Life at the
Nanoscale

ATOMIC FORCE MICROSCOPY
OF LIVE CELLS

Life at the
Nanoscale

ATOMIC FORCE MICROSCOPY OF LIVE CELLS

edited by
Yves Dufrêne

CRC Press
Taylor & Francis Group
Boca Raton London New York

CRC Press is an imprint of the
Taylor & Francis Group, an **informa** business

Life at the Nanoscale: Atomic Force Microscopy of Live Cells

First published 2011 by Pan Stanford Publishing Pte. Ltd.

Published 2018 by CRC Press
Taylor & Francis Group
6000 Broken Sound Parkway NW, Suite 300
Boca Raton, FL 33487-2742

ISBN-13: 978-981-4267-96-0 (hbk)

Visit the Taylor & Francis Web site at
http://www.taylorandfrancis.com

and the CRC Press Web site at
http://www.crcpress.com

British Library Cataloguing-in-Publication Data
A catalogue record for this book is available from the British Library.

Contents

Preface

At the crossroads of life sciences and nanotechnology, the nanoscale analysis of living cells by atomic force microscopy (AFM) is an exciting, rapidly evolving research field. With its ability to observe and force probe cells and biological membranes to molecular resolution and under physiological conditions, AFM offers a wealth of new opportunities in biology and medicine. What is the nanoscale organization of biological membranes? How do cell surfaces remodel during cell growth or incubation with drugs? What is the spatial distribution of cell surface receptors? What are the forces driving cell adhesion processes? What are the adhesive and mechanical properties of cells and of their individual constituents, and how are they related to function? These are some of the pertinent questions that can now be addressed by AFM, thereby contributing to improving our understanding of the structure–function relationships of cell membranes and cell walls.

This book provides an overview of the use of AFM and related techniques for cell analysis, going from the basic principles to the applications. The different chapters, all written by leading experts in their field, cover methodologies for preparing and analyzing membranes and cells of all kinds, discuss the principles of advanced AFM modalities, including high-resolution imaging, high-speed imaging, recognition imaging, single-molecule and single-cell force spectroscopy, mechanical measurements, and highlights recent applications in a variety of fields, including cell biology, microbiology, biophysics, structural biology, physiology and medicine.

The first section of the book covers recent progress in imaging cells and membranes using AFM and related scanning probes. In chapter 1, Milhiet and Le Grimellec explore the nanoscale organization of supported lipid bilayers, with an emphasis on lipid microdomains and membrane proteins. The contribution by the Scheuring team (chapter 2) demonstrates the power of high-resolution AFM imaging for resolving the supramolecular architecture of native membranes. Doktycz *et al.* (chapter 3) review the use of AFM for microbial cell imaging, focusing on sample preparation and imaging conditions, and providing various examples of applications in microbiology. In chapter 4, Malkin describes the unique capabilities of AFM for probing the architecture, assembly and dynamics of bacterial surfaces. Jena (chapter 5) shows how AFM can help us understand cell secretion and membrane fusion processes on the nanoscale. Schillers *et al.* (chapter 6) survey the emerging field of nanophysiology, showing how AFM provides new insights into the dynamics of proteins in plasma membranes, into the membrane mechanodynamics of vascular endothelial cells and into the structural and physical properties of the nuclear envelope. The last three contributions of the section deal with advanced imaging modalities. Chtcheglova *et al.* (chapter 7) first describe the basics of a recent AFM imaging mode named "simultaneous topography and recognition imaging" and its application for mapping cell surface receptors. Uchihashi and Ando (chapter 8) next demonstrate how high-speed AFM imaging is revolutionizing our perception of dynamic biological processes and discuss the potential of the method for observing cell membranes. In the last contribution, van Zanten and Garcia-Parajo (chapter 9) focus on the use of near-field scanning optical microscopy for resolving the clustering of membrane receptors.

In the second section of the book, AFM-based force spectroscopy is used to quantify cellular interactions over scales ranging from whole cells to single molecules. Müller and colleagues (chapter 10) first present an overview of the use of single-cell force spectroscopy for quantifying cell adhesion forces. Rico *et al.* (chapter 11) then survey single-molecule force spectroscopy methods and theories for understanding the binding strength of cell adhesion molecules. In chapter 12, Ikai and Afrin describe advances in the detection and mapping of membrane proteins. Moving into the microbial world, Camesano (chapter 13) explains how to measure bacterial polymer elasticity and bacterial interactions with an AFM, while Lower and Lower (chapter 14) explore the forces and bonds at the interface between microorganisms and minerals. Finally, the Dufrêne team (chapter 15) discusses recent progress in measuring the adhesive and mechanical properties of microbial cell envelope proteins.

The last section of the volume focuses on AFM-based mechanical measurements. In chapter 16, Kasas and Dietler provide a short introduction to elasticity and indentation measurements and highlight relevant publications exploring the nanomechanical properties of biological systems, including viruses and cells. Grandbois and colleagues (chapter 17) combine AFM-based force measurements with fluorescence imaging for the label-free monitoring of cell signalling processes. In chapter 18, Pelling *et al.* use AFM as a tool to deliver localized nanomechanical forces to living mammalian cells, while optically imaging biological responses at the single cell level. In the algae context, Higgins and Wetherbee (chapter 19) explain the role that

AFM has played in advancing our understanding of the morphogenesis and mechanical properties of diatoms. Finally, Sharma and Gimzewski (chapter 20) highlight the potential of AFM techniques in medicine, particularly in cancer diagnostics.

I hope that the book will interest students and researchers from various horizons, whether they are newcomers or well trained in the field. The volume should help them to evaluate the advantages and limitations of AFM techniques in their specific field and to define appropriate procedures and controls that will lead them to successful experiments. I am particularly grateful to all authors for their outstanding contributions, and to people at Pan Stanford Publishing for their invaluable help in publishing the book.

Yves Dufrêne

Chapter 1

OBSERVING THE NANOSCALE ORGANIZATION OF MODEL BIOLOGICAL MEMBRANES BY ATOMIC FORCE MICROSCOPY

Pierre-Emmanuel Milhiet and Christian Le Grimellec
INSERM, Unité 554, Montpellier, France
Université de Montpellier, CNRS, UMR 5048,
Centre de Biochimie Structurale, Montpellier, France
pem@2cbs.cnrs.fr

1.1 INTRODUCTION

Biological membranes are essential to cell life, delineating intracellular compartment or forming a protective barrier as plasma membranes do and being involved in cell communication with the extracellular environment. Lipids are the most important components (in terms of the number of molecules), forming a thin film that provides the basic structure of the membrane. Proteins are peripheral or embedded within the membrane. Lipids are organized as a bilayer with two leaflets with different compositions, i.e. the inner leaflet containing phosphatidylserine and the outer leaflet largely enriched in sphingolipids. In addition, membrane components are very dynamic in-plane, and this phenomenon probably represents the most important driving force of their lateral segregation. A consequence of this segregation is the membrane compartmentalization in microdomains, earlier suggested in 1975.[1] Plasma membranes are now viewed as a mosaic of microdomains, but their size and dynamics are still a matter of debate, and lipid–protein interaction remains poorly understood (for recent reviews see Refs. 2 and 3).

Life at the Nanoscale: Atomic Force Microscopy of Live Cells
Edited by Yves Dufrêne

www.panstanford.com

In this complex context, artificial membranes have been extensively used to mimic membrane organization, using either free-standing membranes like liposomes or planar and supported model membranes.[4] Giant unilamellar vesicles (GUVs) are very useful to study dynamic events and have been widely used to explore lipid domain formation using single-molecule optical microscopy.[5] However, this approach is restricted by diffraction-limited resolution and is therefore not suitable to probe membrane on the mesoscopic scale. Membranes supported on a solid support (supported lipid bilayer, or SLB) are very useful and robust systems that are compatible with most biophysical techniques, including fluorescence microscopy, ellipsometry and atomic force microscopy (AFM). The advantage of AFM, compared with other techniques, is the possibility to image, in real time, the topography of samples with nanometer lateral resolution. AFM, which consists in raster scanning of a sample surface with a sharp tip at the end of a soft cantilever, has been largely used for probing the two-dimensional (2D) organization of model membranes and for elucidating the mechanisms underlying lateral segregation of membrane constituents, especially membrane microdomain formation (for recent reviews see Refs. 6–8). Structural information of membrane proteins incorporated into SLBs with a subnanometer lateral resolution can also be obtained under conditions where proteins are tightly packed.[9,10]

In this chapter, we describe the main strategies to prepare SLBs that are suitable for AFM analysis. After a brief methodological description of AFM imaging in liquid, we review major advances in the exploration of the topology of SLBs, focusing on the study of membrane microdomains and of membrane proteins. Progress in nanobiotechnology and recent technical developments that have improved the time and lateral resolution of AFM are also covered.

1.2 PREPARATION OF ARTIFICIAL SUPPORTED LIPID MEMBRANES

Artificial membranes are generally prepared on chemically inert, hydrophilic and flat solid supports, such as mica, highly oriented pyrolitic graphite, glass, silicon and gold. Different methods have been developed to prepare SLBs, but the most popular technique, first described by McConnel's group,[11] remains the formation of supported membranes by fusion of large unilamellar lipid vesicles (LUVs) on a solid surface. LUVs are generally prepared via sonication or extrusion, and the vesicle solution is then added on top of the support. Vesicles then adsorb on the substrate before rupturing (Fig. 1.1). The composition of the buffer bathing the substrate has to be finely tuned for allowing optimal vesicle–substrate interaction. Divalent cations especially influence the process. For instance, adsorption of negatively charged vesicles made from a mixture of palmitoyl-oleoyl-phosphoglycerol (POPG)/palmitoyl-oleoyl-phosphatidylethanolamine (POPE) lipids is only possible in the presence of calcium chloride.[12] Rupture of intact vesicles can be immediate after vesicle adsorption on the surface or delayed until a critical coverage is reached. It also depends on lipid composition, vesicle concentration and diameter.[8,13] The main drawback of the vesicle fusion method is the symmetry of SLBs that are obtained and that imperfectly mimic biological membranes. Another drawback is the partial loss of membrane dynamics due to strong interaction between the lipid polar heads of the inner leaflet and the substrate, modifying the thickness of the buffer layer trapped between support and SLB.

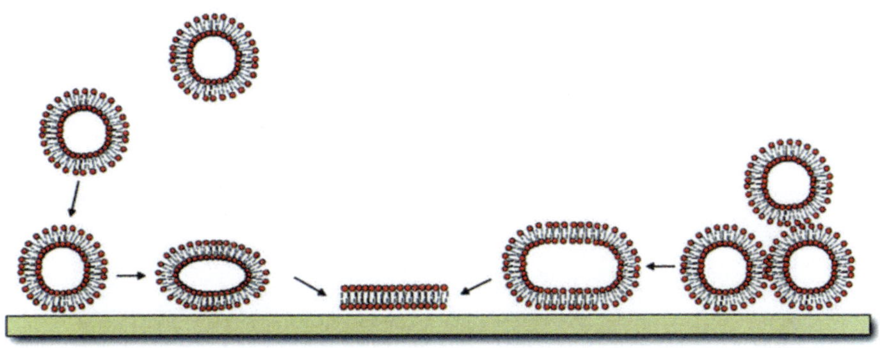

Figure 1.1. Schematic view of the formation of supported lipid bilayers using the fusion of unilamellar vesicles. Single vesicles (lipid polar heads are in red) can adsorb on the surface and rupture to form a supported lipid bilayer (SLB) (left part of the scheme). Alternatively, vesicles can fuse together prior to the rupture (right part of the scheme). A water layer is trapped between the lipids and the support and can act as lubricant.

This thickness largely influences the physical properties of the membrane. It is, for instance, clear that divalent cations can bridge the polar heads of lipids with mica, leading to a large decrease in the interfacial buffer layer as recently observed with SLB composed of neutral phospholipids.[14,15] Similarly, it was described that the way glass coverslips are cleaned largely modulates membrane dynamics and domain formation, probably by changing the viscosity of the water layer trapped between glass and lipid polar heads.[16] More recently, using a POPG/POPE mixture, it was demonstrated that ionic strength largely influences the structure of the water layer, probably by screening the substrate surface charge and by modifying the Debye length.[17] The decreased thickness of the water layer could also explain decoupling of the inner and outer leaflets of SLBs observed in temperature-controlled AFM experiments.[18-20] In addition to the decoupling of the two leaflets, the symmetric versus asymmetric distribution of lipids within SLBs is a question that is not elucidated yet. We and others have observed a symmetric distribution of lipids in the inner and outer membrane leaflets at least for the mixture dioleoyl-phosphatidylcholine/dipalmitoyl-phosphatidylcholine (DOPC/DPPC) and distearoyl-phosphatidylcholine (DSPC)/DPPC,[21,22] whereas the opposite trend has also been noted for the same mixtures.[23,24] An intermediate situation, mixed symmetry, was observed with the DSPC/dilauroyl-phosphatidylcholine (DLPC) mixture. This was explained by a difference in the method of SLB fabrication, i.e. the temperature used for MLV extrusion and fusion that can influence the symmetry of lipid distribution into the bilayers.[24] Further systematic studies are clearly needed to better understand the molecular mechanisms underlying this phenomenon. Finally, it is noteworthy that SLB formation can also be influenced by the roughness of the supporting surface. The shape of gel domains within DOPC-DPPC bilayers formed under the same experimental conditions is completely different when the bilayers are supported on mica (Fig. 1.2a) and on glass (Fig. 1.2b).

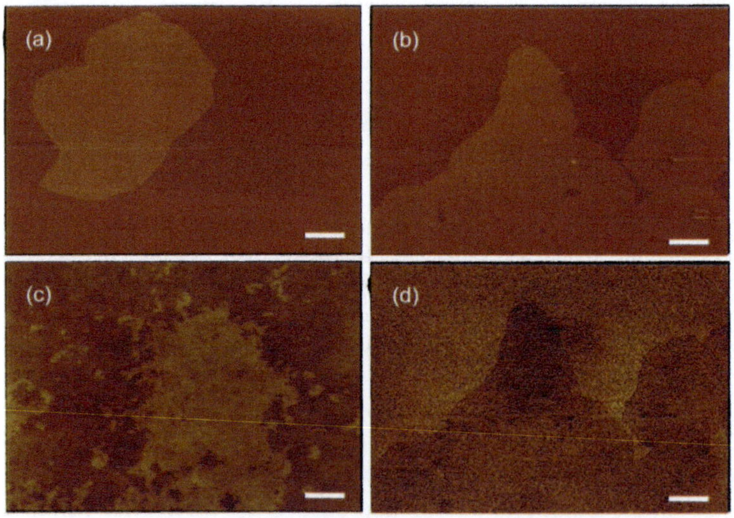

Figure 1.2. AFM imaging of DOPC/DPPC supported lipid bilayers. DOPC-DPPC lipid mixtures (1:1) were used to form SLBs, either on freshly cleaved mica (a, c, d) or on clean glass (b). The shape of the DPPC domains was completely different for the two supports (compare a and b). This difference is probably due to the roughness of glass (~0.2 nm) compared with mica (~ 0.04 nm). (d) is the phase image of the SLB obtained simultaneously with the height image in c. As expected, the phase lag is lower for a gel phase compared with a fluid phase. The *z* scale is 10 nm (a, b, c), and the phase scale is 15° (d). Images were obtained in the tapping mode. Scale bars are 1 μm (a, b) and 0.5 μm (c, d).

Another approach to form SLBs on a solid support is the use of the Langmuir–Blodgett or Langmuir–Schaefer techniques. Both consist in the transfer of a lipid monolayer (inner leaflet) on a hydrophilic support by pulling it vertically through a lipid monolayer at the air–water interface. The outer leaflet is then transferred using either another vertical immersion of the support through the lipid monolayer at the air-water interface or by horizontally dipping the support into the lipid monolayer at the air–water interface. In theory, the advantage of the double transfer methods is that asymmetrical bilayers can be formed. However, it appears that the lipid composition of each leaflet is often very far from the expected composition.[25] Moreover, thinning of the water layer between the mica and the inner leaflet, during the lag time before the second monolayer transfer, often results in a change in the diffusion properties of this inner leaflet.[26] In addition, this technique cannot be used to incorporate transmembrane proteins during bilayer assembly since the protein could be exposed to air during the creation of the second leaflet.

To minimize the membrane–support interaction mentioned above, polymer-supported bilayers (PSBs) have also been developed.[27,28] They can be composed of a soft polymer cushion with typically less than 100 nm thickness to act as a lubricating layer between the support and the bilayer. Alternatively, lipopolymer tethers can also be used to separate membrane components from the support. Generally, PSBs are obtained by the Langmuir–Blodgett technique, vesicle fusion or a combination of both techniques which involves the fusion of LUVs on a pre-deposited monolayer.[29] They have been successfully used for incorporating proteins, preserving their functions, and this technique has now been extended to the biosensors field. However, getting free diffusion of proteins in cushion-supported membranes is not so straightforward, and it seems that protein mobility is strongly dependent on the method of fabrication.[30]

1.3 AFM IMAGING OF SUPPORTED LIPID BILAYERS

Artificial supported membranes are very soft materials, meaning that the tip–sample interaction has to be finely tuned to minimize the force applied during tip scanning, thus preventing the membrane to be swept away (the force between tip and sample can be simplified as a combination of the effects of van der Waals attraction and electrostatic repulsion due to the so-called double layer of counterions).[31] To do so, the spring constant of the cantilever should be low, generally in the 1–100 mN/m range, and force–distance curves should be performed to adjust the force. The pH and buffer (mainly monovalent and divalent ions) conditions are adjusted in such a way as to obtain a mild electrostatic repulsion of the silicon nitride tip by a negatively charged sample.[32]

The contact mode is suitable for imaging flat or weakly corrugated surfaces. During scanning, the tip is always in contact with the sample surface and the force applied by the tip is kept constant (<100 pN) using a z-feedback loop. The highest lateral resolution is actually obtained with this mode (below the nanometer range), as shown below for protein-enriched membranes.[33] A scanning rate of 2–5 Hz is easily achieved with artificial membranes. However, when the sample is weakly adsorbed or corrugated, or when it is difficult to properly tune the tip–sample interaction, the tapping mode (also called intermittent contact mode) is more appropriate and prevents sample damage and tip contamination by reducing shear forces. In this mode, oscillation of the tip can be obtained using acoustic or magnetic excitation, and the oscillation amplitude is used as a feedback signal. Although less performant than the contact mode, the tapping mode yields good lateral and vertical resolutions on artificial membranes.[10] By using this mode the tip can also probe visco-elastic properties of the sample by measuring the phase signal lag, which was successfully used to differentiate fluid and gel phases in SLBs (Fig. 1.2c,d). Generally, the scan rate is ~1 Hz and the oscillation amplitude ~10–100 nm, and the setpoint is adjusted to achieve less than 10% oscillation damping. The group of Hinterdorfer demonstrated that the use of the second harmonic oscillation amplitude as a feedback signal can increase the lateral resolution and the sensitivity to local variations in elasticity.[34] Improvements in temporal resolution and material properties mapping have also been obtained using a torsional harmonic cantilever with an off-axis tip.[35] It is important to notice that tapping mode imaging should be further developed in the coming years because most current AFM developments rely on this mode (see Section 1.7).

1.4 LATERAL MEMBRANE ORGANIZATION

AFM has been extensively used to address the problem of lateral heterogeneity and segregation of lipids and proteins in biological membranes. It has been especially used to characterize rafts microdomains, a subset category of liquid-ordered lipid domains enriched in sphingolipids and cholesterol (Chl), which work as functional platforms in cells. Using lipid mixtures mimicking the composition of the plasma membrane's outer leaflet, essentially mixtures of phosphatidylcholine (PC), sphingomyelin (SM) and Chl, numerous papers have demonstrated the coexistence of the fluid- or liquid-disordered (*ld*) phase with the liquid-ordered (*lo*) or gel phase and tried to rely the physical principles of domain formation in artificial membranes on eukaryotic plasma membrane microdomains (for reviews, see Refs. 3, 7 and 36) or to interpret detergent insolubility of microdomains in cells.[37] From numerous studies, it was confirmed that Chl is a key component of the membrane organization that interacts with saturated fatty acid chains of membrane lipids promoting microdomain formation. However, discrepancies in publications were often observed and could be explained by slight differences in the composition of the lipid mixtures used. As an example, the physical properties of SLBs containing SM can differ depending on whether synthetic lipids or natural (bovine brain) lipids are used. In the latter case, SM fatty acid chains are heterogeneous in length and in saturation. Consequently, a gel-to-fluid transition occurs over a broad range of temperature, including the physiological temperature, and a gel–gel phase separation within SM domains using a SM/DOPC mixture has

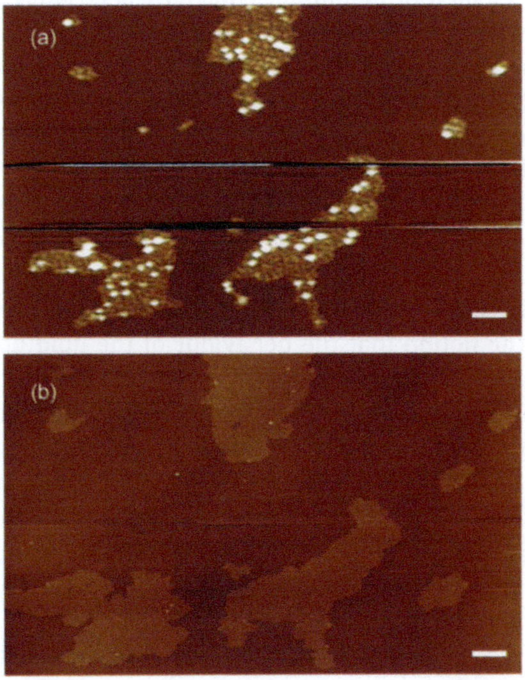

Figure 1.3. Imaging of sphingomyelin domains. (a) SLBs made of a mixture of SM and DOPC (1:1) were observed using AFM contact mode in PBS buffer. SM can form domains of different shapes, protruding from the darker DOPC fluid phase: (a) Corrugated domains formed by closely packed globular structures that protrude 5 nm above the fluid phase; (b) flat domains that protrude 1 nm above the fluid phase. (b) is the same area as the one in (a) observed after the addition of 5 mM $CaCl_2$ in the AFM fluid cell, meaning that the SM domain shape is dependent on divalent cations. The *z* scales in (a) and (b) are 50 and 10 nm, respectively. The scale bars are 1 μm.

been observed.[38] Additional problems of interpretation could come from the fact that natural SM is able to form ripple phases or closely packed globular structures using this binary lipid mixture (Fig. 1.3a). Interestingly, this pattern can be suppressed when divalent cations are added in the phosphate buffer bathing the tip (Fig. 1.3b), thereby strengthening again the importance of a precise tuning of buffer composition for both SLBs formation and imaging. Chl contents as well as the composition of fatty acid chains of the fluid phase are also critical parameters for the organization of membrane microdomains (see "alkaline phosphatase" below), and we think that the *ld* phase of eukaryotic cells should be preferentially mimicked using lipids with asymmetric fatty acid chain, such as POPC, which are more representative of the PC species in plasma membranes.[38]

More recently, attention has been focused on ceramides (Cer), some sphingolipids of eukaryotic membranes that have been lately postulated to be important for membrane structure and function. This lipid can be produced by *de novo* synthesis or through hydrolysis of the SM phosphocholine headgroup by sphingomyelinase. In a DOPC/SM/Chl mixture, replacing SM by Cer induces reorganization of the *lo* phase and formation of a gel phase enriched in Cer,[39,40] confirming that Chl and Cer could compete for SM association. Similarly, in a ternary mixture of POPC, natural Cer and Chl, coexistence of gel-like and *lo* domains was observed up to 20% Chl.[41] Taken together, these studies demonstrate that Cer can induce gel-like domains within membranes (the presence of gel phases in eukaryotic cell membranes is still a matter of debate) and underline the notion that the behaviour of the lipid domain structure as a function of the Chl content is different for membranes containing Cer compared with those containing SM. This lateral organization is, therefore, different from that proposed in the raft hypothesis. Interestingly, when Cer is produced by SM hydrolysis, its effects on the DOPC/SM/Chl mixture's SLB organization is much more pronounced compared with conditions where Cer is directly incorporated in LUVs. It gives large clusters of domains that are heterogeneous, with two distinct heights.[42] This more complex topology could be explained by Cer flip-flop to the lower leaflet, since the presence of this component in model and cell membranes has been described to allow rapid transverse diffusion between inner and outer leaflets.[43] The fact that SLBs are symmetric could influence this phenomenon, compared with native membranes.

Important insights in the distribution of proteins associated with raft have also been obtained by AFM imaging of lipid phase-separated membranes.[44] One of these markers is alkaline phosphatase (AP), a protein expressed at the membrane thanks to a glycosyl-phosphatidylinositol (GPI) anchor which, it has been proposed, is involved in its partitioning in microdomains. In contrast to transmembrane proteins, these proteins can be easily incorporated into SLBs. Two AFM papers using DOPC/SM/Chl supported bilayers have shown that intestine or placental AP (respectively IAP and PLAP) are mainly associated with the *lo* phase.[45,46] However, a study using SLB made of the same lipid mixture and analysed with a combined AFM–fluorescence correlation spectroscopy (FCS) setup showed that PLAP preferentially partitions in the *ld* phase and that only 25% of the proteins are associated with the *lo* phase.[47] The authors proposed that AFM is not suitable for imaging dynamic individual molecules, such as AP inserted in a fluid phase.[47] It is obvious that the scan rate can be a critical parameter to image fast dynamic processes, but, in this case, the diffusion coefficients in fluid and *lo* phases are theoretically comparable (the classical difference of dynamics between these two phases is a factor of 2)[21] and cannot explain the absence of IAP in the fluid phase observed by AFM. Another explanation that can be postulated to describe the different behaviours of IAP and PLAP is the fact that fatty acid chains of the IAP GPI-anchor are more saturated than that of the PLAP enzyme.[48] Our recent data strongly suggest that IAP interacts with the most ordered lipid species present in the gel phases of membranes exhibiting phase separation, probably to prevent any hydrophobic mismatch.[48] Moreover, using a GPI-anchored form of the angiotensin-converting enzyme, it was found that an anchor with a chain length of C18 or longer induces the protein to mainly partition in microdomains enriched in brain SM (the fatty acid chains are mainly C18:0 and C24:1) in DOPC/SM/Chl SLBs. Such a partition was no more observed using synthetic palmitoyl or stearoyl SM.[49]

Taken together these results indicate that particular attention has to be paid to the choice of lipids as well as fatty acid chain composition of GPI-anchored proteins. These different examples also strengthen the potency of AFM to investigate lipid and protein partitioning within SLB but also highlight the difficulties encountered in interpreting and comparing results, even with simple binary or ternary mixtures of lipids. To be accurately compared, model membranes need to be identical in composition and certainly supported on the same material, and, if necessary, proteins should be incorporated using a similar technique. In addition, one has to keep in mind that membrane components are highly dynamic, even in a gel phase, meaning that increasing temporal resolution of AFM is clearly a key issue in this field.

1.5 STRUCTURAL ANALYSIS OF MEMBRANE PROTEINS

Besides its important contribution to the field of lipid microdomains, AFM is also the only microscopy technique that allows images of membrane proteins to be acquired with subnanometer resolution and under physiological conditions. Topographs of the extramembraneous domains of membrane proteins can be acquired with a vertical resolution of 1 Å providing fine details of the protruding protein domains above the membrane. Information about the oligomeric state of a single protein, the organization of individual components within multi-protein assemblies, as well as the individual β-turns and loops connecting transmembrane α-helices can be obtained.[33] In the challenging context of the structural analysis of membrane proteins only ~200 structures are currently available in the protein data bank), AFM is a powerful tool to analyse proteins reconstituted into artificial membranes. Purified and detergent-solubilized membrane proteins can be reconstituted by detergent removal in the presence of additional lipids, at high protein density, allowing the formation of 2D crystals in the plane of the membranes.[50] These protein-enriched membranes can be prepared on mica using appropriate buffers (generally, Tris supplemented with KCl and $MgCl_2$) and imaged at high resolution with AFM. Some remarkable examples include the characterization of the dimeric PufX-containing core complex of *Rhodobacter blasticus* of the bacterial photosynthetic apparatus,[51] the first view of the trimeric structure AmtB, an archetypal member of the ammonium transporter family,[52] or the identification and structure of a putative Ca^{2+}-binding domain at the C terminus of aquaporin 1.[53]

Starting from pure lipid bilayers supported on mica, we have developed in collaboration with Levy's group a new method for the reconstitution of transmembrane proteins. It is based on previous studies of direct incorporation of membrane protein into liposomes for functional studies.[54] The principle relies on the direct incorporation of purified membrane proteins in pre-formed SLBs destabilized by sugar-based detergent. As a main result, the amount of purified proteins per trial is less than 1 picomole, far below the amount requested in 2D or 3D crystallization, thereby allowing the AFM analysis of eukaryotic membrane proteins that are difficult to overexpress.[10] Two large membrane complexes, the light-harvesting 1 (LH1) and LH2 complexes from the bacterial photosynthetic apparatus as well as the bacterial LacY permease, have been successfully incorporated and imaged with a lateral resolution in the nanometer range (Fig. 1.4). This resolution can be achieved because lateral segregation and packing of incorporated transmembrane proteins are possible within SLBs. This also supports the idea that, under appropriate conditions, the lipid polar head–support interaction can be minimized and that the water layer at the inner leaflet–support interface is sufficient for allowing membrane diffusion.

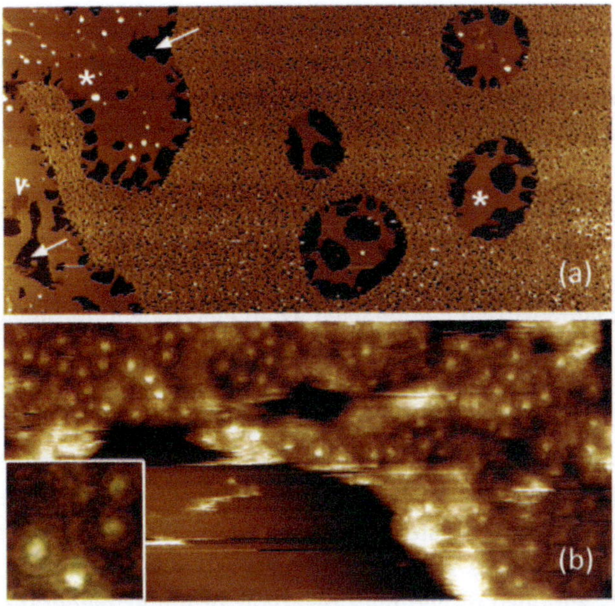

Figure 1.4. Structural analysis of directly incorporated transmembrane proteins. LH1-RC core complexes from *Rhodobacter veldkampi* were directly incorporated at 4°C into a DOPC/DPPC pre-formed bilayer and imaged in contact mode AFM.[14] (a) AFM image (30 μm scan size), where smooth areas alternate with corrugated domains (asterisks are flat lipid protein-free areas, white arrows indicate holes in the SLB and the white arrowhead shows DOPC/DPPC phase separation). (b) A 400 nm zoom in the corrugated domains in which incorporated proteins can be identified and better observed in the inset (LH1-RC is a complex constituted by an LH1 ring of ~10 nm surrounding the RC). The *z* scales in (a) and (b) are 30 and 10 nm, respectively.

Direct addition of membrane proteins to pre-formed SLBs has also been used to study the prepore-to-pore transition mechanism of the cholesterol-dependent cytolysins Perfringolysin O.[9] AFM demonstrated that the prepore-to-pore transition of such pore-forming protein is associated with a dramatic and vertical collapse of its structure, thereby illustrating a new mechanism of membrane insertion.

1.6 APPLYING AFM IN MEMBRANE-BIOINSPIRED NANOTECHNOLOGY

Thanks to their ability to mimic biological membranes and their relative ease to be handled and functionalized, SLBs represent an attractive system in the development of membrane-inspired biosensors. The issue for biosensor applications is to get bilayers separating two compartments for studying the properties of cell membranes such as permeability, active transport or signal transduction by transmembrane proteins. The ultimate goal is to probe single molecules in nano-size compartments. In this section, we highlight a few recent biosensors developments which have used AFM as the main characterization tool, knowing that interest in this field is increasing very quickly (for a recent review, see Ref. 28).

One approach for making biosensors is to form a lipid membrane on top of a planar support that can be used as an electrode, typically a hydrophilic semiconductor or oxide material like gold. Direct fusion of lipid vesicles can occur spontaneously on these surfaces to form a planar bilayer, but, most often, tethered bilayer lipid membranes (tBLMs) are used to generate an additional aqueous space between the support and the membrane. This space can be useful in studying the function of transmembrane proteins.[55] However, AFM imaging of these systems is sparsely documented.[56,57] More recently, porous materials have emerged as good candidates for supporting lipid membranes and also providing a reservoir of buffer below the membrane. Porous silicon obtained by the electrochemical etching of crystalline silicon wafers is especially interesting because it behaves as a photonic crystal reflector and can be used as a label-free optical biosensor. Deposition of a continuous planar phospholipid bilayer using phosphatidylethanolamine/PC/Chl or DOPC/DPPC mixture at the surface of porous silicon has validated the proof of concept.[58,59] Under these conditions, membrane dynamics was well preserved.

Another strategy to make biosensors is to use nano-size holes supporting the lipid membrane. An elegant method of fabrication of SLB on top of porous alumina by vesicle spreading has been developed by Steinem's group.[60] Briefly, a porous alumina surface, having 50 nm hexagonal organized pores, is coated by a gold layer and further functionalized by thiols. Membrane bilayers that include positively charged lipids are formed on the surface and pictured by AFM. Under these conditions, most of the surface is covered by free-standing bilayers over the holes. This system has been used to study the channel activity of several proteins. One problem encountered with aperture-spanning membranes is their low stability in time and their tendency to rupture. A scaffold composed of S-layer proteins of *Bacillus sphaericus*, pre-coated on the support,[61] as well as a gelling solution bathing the membrane,[62] can increase the stability of free-spanning membranes.

Whatever the strategy used to form the lipid bilayer in membrane-inspired biosensors, the main bottleneck remains the incorporation of functional proteins. Spontaneous insertion of ion-channel proteins has been successfully used, but this strategy cannot be applied to the majority of transmembrane proteins. One of the promising ways is to form membrane by proteoliposomes fusion. However, this approach presents important bottlenecks, such as the purification of proteins in large amounts. In addition, proteins generally display two different membrane orientations, except in some detergent-mediated reconstitution approaches.[54] Direct incorporation of membrane proteins, similar to what we have developed for structural analysis purposes, can be useful.

1.7 AFM METHODOLOGICAL DEVELOPMENTS

As we have seen, AFM represents a very powerful tool to explore the structure of SLBs, whether or not containing proteins. However, commercial setups cannot investigate membrane dynamics (lipids diffuse in the membrane with a diffusion coefficient in the $\mu m^2/s$ range) because of the limit of the tip scanning rate (typically between 0.5 and 7 Hz for this type of sample). Recent advances have been made owing to the combination of AFM with FCS, first described in 2005.[63] Structural information and spatial distribution of membrane components, i.e. microdomains, is given by AFM, whereas FCS provides their local dynamic properties. This combination has been applied to studying partitioning of GM1, Cer and AP into SLBs.[39,63] Nevertheless, the FCS lateral resolution is still rather poor compared with that of AFM (the detection area or beam waste diameter used in FCS experiments is larger than 200 nm). A very promising way to get simultaneous dynamics and topography of SLBs is to break the speed limit of AFM. Important progress has been made in cantilevers, scanners and controllers,[64] and two main setups allowing video-rate imaging in liquid seem to be promising in the membrane field. The first setup is a high-speed contact mode AFM developed by the Miles group in Bristol[65] in which the sample is placed on a flexure stage that is aligned with the fast-scanning x direction, and the cantilever is positioned on a piezo tube that is used for slow-scanning y and z directions. This setup allows video-rate acquisition of biological samples, but no images of membranes have been reported so far. The second setup is a high-speed tapping mode AFM developed by the group of Ando in Kanazawa[66] (see Chapter 8). In order to increase the scanning rate in tapping mode, small cantilevers with a high resonance frequency and a low spring constant (150–280 pN/nm and 1.3–1.8 MHz), as well as new AC-to-DC converters, scanners and dynamic PID controllers, have been developed. Different biological systems such as GroES/GroES or myosin motors have been investigated, providing real-time imaging of individual molecule at work,[67] and this setup is already convenient to image biological membranes. Rupture of liposomes on mica and formation of SLB from a ternary mixture of lipids were observed at one image per second,[7] meaning that this setup should be very useful to elucidate membrane phenomena such as microdomain nucleation, diffusion of nanoscale domains and diffusion of membrane components. Recently, high-resolution movies of individual bacteriorhodopsin trimers were acquired at a 100 ms frame and highlighted temporal fluctuation at the crystal edges.[68]

The second main drawback of AFM for imaging complex systems such as biological membranes or model membranes, including several proteins, is the identification of membrane components. One of the solutions is to combine AFM with fluorescence microscopy, even if the lateral resolution in classical fluorescence microscopy (~ 200–300 nm due to the diffraction limit) is very weak compared with that of AFM. In a seminal paper,[69] membrane microdomains were localized on a wide scale by fluorescence in a DOPC/DPPC supported bilayer, whereas AFM provided topography in the mesoscopic scale. This combination is now proposed by most manufacturers. Interestingly, the development of super-resolution fluorescence microscopy should reduce the gap between AFM and fluorescence approaches. Using far-field optical microscopy or nanoscopy such as stimulated emission depletion (STED), photoactivated localization microscopy (PALM) or stochastic optical reconstruction microscopy (STORM), the lateral resolution can reach a few tens of nanometers.[70] Membrane dynamics can also be explored with optical nanoscopy.[71] Combined with AFM, optical nanoscopy might open a new field of exploration of biological membranes that sounds very exciting. However, according to the scanning rate of commercially available microscopes, it is necessary to immobilize membrane components for superimposing optical and AFM images. Combining high-speed AFM and optical nanoscopy could eventually solve this drawback. Another strategy to identify membrane components by AFM is to perform single-molecule recognition imaging.[72] Single-molecule imaging requires tip functionalization with relevant molecules to recognize a specific molecule associated with the membrane. Detection of molecular recognition events can be performed using adhesion force mapping or dynamic recognition force mapping (see Chapter 7). Finally, we can speculate that combining AFM with nanoSIMS (secondary ion mass spectroscopy), a new technique that has been used to identify molecules on top of membranes, should be very interesting. NanoSIMS uses the secondary emitted ions from a bombarded surface to get a mass spectrum and to identify the species present on the surface. The beam scans the surface, and a mass spectrum is made at each pixel. The lateral resolution, which is given by the ion beam diameter, can presently reach 50 nm. The nanoSIMS technique was already able to study the organization of raft-like systems.[73] The ability of AFM to explore biological membranes should also benefit from its coupling to other techniques such as polarized total internal reflection fluorescence microscopy. Using such combination, it was possible to evaluate membrane order parameters and to track changes in lipid headgroup and acyl chain reordering in SLBs, while simultaneously resolving molecular-scale topographical changes.[74]

Progress in the development of non-contact mode AFM, mainly frequency modulation (FM)-AFM, is also expected. Here, the cantilever is excited at fixed amplitude and the topography followed by keeping the frequency modulation constant. FM-AFM imaging in liquid was achieved on SLBs[75] or purple membranes.[76] It could be implemented in commercial setups but still needs further improvement to better control tip–sample interactions, especially with corrugated samples where adhesion events largely perturb the detection of frequency shifts. Another non-destructive approach, called scanning near-field ultrasonic holography or ultrasonic force microscopy, is currently under development. It uses high-frequency oscillators to generate spatially resolved images. In collaboration with Lesniewska's group, we could observe lipid phase separation within SLBs using this approach.

1.8 CONCLUSION

AFM represents a first-choice technique in the study of SLBs, allowing the topography of membrane components to be acquired at (sub)molecular resolution under physiological conditions. Particular attention has to be paid to the choice of lipids to mimic biological membranes as well as to experimental conditions used for lipid vesicles adsorption and fusion. Major advances in the understanding of membrane components partitioning into microdomains have been achieved with AFM. Moreover, this technique can also be useful in delineating the structure of membrane proteins. We can speculate that future technical improvements of AFM techniques will contribute to their broader use in biomembrane research.

Acknowledgements

This work was supported by institutional grants from INSERM and CNRS and by specific grants from the French research agency ANR (PCV08-AFM-MB-PROT, PCV08_343399 and 08-NANO-010).

References

1. Morrisset, J. D., Pownall, H. J., Plumlee, R. T., Smith, L. C., Zehner, Z. E., Esfahani, M., and Wakil, S. J. (1975) Multiple thermotropic phase transitions in *Escherichia coli* membranes and membrane lipids, *J. Biol. Chem.*, **250**, 6969–6976.
2. Jacobson, K., Mouritsen, O. G., and Anderson, R. G. (2007) Lipid rafts: at a crossroad between cell biology and physics, *Nat. Cell Biol.*, **9**, 7–14.
3. Johnston, L. J. (2007) Nanoscale imaging of domains in supported lipid membranes, *Langmuir*, **23**, 5886–5895.
4. Sackmann, E. (1996) Supported membranes: scientific and practical applications, *Science*, **271**, 43–48.
5. Kahya, N., Scherfeld, D., Bacia, K., and Schwille, P. (2004) Lipid domain formation and dynamics in giant unilamellar vesicles explored by fluorescence correlation spectroscopy, *J. Struct. Biol.*, **147**, 77–89.
6. El Kirat, K., Morandat, S., and Dufrene, Y. F. (2010) Nanoscale analysis of supported lipid bilayers using atomic force microscopy, *Biochim. Biophys. Acta.*, **1798**, 750–765.
7. Giocondi, M. C., Yamamoto, D., Lesniewska, E., Milhiet, P. E., Ando, T., and Le Grimellec, C. (2010) Surface topography of membrane domains, *Biochim. Biophys. Acta*, **1798**, 703–718
8. Goksu, E. I., Vanegas, J. M., Blanchette, C. D., Lin, W. C., and Longo, M. L. (2009) AFM for structure and dynamics of biomembranes, *Biochim. Biophys. Acta*, **1788**, 254–266.
9. Czajkowsky, D. M., Hotze, E. M., Shao, Z., and Tweten, R. K. (2004) Vertical collapse of a cytolysin prepore moves its transmembrane beta-hairpins to the membrane, *Embo. J.*, **23**, 3206–3215.
10. Milhiet, P. E., Gubellini, F., Berquand, A., Dosset, P., Rigaud, J. L., Le Grimellec, C., and Levy, D. (2006) High resolution AFM of membrane proteins directly incorporated at high density in planar lipid bilayer, *Biophys. J.*, **91**, 3268–3275.
11. Brian, A. A. and McConnell, H. M. (1984) Allogeneic stimulation of cytotoxic T cells by supported planar membranes, *Proc. Natl. Acad. Sci. USA*, **81**, 6159-6163.
12. Picas, L., Montero, M. T., Morros, A., Cabanas, M. E., Seantier, B., Milhiet, P. E., and Hernandez-Borrell, J. (2009) Calcium-induced formation of subdomains in phosphatidylethanolamine-phosphatidylglycerol bilayers: a combined DSC, 31P NMR, and AFM study, *J. Phys. Chem. B*, **113**, 4648–4655.
13. Richter, R. P., Berat, R., and Brisson, A. R. (2006) Formation of solid-supported lipid bilayers: an integrated view, *Langmuir*, **22**, 3497–3505.
14. Berquand, A., Levy, D., Gubellini, F., Le Grimellec, C., and Milhiet, P. E. (2007) Influence of calcium on direct incorporation of membrane proteins into in-plane lipid bilayer, *Ultramicroscopy*, **107**, 928–933.
15. Garcia-Manyes, S., Oncins, G., and Sanz, F. (2005) Effect of ion-binding and chemical phospholipid structure on the nanomechanics of lipid bilayers studied by force spectroscopy, *Biophys. J.*, **89**, 1812–1826.

16. Seu, K. J., Pandey, A. P., Haque, F., Proctor, E. A., Ribbe, A. E., and Hovis, J. S. (2007) Effect of surface treatment on diffusion and domain formation in supported lipid bilayers, *Biophys. J.*, **92**, 2445–2450.

17. Seeger, H. M., Marino, G., Alessandrini, A., and Facci, P. (2009) Effect of physical parameters on the main phase transition of supported lipid bilayers, *Biophys. J.*, **97**, 1067–1076.

18. Leonenko, Z. V., Finot, E., Ma, H., Dahms, T. E., and Cramb, D. T. (2004) Investigation of temperature-induced phase transitions in DOPC and DPPC phospholipid bilayers using temperature-controlled scanning force microscopy, *Biophys. J.*, **86**, 3783–3793.

19. Keller, D., Larsen, N. B., Moller, I. M., and Mouritsen, O. G. (2005) Decoupled phase transitions and grain-boundary melting in supported phospholipid bilayers, *Phys. Rev. Lett.*, **9402**, 186–189.

20. Charrier, A. and Thibaudau, F. (2005) Main phase transitions in supported lipid single-bilayer, *Biophys. J.*, **89**, 1094–1101.

21. Almeida, P. F., Vaz, W. L., and Thompson, T. E. (1992) Lateral diffusion and percolation in two-phase, two-component lipid bilayers. Topology of the solid-phase domains in-plane and across the lipid bilayer, *Biochemistry*, **31**, 7198–7210.

22. Giocondi, M. C., Vie, V., Lesniewska, E., Milhiet, P. E., Zinke-Allmang, M., and Le Grimellec, C. (2001) Phase topology and growth of single domains in lipid bilayers., *Langmuir*, **17**, 1653–1659.

23. Choucair, A., Chakrapani, M., Chakravarthy, B., Katsaras, J., and Johnston, L. J. (2007) Preferential accumulation of Abeta(1–42) on gel phase domains of lipid bilayers: an AFM and fluorescence study, *Biochim. Biophys. Acta*, **1768**, 146–154.

24. Lin, W. C., Blanchette, C. D., Ratto, T. V., and Longo, M. L. (2006) Lipid asymmetry in DLPC/DSPC-supported lipid bilayers: a combined AFM and fluorescence microscopy study, *Biophys. J.*, **90**, 228–237.

25. Kiessling, V., Wan, C., and Tamm, L. K. (2009) Domain coupling in asymmetric lipid bilayers, *Biochim. Biophys. Acta*, **1788**, 64–71.

26. Hollars, C. W. and Dunn, R. C. (1998) Submicron structure in L-alpha-dipalmitoylphosphatidylcholine monolayers and bilayers probed with confocal, atomic force, and near-field microscopy, *Biophys. J.*, **75**, 342–353.

27. Tanaka, M. and Sackmann, E. (2005) Polymer-supported membranes as models of cell surface, *Nature*, **437**, 656–663.

28. Reimhult, E. and Kumar, K. (2008) Membrane biosensor platforms using nano- and microporous supports, *Trends Biotechnol.*, **26**, 82–89.

29. Kalb, E., Frey, S., and Tamm, L. K. (1992) Formation of supported planar bilayers by fusion of vesicles to supported phospholipid monolayers, *Biochim. Biophys. Acta*, **1103**, 307–316.

30. Merzlyakov, M., Li, E., Gitsov, I., and Hristova, K. (2006) Surface-supported bilayers with transmembrane proteins: role of the polymer cushion revisited, *Langmuir*, **22**, 10145–10151.

31. Israelachvili, J. (2002) Electrostatic forces between surfaces in liquids, in *Intermolecular and Surface Forces*, Academic Press, San Diego, pp. 213–259.
32. Muller, D. J., Amrein, M., and Engel, A. (1997) Adsorption of biological molecules to a solid support for scanning probe microscopy, *J. Struct. Biol.*, **119**, 172–188.
33. Frederix, P. L., Bosshart, P. D., and Engel, A. (2009) Atomic force microscopy of biological membranes, *Biophys. J.*, **96**, 329–338.
34. Preiner, J., Tang, J., Pastushenko, V., and Hinterdorfer, P. (2007) Higher harmonic atomic force microscopy: Imaging of biological membranes in a liquid, *Phys. Rev. Lett.*, **99**, 046102.
35. Sahin, O., Magonov, S., Su, C., Quate, C. F., and Solgaard, O. (2007) An atomic force microscope tip designed to measure time-varying nanomechanical forces, *Nat. Nanotechnol.*, **2**, 507–514.
36. London, E. (2005) How principles of domain formation in model membranes may explain ambiguities concerning lipid raft formation in cells, *Biochim. Biophys. Acta*, **1746**, 203–220.
37. Brown, D. A. and London, E. (1997) Structure of detergent-resistant membrane domains: does phase separation occur in biological membranes? *Biochem. Biophys. Res. Commun.*, **240**, 1–7.
38. Giocondi, M. C., Boichot, S., Plenat, T., and Le Grimellec, C. C. (2004) Structural diversity of sphingomyelin microdomains, *Ultramicroscopy*, **100**, 135–143.
39. Chiantia, S., Kahya, N., Ries, J., and Schwille, P. (2006) Effects of ceramide on liquid-ordered domains investigated by simultaneous AFM and FCS, *Biophys. J.*, **90**, 4500–4508.
40. Johnston, I. and Johnston, L. J. (2006) Ceramide promotes restructuring of model raft membranes, *Langmuir*, **22**, 11284–11289.
41. Fidorra, M., Duelund, L., Leidy, C., Simonsen, A. C., and Bagatolli, L. A. (2006) Absence of fluid-ordered/fluid-disordered phase coexistence in ceramide/POPC mixtures containing cholesterol, *Biophys. J.*, **90**, 4437–4451.
42. Johnston, L. J. (2008) Sphingomyelinase generation of ceramide promotes clustering of nanoscale domains in supported bilayer membranes, *Biochim. Biophys. Acta*, **1778**, 185–197.
43. Contreras, F. X., Villar, A. V., Alonso, A., Kolesnick, R. N., and Goni, F. M. (2003) Sphingomyelinase activity causes transbilayer lipid translocation in model and cell membranes, *J. Biol. Chem.*, **278**, 37169–37174.
44. Giocondi, M. C., Seantier, B., Dosset, P., Milhiet, P. E., and Le Grimellec, C. (2008) Characterizing the interactions between GPI-anchored alkaline phosphatases and membrane domains by AFM, *Pflugers Arch.*, **456**, 179–188.
45. Milhiet, P., Giocondi, M., Baghdadi, O., Ronzon, F., Roux, B., and Le Grimellec, C. (2002) Spontaneous insertion and partitioning of alkaline phosphatase into model lipid rafts, *EMBO Rep.*, **3**, 485–490.
46. Saslowsky, D. E., Lawrence, J., Ren, X. Y., Brown, D. A., Henderson, R. M., and Edwardson, J. M. (2002) Placental alkaline phosphatase is efficiently targeted to rafts in supported lipid bilayers, *J. Biol. Chem.*, **277**, 26966–26970.

47. Chiantia, S., Ries, J., Kahya, N., and Schwille, P. (2006) Combined AFM and two-focus SFCS study of Raft-exhibiting model membranes, *ChemPhysChem*, **7**, 2409–2418.

48. Giocondi, M. C., Besson, F., Dosset, P., Milhiet, P. E., and Le Grimellec, C. (2007) Remodelling of ordered membrane domains by gpi-anchored intestinal alkaline phosphatase, *Langmuir*, **23**, 9358–9364.

49. Garner, A. E., Smith, D. A., and Hooper, N. M. (2007) Sphingomyelin chain length influences the distribution of GPI-anchored proteins in rafts in supported lipid bilayers, *Mol. Membr. Biol.*, **24**, 233–242.

50. Rigaud, J., Chami, M., Lambert, O., Levy, D., and Ranck, J. (2000) Use of detergents in two-dimensional crystallization of membrane proteins, *Biochim. Biophys. Acta*, **1508**, 112–128.

51. Scheuring, S., Busselez, J., and Levy, D. (2005) Structure of the dimeric PufX-containing core complex of Rhodobacter blasticus by in situ atomic force microscopy, *J. Biol. Chem.*, **280**, 1426–1431.

52. Conroy, M. J., Jamieson, S. J., Blakey, D., Kaufmann, T., Engel, A., Fotiadis, D., Merrick, M., and Bullough, P. A. (2004) Electron and atomic force microscopy of the trimeric ammonium transporter AmtB, *EMBO Rep.*, **5**, 1153–1158.

53. Fotiadis, D., Suda, K., Tittmann, P., Jeno, P., Philippsen, A., Muller, D. J., Gross, H., and Engel, A. (2002) Identification and structure of a putative Ca2+-binding domain at the C terminus of AQP1, *J. Mol. Biol.*, **318**, 1381–1394.

54. Rigaud, J. and Levy, D. (2003) Reconstitution of membrane proteins into liposomes, *Methods Enzymol.*, **372**, 65–86.

55. Naumann, C., Prucker, O., Lehmann T, Ruhe, J., Knoll, W., and Frank, C. W. (2002) The polymer-supported phospholipid bilayer: tethering as a new approach to substrate-membrane stabilization, *Biomacromolecules*, **3**, 27–35.

56. Dorvel, B. R., Keiser, H. M., Fine, D., Vuorinen, J., Dodabalapur, A., and Duran, R. S. (2007) Formation of tethered bilayer lipid membranes on gold surfaces: QCM-Z and AFM study, *Langmuir*, **23**, 7344–7355.

57. Jeuken, L. J., Connell, S. D., Henderson, P. J., Gennis, R. B., Evans, S. D., and Bushby, R. J. (2006) Redox enzymes in tethered membranes, *J. Am. Chem. Soc.*, **128**, 1711–1716.

58. Cunin, F., Milhiet, P. E., Anglin, E., Sailor, M. J., Espenel, C., Le Grimellec, C., Brunel, D., and Devoisselle, J. M. (2007) Continuous planar phospholipid bilayer supported on porous silicon thin film reflector, *Ultramicroscopy*, **107**, 1048–1052.

59. Worsfold, O., Voelcker, N. H., and Nishiya, T. (2006) Biosensing using lipid bilayers suspended on porous silicon, *Langmuir*, **22**, 7078–7083.

60. Janshoff, A. and Steinem, C. (2006) Transport across artificial membrane-an analytical perspective, *Anal. Bioanal. Chem.*, **385**, 433–451.

61. Schuster, B. and Sleytr, U. B. (2009) Composite S-layer lipid structures, *J. Struct. Biol.*, **168**, 207–216.

62. Kang, X. F., Cheley, S., Rice-Ficht, A. C., and Bayley, H. (2007) A storable encapsulated bilayer chip containing a single protein nanopore, *J. Am. Chem. Soc.*, **129**, 4701–4705.
63. Burns, A. R., Frankel, D. J., and Buranda, T. (2005) Local mobility in lipid domains of supported bilayers characterized by atomic force microscopy and fluorescence correlation spectroscopy, *Biophys. J.*, **89**, 1081–1093.
64. Hansma, P. K., Schitter, G., Fantner, G. E., and Prater, C. (2006) Applied physics. High-speed atomic force microscopy, *Science*, **314**, 601–602.
65. Picco, L. M., Bozec, L., Ulcinas, A., Engledew, D. J., Antognozzi, M., Horton, M. A., and Miles, M. J. (2007) Breaking the speed limit with atomic force microscopy, *Nanotechnology*, **18**, 044030.
66. Ando, T., Kodera, N., Uchihashi, T., Miyagi, A., Nakakita, R., Yamashita, H., and Matada, K. (2005) High-speed atomic force microscopy for capturing dynamic behavior of protein molecules at work, *e-J. Surf. Sci. Nanotech.*, **3**, 384–392.
67. Ando, T., Uchihashi, T., Kodera, N., Yamamoto, D., Miyagi, A., Taniguchi, M., and Yamashita, H. (2008) High-speed AFM and nano-visualization of biomolecular processes, *Pflugers Arch.*, **456**, 211–225.
68. Casuso, I., Kodera, N., Le Grimellec, C., Ando, T., and Scheuring, S. (2009) Contact-mode high-resolution high-speed atomic force microscopy movies of the purple membrane, *Biophys. J.*, **97**, 1354–1361.
69. Burns, A. R. (2003) Domain structure in model membrane bilayers investigated by simultaneous atomic force microscopy and fluorescence imaging, *Langmuir*, **19**, 8358–8363.
70. Hell, S. W. (2007) Far-field optical nanoscopy, *Science*, **316**, 1153–1158.
71. Shroff, H., Galbraith, C. G., Galbraith, J. A., and Betzig, E. (2008) Live-cell photoactivated localization microscopy of nanoscale adhesion dynamics, *Nat. Methods*, **5**, 417–423.
72. Hinterdorfer, P. and Dufrene, Y. F. (2006) Detection and localization of single molecular recognition events using atomic force microscopy, *Nat. Methods*, **3**, 347–355.
73. McQuaw, C. M., Zheng, L., Ewing, A. G., and Winograd, N. (2007) Localization of sphingomyelin, in cholesterol domains by imaging mass spectrometry, *Langmuir*, **23**, 5645–5650.
74. Oreopoulos, J. and Yip, C. M. (2009) Probing membrane order and topography in supported lipid bilayers by combined polarized total internal reflection fluorescence-atomic force microscopy, *Biophys. J.*, **96**, 1970–1984.
75. Fukuma, T., Higgins, M. J., and Jarvis, S. P. (2007) Direct imaging of lipid-ion network formation under physiological conditions by frequency modulation atomic force microscopy, *Phys. Rev. Lett.*, **98**, 106101, 106104 pp.
76. Hoogenboom, B. W., Frederix, P. L. T. M., Yang, J. L., Martin, S., Pellmont, Y., Steinacher, M., Zäch, S., Langenbach, E., Heimbeck, H.-J., Engel, A., and Hug, H. J. (2005) A Fabry-Perot interferometer for micrometer-sized cantilevers, *Appl. Phys. Lett.*, **86**, 074101–074103.

Chapter 2

HIGH-RESOLUTION ATOMIC FORCE MICROSCOPY OF NATIVE MEMBRANES

Nikolay Buzhynskyy, Lu-Ning Liu, Ignacio Casuso and Simon Scheuring
Institut Curie, U1006 INSERM, 26 rue d'Ulm, Paris, France
Simon.Scheuring@curie.fr

2.1 AFM IN STRUCTURAL BIOLOGY OF MEMBRANE PROTEINS

The atomic force microscope (AFM) has developed into a powerful tool in membrane protein research.[1] Two reasons why AFM is the tool of choice for membrane protein studies are its capability to study single molecules in a sample that needs relatively little prior biochemical treatment and the nativeness of the sample studied.

All techniques, with the exception of AFM, appeal to molecule averaging to obtain structural information (Fig. 2.1). Depending on the signal-to-noise ratio (SNR) provided by a technique, a certain numbers of molecules must be merged to acquire structural information. The averaging methodology varies among techniques and can imply Fourier transformations for techniques where proteins are crystallized (X-ray and electron crystallography), computational overlay of images where individual molecules are imaged at low SNR (single-particle electron microscopy) and averaging of resonances of inter-atom distances of many molecules in solution (nuclear magnetic resonance [NMR] analysis). As detailed in Fig. 2.1, NMR, X-ray and electron crystallography can attain three-dimensional (3D) atomic resolution. However, to obtain atomic resolution datasets these techniques imply merging of billions of molecules. Furthermore, a series of biochemical procedures are needed

Life at the Nanoscale: Atomic Force Microscopy of Live Cells
Edited by Yves Dufrêne

www.panstanford.com

to obtain an analysable sample, including overexpression, solubilization, purification and crystallization. Any of these steps may represent a severe experimental bottleneck. AFM, in contrast, while restricted to a topographical analysis of the sample at a relatively low lateral resolution (~10 Å), provides such a high SNR that single molecules can be structurally analysed.[2] This is the feature that allows the structural analysis of membrane proteins in a native membrane packed with several molecule species.

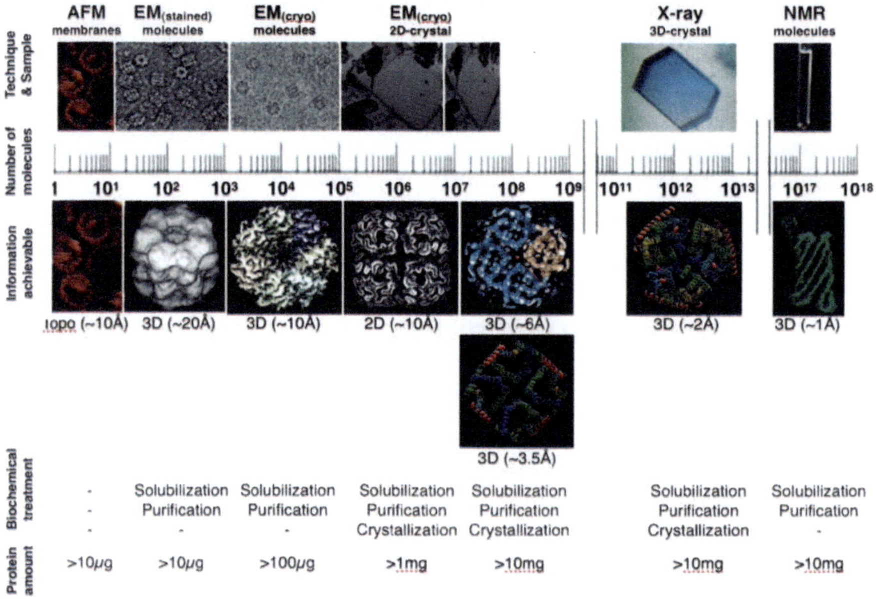

Figure 2.1. Comparison of techniques used to analyse membrane protein structure, taking into consideration the number of molecules to be analysed to obtain wished or achievable structural information, the corresponding required biochemical procedures and an estimated amount of protein needed to obtain an analysable sample.

As detailed in Fig. 2.2, AFM has a second trump that renders the technique particularly attractive for the assessment of structure–function relationships. AFM measures these relationships under physiological conditions, i.e., in a physiological buffer, at room temperature and under atmospheric pressure. This is a major breakthrough compared with other high-resolution techniques that often demand vacuum or low-temperature conditions. Besides this technical advantage, the above-mentioned extraordinary SNR of AFM allows for studying more native samples than what any other techniques can survey (Fig. 2.2). Ultimately, biologists would like to see membrane proteins in native membranes directly on a live cell. This has been impossible to date because of physical reasons, such as the softness of the cell membrane and diffusion dynamics, which seem to occlude imaging of individual membrane molecules on a cell, at least in the near future. To date, we have been able to image native membranes *ex cellula*, which means that minimal biochemical treatments have been employed (cell breakage and density gradient centrifugation) to isolate membranes from other cellular constituents.

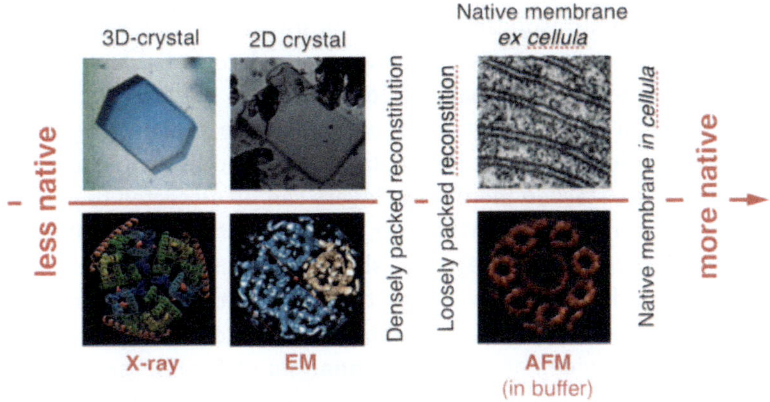

Figure 2.2. The nativeness of studied samples from left (less native) to the right (more native): membrane protein 3D crystal, membrane protein 2D crystal, densely packed reconstitution, loosely packed reconstitution, native membrane *ex cellula* and native membrane *in cellula*.

This chapter focuses on the imaging of proteins in native membranes by AFM, taking advantage of its capability to image single molecules in native samples. We will discuss the examples of AFM where novel information about molecular variability and supramolecular assemblies of membrane proteins, not analysable by any other technique to date, has been presented. The following sections focus on the studies about prokaryotic photosynthetic membranes and on the characterization of eukaryotic eye lens membranes. Finally, the power of the recently introduced high-speed AFM (HS-AFM) for biomembrane research is evidenced.

2.2 HIGH-RESOLUTION AFM IMAGING OF PROKARYOTIC MEMBRANES

The application of AFM in imaging membrane proteins has been successful in two-dimensional (2D) reconstituted systems, i.e., water channel aquaporin Z,[3] potassium channel KirBac3.1,[4] halorhodopsin,[5] outer membrane (OM) porins,[6] ATP synthase (ATPase)[7-12] and light-harvesting complexes.[13-17] Proteins of interest were initially isolated from biological membranes, followed by 2D reconstitution to form regular arrangements or dense packing. However, the shortcoming of this procedure is that the isolation and reconstitution processes may disturb the native state of membrane proteins. Understanding the function of a membrane protein requires its structural study in the native membrane; additionally, this allows the protein of interest to be observed together with other partner proteins.

In addition to high-resolution topographic imaging, another key advantage of AFM is its excellent capacity to nano-manipulate individual membrane proteins by applying additional loading forces to the imaging tip, along with adapted scan rates and feedback parameters to deliberately act on the surface of the biological object (Fig. 2.3a,d). When additional loading forces are relatively high, stacked membrane layers can be dissected to give access to underlying membranes.[18–22] Moreover, individual protein subunits of multi-protein complexes can be dissected at slightly increased forces, allowing for the analysis of underlying protein structures.[17,23,24] At low additional loading forces, individual protein domains can be manipulated. This process is non-destructive and provides access to the analysis of flexible protein surface motifs.[3,22,25] High-resolution imaging and manipulation often benefit each other during the study of membrane protein structure.

2.2.1 Surface Layers

Surface layers (S-layers) are regular, 2D protein networks, functioning as the outermost cell wall layer of many bacteria and archaea.[26,27] These layers withstand non-physiological pH, radiation, temperature, proteolysis, pressure and detergent treatment, thus protecting the cell from such hostile factors.[28,29]

PS2 is the protein that forms the S-layer of *Corynebacterium glutamicum*.[30,31] Native *C. glutamicum* S-layers stick together via their inner surfaces in aqueous condition, with the outer surfaces exposed on opposite sides (Fig. 2.3a). When imaging at a minimal loading force of 100 pN, the outer S-layer surface of the top layer is imaged at high resolution, showing triangle-shaped protrusions with unit cell dimensions of $a = b = 16.0 \pm 0.2$ nm and $\gamma = 60 \pm 1°$ (Fig. 2.3b).[20] The hexameric core between the triangles could be identified as the other side of the flower-shaped hexameric core visible on the inner surface. During image acquisition, the loading force to the tip is increased to 500 pN. This mechanical treatment punctures the top S-layer and gave the AFM tip access to the inner surface of the layer below without severe damage, exposing a flower-shaped surface (Fig. 2.3c). A 3D reconstruction of the S-layer architecture has been calculated from the topographies of both surfaces.[20]

The hexagonally packed intermediate (HPI) layer in *Deinococcus radiodurans* is another typical bacterial S-layer.[32,33] There are six identical protomers that form the HPI layer pore, and unplugged and plugged conformations have been observed. High-resolution AFM images, together with manipulation, exhibit both outer and inner layer surface (Fig. 2.3d). The outer surface of the HPI layer is a hexagonal lattice with a unit cell size of 18 nm (Fig. 2.3e).[34,35] This arrangement presents relatively large openings around the six-fold axes. The inner surface of the HPI layer shows conformational dynamics of pores: a "closed" pore with a central plug and an "open" pore without it (Fig. 2.3f).[35] On both S-layers from *C. glutamicum* and *D. radiodurans* force measurements have been performed.[20,32,36]

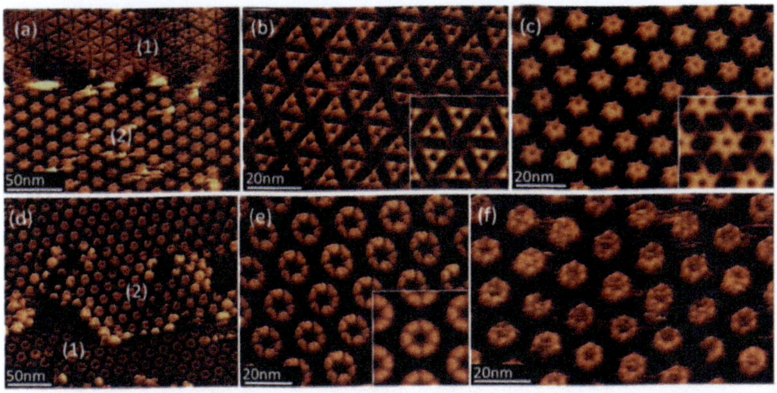

Figure 2.3. (a) Nano-dissection of the top outer surface of the S-layer of *Corynebacterium glutamicum* using AFM (1), providing access to the bottom S-layer and revealing the topography of the inner surface (2) of *C. glutamicum*.[20] (b) High-resolution topograph of the outer S-layer surface of *C. glutamicum* (inset: average). (c) High-resolution topograph of the inner S-layer surface (inset: average). (d) Nano-dissection and high-resolution imaging of S-layers of *Deinococcus radiodurans* using AFM, showing the bottom outer surface of the S-layer (1) and the top S-layer which exposes the inner surface (2) of *D. radiodurans*.[32,33] (e) High-resolution topograph of the outer S-layer surface of *D. radiodurans*. The high imaging contrast allows detection of the substructure on each individual subunit, revealing V-shaped units with a slight left-handed twist (inset: average). (f) High-resolution topograph of the inner S-layer surface of *D. radiodurans*.

Furthermore, the possibility to image the S-layer *in vivo* without invasive sample preparation has been proved.[37] More recently, Dupres and co-workers have made great effort in visualizing S-layer nanoarrays on living *C. glutamicum* bacterial cells.[38] The *in situ* high-resolution imaging is significant in understanding the structure of protein monomolecular arrays in their native state.

2.2.2 Outer Membrane

The OM of gram-negative bacteria protects the cell against bactericidal substances. Passage of nutrients and waste is assured by porins, the β-barrel transmembrane channels in OMs.

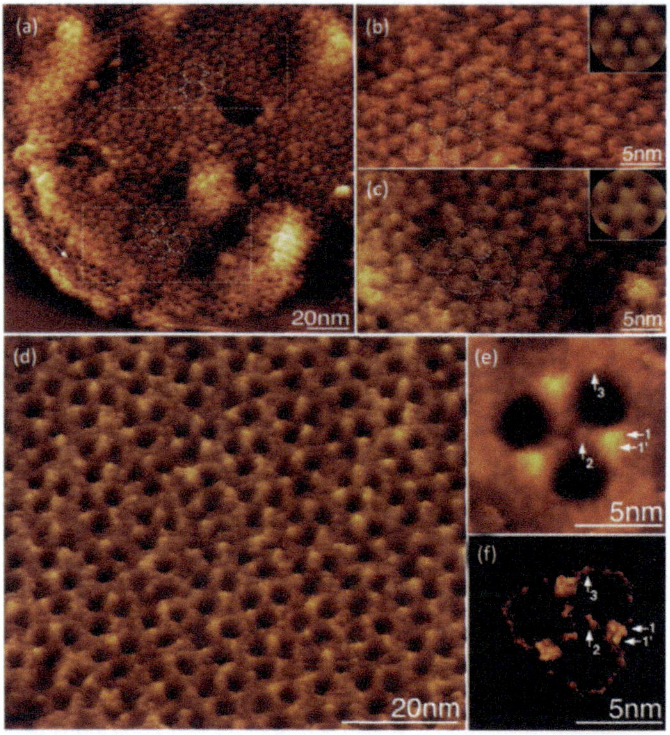

Figure 2.4. AFM images of the native extracellular and the periplasmic OM surfaces of *Roseobacter denitrificans*.[22] (a) Topograph of the top layer exposing the extracellular surface to the AFM tip. The arrow marks a border strip, which exposes the underlying layer showing the periplasmic surface. Encircled are the trimers representing two different conformations of the extracellular domains. (b and c) The two conformations of the extracellular protrusions correlate with their location within the membrane. The "relaxed" conformation (top) was found inside the membrane patch (dashed outlines). The three-fold symmetrized average discloses a well-preserved, central indentation. The "contracted" conformation (bottom) displayed by porin located close to the membrane's border (dashed outlines). (d) High-resolution analysis of the periplasmic surface of the OM. (e) High-resolution average revealing substructure comparable to (f), the X-ray structure of a homologous porin (PDB 1PRN[39]).

The first high-resolution AFM view of a bacterial OM revealed that porins cover 70% of the membrane surface and form locally regular lattices in *Roseobacter denitrificans* (*R. denitrificans*).[22] After nano-manipulation to remove peptidoglycan remnants or stacked layers, both extracellular and periplasmic surfaces were visualized. The extracellular surface of porins exists in two distinct conformations: one with separated protrusions and another with protrusions contracted at the three-fold axis (Fig. 2.4a). Their occurrence was correlated with their position within the membrane: trimers which were positioned in the centre of a membrane displayed the "relaxed" conformation (Fig. 2.4b), whereas the trimers at the membrane edges displayed the "contracted" conformation (Fig. 2.4c), a phenomenon attributed to trapped ions under the central membrane region. The periplasmic surface of the porins exhibited oval-shaped cavities separated by walls crowned by three major protrusions with three-fold axis at the pore brims (Fig. 2.4d). The unit cell in ordered regions is $a = b = 81 \pm 2.5$ Å, $\gamma = 60°$. High-resolution topography averages could be compared to molecular surface representations of the X-ray structure[39] in great detail (Fig. 2.4e,f).

2.3 PHOTOSYNTHETIC MEMBRANE

Owing to its high SNR ratio, AFM has emerged as an indispensable tool that allows for the structural identification of individual membrane proteins not only with regular protein arrangements but also with non-ordered assemblies. A striking breakthrough is the high-resolution AFM imaging of bacterial photosynthetic membranes in contact mode, which has provided the first surface views of the organization of multi-component biological membranes at submolecular resolution.[40,41]

In photosynthesis, light capture and energy transfer are efficiently accomplished by the strong cooperativeness among different photosynthetic components, i.e., light-harvesting complexes (LH1 and LH2), reaction centre (RC), cytochrome (cyt) bc_1 complex and ATPase, which are embedded in the photosynthetic membranes. The shape of bacterial photosynthetic membranes is highly species dependent. They may be arranged in regular parallel layers (*Rhodopseudomonas* [*Rps.*] *viridis*, *Rhodopseudomonas* [*Rps.*] *palustris*, *Rhodospirillum* [*Rsp.*] *photometricum*, *Phaeospirillum* [*Phsp.*] *molischianum*) or form vesicles (*Rhodobactor* [*Rb.*] *spheroides*, *Rhodobactor* [*Rb.*] *blasticus*), or small regular stacks of tubular lamellae (*Rhodospirillum* [*Rsp.*] *fulvum*, *Thiocapsa spp.*). The information concerning protein domain formation, complex assembly and protein heterogeneity, derived from high-resolution AFM imaging, provided important insights into the physiological roles of the photosynthetic machinery.

AFM imaging of native photosynthetic membranes has provided detailed information about the architecture of RC-LH1 core complexes and the LH2 complexes. Indeed, the core-complex architecture varied considerably between species: core complexes in a native membrane of *Rps. viridis*,[24] *Rsp. photometricum*[21,42,43] and *Phsp. molischianum*[44] had $LH1_{16}$-$RC_{L,M,H}$ architecture, topped by a non-membranous tetraheme cytochrome (4Hcyt), that was not visible in *Phsp. molischianum*. In native membranes of *Rb. blasticus*[45] the core complexes had $(PufX_2$-$LH1_{13}$-$RC_{L,M,H})_2$ architecture probably like the *Rba. sphaeroides* core,[46] but the precise Rhodobacter core-complex architecture is still a matter of debate.[45–48] Finally, core complexes in *Rps. palustris* membranes had W-$LH1_{15}$-$RC_{L,M,H}$ architecture.[49] Similarly, high-resolution AFM allowed studying LH2 complex architecture at the single-molecule level and depicting molecular heterogeneity *in situ*. The structure of LH2 complexes in native membranes was found to be rather variable. In the study of LH2 in *Rsp. photometricum* membranes, in addition to normal nonameric LH2 complexes, several different types of complexes were observed:[50] About 10% of the complexes were octameric and another 10% were decameric. This size heterogeneity was attributed to the known spectral heterogeneity of LH2. In addition, some of them presented small rings containing six or seven subunits, open C-shaped complexes, or large complexes containing up to 14 subunits, maybe LH2/LH1 chimeric rings.[50] Such heterogeneous stoichiometry appears to be an inherent feature of LH2, as it has also been observed in *Phsp. molischianum* [44] and *Rps. palustris*.[49]

Beyond the structure analysis of the individual components, the AFM gives the exciting possibility of studying supramolecular non-ordered protein assemblies in the native membrane. Analysis of the distribution of photosynthetic complexes showed clustering of LH2 and RC-LH1 core complexes in native membranes (Fig. 2.5). Clustering of complexes is a functional necessity, as each light-harvesting component must pass its harvested energy to a neighbouring complex and eventually to the RC. Clustering of bacterial photosynthetic complexes has been seen in membranes from all different species studied so far.[41] With the exception of *Rb. sphaeroides*,[51] no regular structural assembly of LH2 and core complexes were observed, and these photosynthetic complexes were not in an ordered arrangement in the native membranes. Core complexes completely surrounded by several LH2 and core complexes making multiple core–core contacts were visualized (Fig. 2.5a)[21,42,43]. However, their organization is far from random. Pair correlation function analysis has shown that there is a most favourable assembly within these membranes, which is core–LH2–

core–LH2 and so on. Additional LH2 synthetized under low-light conditions segregate in antenna domains. The photosynthetic apparatus structure of *Phsp. molischianum* resembles that of *Rsp. photometricum.*[44] LH complexes are segregated into two structurally different types of domains, consisting of a mixture of core and octameric LH2 complexes, as well as paracrystalline, hexagonally packed octameric LH2 rings. One specific example is the photosynthetic membrane of *Rps. viridis* which has no LH2 complexes and shows core complexes forming a hexagonal lattice in the photosynthetic membrane.[24]

Photosynthetic membranes are densely packed with photosynthetic proteins favourable for excitation energy transfer; however, such a dense packing becomes an obstacle for efficient quinone/quinol membrane diffusion between cores and cyt bc_1 complex. Pair correlation function analysis of the entire photosynthetic membrane showed that core complexes influence their molecular environment within a critical radius of 250 Å.[21] Analysis of the molecular environment further indicated that the local environment of core complexes contained more lipid spaces within the membrane, due to the geometric mismatch between cores and LH2. Short-range core interaction and larger lipid space finally establish long-range quinone/quinol pathways through the entire photosynthetic membranes (Fig. 2.5b).[21]

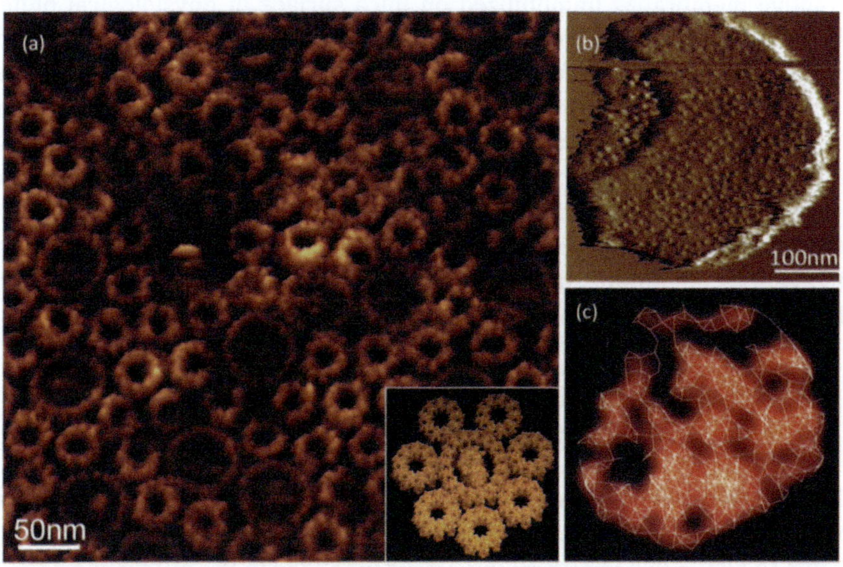

Figure 2.5. (a) AFM topograph of the native photosynthetic membranes of *Rsp. photometricum*, showing the specific assembly of LH2 and core complexes (Inset: structural model of native protein assembly).[52] (b) AFM deflection image of entire native chromatophores of *Rsp. photometricum* and (c) calculated quinone pathways.[21]

2.4 HIGH-RESOLUTION AFM IMAGING OF EUKARYOTIC MEMBRANES

Eukaryotic cells are systems of higher complexity compared with prokaryotes. Eukaryotic cells employ specialized cellular compartments to effectively manage their increased demands in exchange of matter, energy generation and signal transduction with the environment. These specialized substructures are defined by biological membranes; a lipid bilayer forms intracellular compartments and serves as a matrix for numerous membrane proteins. The membrane proteins are known to be of key importance for many vital cellular functions, some of which are mentioned above. High-resolution studies of individual membrane proteins enhanced our understanding of the molecular mechanisms of their functions. However, many membrane proteins serve as a part of supramolecular assemblies, or complex protein machineries, which fulfil ensemble functions. This is in agreement with the emerging concept that eukaryotic membranes are much higher organized than previously assumed.[53] Since many membrane proteins are implicated in important functions, it is also clear that disorders of functional ensembles of membrane proteins may lead to various pathologies.

Since AFM allows for the investigation of the protein complexes directly in native membranes immobilized on mica surface in physiological buffer solution at room temperature and normal pressure, it may become a powerful technique to analyse native and malformed membrane protein assemblies in eukaryotic membranes. Only few studies have so far reported high-resolution views of eukaryotic membranes, but the potential to see individual eukaryotic membrane proteins within their native context is appealing as demonstrated by rhodopsin imaging in disk membranes.[54] Here, we discuss some of the few high-resolution AFM studies on native eukaryotic membranes and one case, the eye lens membrane, of which native and pathological assemblies were distinguishable.

2.4.1 Outer Mitochondrial Membranes: Supramolecular Organization of VDAC

The voltage-dependent anion channel (VDAC) is a 30 kDa protein found in the mitochondrial outer membranes (MOM) of all eukaryotes.[55] VDAC was shown to play a role in several important functions in the mitochondria and also in the cell. It is the most abundant protein in the external membrane of the mitochondria and serves as a major gate for molecules connecting mitochondria and cytoplasm.[56,57]

MOM were purified from yeast as previously described[58] and adsorbed on clean mica surfaces. In medium resolution topographs, corrugated VDAC-containing membrane areas were easily distinguished from the smoother, lipid bilayer regions, allowing the estimation of pore packing densities.[59] Mixed domains contained VDAC pores at low (~20%) density; in other regions the pores were packed at high (~80%) density (Fig. 2.6a,b). Single VDAC molecules were imaged with characteristic pore dimensions of 3.8 × 2.7 nm and were ~2 nm in measured depth.

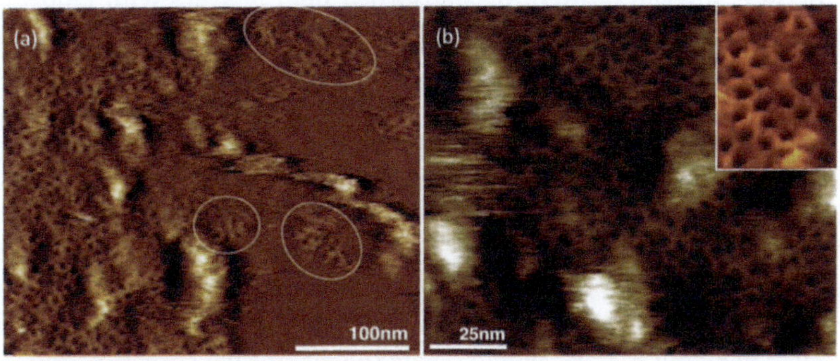

Figure 2.6. Supramolecular organization of VDAC in the mitochondrial outer membranes. (a) Supramolecular organization of VDAC channels. High-density region on the left and low-density region on the right with some protein "islands" of variable size, ranging from two to about twenty VDAC (outlines). (b) High-resolution AFM analysis of densely packed region of VDAC (pore dimensions 3.8 x 2.7 nm; pore depth ~2 nm).

The obtained results were in line with previously suggested mechanisms of VDAC's channel regulation that links VDAC's functionality to its membrane surface density.[58] High mobility of VDAC groups in the observed low-density mixed domains (Fig. 2.6a) could trigger VDAC association in the densely packed domains and thereby act as a regulator of VDAC activity on the cellular level. Such an association–dissociation equilibrium is a simple way of modulating channelling activity. It also points towards possible ways to control VDAC activity with eventual pharmaceutical agents modifying the oligomerization properties of VDAC molecules.

2.4.2 Inner Mitochondrial Membranes: Rows of ATP Synthase Dimers

The ATP synthase functions as a nanometric rotary machine that employs a transmembrane electrochemical gradient to produce ATP.[60] In spite of the recent success in structural studies of most components of the ATP synthase, the supramolecular assembly of ATP synthases in biological membranes remains unclear. A number of indirect studies indicated the importance of subunits *e* and *g* for ATP synthase dimerization.[61,62] Also the so-called inhibitory factor peptide IF1 was shown to interact with ATP synthase dimers.[63,64] AFM was used to investigate native inner mitochondrial membranes from yeasts and, with submolecular resolution, showed the supramolecular organization of ATP synthases.

The AFM images showed mica-adsorbed lipid bilayers with embedded ring-shaped molecules with characteristic diameter of ~8 nm. These objects were identified as ATP synthase c-rings, viewed on the periplasmic surface of the inner mitochondrial membrane (Fig. 2.7a).[65] ATP synthase molecules form dimers with characteristic 15 nm distance between the rotor axes through stereospecific interactions of the membrane-embedded portions of their stators (Figs. 2.7b and 2.7c). According to this model, the surfaces of subunits *e* and *g* are responsible for dimer formation while subunit *b* plays a role in the formation of rows of dimers. Such an organization reinforces the role of the ATP synthase in mitochondrial morphology, where the sterical mismatch of F_0 and F_1 parts would create the membrane curvature and thus contribute to the formation of mitochondrial cristae.[62] Some ATP synthase dimers have 10 nm stalk-to-stalk distance, interpreted as ATP synthases that are accessible to IF1 inhibition. Existence of mitochondrial ATP synthases in functional rows dimers was supported by cryo-tomography studies.[66]

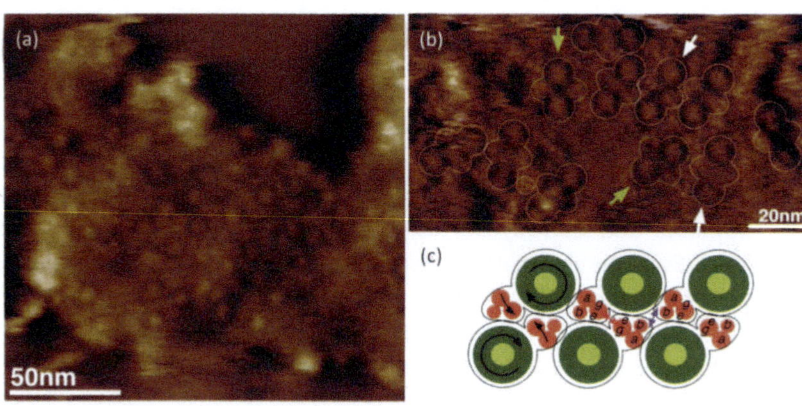

Figure 2.7. Supramolecular organization of ATP synthases in native mitochondrial inner membranes. (a) Topograph of mica-adsorbed inner mitochondrial membrane with rows of ATP synthases. (b) High-resolution image of ATP synthase dimers (outlined). Dimer with intermolecular distances of 15 nm and 10 nm are marked with white and yellows arrows, respectively. (c) Schematic representation of dimer and oligomer assembly.

2.4.3 Eye Lens Membranes: Organization of Protein Junctional Microdomains

The main function of the eye lens is to focus light onto the retina. This underlines the importance of lens transparency, which is achieved by a number of adaptations. To avoid light scattering, the lens tissue is avascular and built of densely packed arrays of fibre cells, maturating from the lens epithelium towards the core. To minimize light scattering, organelles are degraded in the fibre lens cells during cell differentiation and around 90% of the fibre cell protein content presented by crystallins.[67] Lens fibres are held together by junctional protein microdomains, combining thin and gap junctions, connecting cells and permitting fluid flow in a so-called microcirculation system that provides core fibre cells with nutrients and canalizes metabolic waste.[68]

Thin junctions are formed by aquaporin-0 (AQP0), which is the lens-specific mammalian aquaporin. AQP0 comprises up to 60% of overall membrane protein content in lens tissue.[69] The structure of AQP0 is determined to a resolution of 1.9 Å, using reconstituted 2D crystals and electron crystallography,[70] and to 2.2 Å, using X-ray diffraction of 3D crystals.[71] Gap-juctions formed by connexins are the second most abundant membrane proteins; they constitute more than 10% of the total membrane proteins in lenses.[72] Recently, the high-resolution structure of connexin in the gap-junction channel form was solved.[73]

AFM studies have shown that membrane preparations from healthy eye lenses contain large lipid bilayer fragments (Fig. 2.8a) with preserved intercellular junctions that appeared as corrugated protrusions on AFM topographs.[74] Some of the intercellular junctions were dissected during scanning, exposing the extracellular surface of the junctional protein microdomains to the AFM tip (Fig. 2.8b). High-resolution AFM analysis showed that these domains consisted of tetragonally arranged ($a = b =$ 65.5 Å, $\gamma = 90°$) AQP0 tetramers surrounded by densely packed non-ordered gap-junction connexon channels (Fig. 2.8c). It was postulated that connexons act as lineactants inside the membrane and confine AQP0 in the junctional microdomains. These microdomains simultaneously provide adhesion and communication between fibre cells (Fig. 2.8e, left). This first high-resolution view of a multi-component eukaryotic membrane showed membrane proteins self-assembled into functional microdomains.

Cataracts are opacifications of eye lenses and the leading cause of blindness in the world, especially among the senior population, and currently, surgery is the only cure. Besides age, recognized risk factors of cataract are ultraviolet and radiation exposure, hypertension and diabetes diseases, side effect of certain pharmaceuticals. A comparative AFM study of the membrane protein organization in native and cataract-affected eye lens membranes allowed to extend our knowledge about the molecular bases of cataract in an individual patient and to highlight the potential of AFM as a future medical imaging tool.

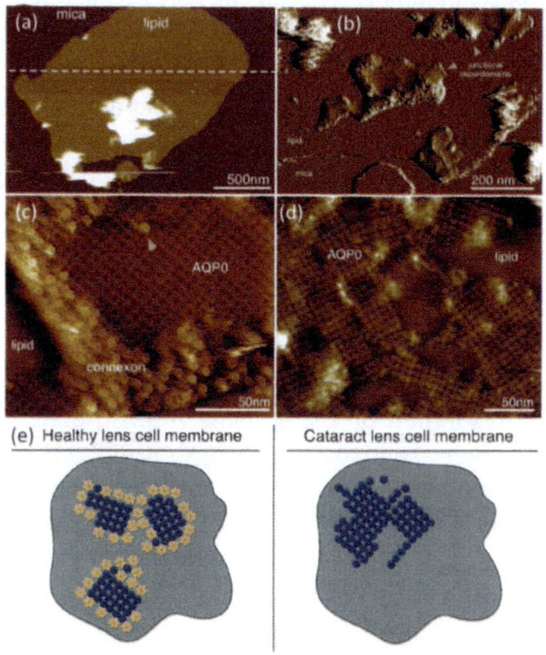

Figure 2.8. Junctional microdomains in lens membranes from healthy and cataract-affected eye lenses. (a) AFM image of a lens fibre cell membrane fragment on mica surface. The membrane thickness measured along the dashed line is about 4.5 nm. (b) The extracellular face of junctional microdomains (marked by arrows). (c) High-resolution AFM image of a junctional microdomain. Aquaporin-0 (AQP0)-formed 2D arrays edged by closed packed connexons. (d) AFM topograph of a junctional microdomain from a cataract-affected eye lens. AQP0 constituted malformed arrays and connexon domains were absent. (e) Models of the supramolecular assembly of junctional microdomains in lens core cell membranes (*left:* healthy; *right:* cataract-affected). In the healthy case, AQP0 form square arrays edged by connexons, providing water and metabolite transport together, and cell adhesion. In the pathological case, connexons were absent and AQP0 assemble in larger and continuous but less-ordered domains. Failure of the lens microcirculation causes clouding of the lens.

Eye lens membranes originating from patients with senile cataract and type II diabetes–induced cataract have been studied.[75,76] In both cases, the membranes contained larger and less structured AQP0 domains arrays compared to the native case. More importantly, in the cataract membranes connexons were absent at the edges of the AQP0 arrays and in the membrane in general (Fig. 2.8d). The absence of connexons in lens membranes obviously led to the fusion and malformation of AQP0 domains and failure of intercellular communication in the tissue (Fig. 2.8e, right). As connexons function as ion, metabolite and waste channels between neighbouring cells, their absence is certainly responsible for the breakdown of the microcirculation system[68] in the lens tissue, thus causing cell death and leading to lens opacification.

2.5 HIGH-SPEED AFM STUDIES OF THE DYNAMICS OF BIOMOLECULES

AFM[77] is a powerful tool for the characterization of biological molecules, providing high-resolution topographic data of biological molecules under physiological buffer conditions at room temperature and ambient pressure. This capability of imaging in conditions where biomolecules are functional makes AFM the ideal technique for characterizing the dynamics of biomolecules. As a matter of fact, only three years after the invention of AFM in 1986 the first observations of the dynamic clotting aggregation processes were performed;[78] few other studies were performed over the next years on antibody binding processes[79] and DNA–protein interactions.[80,81] After that period AFM was perhaps less used to study dynamic processes of biomolecules mainly because of the limitation of the imaging speed of conventional AFMs that is around one minute per image. Such an imaging rate is simply too slow for capturing the dynamics of most biomolecular processes. Therefore, the AFM has mainly been used for structural studies of native proteins or as a force-measurement tool.[1] To expand the use of AFM for the study of the dynamics of biomolecules, a significant increase of the imaging speed is required.

In response to the need for faster imaging rates, a new generation of faster high-speed (HS) atomic force microscopes have been developed in recent years. The key feature of this new generation of AFMs is the increase of the speed of response of the moving components of the AFM setups, in particular the probe and the piezoelectric stage. The increase of speed of the moving components is obtained by reducing their dimensions.[82] Different research groups have been active on this development.[83–88] Best performances on biomolecule imaging to date range around an imaging speed of 30 ms per frame with about 150 x 150 pixels (for more details, see Chapter 8).[89]

The principle of the AFM is based on the force of interaction between the probe and the sample. For successful imaging of biomolecules, the force applied by the probe on the sample must be precisely controlled: biomolecules are of soft nature and get damaged easily when an excessive force (typically higher than 300 pN[90]) is applied. To obtain this precise force control, two strategies have been used in conventional slow AFMs: (1) the contact mode imaging using soft probes of spring constant of 0.01–0.1 N/m—these probes provide high deflection-to-applied-force ratio and thus allow contouring the sample while maintaining the applied forces within a small range of several tens of piconewtons; (2) the oscillatory mode imaging using probes of spring constant of 0.1 N/m—the oscillation minimizes the contact time of the probe and the sample, reducing the friction between probe and sample. Both strategies have yielded similar image quality with a lateral resolution of about 1 nm. Yet, they differ in that the contact mode is around 5 to 10 times faster than the oscillatory mode but limited to samples with low corrugation, while the oscillatory mode can image samples of higher corrugations but at a slower rate.

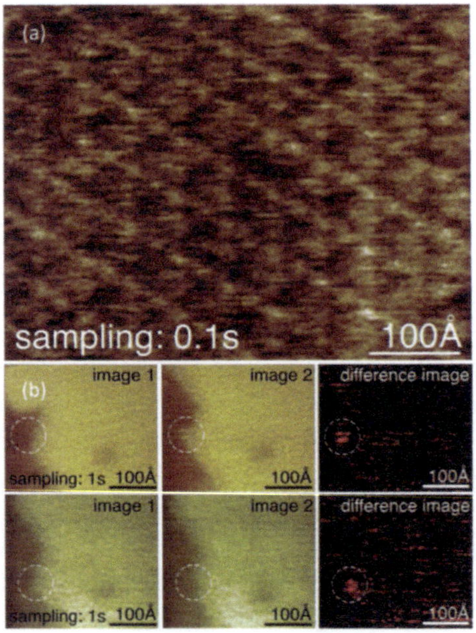

Figure 2.9. High-speed contact mode AFM imaging of purple membrane. (a) Topograph at submolecular resolution at an imaging rate of 10 frames per second. (b) Monitoring association and dissociation of bR trimers from and to the edge of the bR array.

From the two strategies used for the high-resolution imaging of biomolecules with the conventional AFM, only the oscillatory mode has been implemented in the high-speed and high-resolution AFM imaging of biomolecules. The reasons for this are probably the concerns about causing sample damage due to the higher stiffness of the short high-speed probes compared with the probes used in contact mode in conventional AFMs, the instinctive association of high speed with high friction, and the fact that samples mainly studied with the HS-AFM are isolated biomolecules lying on flat substrates, thus creating elevated local corrugations.[91] Only recently, contact mode has been used for high-speed high-resolution AFM imaging. The sample selected was the purple membrane, a well-structured lattice of the bacteriorhodopsin (bR) protein that shows a low corrugation (Fig. 2.9a).[92] The results show that a membrane sample is stable under high-speed contact mode AFM imaging and, furthermore, that the high-speed contact mode imaging results in comparable resolution to imaging of the purple membrane using conventional AFM or HS-AFM in oscillatory mode.[93] Interestingly, the high-speed imaging rate obtained for the contact and oscillatory modes was the same in both cases—10 frames per second. This contrasts with conventional AFM, where the contact mode allows for an imaging rate on the biological membrane that is between 5 to 10 times faster than the oscillatory mode. Probably factors not related to the probe, such as piezoelectric stage or the control electronics, could have limited the speed in contact mode HS-AFM, and future works could show slightly higher imaging speeds in contact mode HS-AFM with respect to the oscillatory mode for imaging low-corrugation samples the same way as in conventional AFM. At lower speed, the imaging contrast was higher. Individual bR trimers associating and dissociating to the edges of the bR array could be monitored (Fig. 2.9b).[94] Similar results have been acquired using oscillating mode HS-AFM.[93]

Finally, it is important to highlight the role of the apex of the AFM probe on the quality of AFM imaging. An AFM probe comes in contact of the sample only through the apex of the microfabricated pyramid that is located at the end of the cantilever. The shape of the tip of the apex determines the imaging resolution and the pyramid the distribution of the applied force over the sample. In biological samples with low corrugation, such as the purple membrane, it has been shown by using conventional slow AFM that a blunt apex presenting some nanometric protrusion can provide the optimal imaging conditions for contact mode imaging. The reason is that the force applied by the probe is distributed over a larger area, but that the protrusion is sharp enough to provide a local sensitivity and obtain high-resolution images.[90] For the high-speed contact mode AFM studies[92] the blunt probes were also used and provided optimal imaging conditions. In contrast, probes with a sharper apex tended to easily damage the samples in contact mode HS-AFM. On the other hand, previous high-speed studies in oscillatory mode typically used a sharpened apex. The high corrugation of the isolated samples studied required sharp, high–aspect ratio probes to minimize the convolution and enhance the resolution.[91]

Acknowledgements

The work in our team was supported by the Institut Curie, the Institut National de la Santé et Recherche Médicale (INSERM), the Centre National de la Recherche Scientifique (CNRS), the Agence Nationale de la Recherche (ANR), and the City of Paris.

References

1. Engel, A., and Gaub, H. E. (2008) Structure and mechanics of membrane proteins, *Annu. Rev. Biochem.*, **77**, 127–148.
2. Fechner, P., Boudier, T., Mangenot, S., Jarosławski, S., Sturgis, J. N., and Scheuring, S. (2009) Structural information, resolution, and noise in high-resolution atomic force microscopy topographs, *Biophys. J.*, **96**, 3822–3831.
3. Scheuring, S., Ringler, P., Borgnia, M., Stahlberg, H., Müller, D. J., Agre, P., and Engel, A. (1999) High resolution AFM topographs of the *Escherichia coli* water channel aquaporin Z, *EMBO J.*, **18**, 4981–4987.
4. Jarosławski, S., Zadek, B., Ashcroft, F., Venien-Bryan, C., and Scheuring, S. (2007) Direct visualization of KirBac3.1 potassium channel gating by atomic force microscopy, *J. Mol. Biol.*, **374**, 500–505.
5. Persike, N., Pfeiffer, M., Guckenberger, R., Radmacher, M., and Fritz, M. (2001) Direct observation of different surface structures on high-resolution images of native halorhodopsin, *J. Mol. Biol.*, **310**, 773–780.
6. Müller, D. J., and Engel, A. (1999) Voltage and pH-induced channel closure of porin OmpF visualized by atomic force microscopy, *J. Mol. Biol.*, **285**, 1347–1351.
7. Müller, D. J., Engel, A., Matthey, U., Meier, T., Dimroth, P., and Suda, K. (2003) Observing membrane protein diffusion at subnanometer resolution, *J. Mol. Biol.*, **327**, 925–930.
8. Arechaga, I., and Fotiadis, D. (2007) Reconstitution of mitochondrial ATP synthase into lipid bilayers for structural analysis, *J. Struct. Biol.*, **160**, 287–294.
9. Matthies, D., Preiss, L., Klyszejko, A. L., Müller, D. J., Cook, G. M., Vonck, J., and Meier, T. (2009) The c13 ring from a thermoalkaliphilic ATP synthase reveals an extended diameter due to a special structural region, *J. Mol. Biol.*, **388**, 611–618.
10. Pogoryelov, D., Yu, J., Meier, T., Vonck, J., Dimroth, P., and Müller, D. J. (2005) The c15 ring of the *Spirulina platensis* F-ATP synthase: F1/F0 symmetry mismatch is not obligatory, *EMBO Rep.*, **6**, 1040–1044.
11. Stahlberg, H., Müller, D. J., Suda, K., Fotiadis, D., Engel, A., Meier, T., Matthey, U., and Dimroth, P. (2001) Bacterial Na(+)-ATP synthase has an undecameric rotor, *EMBO Rep.*, **2**, 229–233.

12. Seelert, H., Poetsch, A., Dencher, N. A., Engel, A., Stahlberg, H., and Müller, D. J. (2000) Proton-powered turbine of a plant motor, *Nature*, **405**, 418–419.

13. Bahatyrova, S., Frese, R. N., van der Werf, K. O., Otto, C., Hunter, C. N., and Olsen, J. D. (2004) Flexibility and size heterogeneity of the LH1 light harvesting complex revealed by atomic force microscopy: functional significance for bacterial photosynthesis, *J. Biol. Chem.*, **279**, 21327–21333.

14. Scheuring, S., Seguin, J., Marco, S., Lévy, D., Breyton, C., Robert, B., and Rigaud, J. L. (2003) AFM characterization of tilt and intrinsic flexibility of *Rhodobacter sphaeroides* light harvesting complex 2 (LH2), *J. Mol. Biol.*, **325**, 569–580.

15. Scheuring, S., Reiss-Husson, F., Engel, A., Rigaud, J. L., and Ranck, J. L. (2001) High-resolution AFM topographs of *Rubrivivax gelatinosus* light-harvesting complex LH2, *EMBO J.*, **20**, 3029–3035.

16. Gonçalves, R. P., Busselez, J., Lévy, D., Seguin, J., and Scheuring, S. (2005) Membrane insertion of *Rhodopseudomonas acidophila* light harvesting complex 2 investigated by high resolution AFM, *J. Struct. Biol.*, **149**, 79–86.

17. Liu, L. N., Aartsma, T. J., and Frese, R. N. (2008) Dimers of light-harvesting complex 2 from *Rhodobacter sphaeroides* characterized in reconstituted 2D crystals with atomic force microscopy, *FEBS J.*, **275**, 3157–3166.

18. Hoh, J. H., Lal, R., John, S. A., Revel, J. P., and Arnsdorf, M. F. (1991) Atomic force microscopy and dissection of gap junctions, *Science*, **253**, 1405–1408.

19. Schabert, F. A., Henn, C., and Engel, A. (1995) Native *Escherichia coli* OmpF porin surfaces probed by atomic force microscopy, *Science*, **268**, 92–94.

20. Scheuring, S., Stahlberg, H., Chami, M., Houssin, C., Rigaud, J. L., and Engel, A. (2002) Charting and unzipping the surface layer of *Corynebacterium glutamicum* with the atomic force microscope, *Mol. Microbiol.*, **44**, 675–684.

21. Liu, L. N., Duquesne, K., Sturgis, J. N., and Scheuring, S. (2009) Quinone pathways in entire photosynthetic chromatophores of *Rhodospirillum photometricum*, *J. Mol. Biol.*, **393**, 27–35.

22. Jarosławski, S., Duquesne, K., Sturgis, J. N., and Scheuring, S. (2009) High-resolution architecture of the outer membrane of the Gram-negative bacteria *Roseobacter denitrificans*, *Mol. Microbiol.*, **74**, 1211–1222.

23. Fotiadis, D., Müller, D. J., Tsiotis, G., Hasler, L., Tittmann, P., Mini, T., Jeno, P., Gross, H., and Engel, A. (1998) Surface analysis of the photosystem I complex by electron and atomic force microscopy, *J. Mol. Biol.*, **283**, 83–94.

24. Scheuring, S., Seguin, J., Marco, S., Lévy, D., Robert, B., and Rigaud, J. L. (2003) Nanodissection and high-resolution imaging of the *Rhodopseudomonas viridis* photosynthetic core complex in native membranes by AFM, *Proc. Natl. Acad. Sci. U S A*, **100**, 1690–1693.

25. Müller, D. J., Büldt, G., and Engel, A. (1995) Force-induced conformational change of bacteriorhodopsin, *J. Mol. Biol.*, **249**, 239–243.

26. Sleytr, U. B., and Sara, M. (1997) Bacterial and archaeal S-layer proteins: structure-function relationships and their biotechnological applications, *Trends Biotechnol.*, **15**, 20–26.

27. Baumeister, W., Wildhaber, I., and Phipps, B. M. (1989) Principles of organization in eubacterial and archaebacterial surface proteins, *Can. J. Microbiol.*, **35**, 215–227.

28. Engelhardt, H., and Peters, J. (1998) Structural research on surface layers: a focus on stability, surface layer homology domains, and surface layer-cell wall interactions, *J. Struct. Biol.*, **124**, 276–302.

29. Gonçalves, R. P., and Scheuring, S. (2006) Manipulating and imaging individual membrane proteins by AFM, *Surf. Interface Anal.*, **38**, 1413–1418.

30. Peyret, J. L., Bayan, N., Joliff, G., Gulik-Krzywicki, T., Mathieu, L., Schechter, E., and Leblon, G. (1993) Characterization of the cspB gene encoding PS2, an ordered surface-layer protein in *Corynebacterium glutamicum*, *Mol. Microbiol.*, **9**, 97–109.

31. Chami, M., Bayan, N., Dedieu, J., Leblon, G., Shechter, E., and Gulik-Krzywicki, T. (1995) Organization of the outer layers of the cell envelope of *Corynebacterium glutamicum*: a combined freeze-etch electron microscopy and biochemical study, *Biol. Cell.*, **83**, 219–229.

32. Karrasch, S., Hegerl, R., Hoh, J. H., Baumeister, W., and Engel, A. (1994) Atomic force microscopy produces faithful high-resolution images of protein surfaces in an aqueous environment, *Proc. Natl. Acad. Sci. U S A*, **91**, 836–838.

33. Müller, D. J., Baumeister, W., and Engel, A. (1996) Conformational change of the hexagonally packed intermediate layer of *Deinococcus radiodurans* monitored by atomic force microscopy, *J. Bacteriol.*, **178**, 3025–3030.

34. Müller, D. J., Baumeister, W., and Engel, A. (1999) Controlled unzipping of a bacterial surface layer with atomic force microscopy, *Proc. Natl. Acad. Sci. U S A*, **96**, 13170–13174.

35. Oesterhelt, F., and Scheuring, S. (2006) High-resolution imaging and force measurement of individual membrane proteins by AFM, *Curr. Nanosc.*, **2**, 329–335.

36. Goncalves, R. P., Agnus, G., Sens, P., Houssin, C., Bartenlian, B., and Scheuring, S. (2006) Two-chamber AFM: probing membrane proteins separating two aqueous compartments, *Nat. Methods*, **3**, 1007–1012.

37. Lister, T. E., and Pinhero, P. J. (2001) *In vivo* atomic force microscopy of surface proteins on *Deinococcus radiodurans*, *Langmuir*, **17**, 2624–2628.

38. Dupres, V., Alsteens, D., Pauwels, K., and Dufrene, Y. F. (2009) *In vivo* imaging of S-layer nanoarrays on *Corynebacterium glutamicum*, *Langmuir*, **25**, 9653–9655.

39. Kreusch, A., and Schulz, G. (1994) Refined structure of the porin from Rhodopseudomonas blastica. Comparison with the porin from Rhodobacter capsulatus, *J. Mol. Biol.*, **243**, 891–905.

40. Sturgis, J. N., Tucker, J. D., Olsen, J. D., Hunter, C. N., and Niederman, R. A. (2009) Atomic force microscopy studies of native photosynthetic membranes, *Biochemistry*, **48**, 3679–3698.

41. Scheuring, S., and Sturgis, J. N. (2009) Atomic force microscopy of the bacterial photosynthetic apparatus: Plain pictures of an elaborate machinery, *Photosynth. Res.*, **102**, 197–211.

42. Scheuring, S., Sturgis, J. N., Prima, V., Bernadac, A., Levy, D., and Rigaud, J. L. (2004) Watching the photosynthetic apparatus in native membranes, *Proc. Natl. Acad. Sci. U S A*, **101**, 11293–11297.

43. Scheuring, S., and Sturgis, J. N. (2005) Chromatic adaptation of photosynthetic membranes, *Science*, **309**, 484–487.

44. Gonçalves, R. P., Bernadac, A., Sturgis, J. N., and Scheuring, S. (2005) Architecture of the native photosynthetic apparatus of *Phaeospirillum molischianum*, *J. Struct. Biol.*, **152**, 221–228.

45. Scheuring, S., Busselez, J., and Lévy, D. (2005) Structure of the dimeric PufX-containing core complex of *Rhodobacter blasticus* by in situ atomic force microscopy, *J. Biol. Chem.*, **280**, 1426–1431.

46. Scheuring, S., Francia, F., Busselez, J., Melandri, B. A., Rigaud, J. L., and Lévy, D. (2004) Structural role of PufX in the dimerization of the photosynthetic core complex of *Rhodobacter sphaeroides*, *J. Biol. Chem.*, **279**, 3620–3626.

47. Qian, P., Bullough, P. A., and Hunter, C. N. (2008) Three-dimensional reconstruction of a membrane-bending complex: The RC-LH1-PufX core dimer of *Rhodobacter sphaeroides*, *J. Biol. Chem.*, **283**, 14002–14011.

48. Qian, P., Hunter, C. N., and Bullough, P. A. (2005) The 8.5 Å projection structure of the core RC-LH1-PufX dimer of *Rhodobacter sphaeroides*, *J. Mol. Biol.*, **349**, 948–960.

49. Scheuring, S., Gonçalves, R. P., Prima, V., and Sturgis, J. N. (2006). The photosynthetic apparatus of *Rhodopseudomonas palustris*: structures and organization, *J. Mol. Biol.*, **358**, 83–96.

50. Scheuring, S., Rigaud, J. L., and Sturgis, J. N. (2004) Variable LH2 stoichiometry and core clustering in native membranes of *Rhodospirillum photometricum*, *EMBO J.*, **23**, 4127–4133.

51. Bahatyrova, S., Frese, R. N., Siebert, C. A., Olsen, J. D., Van Der Werf, K. O., van Grondelle, R., Niederman, R. A., Bullough, P. A., and Hunter, C. N. (2004) The native architecture of a photosynthetic membrane, *Nature*, **430**, 1058–1062.

52. Scheuring, S., Boudier, T., and Sturgis, J. N. (2007) From high-resolution AFM topographs to atomic models of supramolecular assemblies, *J. Struct. Biol.*, **159**, 268–276.

53. Engelman, D. (2005) Membranes are more mosaic than fluid, *Nature*, **438**, 578–580.

54. Fotiadis, D., Liang, Y., Filipek, S., Saperstein, D. A., Engel, A., and Palczewski, K. (2003) Atomic-force microscopy: Rhodopsin dimers in native disc membranes, *Nature*, **421**, 127–128.

55. Colombini, M. (1980) Structure and mode of action of a voltage-dependent anion-selective channel (VDAC) located in the outer mitochondrial membrane, *Ann. N Y Acad. Sci.*, **341**, 552–563.

56. Colombini, M., Blachly-Dyson, E., and Forte, M. (1996) VDAC, a channel in the outer mitochondrial membrane, *Ion Channels*, **4**, 169–202.

57. Rostovtseva, T., and Colombini, M. (1996) ATP flux is controlled by a voltage-gated channel from the mitochondrial outer membrane, *J. Biol. Chem.*, **271**, 28006–28008.

58. Mannella, C. A. (1998) Conformational changes in the mitochondrial channel protein, VDAC, and their functional implications, *J. Struct. Biol.*, **121**, 207–218.

59. Gonçalves, R. P., Buzhynskyy, N., Prima, V., Sturgis, J. N., and Scheuring, S. (2007) Supramolecular assembly of VDAC in native mitochondrial outer membranes, *J. Mol. Biol.*, **369**, 413–418.

60. Boyer, P. (1997) The ATP synthase—a splendid molecular machine, *Annu. Rev. Biochem.*, **66**, 717–749.

61. Arnold, I., Pfeiffer, K., Neupert, W., Stuart, R., and Schagger, H. (1998) Yeast mitochondrial F1F0-ATP synthase exists as a dimer: identification of three dimer-specific subunits, *EMBO J.*, **17**, 7170–7178.

62. Paumard, P., Vaillier, J., Coulary, B., Schaeffer, J., Soubannier, V., Mueller, D. M., Brèthes, D., di Rago, J.-P., and Velours, J. (2002) The ATP synthase is involved in generating mitochondrial cristae morphology, *EMBO J.*, **21**, 221–230.

63. Cabezon, E., Montgomery, M. G., Leslie, A. G. W., and Walker, J. E. (2003) The structure of bovine F1-ATPase in complex with its regulatory protein IF1, *Nat. Struct. Mol. Biol.*, **10**, 744–750.

64. Cabezon, E., Runswick, M., Leslie, A., and Walker, J. (2001) The structure of bovine IF(1), the regulatory subunit of mitochondrial F-ATPase, *EMBO J.*, **20**, 6990–6996.

65. Buzhynskyy, N., Sens, P., Prima, V., Sturgis, J. N., and Scheuring, S. (2007) Rows of ATP synthase dimers in native mitochondrial inner membranes, *Biophys. J.*, **93**, 2870–2876.

66. Strauss, M., Hofhaus, G., Schroeder, R. R., and Kuhlbrandt, W. (2008) Dimer ribbons of ATP synthase shape the inner mitochondrial membrane, *EMBO J.*, **27**, 1154–1160.

67. Andley, U. (2007) Crystallins in the eye: Function and pathology, *Prog. Retin. Eye. Res.*, **25**, 78–98.

68. Donaldson, P., Kistler, J., and Mathias, R. T. (2001) Molecular solutions to mammalian lens transparency, *News Physiol. Sci.*, **16**, 118–123.

69. Alcala, J., Lieska, N., and Maisel, H. (1975) Protein composition of bovine lens cortical fiber cell membranes, *Exp. Eye Res.*, **21**, 581–595.

70. Gonen, T., Cheng, Y., Sliz, P., Hiroaki, Y., Fujiyoshi, Y., Harrison, S. C., and Walz, T. (2005) Lipid-protein interactions in double-layered two-dimensional AQP0 crystals, *Nature*, **438**, 633–638.

71. Harries, W. E., Akhavan, D., Miercke, L. J., Khademi, S., and Stroud, R. M. (2004) The channel architecture of aquaporin 0 at a 2.2-Å resolution, *Proc. Natl. Acad. Sci. U S A*, **101**, 14045–14050.

72. Fleschner, C., and Cenedella, R. (1991) Lipid composition of lens plasma membrane fractions enriched in fiber junctions, *J. Lipid. Res.*, **32**, 45–53.

73. Maeda, S., Nakagawa, S., Suga, M., Yamashita, E., Oshima, A., Fujiyoshi, Y., and Tsukihara, T. (2009) Structure of the connexin 26 gap junction channel at 3.5 Å resolution, *Nature*, **458**, 597–602.

74. Buzhynskyy, N., Hite, R. K., Walz, T., and Scheuring, S. (2007) The supramolecular architecture of junctional microdomains in native lens membranes, *EMBO Rep.*, **8**, 51–55.

75. Mangenot, S., Buzhynskyy, N., Girmens, J. F., and Scheuring, S. (2009) Malformation of junctional microdomains in cataract lens membranes from a type II diabetes patient, *Pflugers Arch.*, **457**, 1265–1274.

76. Buzhynskyy, N., Girmens, J. F., Faigle, W., and Scheuring, S. (2007) Human cataract lens membrane at subnanometer resolution, *J. Mol. Biol.*, **374**, 162–169.

77. Binnig, G., Quate, C. F., and Gerber, C. (1986) Atomic force microscope, *Phys. Rev. Lett.*, **56**, 930–933.

78. Drake, B., Prater, C. B., Weisenhorn, A. L., Gould, S. A. C., Albrecht, T. R., Quate, C. F., Cannell, D. S., Hansma, H. G., and Hansma, P. K. (1989) Imaging crystals, polymers, and processes in water with the atomic force microscope, *Science*, **243**, 1586–1588.

79. Ohnesorge, F., Heckl, W. M., Häberle, W., Pum, D., Sara, M., Schindler, H., Schlicher, K., Kiener, A., Smith, D. P. E., Sleytr, U. B., and Binnig, G. (1992) Scanning force microscopy studies of the S-layers from *Bacillus coagulans* E38-66, *Bacillus sphaericus* CCM2177 and of an antibody binding process, *Ultramicroscopy*, **42–44**, 1236–1242.

80. Bezanilla, M., Drake, B., Nudler, E., Kashlev, M., Hansma, P. K., and Hansma, H. G. (1994) Motion and enzymatic degradation of DNA in the atomic force microscope, *Biophys. J.*, **67**, 2454–2459.

81. Guthold, M., Bezanilla, M., Erie, D. A., Jenkins, B., Hansma, H. G., and Bustamante, C. (1994) Following the assembly of RNA polymerase DNA complexes in aqueous solutions with the scanning force microscope, *Proc. Natl. Acad. Sci. U S A*, **91**, 12927–12931.

82. Viani, M. B., Schäfer, T. E., Chand, A., Rief, M., Gaub, H., and Hansma, P. K. (1999) Small cantilevers for force spectroscopy of single molecules, *J. Appl. Phys.*, **86**, 2258–2262.

83. Viani, M. B., Pietrasanta, L. I., Thompson, J. B., Chand, A., Gebeshuber, I. C., Kindt, J. H., Richter, M., Hansma, H. G., and Hansma, P. K. (2000) Probing protein-protein interactions in real time, *Nat. Struct. Biol.*, **7**, 644–647.

84. Hansma, P. K., Schitter, G., Fantner, G. E., and Prater, C. (2006) High-speed atomic force microscopy, *Science* **314**, 601–602.

85. Humphris, A. D. L., Miles, M. J., and Hobbs, J. K. (2005) A mechanical microscope: High-speed atomic force microscopy, *Appl. Phys. Lett.*, **86**, 34106–34108.

86. Picco, L. M., Bozec, L., Ulcinas, A., Engledew, D. J., Antognozzi, M., Horton, M. A., and Miles, M. J. (2007) Breaking the speed limit with atomic force microscopy, *Nanotechnology*, **18**, 44030–44033.

87. Ando, T., Kodera, N., Takai, E., Maruyama, D., Saito, K., and Toda, A. (2001) A high-speed atomic force microscope for studying biological macromolecules, *Proc. Natl. Acad. Sci. USA.*, **98**, 12468–12472.

88. Kodera, N., Kinoshita, T., Ito, T., and Ando, T. (2003) High-resolution imaging of myosin motor in action by a high-speed atomic force microscope, *Adv. Exp. Med. Biol.*, **538**, 119–127.

89. Ando, T., Uchihashi, T., Kodera, N., Yamamoto, D., Miyagi, A., Taniguchi, M., and Yamashita, H. (2008) High–speed AFM and nano–visualization of biomolecular processes, *Pflugers Arch. Eur. J. Physiol.*, **456**, 211–225.

90. Müller, D. J., Fotiadis, D., Scheuring, S., Müller, S. A., and Engel, A. (1999) Electrostatically balanced subnanometer imaging of biological specimens by atomic force microscope, *Biophys. J.*, **76**, 1101–1111.

91. Ando, T., Uchihashi, T., Kodera, N., Miyagi, A., Nakakita, R., Yamashita, H., and Sakashita, M. (2006) High-speed atomic force microscopy for studying the dynamic behavior of protein molecules at work,. *Jpn. J. Appl. Phys.*, **45**, 1897–1903.

92. Casuso, I., Kodera, N., Le Grimellec, C., Ando, T., and Scheuring, S. (2009) Contact-mode high-resolution high-speed atomic force microscopy movies of the purple membrane, *Biophys. J.*, **97**, 1354–1361.

93. Yamashita, H., Voitchovsky, K., Uchihashi, T., Contera, S. A., Ryan, J. F., and Ando, T. (2009) Dynamics of bacteriorhodopsin 2D crystal observed by high-speed atomic force microscopy, *J. Struct. Biol.*, **167**, 153–158.

94. Casuso, I., Sens, P., and Scheuring, S. (2010) Experimental evidence for membrane-mediated protein-protein interaction, Biophys. J, 99, L47-49.

Chapter 3

MICROBIAL CELL IMAGING USING ATOMIC FORCE MICROSCOPY

Mitchel J. Doktycz,[a] Claretta J. Sullivan,[b] Ninell Pollas Mortensen[a,c] and David P. Allison[a,c]

[a] *Biological and Nanoscale Systems Group, Biosciences Division, Oak Ridge National Laboratory, Oak Ridge, Tennessee 37831–6445, USA*

[b] *Eastern Virginia Medical School, Department of Surgery P.O. Box 1980 Norfolk, VA 23501, USA*

[c] *Department of Biochemistry and Cellular and Molecular Biology, University of Tennessee, Knoxville, Tennessee, 37996–0840, USA*

doktyczmj@ornl.gov

3.1 INTRODUCTION

Atomic force microscopy (AFM) is finding increasing application in a variety of fields including microbiology. Until the emergence of AFM, techniques for investigating processes in single microbes were limited. From a biologist's perspective, the fact that AFM can be used to generate high-resolution images in buffers or media is its most appealing feature as live-cell imaging can be pursued. Imaging living cells by AFM allows dynamic biological events to be studied, at the nanoscale, in real time. Few areas of biological research have as much to gain as microbiology from the application of AFM. Whereas the scale of microbes places them near the limit of resolution for light microscopy, AFM is well suited for the study of structures on the order of a micron or less. Although electron microscopy techniques have been the standard for high-resolution imaging of microbes, AFM is quickly gaining favour for several reasons. First, fixatives that impair biological activity are not required. Second, AFM is capable of detecting forces in the pN range,

Life at the Nanoscale: Atomic Force Microscopy of Live Cells
Edited by Yves Dufrêne

www.panstanford.com

and precise control of the force applied to the cantilever can be maintained. This combination facilitates the evaluation of physical characteristics of microbes. Third, rather than yielding the composite, statistical average of cell populations, as is the case with many biochemical assays, the behaviour of single cells can be monitored.

Despite the potential of AFM in microbiology, there are several limitations that must be considered. For example, the time required to record an image allows for the study of gross events such as cell division or membrane degradation from an antibiotic but precludes the evaluation of biological reactions and events that happen in just fractions of a second. Additionally, the AFM is a topographical tool and is restricted to imaging surfaces. Therefore, it cannot be used to look inside cells as with optical and transmission electron microscopes. Other practical considerations are the limitation on the maximum scan size (roughly 100×100 μm) and the restricted movement of the cantilever in the *Z* (or height) direction. In most commercial AFMs, the *Z* range is restricted to roughly 10 μm such that the height of cells to be imaged must be seriously considered. Nevertheless, AFM can provide structural–functional information at nanometre resolution and do so in physiologically relevant environments. Further, instrumentation for scanning probe microscopy continues to advance. Systems for high-speed imaging are becoming available,[1–3] and techniques for looking inside the cells are being demonstrated.[4] The ability to combine AFM with other imaging modalities is likely to have an even greater impact on microbiological studies.

AFM studies of intact microbial cells started to appear in the literature in the 1990s. For example, AFM studies of *Saccharomyces cerevisiae* examined budding scars after cell division and detailed changes related to cell growth processes.[5,6] Also, the first AFM studies of bacterial biofilms appeared.[7] In the late 1990s, AFM studies of intact fungal spores described clear changes in spore surfaces upon germination, and studies of individual bacterial cells were also described.[8–10] These early bacterial imaging studies examined changes in bacterial morphology due to antimicrobial peptides exposure and bacterial adhesion properties.[8,11]

The majority of these early studies were carried out on dried samples and took advantage of the resolving power of AFM. The lack of cell mounting procedures presented an impediment for cell imaging studies. Subsequently, several approaches to mounting microbial cells have been developed, and these techniques are described later. Also highlighted are general considerations for microbial imaging and a description of some of the various applications of AFM to microbiology.

3.2 IMMOBILIZATION OF MICROBIAL CELLS FOR AFM IMAGING

The problem of cell immobilization has come to be recognized as a significant barrier to the application of AFM to microbiology. As indicated by the number of papers that continue to be published on the topic, it is also a persistent one. Immobilization is necessary to prevent displacement of the cell by the scanning tip. Because the shape, size, rigidity and chemical properties of cells can differ dramatically, so must strategies for their immobilization. Drying bacteria onto the substrate before imaging is a popular choice for immobilization.[12-15] Unfortunately, drying the cells can result in cell dehydration and a flattened or collapsed appearance in the resulting AFM images.[16,17] Moreover, cells immobilized this way may not be viable and are frequently not stable when imaged in liquid. Imaging in liquid is a requirement for live-cell imaging but adds additional challenges because of the tendency of the tip to more easily disturb hydrated bacteria. Nevertheless, these challenges must be confronted if imaging dynamic processes are to be realized.

Bacterial surfaces vary because of differences in proteins, saccharides and appendages (pili, fimbriae, flagella) as determined by the genetics of the strain. Competition between these surface constituents and media components for binding sites on the substrate can prevent immobilization. This point is demonstrated by an AFM imaging study of purple membranes wherein Müller and colleagues enhanced immobilization by optimizing the ion content and pH of the imaging buffer.[18] Cellular imaging, however, requires an appreciation of how imaging solutions impact the physiology of the cells. Depending on the focus of the study, changes in properties such as osmotic potential or metabolism of the microbe may be undesirable. For example, force–distance measurements of *Pseudomonas aeruginosa* carried out in water and media show significant differences in bacterial spring constants.[19] Also, the pH of the liquid has been shown to have a significant influence on the nanomechanical properties on *Shewaenella putrefaciens.*[20] Imaging living microbes in liquid requires careful consideration of the immobilization technique used to ensure that the physiology of the bacteria is not compromised by the immobilization and imaging conditions. Therefore, immobilization strategies must be developed and systematically tested for individual organisms. Several approaches for mounting and immobilizing microbes have been described. Successful approaches generally fall into one of two categories: physical entrapment or chemical attachment (Fig. 3.1).

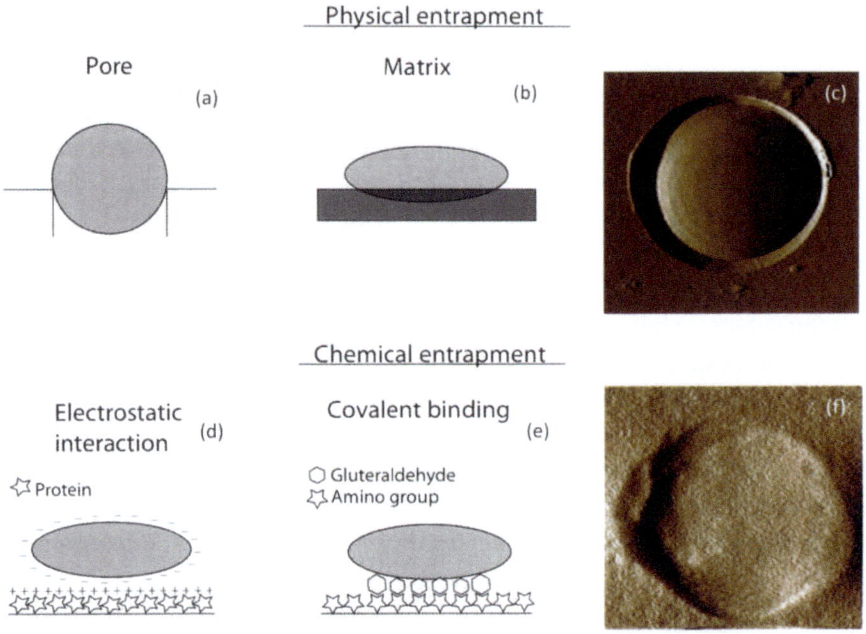

Figure 3.1. Schematics of physical entrapment and chemical attachment of microbial cells. In (a) and (b), two types of physical entrapment are shown. In (c), a 6 μm × 6 μm deflection image in liquid of a spherical yeast cell trapped in a filter pore is shown (image courtesy of Y. Dufrêne). In (d), immobilization by electrostatic interactions occurs between the negatively charged cell surface and positively charged imaging substrate. In (e), covalent attachment occurs by cross linking amino groups on the cell surface with glutaraldehyde to amino groups covering the substrate surface. The image in (f) shows a 1 μm × 1 μm amplitude image taken in liquid of an *E. coli* spheroplast covalently linked to an APTES covered surface.

3.2.1 Physical Entrapment

A successfully used method for immobilizing rigid, spherical bacteria, as well as yeast and fungal spores, is to use polymer membranes with a pore size comparable to the dimensions of the cell (Fig. 3.1a). The cells become mechanically trapped in the pores, allowing repeated imaging without cell detachment or damage.[6,9,21-24] Most commonly, spherical microbial cells have been examined (Fig. 3.1c) using the trapping strategy. The rigidity provided by the peptidoglycan layer in the cell assists in maintaining the spherical conformation of the bacteria as well as holding the cells in the pores. Although the strategy is generally not applicable for immobilizing rod-shaped bacteria or cell wall deficient forms of bacteria, at least one report applied the technique for imaging rod-shaped mycobacteria.[25] To accommodate other bacterial shapes, lithographic patterning of silicon wafers to purposefully define pores suitable for microbial immobilization has been described.[26] However, entrapment within pores is not without risk of mounting artifacts. Mendez-Vilas *et al.* evaluated mechanically trapped *Staphylococcus epidermidis* strains and found that the friction between the spherical bacterial cell and the sides of the filter pore can lead to accumulation of extracellular polymeric material deposited on the exposed surface of the cell.[22]

3.2.2 Chemical Attachment

Chemically attaching the cells to the surface is yet another approach to immobilizing microbial cells for AFM imaging in liquid. One variation is to modify the substrate in such a way that it facilitates adsorption of the cells to the surface. This approach typically takes advantages of the negatively charged surface of most bacteria. Hence, cationic substrate modifications are effective at immobilizing a wide variety of cells.[17,27–34]

One material that is amenable to cell adsorption is gelatin (Fig. 3.1d). It is suitable for immobilizing both Gram-negative and Gram-positive bacteria and involves briefly incubating a bacterial suspension with a gelatin-coated mica surface.[17,35] The bacteria are typically suspended in water and allowed to stand on the gelatin-coated surface. Afterwards, the sample is rinsed for several minutes with water. The immobilized bacteria on the surface can be imaged in liquid without further alteration. The technique results in isolated bacteria, distributed throughout the sample, thereby reducing the time required to find a region of interest. Apparently, not all commercially available gelatins can be effectively used in this technique. Several bovine gelatins purchased from Sigma did not immobilize the bacteria while two porcine gelatins purchased from Sigma (G-2624, G6144) did.[17] Bacterial adhesion to gelatin is believed to occur in two stages: the initial, reversible attachment and the more durable, irreversible attachment which follows.[36,37] Rinsing has been shown to displace bacteria in the earlier stages of adhesion while irreversible attachment occurs within minutes.[38] Bacteria immobilized on gelatin are stably attached and can withstand rinsing under a stream of liquid for several minutes. Gelatin is denatured collagen, and several bacterial species have been shown to bind collagen via specific binding sites on the bacterial surface.[39,40] It is likely that these binding sites, along with electrostatic and hydrophobic interactions, contribute to retaining bacteria on gelatin-coated substrates. Gelatin-coated mica has been effective for immobilizing a number of different bacteria including *Escherichia coli, P.*

aeruginosa, Listeria monocytogenes and *Bacillus atrophaeus.*[28,30,31,41] It has also been useful for immobilizing and imaging of diatoms.[42] Although gelatin-coated mica has been used successfully for a number of microbial imaging studies, not all bacteria have surfaces compatible with immobilization on gelatin. Even within a species, variations in surface characteristics can decrease the affinity of a microbe for the substrate. Further, the presence of rich media or buffer salts can interfere with bacterial adhesion to gelatin-coated surfaces.

Other cationic surface coatings have been prepared using amino-containing silane reagents, poly-*L*-lysine and other cationic polymers to electrostatically immobilize bacteria. For example, poly-*L*-lysine has been used to immobilize Gram-negative bacteria, including *E. coli, Shewanella oneidensis, Burkholderia cepacia, P. aeruginosa, Geobacter sulfurreducens* and Gram-positive *Listeria ivanovii.*[16,43–45] Glass slides coated with polyethylenimine have been useful for immobilization and imaging *E. coli* in liquid.[46,47] Substrates coated with 3-aminopropyltriethoxy silane (APTES) have been used to immobilize both Gram-negative and Gram-positive bacteria, and polycoated vinyl plastic has proven useful for immobilizing bacterial spores such as *B. atrophaeus* in liquid.[31,48,49]

In some cases, the interactions between the microbial cell and an untreated substrate are strong enough to ensure immobilization. For example, the actinomycete *Streptomyces coelicolor* was imaged in liquid after immobilization on freshly cleaved mica.[50] In biofilm research, simply growing the cells on the imaging substrate has also been used to immobilize bacteria for AFM imaging.[51–53] Increased time (hours to days) is generally required to prepare a sample in this manner as the bacteria synthesize polymers to condition the substrate to make it conducive for attachment and growth. Whereas researchers studying biofilm characteristics may prefer the close packing of cells in biofilms, others favour cells that are somewhat dispersed on the substrate so that morphological distortions and constraints due to tight packing are avoided.

For immobilization techniques that rely primarily on electrostatic interactions between the microbe and the substrate, buffer salts need to be evaluated. In some cases, the addition of divalent cations to the buffer can facilitate binding.[54] Successful imaging in water has been reported, but one must consider that water can create large, detrimental osmotic pressures on some bacteria. Therefore, using an isotonic solution like 0.25 M sucrose is advisable.[30,33,55] Because of these limitations, even stronger binding has been sought. Alternatively, covalent bonding strategies have been used in which substrates modified with amino groups were subsequently cross-linked to cells using glutaraldehyde (Figs. 3.1e,f).[33] This method has been used for imaging Gram-negative *B. cepacia*, *Pseudomonas stutzeri*, *E. coli*, *Pseudomonas putida* and Gram-positive *Bacillus subtilis* and *Micrococcus luteus*.[56,57] Immobilization by covalent attachment depends on favourable microbe-to-substrate contact, and any repulsion forces that might prevent this contact must be overcome. This immobilization technique has been used successfully for AFM imaging with various buffers.[21,23,24,48,49]

3.3 GENERAL CONSIDERATIONS FOR AFM IMAGING OF MICROBIAL CELLS

In addition to cell mounting procedures, other considerations related to microbial cell imaging should be considered. In general, imaging bacteria in liquid shows that hydrated cells have a smooth surface with greater heights compared with bacteria imaged in air. However, the imaging mode used, either contact or non-contact, can also affect the image. For example, the morphology of *P. aeruginosa* treated with the antimicrobial peptide colistin appears remarkably different when imaged in MACmode when compared with images taken in contact mode.[30] After 3 hours of colistin treatment, the cell surface changes and appears rough in MACmode images. In contrast, contact mode images result in a wavy morphology (Fig. 3.2). Even though both imaging modes indicate that colistin strongly affects the bacterial envelope, the actual morphology of the bacterial surface after treatment with colistin is difficult to ascertain. Contact mode images of bacterial spheroplasts reinforce the importance of selecting the appropriate imaging mode for the probed sample. When untreated spheroplasts were imaged in contact mode using a cantilever with a relatively low spring constant (0.01 nN/nm), they conformed to the shape of the tip (Fig. 3.3). Intermittent contact imaging (e.g. MACmode, acoustic, tapping), which applies a lower force, prevents these tip artifacts and allows for imaging soft bacterial cell surfaces where the cell wall has been removed.

Although liquid imaging is preferred for the reasons mentioned earlier, imaging in air generally results in better resolution of fine structures (Fig. 3.4). Presumably, the fluidity of the microbial surface and its various appendages are reduced as the surface is dried and appendages become immobilized on the surface. This allows for routine resolution of bacterial flagella, pili and changes to surface ultrastructure. Nevertheless, imaging and interpreting dried samples will have to account for artifacts that result from the drying process.

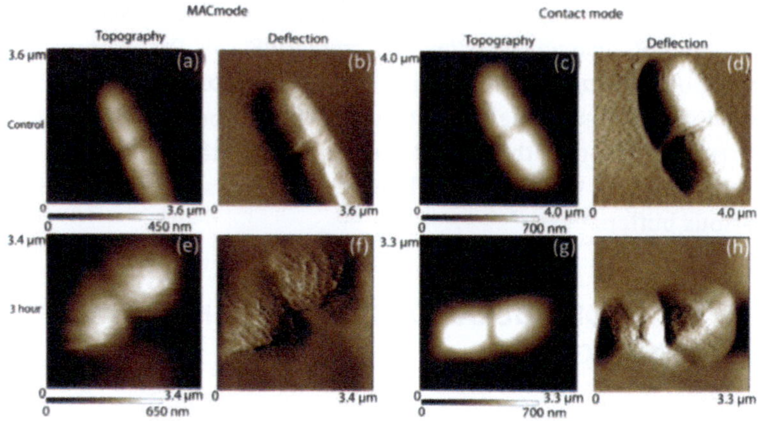

Figure 3.2. Mid-exponential *P. aeruginosa* (PA14) cells treated with 10 µg/ml colistin for 3 hours and imaged in 0.25 M sucrose. (a–b, e–f) are MACmode images while (c–d, g–h) are contact mode images. In the untreated controls, (a) topography and (b) amplitude are Macmode images, while (c) topography and (d) deflection are contact mode images. Cells treated with colistin and imaged in MACmode are shown in (e) and (f), while contact mode images are shown in (g) and (h). The surface structures in MACmode appear "spiky," while contact mode images appear wavy.

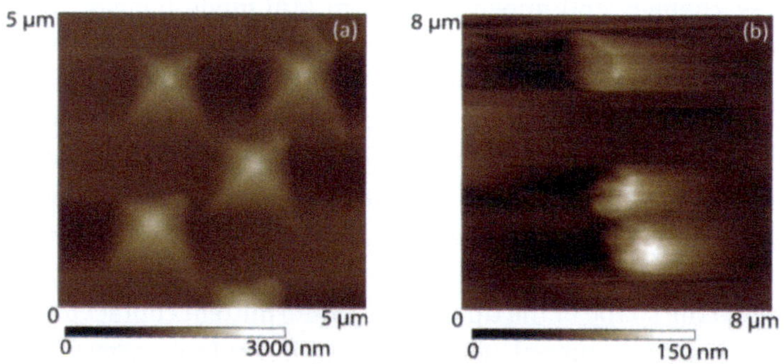

Figure 3.3. Immobilized *E. coli* spheroplasts imaged in liquid in contact and non-contact modes. (a) Contact mode imaging imposes the shape of the probe into the image as it contacts the pliable spheroplasts. (b) Non-contact mode (MACmode) imaging causes less disturbance of the cell shape.

Other imaging artifacts can result from the shape of the probe tip and the forces exerted by it as first described by Velegol *et al.*[47] The relatively tall microbial cell requires slow scanning speeds and careful optimization of gain settings to enable the tip to track over the cell without significantly disturbing it. Further, in the early AFM studies of bacteria in liquid, it was speculated that material was being scraped off the cell surface resulting in the piling up of "material" with repeated scanning. This observation has been described for both bacteria and for bacterial spores when sample height exceeded 1 µm.[47] Velegol *et al.* observed that the orientation of the "material" was the same despite changes in the scan direction, and they noticed a consistent angle of ±27° to the scan direction, thereby showing that the "material" really was an imaging artifact.[47]

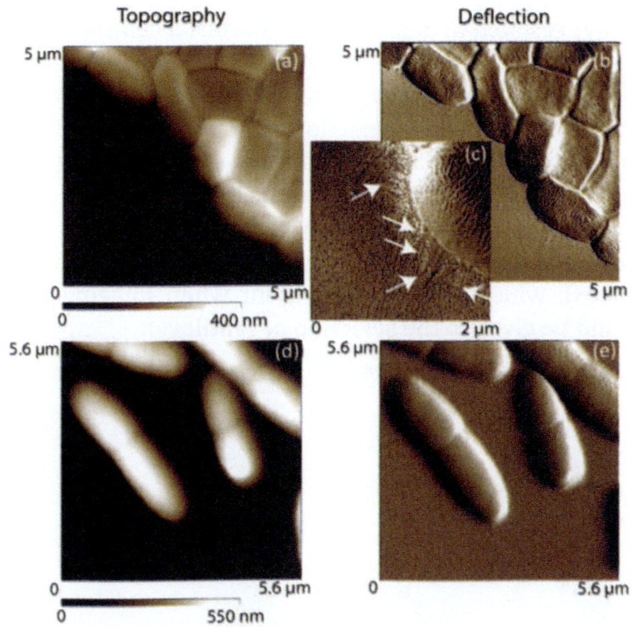

Figure 3.4. Contact mode AFM images in air (a–c) and in water (d–e) show differences in resolution of fine structure. Panels (a–b), respectively, are topographic and deflection images while (c) is a friction image of enteroaggregative *E. coli* 042. The white arrows in the friction image point to fimbriae, which are clearly visible in image (c). In (d) and (e), images of *P. aeruginosa* PA14 taken in water are shown. The reduced level of details showing fimbriae and cellular surface structure is especially obvious when comparing the deflection mode images taken in air and water.

The shape of the artifact is explained as being a combination of the object being scanned, the geometry of the tip, the angle of tip tilt and the scan direction.[47] The choice of cantilever is an important one because of the available variations in material, size, shape and tip geometry. Collectively, these features determine the mechanical forces imposed on the sample during imaging. Commercially available silicon cantilevers are stiffer than their silicon nitride counterparts. High aspect ratio tips which have a smaller radius of curvature are better for imaging rigid samples with tall features, but they are costlier and more fragile than low aspect ratio tips. On the other hand, the blunter silicon nitride tips may avoid puncturing fragile, tall samples such as cells. To illustrate the importance of cantilever selection, intact *E. coli* were imaged using a silicon cantilever. According to the manufacturer, this cantilever has an estimated spring constant of 0.6 nN/nm, and the cantilever tip radius of curvature is said to be less than 10 nm. As shown in noncontact MACmode™ imaging provides acceptable images of intact *E. coli* (Fig. 3.5a,b). However, repeated contact and retraction in a chosen location (movement in the *Z* plane only) resulted in damage to the bacterial cell surface as indicated by subsequent images (Fig. 3.5c). When a second location on the bacteria was similarly treated, more damage was observed (Fig. 3.5d). This is only one example of how cantilever selection may impact experimental results. The other possibility is selecting cantilevers too soft for the condition of the experiment. Specifically, it can be difficult to approach the surface in air if cantilevers with very low spring constants are used. The difficulty arises because softer cantilevers are more vulnerable to forces of adhesion (capillary forces), which are always present when imaging in air. Careful attention should be given to the selection of cantilever.

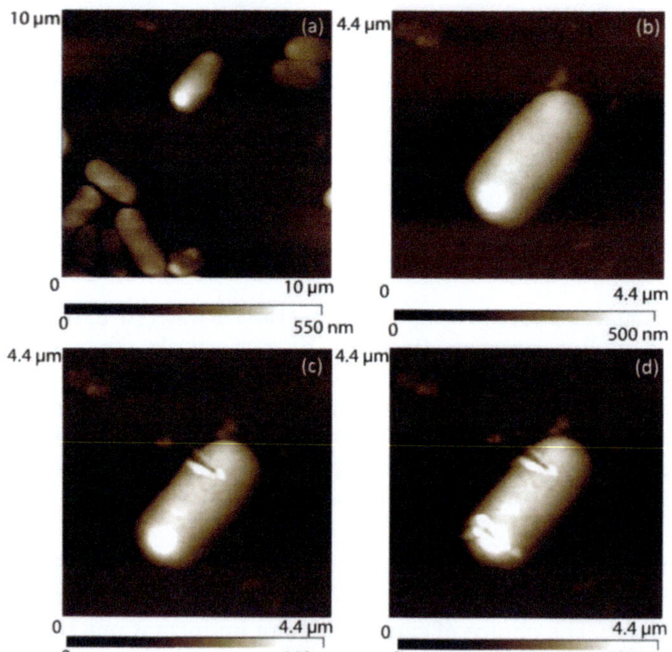

Figure 3.5. AFM provides the ability to both image and manipulate single microbes. (a, b) *E. coli* imaged in MACmode using a silicon cantilever. (c) Repeatedly contacting the surface with a silicon tip in the same location scars the cell surface. (d) The same cell after a second location was similarly treated. All images are topography images.

A continuing challenge with AFM images is distinguishing structures or molecules of interest on the cell surface. In electron microscopy, immunolabelled nanoparticles are often used, and this technique has also been implemented in AFM under dry conditions and in liquid. Plomp and Malkin used immunolabelled gold nanoparticles to target epitopes on the surface of *Bacillus* spores.[58] By using both monoclonal and polyclonal antibodies, they successfully targeted epitopes on both the spore coat and the underlying exosporium. Using nanoparticles coated with secondary antibodies to anti-lipoarabinomannan, Alsteens *et al.* showed that the four drugs they investigated led to exposure of hydrophilic lipoarabinomannan on the surface of mycobacteria.[59]

3.4 APPICATIONS OF AFM TO MICROBIOLOGY

The potential applications of AFM in microbiology are numerous and diverse. Published reports based on AFM imaging of whole cells have exploited AFM's spatial resolution capabilities and ability to operate in liquid environments. Characterizing a microbial cell's response to chemicals is a common application in which images before and after chemical treatment are compared. Microbial imaging is contributing to the understanding of morphological characteristics of single cells, their extracellular structures, bacterial communities such as biofilms and microbial responses to everything from growth conditions to antibiotics and nanoparticle exposure. Considering the wide variety of cells and the diverse array of chemical environments, the use of AFM for such applications will likely continue to grow. The capability to study dynamic biological processes at the nanoscale is a unique attribute of AFM that continues to evolve. Studies using AFM for this purpose have been reported and were somewhat dependent on improvements in sample preparation and mounting techniques. Other applications such as the measurement of interaction forces on living cells and cellular elasticity are being described in other chapters in this volume. Therefore, imaging-based applications of AFM are selected for review in the following section. The reader is referred to other chapters for discussions concerning force spectroscopy applications.

3.4.1 AFM Studies of Microbial Response to Chemical Changes

AFM has been used to study microbial response to different growth conditions and to exposure to various antimicrobial treatments. Most of these studies have been performed on dry samples, but a few are dynamic studies of living cells that have been carried out in liquid environments.

In one study, the microbial response to differences in growth conditions shows that the organization of *Corynebacterium glutamicum* S-layer is dependent on the growth media. This work showed that AFM images of *C. glutamicum* grown in nutrient-rich media had a smooth surface with no ordered S-layer visible, while bacteria grown in brain-heart infusion media showed a highly ordered hexagonal lattice S-layer.[60] An early AFM-based study in air examined the response of *E. coli* to different concentrations of ethylenediaminetetraacetic acid (EDTA). AFM images of dried samples revealed changes in bacterial surface morphology and collapsed cells due to the EDTA treatment. The study also showed that metal depletion caused irregularly shaped pits in the cell's outer membrane.[11] AFM has also been used to visualize morphological changes in *E. coli* after bacteriophage infection. AFM images taken in air showed increased smoothness and decreased bacterial height.[61]

Several reports have established the utility of AFM in understanding the mechanism of antibiotic action. AFM has been used to examine the mechanism of action for a range of antimicrobial peptides including PGLa,[62,63] magainin 2, melittin,[63] SB006[64] and the so-called sushi peptides.[65] In all cases, AFM images in air show that the antimicrobial peptides caused damage to the bacterial cell envelope. AFM has also been used to study in air the effects of exposure of various bacteria to the antibiotic cefodizime,[8] a polymeric drug risug,[66] nanoparticles bound to lysozyme[67] and nitric oxide (NO).[68] In both cefodizime- and risug-treated *E. coli*, the centre of the cells was observed to be collapsed when imaged in air, and a knobby and irregular appearance to the bacterial surface was also seen. Silver nanoparticles coupled with lysozyme showed strong antimicrobial properties on *Proteus mirabilis* and *E. coli* where AFM imaging in air revealed a partially destroyed cell envelope.[67] *E. coli and P. aeruginosa* exposed to NO showed morphology similar to bacteria exposed to the antibiotic amoxicillin, indicating that NO leads to cell envelope deterioration.[68] AFM revealed that bacitracin, a metal-dependent dodecapeptide, inhibited cell growth and division in *Staphylococcus aureus* and that the effect of bacitracin was increased when the peptide was coupled to metal ions.[69] The antimicrobial effect of chitosan was investigated on both vegetative bacteria and its spores of *Bacillus cereus*. AFM showed that the polymer forms a film that surrounds the cells.[13] A study of the antimicrobial peptide colistin performed on living *P. aeruginosa* in liquid showed that the cell morphology changed to a wavy phenotype and increased stiffness of the cell surface.[30]

As a natural extension to studying single cells, bacterial communities such as biofilms have also been investigated.[57,70] AFM has been used to study biofilm organization, response to chemical exposure and bacterial predation. Most studies have examined the early stages of biofilm formation and are carried out in air. Hydrated, mature biofilms have dramatic topographies and gel-like characteristics, which make them difficult to image with scanning probes. One study of biofilm organization, imaged in air, examined a number of surface protein mutants of *Streptococcus mutans*, the primary cause of tooth decay in humans.[70] Investigations of the biofilm matrix components curli (which are amyloid fibres), cellulose and the cell surface protein BapA in biofilm and colony morphology of *Salmonella typhimurim* indicate that curli and cellulose but not BapA have an impact on the formation and the morphology of a biofilm.[71] When compared with cellulose, curli appears to be more important for the formation of cell aggregates.[71]

The adhesive properties and spring constants of biofilms derived from four different bacteria (*P. putida, E. coli, M. luteus and B. subtilis)* grown on glass were examined in liquid.[57] This study showed that the spring constant of bacteria in a biofilm was higher than that for the same strain that was grown in liquid.[57] A study examining the effect of a range of inorganic compounds on *S. epidermidis* biofilm formation revealed that, while the compounds did not significantly affect the growth of an already established biofilm, they had a strong inhibitory effect on initial bacterial adhesion.[72]

The survival of biofilms exposed to *Bdellovibrio bacteriovorus,* a Gram-negative bacterium that preys on other Gram-negative bacteria, was found to correlate with the nutritional level of the growth media. In nutrition poor media *Bdellovibrio* completely killed the biofilm, whereas in rich media some *E. coli* would remain.[73] In an earlier study, the same authors proposed that predation behaviour at interfaces (air–solid or liquid–solid) differed from those in solution. They simulated the air–solid interface by growing the bacteria on sterilized, small-pore filters placed on agar plates. Over a period of a few days, several life cycles of *Bdellovibrio* were imaged in air by AFM. When *Bdellovibrio* invaded *E. coli,* they utilized the prey's macromolecules for growth. During this period, the combined organism is called a bdelloplast. In electron microscopy studies, unaffected prey could be distinguished based on changes in their two-dimensional shape, but this study provided evidence that there were also changes in the height of the structures. Specifically, bdelloplasts were shown to be "rounded up" and smoother when compared with uninvaded cells. The ability to image these cells without fixation was important to the findings in this study because it allowed the researchers to distinguish between true features and electron microscopy artefacts.[74]

A related application is the use of AFM to evaluate cell–cell interactions between microbes and eukaryotic cells. The interaction of *E. coli* with mouse bladder tissue has been studied by AFM in liquid and showed clusters of bacteria trapped in the intermediate filaments of the urothelial cells.[75] The AFM images in liquid show *E. coli* in early stages of the infection to be loosely organized in the bladder, but later bacteria become more tightly packed in biofilm-like clusters among the filaments of the bladder tissue.

3.4.2 AFM Studies of Dynamic Biological Processes

One of the most attractive prospects of AFM for biologists is the possibility to study dynamic biological processes, such as bacterial growth, in an environment simulating the natural one. Studying dynamic biological processes is in several aspects more challenging than studying static processes. If growth is the focus of the study, biological parameters like temperature and growth media have to be kept constant and favourable, while maintaining stable immobilization during the study. Because cell surface remodelling may occur during dynamic processes, immobilization remains a key challenge. Nevertheless, studying dynamic biological processes at high resolution can be enabled by AFM studies. Growth and septum formation of *S. aureus* have been visualized using AFM, and the results were in good correlation with images of cell division obtained by transmission electron microscopy.[23] The growth of bacteria immobilized on patterned surfaces has also been observed.[26]

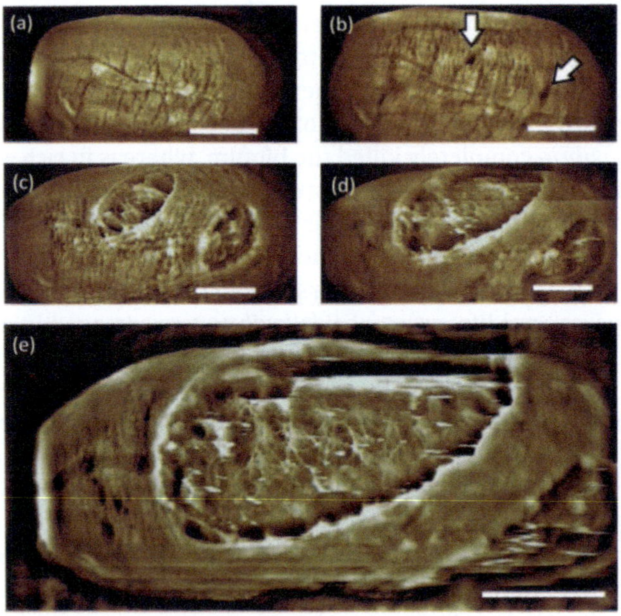

Figure 3.6. Dynamic AFM imaging was used to study germination of *B. atrophaeus* spores. In this study, spores were immobilized on polycoated vinyl plastic surfaces for imaging in liquid at 37°C. Spores were first imaged in water (a) followed by the addition of germination solution. Onset of germination (b) is first observed by cracks (arrows) appearing in the spore surface. These cracks continue to expand (c–e), resulting in the final release of vegetative cells. This process can take 2–15 hours. Scale bars: 500 nm. Image courtesy of M. Plomp and A. J. Malkin. Reproduced with permission from Ref. 31. Copyright 2007, National Academy of Sciences, U.S.A.

In other reports, AFM was used to examine the development and structure of various bacterial spores (see also Chapter 4).[9,13,14,52,76,77] Images of the dynamic process of germination of *B. atrophaeus* spores revealed germination-induced changes in spore coat topography and structure.[31] These imaging studies began in water (non-inducing conditions) and later changed to a buffer designed to induce germination (Fig. 3.6). Imaging was conducted at 37°C and continued for several hours. Ultimately, the authors found that outer rodlet structures were disrupted by small defects that formed perpendicular to the rodlet array. These defects expanded and coalesced over time until vegetative cells emerged. A previously unrecognized ordered layer was also revealed as a result of the impressive images in this report. The same group also followed spore germination of *Clostridium novyi* in liquid.[14]

AFM has been used to visualize other dynamic processes of microbes. For example, the enzymatic digestion of a bacterial cell wall has been documented by time-series images of a single bacterial cell during enzyme exposure. When the cell wall of *S. aureus* was treated with the enzyme lysostaphin, the cell surface roughness increased with time.[21] An example from virology illustrates the benefit of using AFM for simulating the dynamic environment encountered by biological systems *in vivo*. Kienberger *et al.* used MACmode® to monitor the release of RNA from a type of human rhinovirus (HRV2).[54] High-resolution images of the immobilized viruses yielded height measurements around 30 nm, which correlated well with those previously reported in literature. It was known before their study that RNA is released from the HRV2 capsid in low pH conditions *in vivo*. The authors were able to simulate this condition by reducing the pH of the imaging buffer. After a period of time in the low pH buffer, the extrusion of RNA molecules was visualized, illustrating the resolving power of AFM in dynamic environments. The presence of RNA was confirmed by comparing these images with those taken in buffer supplemented with RNase A. In the latter case, the fibres believed to be RNA were not present. On the basis of the presence of fork-like structures at the end of fully released RNA, the authors were able to speculate about the initial orientation of the RNA during extrusion. The ability to alter buffer conditions during imaging permits visualization of dynamic processes and is an important advantage of AFM.

3.4.3 AFM Studies of Microbial Cell Substructures

Although microbes do not have organelles in the classical sense, some have subcellular macromolecular structures that are important to their physiology. Such components are impossible to visualize using optical microscopy without fluorescent constructs. Extracellular structures of microbes can be investigated by AFM as demonstrated by a study examining DNA transfer in *E. coli*.[32] During conjugation, the F-plasmid of donor cells encodes proteins to prepare specialized pili that bind to the surface of recipient cells. After gaining access to the recipient cell's cytoplasm, the pilus involved in the exchange is depolymerized, bringing the mating pair into intimate contact. Whether the DNA exchange occurs through the extended pilus or during contact between the cells is debated. Shu *et al.*[32] imaged mating pairs of *E. coli* sharing an extended pilus. After using the AFM tip to sever the extended pilus, the area was probed with a tip functionalized with an antibody to ssDNA. The authors assert that the DNA was being transferred through the pilus at the time of dissection, accounting for the recognition by the functionalized tip. As underscored by this study, the ability to image in liquid and to manipulate a localized region of the sample is a distinct advantage of AFM.

In a study by Dennis *et al.*, storage inclusions from the bacterium *Cupriavidus necator* were examined.[78] Owing to the amount of sonication used during the preparation, a mixed population of inclusions was isolated. Using surface morphology as selection criteria, the authors examined the properties of each group and determined that each form presented a different layer of the storage inclusion. Importantly, a previously unidentified protein network about 2–4 nm thick was revealed in the study. Biochemical studies of the PhaP protein had shown that the protein was associated with storage inclusions but not until the AFM study was the protein localized to the space beneath the inclusion envelope.

Touhami and colleagues generated high-resolution AFM images to characterize the pili of a laboratory strain of *P. aeruginosa* (PA01).[79] In addition, the authors immobilized PA01 to the cantilever and approached the mica substrate, giving pili on the bacteria the opportunity to adhere to the mica. This configuration permitted the authors to calculate the forces required to either dissociate the pili from the mica or to break the pili. Although the two scenarios could not be distinguished in their experiments, the sensitivity of AFM for force measurements is nevertheless highlighted.

The cell walls of Gram-positive bacteria include a thick peptidoglycan to which teichoic acids are attached, whereas Gram-negative cell walls have a much thinner peptidoglycan and a lipopolysaccharide-containing outer membrane.[80,81] In both instances, these cell wall components obstruct access to the cytoplasmic membrane in AFM studies. Cell walls can be removed enzymatically using established protocols to provide access to the various membrane proteins that reside there.[82,83] The resulting spheroplasts are extremely soft and can be imaged in liquid in a non-destructive way.[11,55] Sullivan *et al.* used AFM to study live *E. coli* spheroplasts and compare their nanomechanical properties with intact *E. coli*.[33] The study showed that non-contact imaging of live spheroplasts provided the best images. Furthermore, despite using a soft cantilever, the pliability of the untreated spheroplast surface prevented measurement of its elastic properties.

Diatoms are single-cell photosynthetic algae that are found in all bodies of water. There are tens of thousands of species of diatoms that are differentiated by the silica skeletons which are derived from the minute concentrations of silicic acid found in the water. From an ecological standpoint, diatoms are critical components of the ecosystem. From a materials standpoint, they serve as an ideal model system for understanding natural strategies to the synthesis and patterning of hard materials. As revealed by AFM, formation of the silica skeleton can begin with fusion of silica beads as small as 40 nm to create highly ordered micron-scale structures.[84] Determining how these skeletons are produced with accuracy spanning the nano to the macroscale exceeds our synthetic capabilities. The accuracy of the diatom's biosynthetic capability, to reproduce its silica skeleton, is illustrated in the example of the AFM images of the large diatom *Gyrosigma balticum* shown in Figure 3.7. The AFM analysis of diatoms, not only in terms of structure but also adhesive and mechanical properties, is an active area of research that can be further accessed in reviews[85–87] as well as in Chapter 19.

3.5 FUTURE DIRECTIONS

Earlier efforts in applying AFM to microbiology had a necessary focus on demonstrating the applicability of AFM to such studies and on addressing technical concerns related to the mounting and imaging of microbial cells. In the process, AFM has proven valuable for understanding the physical and morphological responses of the cell to hydration and chemical treatment. Elucidating the physical consequences of antibiotic treatment will likely be a topic of continued interest.

The spatial resolution afforded by AFM allows examination of structural changes related to such exposures as well as detailed investigation of extracellular structures and biological processes such as cell division and sporulation. Although high-resolution imaging of living microbial cells will continue to be a defining attribute of AFM imaging, the field is transitioning from the study of static biological samples to dynamic measurements of living systems. Refining capabilities to characterize dynamic processes will be essential. Such advances will likely benefit from refinements in procedures for mounting and imaging but also from improvements in instrumentation.

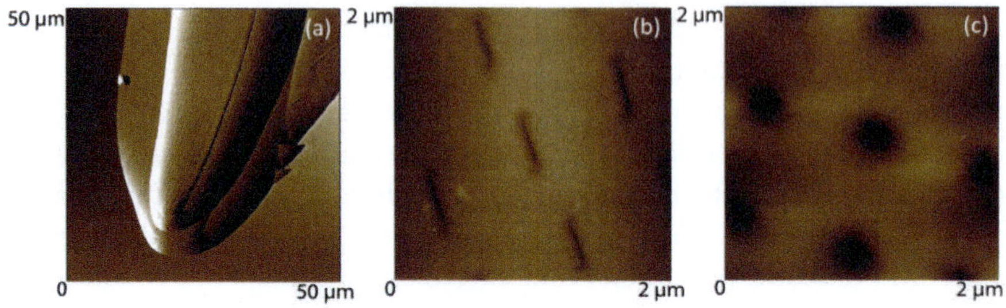

Figure 3.7. *G. balticum* is a large marine diatom 350–450 µm in length with a width of roughly 20 µm. When the gelatinous material that covers diatoms is removed, an intricate silica structure is revealed. Although it is not possible to scan the entire length of this large diatom by AFM, the AFM image (a) partially reveals the surfboard shape of the valve of this diatom. Nanometre scale structure is revealed on the outer (distal) valve surface (b) where S-shaped slits 400 nm in length with 50 nm widths are seen evenly spaced along the entire valve surface. On the inner (proximal) valve surface (c), instead of slits, corresponding holes 250 nm in diameter are spaced roughly 850 nm apart in equally spaced rows.

High-speed imaging capabilities are emerging and are being applied to biomolecule characterization.[1-3] Such systems will also facilitate the measurement of dynamic events at the cellular scale. Chemical identification will continue to be a challenge. Beyond physical tagging approaches, various techniques for chemical identification based on tip functionalization have been described and are beginning to be applied to the characterization of live cells.[88-93] Instrumentation advances that allow characterizations beneath the cell surface are also being developed.[4] The exciting possibility of molecular resolution of intracellular species by scanning probe-based tools may soon be available.[4] In the meantime, the combination of AFM with other imaging modalities is commonplace.[94-96] For example, fluorescence microscopy techniques are often joined with AFM and can aid in correlating molecular events that occur at the intracellular and cell surface levels.[53,97,98] Although descriptions of the application of such systems to microbial cells are not common, the advantages of combined imaging tools will likely be realized in future microbial imaging studies. The maturing capabilities in mounting and imaging cells and exciting developments in instrumentation are leading the way in addressing a wide variety of problems in microbiology. The continued application of AFM to microbiology promises big advances for investigations into the small world of microbial systems.

Acknowledgements

The authors acknowledge research support from the US DOE Office of Biological and Environmental Sciences. Oak Ridge National Laboratory is managed by UT-Battelle, LLC, for the US Department of Energy under Contract no. DEAC05-00OR22725. Ninell Pollas Mortensen would like to thank Lundbeck Fonden for financial support.

References

1. Ando, T., Kodera, N., Takai, E., Maruyama, D., Saito, K., and Toda, A. (2001) A high-speed atomic force microscope for studying biological macromolecules, *Proc. Natl. Acad. Sci. USA*, **98**, 12468–12472.
2. Carberry, D. M., Picco, L., Dunton, P. G., and Miles, M. J. (2009) Mapping real-time images of high-speed AFM using multitouch control, *Nanotechnology*, **20**, 434018.
3. Casuso, I., Kodera, N., Le Grimellec, C., Ando, T., and Scheuring, S. (2009) Contact-mode high-resolution high-speed atomic force microscopy movies of the purple membrane, *Biophys. J.*, **97**, 1354–1361.
4. Tetard, L., Passian, A., Venmar, K. T., Lynch, R. M., Voy, B. H., Shekhawat, G., Dravid, V. P., and Thundat, T. (2008) Imaging nanoparticles in cells by nanomechanical holography, *Nat. Nanotechnol.*, **3**, 501–505.
5. Gad, M., and Ikai, A. (1995) Method for immobilizing microbial cells on gel surface for dynamic AFM studies, *Biophys. J.*, **69**, 2226–2233.
6. Kasas, S., and Ikai, A. (1995) A method for anchoring round shaped cells for atomic-force microscope imaging, *Biophys. J.*, **68**, 1678–1680.
7. Steele, A. A., Goddard, D. T., and Beech, I. B. (1994) An atomic force microscopy study of the biodeterioration of stainless steel in the presence of bacterial biofilms, *Int. Biodeterior. Biodegradation*, **34**, 35–46.
8. Braga, P. C., and Ricci, D. (1998) Atomic force microscopy: application to investigation of *Escherichia coli* morphology before and after exposure to cefodizime, *Antimicrob. Agents Chemother.*, **42**, 18–22.
9. Dufrene, Y. F., Boonaert, C. J. P., Gerin, P. A., Asther, M., and Rouxhet, P. G. (1999) Direct probing of the surface ultrastructure and molecular interactions of dormant and germinating spores of *Phanerochaete chrysosporium*, *J. Bacteriol.*, **181**, 5350–5354.
10. van der Aa, B. C., Michel, R. M., Asther, M., Zamora, M. T., Rouxhet, P. G., and Dufrene, Y. F. (2001) Stretching cell surface macromolecules by atomic force microscopy, *Langmuir*, **17**, 3116–3119.

11. Amro, N. A., Kotra, L. P., Wadu-Mesthrige, K., Bulychev, A., Mobashery, S., and Liu, G. Y. (2000) High-resolution atomic force microscopy studies of the *Escherichia coli* outer membrane: structural basis for permeability, *Langmuir*, **16**, 2789–2796.
12. Canetta, E., Adya, A. K., and Walker, G. M. (2006) Atomic force microscopic study of the effects of ethanol on yeast cell surface morphology, *FEMS Microbiol. Lett.*, **255**, 308–315.
13. Fernandes, J. C., Eaton, P., Gomes, A. M., Pintado, M. E., and Malcata, F. X. (2009) Study of the antibacterial effects of chitosans on *Bacillus cereus* (and its spores) by atomic force microscopy imaging and nanoindentation, *Ultramicroscopy*, **109**, 854–860.
14. Plomp, M., McCaffery, J. M., Cheong, I., Huang, X., Bettegowda, C., Kinzler, K. W., Zhou, S., Vogelstein, B., and Malkin, A. J. (2007) Spore coat architecture of *Clostridium novyi* NT spores, *J. Bacteriol.*, **189**, 6457–6468.
15. Sahu, K., Bansal, H., Mukherjee, C., Sharma, M., and Gupta, P. K. (2009) Atomic force microscopic study on morphological alterations induced by photodynamic action of Toluidine Blue O in *Staphylococcus aureus* and *Escherichia coli*, *J. Photochem. Photobiol. B,Biol.*, **96**, 9–16.
16. Bolshakova, A. V., Kiselyova, O. I., and Yaminsky, I. V. (2004) Microbial surfaces investigated using atomic force microscopy, *Biotechnol. Prog.*, **20**, 1615–1622.
17. Doktycz, M. J., Sullivan, C. J., Hoyt, P. R., Pelletier, D. A., Wu, S., and Allison, D. P. (2003) AFM imaging of bacteria in liquid media immobilized on gelatin coated mica surfaces, *Ultramicroscopy*, **97**, 209–216.
18. Müller, D. J., Amrein, M., and Engel, A. (1997) Adsorption of biological molecules to a solid support for scanning probe microscopy, *J. Struct. Biol.*, **119**, 172–188.
19. Yao, X., Walter, J., Burke, S., Stewart, S., Jericho, M. H., Pink, D., Hunter, R., and Beveridge, T. J. (2002) Atomic force microscopy and theoretical considerations of surface properties and turgor pressures of bacteria, *Colloids Surf. B Biointerfaces*, **23**, 213–230.
20. Gaboriaud, F., Bailet, S., Dague, E., and Jorand, F. (2005) Surface structure and nanomechanical properties of *Shewanella putrefaciens* bacteria at two pH values (4 and 10) determined by atomic force microscopy, *J. Bacteriol.*, **187**, 3864–3868.
21. Francius, G., Domenech, O., Mingeot-Leclercq, M. P., and Dufrene, Y. F. (2008) Direct observation of *Staphylococcus aureus* cell wall digestion by lysostaphin, *J. Bacteriol.*, **190**, 7904–7909.
22. Mendez-Vilas, A., Gallardo-Moreno, A. M., and Gonzalez-Martin, M. L. (2007) Atomic force microscopy of mechanically trapped bacterial cells, *Microsc. Microanal.*, **13**, 55–64.
23. Touhami, A., Jericho, M. H., and Beveridge, T. J. (2004) Atomic force microscopy of cell growth and division in *Staphylococcus aureus*, *J. Bacteriol.*, **186**, 3286–3295.

24. van der Mei, H. C., Busscher, H. J., Bos, R., de Vries, J., Boonaert, C. J. P., and Dufrene, Y. F. (2000) Direct probing by atomic force microscopy of the cell surface softness of a fibrillated and nonfibrillated oral streptococcal strain, *Biophys. J.*, **78**, 2668–2674.

25. Dupres, V., Menozzi, F. D., Locht, C., Clare, B. H., Abbott, N. L., Cuenot, S., Bompard, C., Raze, D., and Dufrene, Y. F. (2005) Nanoscale mapping and functional analysis of individual adhesins on living bacteria, *Nat. Methods*, **2**, 515–520.

26. Kailas, L., Ratcliffe, E. C., Hayhurst, E. J., Walker, M. G., Foster, S. J., and Hobbs, J. K. (2009) Immobilizing live bacteria for AFM imaging of cellular processes, *Ultramicroscopy*, **109**, 775–780.

27. Arce, F. T., Carlson, R., Monds, J., Veeh, R., Hu, F. Z., Stewart, P. S., Lal, R., Ehrlich, G. D., and Avci, R. (2009) Nanoscale structural and mechanical properties of nontypeable *Haemophilus influenzae* biofilms, *J. Bacteriol.*, **191**, 2512–2520.

28. Beckmann, M. A., Venkataraman, S., Doktycz, M. J., Nataro, J. P., Sullivan, C. J., Morrell-Falvey, J. L., and Allison, D. P. (2006) Measuring cell surface elasticity on enteroaggregative *Escherichia coli* wild type and dispersin mutant by AFM, *Ultramicroscopy*, **106**, 695–702.

29. Kuznetsov, Y., Gershon, P. D., and McPherson, A. (2008) Atomic force microscopy investigation of vaccinia virus structure, *J. Virol.*, **82**, 7551–7566.

30. Mortensen, N. P., Fowlkes, J. D., Sullivan, C. J., Allison, D. P., Larsen, N. B., Molin, S., and Doktycz, M. J. (2009) Effects of colistin on surface ultrastructure and nanomechanics of *Pseudomonas aeruginosa* cells, *Langmuir*, **25**, 3728–3733.

31. Plomp, M., Leighton, T. J., Wheeler, K. E., Hill, H. D., and Malkin, A. J. (2007) In vitro high-resolution structural dynamics of single germinating bacterial spores, *Proc. Natl. Acad. Sci. USA*, **104**, 9644–9649.

32. Shu, A. C., Wu, C. C., Chen, Y. Y., Peng, H. L., Chang, H. Y., and Yew, T. R. (2008) Evidence of DNA transfer through F-pilus channels during *Escherichia coli* conjugation, *Langmuir*, **24**, 6796–6802.

33. Sullivan, C. J., Venkataraman, S., Retterer, S. T., Allison, D. P., and Doktycz, M. J. (2007) Comparison of the indentation and elasticity of *E. coli* and its spheroplasts by AFM, *Ultramicroscopy*, **107**, 934–942.

34. Suo, Z., Avci, R., Yang, X., and Pascual, D. W. (2008) Efficient immobilization and patterning of live bacterial cells, *Langmuir*, **24**, 4161–4167.

35. Venkataraman, S., Allison, D. P., Qi, H., Morrell-Falvey, J. L., Kallewaard, N. L., Crowe, J. J. E., and Doktycz, M. J. (2006) Automated image analysis of atomic force microscopy images of rotavirus particles, *Ultramicroscopy*, **106**, 829–837.

36. An, Y. H., and Friedman, R. J. (1998) Concise review of mechanisms of bacterial adhesion to biomaterial surfaces, *J. Biomed. Mater. Res.*, **43**, 338–348.

37. Dunne, W. M. (2002) Bacterial adhesion: seen any good biofilms lately? *Clin. Microbiol. Rev.*, **15**, 155-166.

38. Palmer, J., Flint, S., and Brooks, J. (2007) Bacterial cell attachment, the beginning of a biofilm, *J. Ind. Microbiol. Biotechnol.*, **34**, 577–588.

39. Shen, W., and Ljungh, A. (1993) Collagen-binding to *Escherichia coli* strain Ng7c, *Curr. Microbiol.*, **27**, 311–316.

40. Visai, L., Speziale, P., and Bozzini, S. (1990) Binding of collagens to an enterotoxigenic strain of *Escherichia coli*, *Infect. Immun.*, **58**, 449–455.

41. Park, B.-J., Haines, T., and Abu-Lail, N. I. (2009) A correlation between the virulence and the adhesion of *Listeria monocytogenes* to silicon nitride: an atomic force microscopy study, *Colloids Surf. B Biointerfaces*, **73**, 237–243.

42. Hildebrand, M., Holton, G., Joy, D. C., Doktycz, M. J., and Allison, D. P. (2009) Diverse and conserved nano- and mesoscale structures of diatom silica revealed by atomic force microscopy, *J. Microsc.*, **235**, 172–187.

43. Micic, M., Hu, D., Suh, Y. D., Newton, G., Romine, M., and Lu, H. P. (2004) Correlated atomic force microscopy and fluorescence lifetime imaging of live bacterial cells, *Colloids Surf. B Biointerfaces*, **34**, 205–212.

44. Razatos, A., Ong, Y.-L., Sharma, M. M., and Georgiou, G. (1998) Molecular determinants of bacterial adhesion monitored by atomic force microscopy, *Proc. Natl. Acad. Sci. USA*, **95**, 11059–11064.

45. Robichon, D., Girard, J.-C., Cenatiempo, Y., and Cavellier, J.-F. (1999) Atomic force microscopy imaging of dried or living bacteria, *Comptes Rendus de l'Académie des Sciences. Series III, Sciences de la Vie*, **322**, 687–693.

46. Velegol, S. B., and Logan, B. E. (2002) Contributions of bacterial surface polymers, electrostatics, and cell elasticity to the shape of AFM force curves, *Langmuir*, **18**, 5256–5262.

47. Velegol, S. B., Pardi, S., Li, X., Velegol, D., and Logan, B. E. (2003) AFM imaging artifacts due to bacterial cell height and AFM tip geometry, *Langmuir*, **19**, 851–857.

48. Cerf, A., Cau, J.-C., Vieu, C., and Dague, E. (2009) Nanomechanical properties of dead or alive single-patterned bacteria, *Langmuir*, **25**, 5731–5736.

49. Dorobantu, L. S., Bhattacharjee, S., Foght, J. M., and Gray, M. R. (2008) Atomic force microscopy measurement of heterogeneity in bacterial surface hydrophobicity, *Langmuir*, **24**, 4944–4951.

50. Bagchi, S., Tomenius, H., Belova, L. M., and Ausmees, N. (2008) Intermediate filament-like proteins in bacteria and a cytoskeletal function in Streptomyces, *Mol. Microbiol.*, **70**, 1037–1050.

51. Ahimou, F., Semmens, M. J., Novak, P. J., and Haugstad, G. (2007) Biofilm cohesiveness measurement using a novel atomic force microscopy methodology, *Appl. Environ. Microbiol.*, **73**, 2897–2904.

52. Del Sol, R., Armstrong, I., Wright, C., and Dyson, P. (2007) Characterization of changes to the cell surface during the life cycle of Streptomyces coelicolor: atomic force microscopy of living cells, *J. Bacteriol.*, **189**, 2219–2225.

53. Mangold, S., Harneit, K., Rohwerder, T., Claus, G., and Sand, W. (2008) Novel combination of atomic force microscopy and epifluorescence microscopy for visualization of leaching bacteria on pyrite, *Appl. Environ. Microbiol.*, **74**, 410–415.

54. Kienberger, F., Zhu, R., Moser, R., Blaas, D., and Hinterdorfer, P. (2004) Monitoring RNA release from human rhinovirus by dynamic force microscopy, *J. Virol.*, **78**, 3203–3209.

55. Sullivan, C. J., Morrell, J. L., Allison, D. P., and Doktycz, M. J. (2005) Mounting of *Escherichia coli* spheroplasts for AFM imaging, *Ultramicroscopy*, **105**, 96–102.

56. Abu-Lail, N. I., and Camesano, T. A. (2002) Elasticity of *Pseudomonas putida* KT2442 surface polymers probed with single-molecule force microscopy, *Langmuir*, **18**, 4071–4081.

57. Volle, C. B., Ferguson, M. A., Aidala, K. E., Spain, E. M., and Núñez, M. E. (2008) Spring constants and adhesive properties of native bacterial biofilm cells measured by atomic force microscopy, *Colloids Surf. B Biointerfaces*, **67**, 32–40.

58. Plomp, M., and Malkin, A. J. (2009) Mapping of proteomic composition on the surfaces of Bacillus spores by atomic force microscopy-based immunolabeling, *Langmuir*, **25**, 403–409.

59. Alsteens, D., Verbelen, C., Dague, E., Raze, D., Baulard, A. R., and Dufrene, Y. F. (2008) Organization of the mycobacterial cell wall: a nanoscale view, *Pflugers Arch.*, **456**, 117–125.

60. Dupres, V., Alsteens, D., Pauwels, K., and Dufrene, Y. F. (2009) In vivo imaging of S-layer nanoarrays on *Corynebacterium glutamicum*, *Langmuir*, **25**, 9653–9655.

61. Chen, Y.-Y., Wu, C.-C., Hsu, J.-L., Peng, H.-L., Chang, H.-Y., and Yew, T.-R. (2009) Surface rigidity change of *Escherichia coli* after filamentous bacteriophage infection, *Langmuir*, **25**, 4607–4614.

62. da Silva, A., Jr., and Teschke, O. (2003) Effects of the antimicrobial peptide PGLa on live *Escherichia coli*, *Biochim. Biophys. Acta*, **1643**, 95–103.

63. Meincken, M., Holroyd, D. L., and Rautenbach, M. (2005) Atomic force microscopy study of the effect of antimicrobial peptides on the cell envelope of *Escherichia coli*, *Antimicrob. Agents Chemother.*, **49**, 4085–4092.

64. Rossetto, G., Bergese, P., Colombi, P., Depero, L. E., Giuliani, A., Nicoletto, S. F., and Pirri, G. (2007) Atomic force microscopy evaluation of the effects of a novel antimicrobial multimeric peptide on *Pseudomonas aeruginosa*, *Nanomedicine*, **3**, 198–207.

65. Li, A., Lee, P. Y., Ho, B., Ding, J. L., and Lim, C. T. (2007) Atomic force microscopy study of the antimicrobial action of Sushi peptides on Gram negative bacteria, *Biochim. Biophys. Acta*, **1768**, 411–418.

66. Sharma, S., Sen, P., Mukhopadhyay, S. N., and Guha, S. K. (2003) Microbicidal male contraceptive-Risug induced morphostructural damage in E. coli, *Colloids Surf. B Biointerfaces*, **32**, 43–50.

67. Eby, D. M., Schaeublin, N. M., Farrington, K. E., Hussain, S. M., and Johnson, G. R. (2009) Lysozyme catalyzes the formation of antimicrobial silver nanoparticles, *ACS Nano*, **3**, 984–994.

68. Deupree, S. M., and Schoenfisch, M. H. (2009) Morphological analysis of the antimicrobial action of nitric oxide on Gram-negative pathogens using atomic force microscopy, *Acta Biomater.*, **5**, 1405–1415.

69. Qi, Z.-D., Lin, Y., Zhou, B., Ren, X.-D., Pang, D.-W., and Liu, Y. (2008) Characterization of the mechanism of the *Staphylococcus aureus* cell envelope by bacitracin and bacitracin-metal ions, *J. Membr. Biol.*, **225**, 27–37.

70. Cross, S. E., Kreth, J., Zhu, L., Qi, F. X., Pelling, A. E., Shi, W. Y., and Gimzewski, J. K. (2006) Atomic force microscopy study of the structure-function relationships of the biofilm-forming bacterium *Streptococcus mutans*, *Nanotechnology*, **17**, S1-S7.

71. Jonas, K., Tomenius, H., Kader, A., Normark, S., Romling, U., Belova, L., and Melefors, O. (2007) Roles of curli, cellulose and BapA in Salmonella biofilm morphology studied by atomic force microscopy, *BMC Microbiol.*, **7**, 70.

72. Qin, Z., Zhang, J., Hu, Y., Chi, Q., Mortensen, N. P., Qu, D., Molin, S., and Ulstrup, J. (2009) Organic compounds inhibiting *S. epidermidis* adhesion and biofilm formation, *Ultramicroscopy*, **109**, 881–888.

73. Nunez, M. E., Martin, M. O., Chan, P. H., and Spain, E. M. (2005) Predation, death, and survival in a biofilm: Bdellovibrio investigated by atomic force microscopy, *Colloids Surf. B Biointerfaces*, **42**, 263–271.

74. Nunez, M., Martin, M. O., Duong, L. K., Ly, E., and Spain, E. M. (2003) Investigations into the life cycle-of the bacterial predator *Bdeliovibrio bacteriovorus* 109J at an interface by atomic force microscopy, *Biophys. J.*, **84**, 3379–3388.

75. Kreplak, L., Wang, H., Aebi, U., and Kong, X.-P. (2007) Atomic force microscopy of mammalian urothelial surface, *J. Mol. Biol.*, **374**, 365–373.

76. Boonaert, C. J. P., Rouxhet, P. G., and Dufrene, Y. F. (2000) Surface properties of microbial cells probed at the nanometre scale with atomic force microscopy, *Surface Interface Anal.*, **30**, 32–35.

77. Carroll, A. M., Plomp, M., Malkin, A. J., and Setlow, P. (2008) Protozoal digestion of coat-defective *Bacillus subtilis* spores produces "rinds" composed of insoluble coat protein, *Appl. Environ. Microbiol.*, **74**, 5875–5881.

78. Dennis, D., Sein, V., Martinez, E., and Augustine, B. (2008) PhaP is involved in the formation of a network on the surface of polyhydroxyalkanoate inclusions in *Cupriavidus necator* H16, *J. Bacteriol.*, **190**, 555–563.

79. Touhami, A., Jericho, M. H., Boyd, J. M., and Beveridge, T. J. (2006) Nanoscale characterization and determination of adhesion forces of *Pseudomonas aeruginosa* pili by using atomic force microscopy, *J. Bacteriol.*, **188**, 370–377.

80. Beveridge, T. J. (1988) The bacterial surface—general-considerations towards design and function, *Can. J. Microbiol.*, **34**, 363–372.

81. White, D. (2000) *The Physiology and Biochemistry of Prokaryotes*, Oxford University Press, New York.

82. Birdsell, D. C., and Cota-Robles, E. H. (1967) Production and ultrastructure of lysozyme and ethylenediaminetetraacetate-lysozyme spheroplasts of *Escherichia coli*, *J. Bacteriol.*, **93**, 427–437.

83. Miller, I. L., Zsigray, R. M., and Landman, O. E. (1967) The formation of protoplasts and quasi-spheroplasts in normal and chloramphenicol-pretreated *Bacillus subtilis*, *J. Gen. Microbiol.*, **49**, 513–525.

84. Hildebrand, M., York, E., Kelz, J. I., Davis, A. K., Frigeri, L. G., Allison, D. P., and Doktycz, M. J. (2006) Nanoscale control of silica morphology and three-dimensional structure during diatom cell wall formation, *J. Mater. Res.*, **21**, 2689–2698.

85. Allison, D. P., Dufrene, Y. F., Doktycz, M. J., and Hildebrand, M. (2008) Biomineralization at the nanoscale learning from diatoms, *Methods Cell Biol.*, **90**, 61–86.

86. Hildebrand, M., Doktycz, M., and Allison, D. (2008) Application of AFM in understanding biomineral formation in diatoms, *Pflügers Arch.*, **456**, 127–137.

87. Kroger, N., and Poulsen, N. (2008) Diatoms—from cell wall biogenesis to nanotechnology, *Annu. Rev. Genet.*, **42**, 83–107.

88. Hinterdorfer, P., and Dufrêne, Y. F. (2006) Detection and localization of single molecular recognition events using atomic force microscopy, *Nat. Methods*, **3**, 347 -355.

89. Gad, M., Itoh, A., and Ikai, A. (1997) Mapping cell wall polysaccharides of living microbial cells using atomic force microscopy, *Cell Biol. Int.*, **21**, 697–706.

90. Hinterdorfer, P., Baumgartner, W., Gruber, H. J., Schilcher, K., and Schindler, H. (1996) Detection and localization of individual antibody-antigen recognition events by atomic force microscopy, *Proc. Natl. Acad. Sci. USA*, **93**, 3477–3481.

91. Preiner, J., Ebner, A., Chtcheglova, L., Zhu, R., and Hinterdorfer, P. (2009) Simultaneous topography and recognition imaging: physical aspects and optimal imaging conditions, *Nanotechnology*, **20**, 215103.

92. Vadillo-Rodriguez, V., Busscher, H. J., van der Mei, H. C., de Vries, J., and Norde, W. (2005) Role of lactobacillus cell surface hydrophobicity as probed by AFM in adhesion to surfaces at low and high ionic strength, *Colloids Surf. B Biointerfaces*, **41**, 33–41.

93. Weisenhorn, A. L., Schmitt, F. J., Knoll, W., and Hansma, P. K. (1992) Streptavidin binding observed with an atomic force microscope, *Ultramicroscopy*, **42–44** (Pt B), 1125–1132.

94. Chiantia, S., Ries, J., Chwastek, G., Carrer, D., Li, Z., Bittman, R., and Schwille, P. (2008) Role of ceramide in membrane protein organization investigated by combined AFM and FCS, *Biochim. Biophys. Acta.*, **1778**, 1356–1364.

95. Gumpp, H., Stahl, S. W., Strackharn, M., Puchner, E. M., and Gaub, H. E. (2009) Ultrastable combined atomic force and total internal fluorescence microscope, *Rev. Sci. Instrum.*, **80**, 063704.

96. Oreopoulos, J., and Yip, C. M. (2009) Combinatorial microscopy for the study of protein-membrane interactions in supported lipid bilayers: order parameter measurements by combined polarized TIRFM/AFM, *J. Struct. Biol.*, **168**, 21–36.

97. Frankel, D. J., Pfeiffer, J. R., Surviladze, Z., Johnson, A. E., Oliver, J. M., Wilson, B. S., and Burns, A. R. (2006) Revealing the topography of cellular membrane

domains by combined atomic force microscopy/fluorescence imaging, *Biophys. J.*, **90**, 2404–2413.

98. Madl, J., Rhode, S., Stangl, H., Stockinger, H., Hinterdorfer, P., Schutz, G. J., and Kada, G. (2006) A combined optical and atomic force microscope for live cell investigations, *Ultramicroscopy*, **106**, 645–651.

Chapter 4

RESOLVING THE HIGH-RESOLUTION ARCHITECTURE, ASSEMBLY AND FUNCTIONAL REPERTOIRE OF BACTERIAL SYSTEMS BY *IN VITRO* ATOMIC FORCE MICROSCOPY

Alexander J. Malkin
Physical and Life Sciences Directorate, Lawrence Livermore National Laboratory, Livermore, CA 94551, USA
malkin1@llnl.gov

4.1 PROBING THE SPORE COAT HIGH-RESOLUTION STRUCTURE AND ASSEMBLY

When starved for nutrients, *Bacillus* and *Clostridium* cells initiate a series of genetic, biochemical and structural events that results in the formation of a metabolically dormant endospore.[1] Bacterial spores can remain dormant for extended time periods and possess a remarkable resistance to a wide range of environmental insults, including heat, radiation, pH extremes and toxic chemicals.[1] Their unique structure, including a protective multilayer spore coat, plays a major role in the maintenance of spore environmental resistance, dormancy and germination.[1–3]

The *Bacillus* bacterial spore structure (Fig. 4.1) consists,[3] starting from the centre, of an inner core surrounded by the inner cytomembrane, a cortex, outer membrane and an exterior spore coat. In some bacterial species (i.e. *Bacillus thuringiensis* and *Bacillus anthracis*), the coat is surrounded by a loosely attached exosporium. The spore core contains DNA and dipicolinic

Life at the Nanoscale: Atomic Force Microscopy of Live Cells
Edited by Yves Dufrêne

www.panstanford.com

acid, which is associated predominantly with Ca^{2+}. The major role of the spore cortex, which consists of a thick layer of species-dependent peptidoglycan, is to maintain spore heat resistance and dormancy.

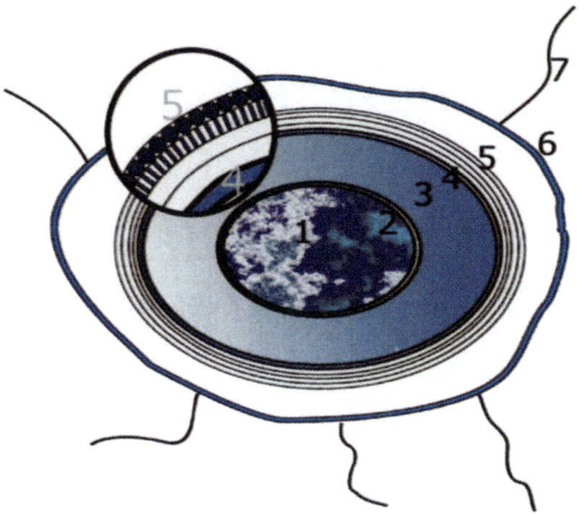

Figure 4.1. Structure of a *Bacillus* spore: spore core (1); inner membrane (2); cortex (3); outer membrane (4); spore coat (5); exosporium (6); and appendages (7). Insert: spore coat with two crystalline layers of the outer spore coat.

One of the current scientific challenges at the intersection of life and physical sciences is to define the biophysical pathways of cellular life and, in particular, to elucidate the complex molecular machines that carry out cellular and microbial function and propagate the disease. The present technological and scientific challenges are to unravel the relationships between the organization and function of protein complexes at cell and microbial surfaces, to understand how these complexes evolve during the bacterial and cellular life cycles and how they respond to environmental changes, chemical stimulants and therapeutics.

Development of atomic force microscopy (AFM) for probing the architecture and assembly of single microbial and cellular surfaces at a nanometre scale under native conditions, and unravelling of its structural dynamics during their life cycle, and in response to changes in the environment, has the capacity to significantly enhance the current insight into molecular architecture and structural and environmental variability of cellular and microbial systems.

In this chapter, we will demonstrate, focusing on the work conducted in our group in the past several years, the capabilities of AFM in probing the architecture and assembly of bacterial surfaces and integument structures, and their evolvement during bacterial life cycles, as well as in response to environmental changes.

4.1.1 Spore Morphology

AFM images of various species of air-dried *Bacillus*[4–7] and *Clostridium novyi-NT*[8] spores are presented in Fig. 4.2. As illustrated in Fig. 4.2, *B. thuringiensis* and *B. anthracis* native spores are enclosed within an exosporium sacculus (indicated with the letter *E* in Fig. 4.2a), which is either larger than the dimensions of the spore body (Fig. 4.2a) or tightly attached to the spore coat as in the case of *B. anthracis* spores (Fig. 4.2b). *C. novyi-NT* spores were found to be encased in amorphous shells composed by irregular amorphous material (Fig. 4.2d), with many spores exhibiting ~200 nm thick "shell tails" at their poles.

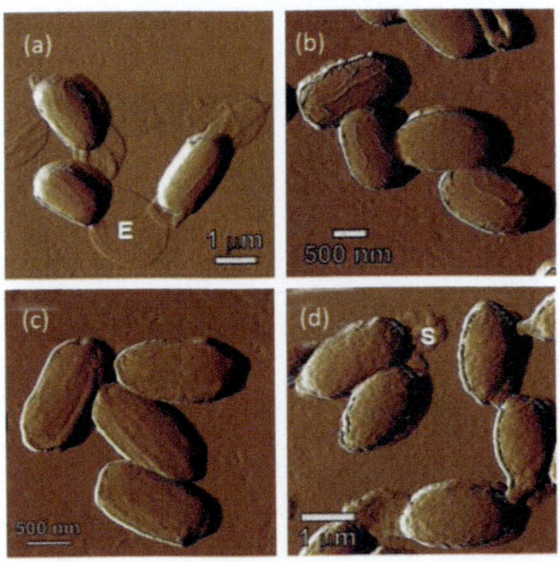

Figure 4.2. AFM images of air-dried bacterial spores. (a) *B. thuringiensis*, (b) *B. anthracis*, (c) *B. atrophaes* and (d) *C. novyi-NT* spores. Exosporium is indicated with E in (a). "Shell tail" in (d) is indicated with S in (d).

4.1.2 Spore Coat Architecture and Assembly

More than 50 *Bacillus* spore coat proteins have been identified by genomic and proteomic analysis.[1–3,9] Despite the recent advances in biochemical and genetic studies,[9] spore coat self-assembly is still poorly understood. In particular, it is not clear which spore coat proteins form the various spore coat layers, what their roles are in the coat assembly and, finally, which proteins are surface-exposed and which ones are embedded beneath the surface. The elucidation of bacterial spore coat architecture and structure–

function relationships is critical to determining mechanisms of pathogenesis, environmental resistance, immune response and spore's physicochemical properties. Thus, the development and application of high-resolution imaging techniques, which could address spatially explicit bacterial spore coat protein architecture at nanometre resolution under physiological conditions, are of considerable importance.

We have directly visualized species-specific high-resolution native spore coat structures of bacterial spores including the exosporium and crystalline layers of the spore coat (Fig. 4.3) of various *Bacillus*[4–7,10–12] and *C. novyi-NT*[8] species in their natural environment, namely, air and fluid.

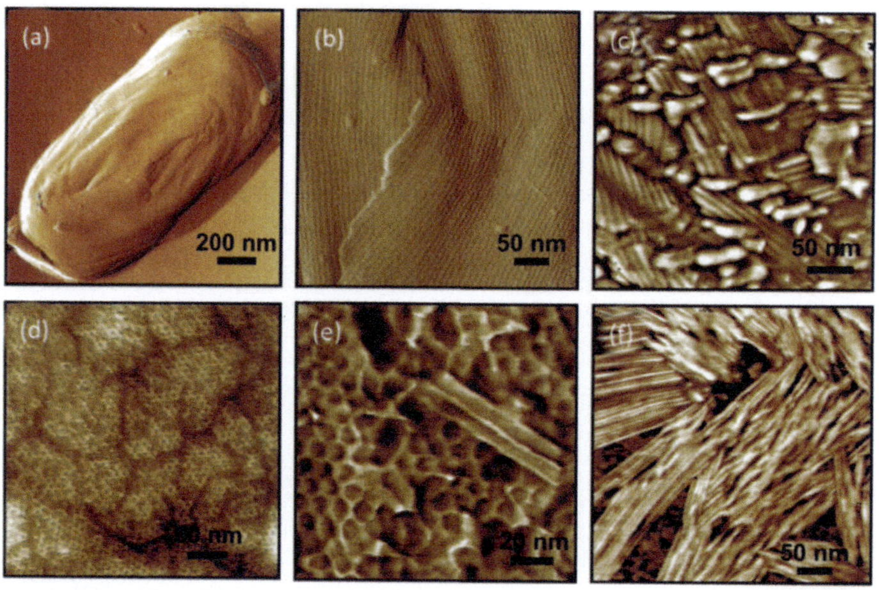

Figure 4.3. High-resolution spore coat structures of *Bacillus* spores. The outer spore coats of *B. atrophaeus* (a,b), *B. cereus* (c) and *B. thuringiensis* (d) consist of crystalline layers rodlet and honeycomb structures. *B. cereus* spores contain a crystalline honeycomb structure (e) beneath the exterior rodlet layer (c). *B. thuringiensis* spore coats do not contain rodlet structures. Rodlet assemblies can be seen adsorbed to the substrate (f). Images reproduced, with permission from Ref. 4. © (2005) Biophysical Society, USA.

For *Bacillus atrophaeus* (Fig. 4.3a,b), the outer spore coat was composed of a crystalline rodlet layer with a periodicity of ~8 nm. In case of *Bacillus subtilis* spores, the rodlet structure was typically completely or partially covered by the amorphous layer (Fig. 4.4a). Patches of amorphous layer were also occasionally seen on *B. atrophaeus* spores. Removal of the *Bacillus cereus* and *B. thuringiensis* exosporium by sonication[4] or single-cell French Press treatment[8] revealed crystalline rodlet (Fig. 4.3c) and hexagonal honeycomb (Fig. 4.3d) outer spore coat structures, respectively.

As seen in Fig. 4.3c, the ~10 nm thick rodlet layer of *B. cereus* spores is formed by multiple randomly oriented domains, comprising parallel subunits with a periodicity of ~8 nm. The size of the domains is typically 100–200 nm. In contrast to the multi-domain rodlet structure of the *B. cereus* spore coat, typically only a single continuous domain or several domains were found to be present on the outer coat of *B. atrophaeus* (Fig. 4.3b) and *B. subtilis* (Fig. 4.4a) spores. Complete removal of the exterior *B. cereus* rodlet layer by sonication revealed an underlying honeycomb structure (Fig. 4.3e) similar to the exterior spore coat layer of *B. thuringiensis* (Fig. 4.3d). For both species, the lattice parameter for the honeycomb structure is ~9 nm, with ~5–6 nm holes/pits (Fig. 4.3d,e). In case of *B. thuringiensis* spores, rodlet structures were not observed as an integral component of the spore coat[4,6]; however, as illustrated in Fig. 4.3f, patches of extrasporal rodlet structures were observed adsorbed to the substrate.[4,6] Rodlet width and thickness (Fig. 4.3.f) were similar to those observed for *B. atrophaeus*, *B. subtilis* and *B. cereus* spore coat structures, which indicates that the similar rodlet proteins could be present during the sporulation in these three species of *Bacillus* spores.

Similar rodlet and honeycomb crystalline structures to those seen in Fig. 4.3 were observed in freeze-etching electron microscopy (EM) studies of several species of *Bacillus* spores[13] and AFM studies of fungal spores.[14] Note that in the case of *B. thuringiensis* spore coat rodlet structures were not observed in freeze-etching EM.[13]

The assembly, physical properties and proteomic nature of these bacterial spore rodlets are poorly understood. The closest structural and functional orthologs to the *Bacillus* species rodlet structure (*not* its protein sequence) are found outside the *Bacillus* genus. Several classes of proteins, with divergent primary sequences, were found to form similar rodlet structures on the surfaces of cells of Gram-negative *Escherichia coli* and *Salmonella enterica*, as well as on spores of Gram-positive streptomycetes and various fungi (for a review, see Ref. 15). Hydrophobins, a new class of structural proteins,[16] were shown to be an integral component of rodlet fungal spore surface structures. However, it has not been possible to identify orthologs of hydrophobin-like proteins in bacterial spores.[17] Similarities in crystalline outer coat layer motifs found in prokaryotic and eukaryotic spore types are a striking and unexpected example of the convergent evolution of critical biological structures. Further investigation is required to determine the molecular composition of prokaryotic endospore rodlets and their evolutionary relationship to eukaryotic rodlet structures.

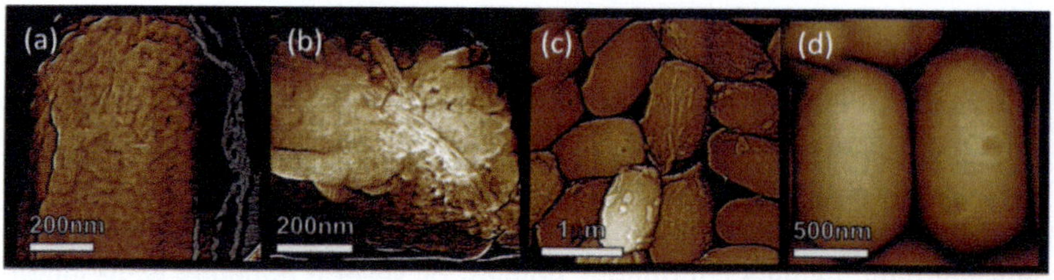

Figure 4.4. AFM images of *B. subtilis* spores of different strains. The spores analyzed were wild type (a), *cotE* (b), *gerE* (c) and *cotEgerE* (d). Images reproduced, with permission from Refs. 11 and 12. © (2008) American Society for Microbiology.

Proper assembly of a multilayer spore coat of *Bacillus* spores is dependent on a number of coat proteins.[1] Loss of any of those proteins could alter significantly the mechanisms of the spore coat assembly and the final spore coat structure. Indeed, as demonstrated in Fig. 4.4, deletion of a single spore coat protein could result in pronounced changes in the spore coat architecture.[11,12] Thus, the AFM analysis demonstrated that intact wild-type *B. subtilis* spores are completely or partially covered by a thin amorphous layer lacking defined structure (Fig. 4.4a). Directly below the amorphous layer is a rodlet crystalline layer (Fig. 4.4a), which has parameters similar to the *B. atrophaeus* rodlet spore coat layer.[6] *CotE* is a major spore coat morphogenetic protein, and in its absence, the outer coat fails to assemble properly.[18] Indeed, we have demonstrated that for most *cotE* spores the outermost structure is formed by 3–5 crystalline layers, each of which is ~6 nm thick (Fig. 4.4b), which likely correspond to the inner coat layers.

Furthermore, surfaces of some *cotE* mutant spores exhibit patches or large regions covering the spore of a hexagonal crystalline layer (located between the rodlet layer and the inner coat multilayer structure) (Fig. 4.4b). Surfaces of *gerE* spores were found to lack completely both amorphous and rodlet structures, being encased in several inner spore coat layers.[7] Finally, spores lacking both *CotE* and *gerE* proteins (*cotEgerE* spores) were found to lack all outer and inner coat structures, with the spore cortex being the outermost structure.[12] Our recent comprehensive analysis of a wide range of *B. subtilis* mutants, which lack various spore coat proteins,[19] has provided improved understanding of the spore coat architecture, assembly and function of coat proteins.

To observe the structure of the *C. novyi-NT* spore coat beneath the amorphous shell, we developed procedures to remove the shells by chemical treatment with various reducing agents and detergents or by physical treatment using a French Press.[8] When either French Press or chemical treatments were used, the majority of the exposed spore coat surface is formed by a ~8–10 nm thick honeycomb layer with a periodicity of 8.7 ± 1 nm (Fig. 4.5).

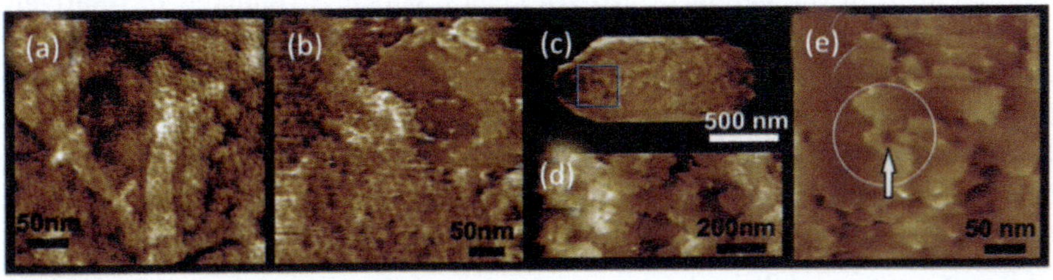

Figure 4.5. *C. novyi-NT* spore coats—high-resolution AFM height images. (a) Removal of the amorphous shell by physico-chemical treatments reveals the underlying honeycomb layer. (b) Most of the honeycomb layers disappeared from the spores within ~1 hour during the germination process. Remaining honeycomb patches (left, lower sides) could be easily removed by scanning with increased force. Below the honeycomb layer, several underlying coat layers (upper right) are revealed. (c–e). Typical growth patterns seen on *C. novyi-NT* spore surface after removing the honeycomb layers. (c) Whole spore with several ~6 nm thick layers exposed on the surface. (d) Zoom-in of the centre of (c) showing that spore coat layers originate at screw dislocations. (e) Zoom-in of the area indicated in (c). The circle in (e) denotes a fourfold screw axis. Many dislocation centres show depressions reminiscent of hollow cores (arrow). Images reproduced, with permission from Ref. 8. © (2007) American Society for Microbiology.

As seen in Fig. 4.5b–e, the removal of the honeycomb layer revealed a multilayer structure formed by ~6 nm thick smooth layers. Typically, there were 3–6 layers exposed on the spore surface, similar to ones observed for *B. subtilis* (Fig. 4.4) and *B. anthracis* (data are not shown here) spores. The spore coat surface patterns (Fig. 4.5b–e) were very similar to ones observed on the surfaces of inorganic, organic and macromolecular crystals.[20-23]

These patterns include steps and growth spirals originating from screw dislocations, such as those previously described in studies of the crystallization of semiconductors,[24] salts[25] and biological macromolecules.[22,23] In the middle of the growth centres, the dislocations cause depressions, typically <15 nm, which are known as *hollow cores* in crystal growth theory and are formed by the stress associated with the dislocations.[26]

Thus, the presence of the aforementioned growth patterns confirms the crystalline nature of the coat layers. However, while AFM resolution is typically sufficient for visualization of crystal lattices on a molecular scale for a wide range of protein crystals,[23,27] we were not able to resolve a regular, crystalline lattice on the spore coat layers. In this case, the lattice periodicity is assumed to be smaller than ~1 nm, which is the resolution associated with the sharpest AFM tips used. Such a periodicity would be small compared with the ~6 nm thickness of individual spore coat layers. In the case of globular proteins, lateral lattice parameters typically do not differ to such an extent from the height of growth layers, which is reflected in relatively small differences between lateral and perpendicular crystallographic unit cell parameters.[27] Thus, the proteins forming the spore coat layers are likely not globular, but rather may be stretched peptides "standing upright" in the layers. This construction, which is found in fat crystals,[28] results in a crystal class with relatively strong, hydrophobic interaction forces between the long neighbouring units (here peptides) and weak interaction forces between the different crystalline layers. Such a crystal type, with tightly packed, strongly interacting longitudinal peptides within a layer, would help explain the toughness associated with bacterial spore coats.[1–3,13] It may also explain why spore coat proteins are difficult to dissolve,[1,13,29] as this type of packing involves hydrophobic interactions and hence a high proportion of hydrophobic amino acids.

In addition to enabling the nucleation and growth of new coat layers during sporulation, the screw dislocations also pin several of these layers together, thereby making the spore coat an interconnected, cohesive entity, rather than a set of separate layers loosely deposited on top of each other. This, combined with the strong in-layer bonds, and possible cross-linking between the coat proteins, likely contributes to the resilient nature of the spore coat.

In biology, crystallization is most often associated with biomineralization, where protein-directed crystallization leads to calcious bone[30] and shell formation.[31,32] Screw dislocations and ensuing spiral growth have been observed for shell formation.[33,34] High-resolution scanning electron probe X-ray microanalysis[35,36] and nanometre-scale secondary ion mass spectrometry[37] studies have demonstrated that the proteinaceous coat of several bacterial spore species is essentially devoid of divalent mineral cations such as calcium, magnesium and manganese. This indicates that *C. novyi-NT* spores could present the first case of non-mineral dislocation growth patterns being revealed for a biological organism.

4.1.3 Formulation-Specific Spore Coat Assembly

The implication of observed crystalline nature of *Bacillus* and *Clostridium* spore coat layers for bacterial spore coat assembly is that, while the proteineous building blocks are produced via biochemical pathways directed by various enzymes and factors,[1] the actual construction of these building blocks into spore coat layers is a self-assembly crystallization process. Similarly, the striking differences in native rodlet motifs seen in *B. atrophaeus* (one major domain for each spore), *B. cereus* (a patchy multi-domain motif) and *B. thuringiensis* (extrasporal rodlets) appear to be a consequence of species-specific nucleation and crystallization mechanisms that regulate the assembly of the outer spore coat. In the case of *B. cereus* outer coat assembly, the surface free energy[38] for crystalline phase nucleation appears to be low enough to allow the formation of multiple rodlet domains resulting in cross-patched and layered assemblies. During the assembly of the outer coat of *B. atrophaeus* spores, the surface free energy may be considerably higher, reducing nucleation to the point that only one major domain is formed covering the entire spore surface. In addition to the possible differences in the surface free energies of the underlying inner coat, the pronounced difference in the nucleation rate of the outer coat rodlet layers for different *Bacillus* species could be caused by different supersaturation levels of the sporulation media during rodlet self-assembly. Since the molecular mechanisms of self-assembly of spore coat structural layers appear to be very similar to those described for nucleation and crystallization of inorganic and macromolecular single crystals,[22,23,38] fundamental and applied concepts developed for the nucleation and growth of inorganic and protein crystals can be applied successfully to understanding the assembly of the spore coat. Thus, based on experimentally observed rodlet structural properties, we have developed a model for rodlet spore surface assembly, which was derived from well-developed molecular-scale crystallization/self-assembly mechanisms.[6]

The consequence of spore coat crystalline assembly process is that similar to inorganic and macromolecular crystallization, and conditions during sporulation such as salt concentration, pH, the presence of impurities, nucleation rates of crystalline self-assembly of spore coat layers and random variations in the number of screw dislocations on spores could change the growth rate and hence the thickness of the spore coat.

Furthermore, these observations suggest that spore coat architecture and assembly are not purely genetically determined but could also be strongly influenced by the modifications of sporulation media, which in turn could affect spore germination competence and physicochemical properties. However, the effects of environmental and chemical perturbations on spore coat structure have not been investigated before.

By observing spore coat high-resolution structures, AFM analysis could be utilized to reconstruct the environmental conditions that were present during spore formation. Thus, we have demonstrated for the first time the pronounced differences in the spore coat architecture of *B. thuringiensis* spores grown under different sporulation conditions. Thus, for spores grown in NB medium, only honeycomb crystalline layers were seen on the spore coat (Fig. 4.6a,b) accompanied by the extrasporal rodlets. However, for spores grown in G medium, patches of rodlet structure were visualized on the spore coat (Fig. 4.6c,d). These data establish that outer coat structural motifs are directly correlated to differences in the medium conditions during sporulation.

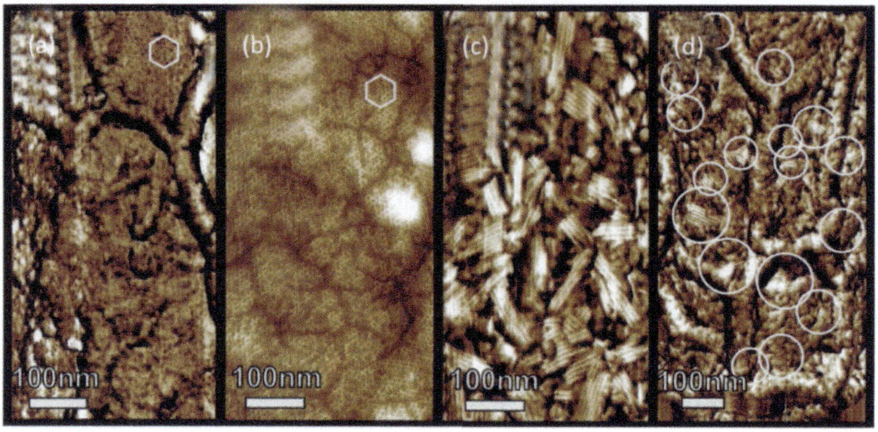

Figure 4.6. AFM images showing the outer coat structure of *B. thuringiensis* spores grown in NB medium (a,b) and G medium (c,d). Honeycomb and rodlet crystalline structures are indicated with hexagons (a,b) and circles (d), respectively.

These findings validate that AFM can identify formulation-specific structural attributes that could be used in bioforensics to reconstruct spore formulation conditions. We have recently successfully demonstrated this approach for probing the formulation-dependent spore coat structures of *B. anthracis* spores.[39]

4.1.4 AFM-Based Immunolabelling of the Proteomic Structures

AFM provides high-resolution topographical information about the spatial and temporal distribution of macromolecules in biological samples. However, simultaneous near-molecular resolution topographical imaging of biological structures and specific recognition of the proteins forming these structures is currently lacking. Of particular importance is the identification of the protein composition of pathogen and microbial surfaces. Pathogen outer surface structures (e.g. virus membranes and capsids, as well as bacterial cell walls, spore coats and exosporia) typically contain multiple proteins. While it is known to a certain degree which proteins are expressed for these surface structures, it is often unknown which of these are exposed on the outside of these structures and which are embedded within the structures.

Detection of surface-exposed proteins is paramount for improving the fundamental understanding of their functional properties as well as for the development of detection, attribution and medical countermeasures against these pathogens.

In the past several years, considerable progress, in particular towards probing of microbial and cellular systems, has been made in identification and mapping of specific receptors and ligands on the biological surfaces using adhesion force mapping and dynamic recognition force mapping (for reviews, see Refs. 40 and 41). EM-based immunolabelling techniques have become an important tool for the elucidation of biological structure and function.[42,43] AFM immunogold markers were utilized in the past for imaging of proteins and macromolecular ensembles.[44–48]

We have recently utilized[10] AFM-based immunochemical labelling procedures for visualization and mapping of the binding of antibodies, conjugated with nanogold particles, to specific epitopes on the surfaces of *Bacillus* spores. We have established the immunospecificity of labelling, through the utilization of specific anti-*B. atrophaeus and B. anthracis* polyclonal and monoclonal antibodies, which were targeted to spore coat and exosporium epitopes (Fig. 4.7). In particular, we have confirmed that bclA glycoprotein is the immuno-dominant epitope on the surface of *B. anthracis* spores.[10]

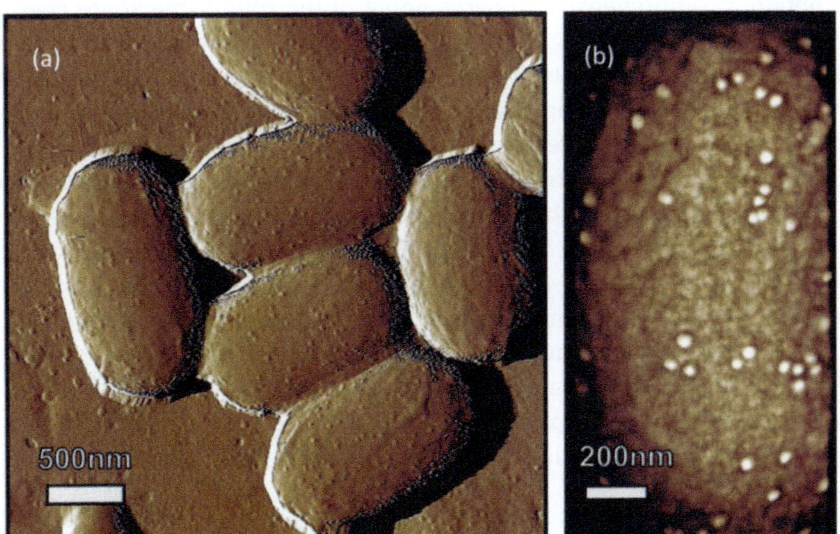

Figure 4.7. AFM images of specific binding of anti-*B. anthracis* gold-labelled polyclonal antibodies to the *B. anthracis* spore exosporia. Images reproduced with permission from Ref. 10. © (2009) American Chemical Society.

4.2 PROBING THE STRUCTURAL DYNAMICS OF SINGLE GERMINATING SPORES

Upon exposure to favourable conditions, metabolically dormant *Bacillus* and *Clostridium* spores break dormancy through the process of germination[49–51] and eventually reenter the vegetative mode of replication. A comprehensive understanding of the mechanisms controlling spore germination is of fundamental importance both for practical applications related to the prevention of a wide range of diseases by spore-forming bacteria as well as for fundamental studies of cell development.

Germination involves an ordered sequence of chemical, biosynthetic and genetic events.[49–51] Spore coat structure regulates the permeation of germinant molecules.[51] However, while significant progress has been made in understanding the biochemical and genetic bases for the germination process,[49] the role of the spore coat in the germination remains unclear.[49–51]

We have utilized[7,8] *in vitro* AFM methods for molecular-scale examination of spore coat and germ cell wall dynamics during spore germination and outgrowth.

4.2.1 Germination-Induced Spore Coat Disassembly

To obtain a comprehensive understanding of the role of the spore coat in germination, AFM imaging on a nanometre scale is required. At this scale, the outer layer of the *B. atrophaeus* spore coat is composed of a crystalline rodlet array (Fig. 4.8a,b; Fig. 4.3a,b) containing a small number of point and planar (stacking fault) defects.[6] Upon exposure to the germination solution, disassembly of the rodlet structures was observed.[7] During the initial stages of germination, the disassembly was initiated through the formation of 2–3 nm wide micro etch pits in the rodlet layer (Fig. 4.8b). Subsequently, the etch pits formed fissures (Fig. 4.8b–d) that were, in all cases, oriented perpendicular to the rodlet direction. Simultaneously, etching commenced on the stacking faults (Fig. 4.8e–f), revealing an underlying hexagonal inner spore coat layer (Fig. 4.8g). During later stages of germination, further disintegration of the rodlet layer (Fig. 4.8e–f) proceeded by coalescence of existing fissures, by their autonomous elongation and widening and by continued formation of new fissures.

Currently, it is unclear what causes this breakdown of the rodlet layer. We have proposed[7] that rodlet structure degradation is caused by specific hydrolytic enzyme(s), located within the spore integument and activated during the early stages of germination. The highly directional rodlet disassembly process suggests that coat-degrading enzymes could be

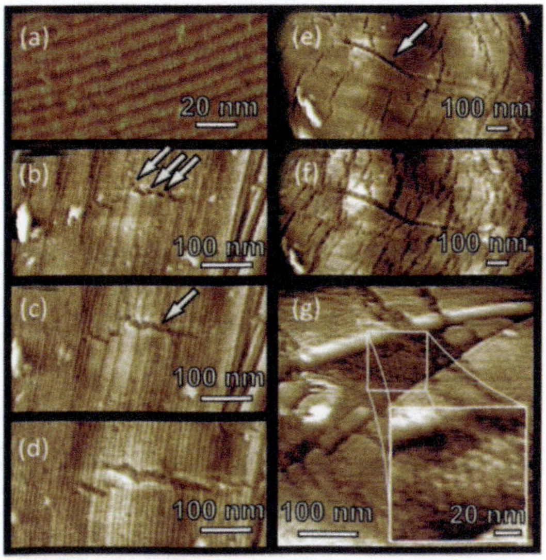

Figure 4.8. Disintegration of the spore coat rodlet layer. (a) The intact rodlet layer covering the outer coat of dormant *B. atrophaeus* spores is ~11 nm thick and has a periodicity of ~8 nm.[6] (b -d) Series of AFM height images tracking the initial changes of the rodlet layer after (b) 13 min., (c) 113 min., and (d) 295 min. of exposure to germination solution. Small etched pits (indicated with arrows in (b) evolve into fissures, indicated with an arrow in (c), perpendicular to the rodlet direction. The fissures expand both in length and width. (e, f) AFM images showing another germinating spore. The spore long axis, as well as major rodlet orientation is left-right. Enhanced etching at stacking faults running from left to centre and indicated with an arrow in (e), as well as increased etching at the perpendicular gaps were visible following (e) 135 min. and (f) 240 min. of germination. Fissure width and length increased from 10-15 nm and 100-200 nm (135 min.) to 15 -30 nm and 125-250 nm (240 min.), respectively. (g) Etching and/or fracture of the rodlet layer at a stacking fault revealed the underlying hexagonal layer of particles with a 10-13 nm lattice period. Images reproduced with permission from Ref. 7 © (2007) National Academy of Sciences, U.S.A.

localized at the etch pits and either recognize their structural features or the etch pits are predisposed to structural deformation during early stages of spore coat disassembly. The gradual elongation of the fissures suggests that once hydrolysis is initiated at an etch pit, processive hydrolysis propagates perpendicular to the rodlet direction and to neighbouring rodlets.

The locations of the small etch pits may coincide with point defects in the rodlet structure. These point defects could be caused by misoriented rodlet monomers or by the incorporation of impurities into the crystalline structure. In both cases, point defects could facilitate access of degradative enzymes to their substrate in an otherwise tightly packed structure.

Disassembly of the higher-order rodlet structure began prior to the outgrowth stage of germination (Fig. 4.9). Disaggregation of the rodlet layer occurred perpendicular to the orientation of individual rodlets, resulting in the formation of banded remnants (Fig. 4.9). Further structural disruption led to the formation of extended, 2–3 nm wide, fibrils (indicated with arrows in Fig. 4.9e), which were also oriented perpendicular to the rodlet direction.

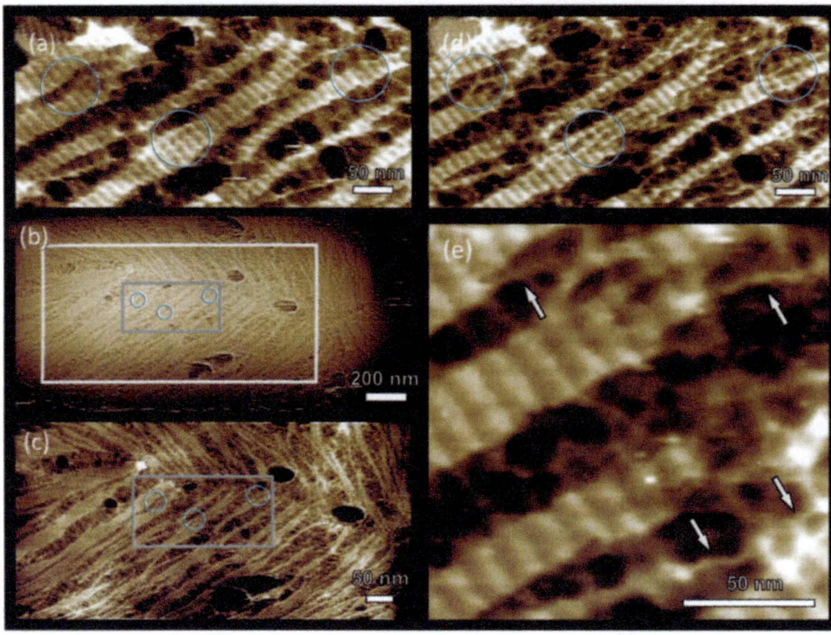

Figure 4.9. (a–d) Series of AFM height images showing the progress of rodlet disassembly. In the circled regions, banded remnants of rodlet structure (a) disassemble into thinner fibrous structures (d). Time between images was 36 min. (a)–(b); 3 min. (b)–(c); and 6 min. (c)–(d), for a total time between (a) and (d) of 45 min. In (b), the area imaged in (c) is indicated with a light grey box. In (b) and (c), the area imaged in (a) and (d) is indicated with a dark grey box. In (e), which is an enlarged part of (d), arrows indicate the end point of rodlet disruption, i.e. fibrils with a diameter of 2–3 nm, oriented roughly perpendicular to the rodlets. Images reproduced, with permission from Ref. 7. © (2007) National Academy of Sciences, USA.

Several classes of proteins, with divergent primary sequences, were found to form similar rodlet structures on the surfaces of cells of Gram-negative *E. coli* and *S. enterica* as well as spores of Gram-positive streptomycetes and various fungi.[15] These rodlets were shown to be structurally highly similar to amyloid fibrils.[15] Amyloids possess a characteristic cross-β structure and have been associated with neural degenerative diseases (i.e. Alzheimer's and prion diseases).[52] Amyloid fibrils or rodlets form microbial surface layers and play important roles in microbial attachment, dispersal and pathogenesis.[15] We have proposed[7] that the structural similarity of *B. atrophaeus* spore coat rodlets and the amyloid rodlets found on other bacterial and fungal spores suggests that *Bacillus* rodlets have an amyloid structure. AFM characterization of the nanoscale properties of individual amyloid fibrils has revealed that these self-assembled structures can have a strength and stiffness comparable with structural steel.[53] The extreme physical, chemical and thermal resistance of *Bacillus* spores suggests that evolutionary forces have captured the mechanical rigidity and resistance of these amyloid self-assembling biomaterials to structure the protective outer spore surface.

Structural studies of amyloids have identified an array of possible rodlet assemblies, each consisting of several (2 or 4) individual cross-β sheet fibrils, which are often helically intertwined.[15] The number of fibrils determines the diameter of the rodlet. Most amyloids resulting from protein folding diseases, and some naturally occurring amyloids, form individual fibrils or disorganized rodlets networks.

In spore coats of *B. atrophaeus*, the higher-order rodlet structure is organized as one major domain of parallel rodlets covering the entire spore surface.[3] Rodlet domain formation requires the periodic bonds in the rodlet direction ("parallel bonds") as well in the direction perpendicular to it ("perpendicular bonds").[54] In the case of amyloid-like rodlets, the *intra*-rodlet, parallel bonds are known and consist primarily of hydrogen bonds associated with the cross-β sheets that form the backbone of the rodlet fibrils. However, the nature of the perpendicular bonds, i.e. the *inter*-rodlet bonds that keep the rodlets tightly packed, is unknown.

On the basis of these rodlet features, one might expect that during germination individual rodlets would detach or erode, leaving a striated pattern *parallel* to the rodlet direction. Surprisingly, striations *perpendicular* to the rodlet direction were observed (Fig. 4.9), and 2–3 nm wide fibrils *perpendicular* to the rodlet direction (Fig. 4.9e) were the culmination product of coat degradation. This result indicates that during germination, perpendicular rodlet bonds are stronger, or are more resistant to hydrolysis, than bonds parallel to the rodlet direction. Second, and most surprisingly, these perpendicular structures facilitate the formation of 200–300 nm long fibres *perpendicular* to the rodlet direction.

It is unclear how microbial amyloid fibres form these perpendicular structures. One possibility is that during the formation of the rodlet layer, both intra-rodlet parallel bonds and inter-rodlet perpendicular bonds form, similar in strength and leading to tightly packed rodlets domains held together by a checkerboard-like bonding pattern. During germination, the intra-rodlet parallel bonds are hydrolyzed, while the inter-rodlet perpendicular bonds remain intact over longer time periods. Spore coat hydrolytic enzymes could target a specific residue or structure (in this case, that of the cross-β sheets) and leave other (here, perpendicular) residues or structures intact. Identification of the gene(s) encoding the rodlet structure and the enzymes responsible for rodlet degradation are important areas for future research.

4.2.2 Emergence of Vegetative Cells

Etch pits were the initiation sites for early germination-induced spore coat fissure formation. During intermediate stages of germination, small spore coat apertures developed that were up to 70 nm in depth (Fig. 4.10a,b). During late stages of germination, these apertures dilated (Fig. 4.10c–e), allowing vegetative cell emergence (data not shown).

In vitro AFM visualization of germling emergence allowed high-resolution visualization of nascent vegetative cell surface structure (Fig. 4.10e–g). Vegetative cell wall structure could be recognized through the apertures approximately 30–60 minutes prior to germ cell emergence. During the release of vegetative cells from the spore integument, the entire cell surface consisted of a porous fibrous network (Fig. 4.10g).

To compare the cell wall structure of germling and mature vegetative cells, we carried out separate experiments in which cultured vegetative *B. atrophaeus* cells were adhered to a gelatin-coated surface[4] and imaged with AFM in water. As seen in Fig. 4.10h, the cell wall of mature vegetative cells contained a porous, fibrous structure similar to the structure observed on the surface of germling cells (Fig. 4.10g).

The bacterial cell wall consists of long chains of peptidoglycan that are cross-linked via flexible peptide bridges.[55] While the composition and chemical structure of the peptidoglycan layer vary among bacteria, its conserved function is to allow bacteria to withstand high internal osmotic pressure.[55] The length of peptidoglycan strands varies from 3–10 disaccharide units in *Staphylococcus aureus* to ~100 disaccharide units in *B. subtilis*, with each unit having a length and diameter of ~1 nm.[56] The fibrous network observed on the germ cell surface with 5–100 nm pores (Fig. 4.10e,g) and the fibrous network observed on mature vegetative cells with 5–50 nm pores (Fig. 4.10h) appear to represent the nascent peptidoglycan architecture of newly formed and mature cell wall, respectively, and is composed of either individual or several intertwined peptidoglycan strands. The cell wall density of mature cells appears to be higher with, on average, smaller pores and more fibrous material, as compared with the germ cells. These results are consistent with murein growth models whereby new peptidoglycan is inserted as single strands and subsequently cross-linked with preexisting murein.[57] The AFM-resolved pore structure of the nascent *B. atrophaeus* germ and vegetative cell surfaces is similar to the honeycomb structure of peptidoglycan oligomers determined by NMR.[55]

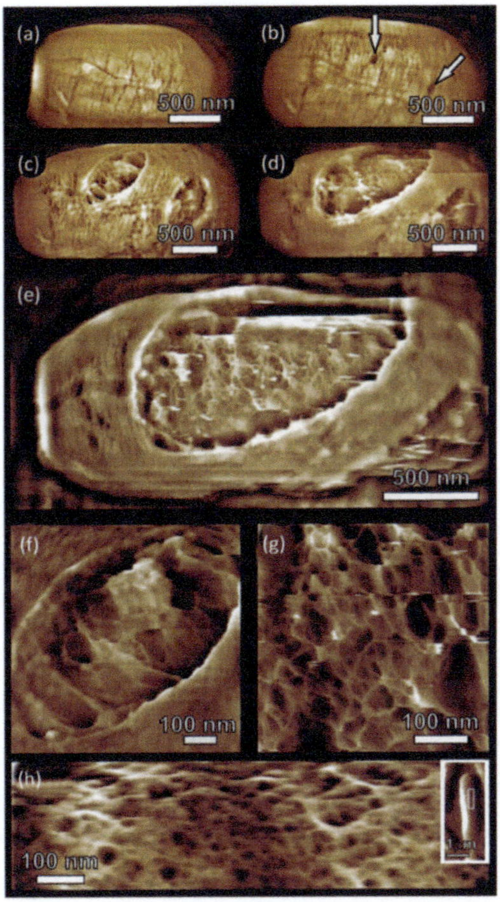

Figure 4.10. Emergence of vegetative cells. (a–g) Series of AFM height images showing 60–70 nm deep apertures in the rodlet layer (indicated with arrows in (b)) that gradually enlarged (c–d) and subsequently eroded the entire spore coat (e). Germ cells emerged from these apertures. (e) Prior to germ emergence from the spore coat, the peptidoglycan cell wall structure was evident. (f) At an early stage of emergence, the cell wall was still partly covered by spore remnants, while (g) immediately prior to cell emergence, the cell wall was free of spore integument debris. The germ cell surface contained 1–6 nm fibres forming a fibrous network enclosing pores of 5–100 nm. Images in (a–g) were collected on the same spore as those shown in Figure 4.8e,f. Elapsed germination time (in hr:min) was (a) 3:40, (b) 5:45, (c) 7:05, (d) 7:30, (e) 7:45, (f) 7:15, (g) 7:50. (h) In separate experiments, cultured vegetative *B. atrophaeus* cells were adhered to gelatin surfaces and imaged in water. AFM height images show a slightly denser, similar fibrous network compared with the germ cell network structure (g), with 5–50 nm pores. In the inset, the imaged part (h) of the entire cell is indicated with a white rectangle. Images reproduced, with permission from Ref. 7. © (2007) National Academy of Sciences, USA.

The structural dynamics of *C. novyi-NT*[8] and *B. atrophaeus* germinating spores appears to be similar. Thus, at later stages of the germination process, the *C. novyi-NT* spore coat layers seen in Fig. 4.11, which are exposed at early stages of germination, start to dissolve. Thus, this process was initiated by the formation of fissures (Fig. 4.11a), which subsequently widened and elongated (Fig. 4.11b–e), resulting in isolated islands of remnant coat layers (Fig. 4.11e,f).

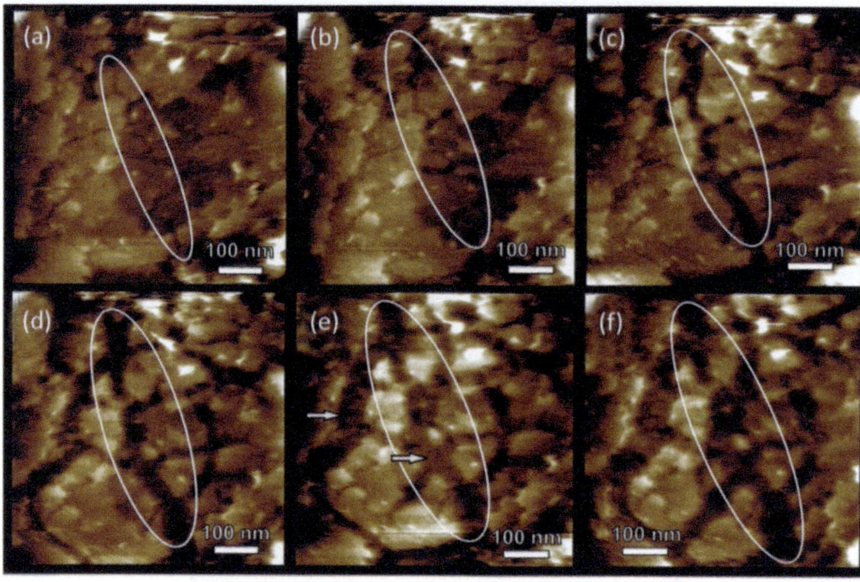

Figure 4.11. Dynamic AFM height imaging of degrading *C. novyi-NT* spore coat layers. Fissures first appeared (a,b), then laterally expanded into wide gaps (c–e) and eventually resulted in the removal of whole layers, exposing the underlying layer (e,f, arrows in (e)). One expanding fissure is indicated with a white oval in (a–f). Time in germination medium in hr:min was 0:45 (a), 0:50 (b), 0:55 (c), 1:00 (d), 1:05 (e), 1:10 (f). Images reproduced, with permission from Ref. 8. © (2007) American Society for Microbiology.

Similarly to *B. atrophaeus* spore germination mechanisms described earlier, coat degradation likely occurs under the influence of germination-activated lytic enzymes. In fact, such lytic enzymes are known to be encoded within the *C. novyi-NT* genome.[58] Interestingly, *C. novyi-NT* spores contain mRNA, and these mRNA molecules are enriched in proteins that could assist with cortex and other degradation.[58]

At the final stages of germination, the coat layers dissolved completely (Fig. 4.12a), fully exposing the ~20–25 nm thick undercoat layer. In the following stage of germination, this layer also disintegrated. This proceeded through the formation and slow expansion of ~25 nm deep flat-bottomed apertures (Fig. 4.12a–f). The cortex was fully lysed by the time spore coat layers dissolved. Hence, the flat-bottomed apertures in this undercoat layer show the underlying cell wall of the emerging *C. novyi-NT* vegetative cell, which, based on its lighter AFM phase contrast (Fig. 4.12f), has different physicochemical properties or/and hence composition than the surrounding coat remnants. The nascent surface of the emerging germ cell appears to be formed by a porous network (Fig. 4.12e–f) of peptidoglycan fibres, similar to one described earlier for *B. atrophaeus* vegetative cells.

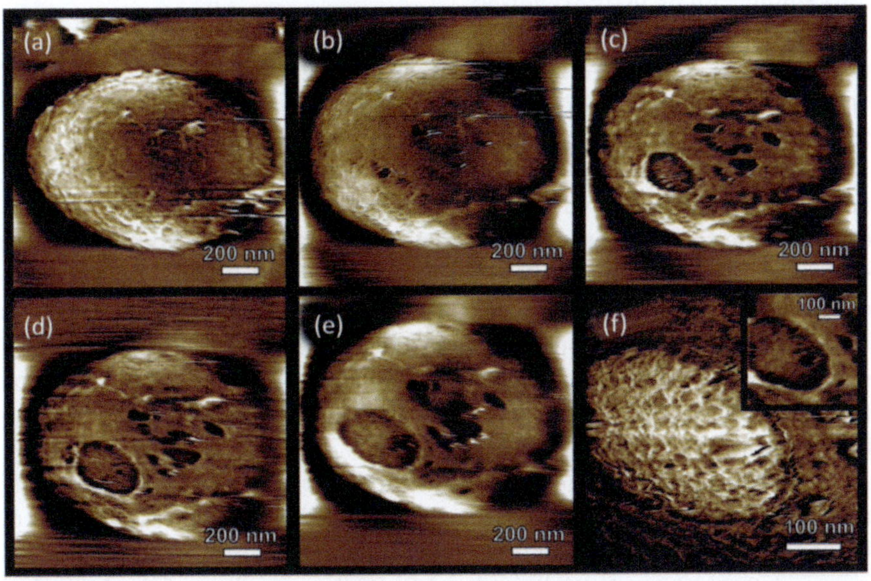

Figure 4.12. (a–e) AFM height images of the final outgrowth stage. (a) After the ~6 spore coat layers were largely dissolved, the underlying structural layer was exposed. (b–e) In this layer, 25 nm deep apertures appeared and grew laterally. (f) Phase image zoom-in of the largest aperture depicted in (c–e), showing the pronounced phase contrast, indicating the different material properties of the emerging cell wall (light) and remaining spore layer (dark). Inset in (f) is the concurrent height image, showing the 25 nm deeper position of the cell wall with respect to the surrounding spore layer. Time in germination medium in hr:min was 1:40 (a), 2:15 (b), 2:50 (c), 3:35 (d), 3:50 (e), 3:55 (f). Images reproduced, with permission from Ref. 8. ©(2007) American Society for Microbiology.

Note that the spore coat degradation process presented in Figs. 4.8–4.12 appears not to be affected by the scanning AFM tip.[7,8] The shapes of fissures and apertures remained unaltered after repeated scanning. Furthermore, when we zoomed out to a larger previously non-scanned area after prolonged scanning on a smaller spore area, the initially scanned area did not display any tip-induced alterations (such as a larger degree of coat degradation). Finally, when we did not image spores for more than an hour between two scans, the coat degradation pattern had developed similarly when compared with spores that were scanned continuously.

Spore germination provides an attractive experimental model system for investigating the genesis of the bacterial peptidoglycan structure. Dormant spore populations can be chemically cued to germinate with high synchrony,[49] allowing the generation of homogenous populations of emergent vegetative cells suitable for structural analysis.

Proposed models for the bacterial cell wall structure posit that peptidoglycan strands are arranged either parallel (planar model) or orthogonal (scaffold model) to the cell membrane.[55] Existing experimental techniques are unable to confirm either the planar or the orthogonal model. The experiments described here do not contain sufficient high-resolution data, in particular of individual peptidoglycan strands, to deduce with certainty the tertiary three-dimensional peptidoglycan structure. The pore structures (Figs. 4.10 and 4.12) of the emergent germ and mature vegetative cell wall — an array of pores — suggest a parallel orientation of glycan strands with peptide stems positioned in stacked orthogonal planes.[55] More detailed studies of germ cell surface architecture and morphogenesis will be required to confirm this peptidoglycan architecture and to investigate whether glycan biosynthesis precedes peptide cross-linking.

The results presented here demonstrate that *in vitro* AFM has the capacity to provide important insight into the time-dependent structural dynamics of individual germinating spores and cell wall high-resolution architecture. This approach could be potentially utilized for the unravelling of the biological role of the cell wall in critical cellular processes and antibiotic resistance.

4.3 PROBING THE BACTERIAL–MINERAL INTERACTIONS ON THE SURFACES OF METAL-RESISTANT BACTERIA

We are currently conducting studies on the elucidation of bioremediation mechanisms of *Arthrobacter oxydans* metal-resistant bacteria. *A. oxydans* is a Gram-positive and chromium (VI)-resistant bacterium, which can reduce highly mobile, carcinogenic, mutagenic and toxic hexavalent chromium to less mobile and much less toxic trivalent chromium. Toxic compounds and heavy metals can be removed from contaminated sites or waste by chemical and physical techniques, which are both difficult and expensive. The extraordinary ability of indigenous microorganisms, like metal-resistant bacteria, for biotransformation of toxic compounds is of considerable interest for the emerging area of environmental bioremediation. However, the underlying mechanisms by which metal-resistant bacteria transform toxic compounds are currently unknown and await elucidation. Stress response pathways are sure to play an important role in the niche definition of metal-resistant bacteria and their effect on the biogeochemistry of many contaminated environments.

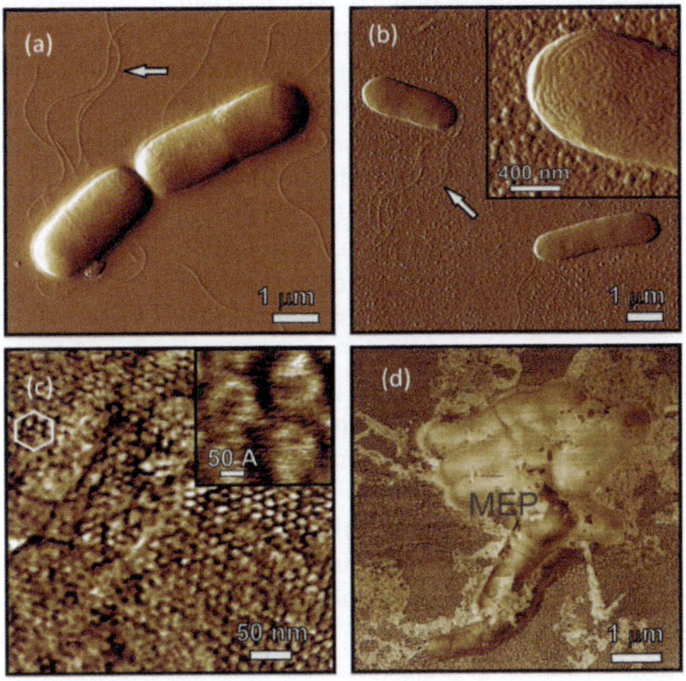

Figure 4.13. AFM images of *A. oxydans* bacteria. (a,b) growth-dependent morphologies; (c) stress-induced supramolecular crystalline hexagonal layer on the bacterial surface; (d) stress-induced microbial extracellular polymer (MEP) layer covering a microbial colony.

We have visualized air-dried *A. oxydans* bacteria and revealed the differences in surface morphology and flagella arrangements during different stages of bacterial growth. Thus, bacteria during the exponential stage growth (Fig. 4.13a) appear to have a rather smooth surface and show a peritrichous flagellar arrangement with flagella seen over the entire cellular surface. The surface of air-dried bacteria grown during the stationary phase (Fig. 4.13b) appears to be tubular (Fig. 4.13, insert), and these bacteria show the lophotrichous flagellar arrangement, with several flagella seen only at one pole of the cell.

We have further visualized for the first time stress responses of *A. oxydans* bacteria in response to the exposure to the toxic environment. Thus, as illustrated in Fig. 4.13c, the formation of a supramolecular crystalline hexagonal structure on the surface of *A. oxydans* bacteria exposed to 35–50 ppm Cr(VI) was observed. Since similar crystalline layers are not seen on control samples (data are not shown here), this structure appears to be stress induced in response to Cr(VI) exposure. At higher Cr(VI) concentrations, we have observed the formation of microbial extracellular polymers, which are seen in Fig. 4.13d, to cover a small microbial colony.

Our AFM observations of the appearance of stress-induced layers on the surfaces of *A. oxydans* bacteria exposed to Cr(VI) are consistent with biochemical studies of stress responses of *A. oxydans* bacteria. Thus, it was reported that *A. oxydans* grown with chromate concentrations above 40 mg/L significantly increased the production of a cell wall protein that had an apparent molecular mass of 60 kDa.[59] Presumably, this protein could form a highly organized particulate layer on the surface of *A. oxydans* bacteria exposed to Cr(VI). The hexagonal stress-induced structure (Fig. 4.13c) is formed by a protein with the size of ~10–11 nm. High-resolution images (Fig. 4.13c, insert) reveal that these particles are oligomers, composed of monomers with a size of ~5 nm. Assuming the globular shape of the protein, this size corresponds well to the molecular mass of ~60 KDa.

It was suggested that reduction of Cr(VI) proceed on the cell wall.[60] This 60 kDa protein could be potentially involved in the reduction of Cr(VI). We are currently developing procedures for *in vitro* high-resolution AFM characterization of the surface architecture and structural dynamics of metal-resistant bacteria in response to changes in the environment and various chemical stimuli. It is expected that these experiments will improve the fundamental understanding of bioremediation mechanisms.

The present technological and scientific challenges are to elucidate the relationships between the stress-induced organization and function of protein and polymer complexes at bacterial cell wall surfaces, to understand how these complexes respond to environmental changes and chemical stimulants and to predict how they guide the formation of biogenic metal phases on the cell surface.

The results presented here demonstrate that *in vitro* AFM is a powerful tool for revealing the structural dynamics and architectural topography of the microbial and cellular systems. AFM allows new approaches to high-resolution real-time dynamic studies of single microbial cells under native conditions. Environmental parameters (e.g. temperature, chemistry or gas phase) can be easily changed during the course of AFM experiments, allowing dynamic environmental and chemical probing of microbial surface reactions. Further incorporation of AFM-based immunolabelling techniques could allow the identification of spore coat proteins that play a role in spore germination and provide a structural understanding of how these proteins regulate spore survival, germination and disease.

Acknowledgements

The author thanks M. Plomp for his critical contributions to this work and B. Vogelstein, T.J. Leighton, P. Setlow and H.-Y. Holman for spore and bacteria preparations, helpful discussions and encouragement. This work was performed under the auspices of the US Department of Energy by Lawrence Livermore National Laboratory under contract number DE-AC52-07NA27344. This work was supported by the Lawrence Livermore National Laboratory through Laboratory Directed Research and Development Grants 04-ERD-002 and 08-LW-027.

References

1. Driks, A. (1999) *Bacillus subtilis* spore coat, *Microbiol. Mol. Biol. Rev.*, **63**, 1–21.
2. Driks, A., and Setlow, P. (2000) Morphogenesis and properties of the bacterial spore, in *Prokaryotic Development* (ed. Brun, Y. V., and Shimkets, L. J.), American Society for Microbiology, Washington, DC, pp. 191–218.
3. Driks, A. (2002) The *Bacillus subtilis* spore coat, in *Bacillus subtilis and its Relatives: From Genes to Cells* (ed. Sonenshein, A. L., Hoch, J. A., and Losick, R. M.), American Society for Microbiology, Washington, DC, pp. 527–536.
4. Plomp, M., Leighton, T. J., Wheeler, K. E., and Malkin, A. J. (2005) The high-resolution architecture and structural dynamics of *Bacillus* spores, *Biophys. J.*, **88**, 603–608.
5. Plomp, M., Leighton, T. J., Wheeler, K. E., and Malkin, A. J. (2005) Architecture and high-resolution structure of *Bacillus thuringiensis* and *Bacillus cereus* spore coat surfaces, *Langmuir*, **21**, 7892–7898.
6. Plomp, M., Leighton, T. J., Wheeler, K. E., Pitesky, M. E., and Malkin, A. J. (2005) *Bacillus atrophaeus* outer spore coat assembly and ultrastructure, *Langmuir*, **21**, 10710–10716.
7. Plomp, M., Leighton, T. J., Wheeler, K. E., Hill, H. D., and Malkin, A. J. (2007) *In vitro* high-resolution structural dynamics of single germinating bacterial spores, *Proc. Natl. Acad. Sci. USA*, **104**, 9644–9649.
8. Plomp, M., McCaffery, J. M., Cheong, I., Huang, X., Bettegowda, C., Kinzler, K. W., Zhou, S., Vogelstein, B., and Malkin, A. J. (2007) Spore coat architecture of *Clostridium novyi-NT* spores, *J. Bacteriol.*, **189**, 6457–6468.

9. Kim, H., Hahn, M., Grabowski, P., McPherson, D. C., Otte, M. M., Wang, R., Ferguson, C. C., Eichenberger, P., and Driks, A. (2006) The *Bacillus subtilis* spore coat protein interaction network, *Mol. Microbiol.*, **59**, 487–502.
10. Plomp, M., and Malkin, A. J. (2009) Mapping of proteomic composition on the surfaces of *Bacillus* spores by atomic force microscopy-based immunolabeling, *Langmuir*, **25**, 403–409.
11. Monroe, C., Plomp, M., Malkin, A. J., and Setlow, P. (2008) Protozoal digestion of coat-defective *Bacillus subtillis* spores produces "rinds" composed of insoluble coat protein, *Appl. Environ. Microbiol.*, **74**, 5875–5581.
12. Ghoshi, S., Setlow, B., Wahone, P. G., Cowan, A. E., Plomp, M., Malkin, A. J., and Setlow, P. (2008) Characterization of spores of *Bacillus subtilis* that lack most coat layers, *J. Bacteriol.*, **190**, 6741–6748.
13. Aronson, A. I., and Fitz-James, P. (1976) Structure and morphogenesis of the bacterial spore coat, *Bact. Rev.*, **40**, 360–402.
14. Dufrêne, Y. F., Boonaert, C. J. P., Gerin, P. A., Asther, M., and Rouxhet, P. G. (1999) Direct probing of the surface ultrastructure and molecular interactions of dormant and germinating spores of *Phanerochaete chrysosporium*, *J. Bacteriol.*, **181**, 5350–5354.
15. Gebbink, M. F. B. G., Claessen, D., Bouma, B., Dijkhuizen, L., and Wosten, H. A. B. (2005) Amyloids — a functional coat for microorganisms, *Nat. Rev. Microbiol.*, **3**, 333–341.
16. Wessels, J. G. H. (1998) Hydrophobins: proteins that change the nature of the fungal surface, *Adv. Microb. Physiol.*, **38**, 1–45.
17. Leighton, T. J., personal communication.
18. Little, S., and Driks, A. (2001) Functional analysis of the morphogenetic spore coat *Bacillus subtilis* protein CotE, *Mol. Microbiol.*, **42**, 1107–1120.
19. Plomp, M., Monroe, C., Setlow, P., and Malkin, A. J. (2010) Spore coat assembly of *Bacillus subtilis* spores. In preparation.
20. DeYoreo, J. J., Land, T. A., and Dair, B. (1994) Growth morphology of vicinal hillocks on the (101) face of KH_2PO_4 from step-flow to layer-by-layer growth, *Phys. Rev. Lett.*, **73**, 838–841.
21. Plomp, M., van Enckevort, W. J. P., van Hoof, P. J. C. M., and van de Streek, C. J. (2003) Morphology and dislocation movement in n-$C_{40}H_{82}$ paraffin crystals grown from solution, *J. Cryst. Growth*, **249**, 600–613.
22. Malkin, A. J., Kuznetsov, Yu. G., Land, T. A., DeYoreo, J. J., and McPherson, A. (1995) Mechanisms of growth for protein and virus crystals, *Nat. Struct. Biol.*, **2**, 956–959.
23. Malkin, A. J., and McPherson, A. (2004) Probing of crystal interfaces and the structures and dynamic properties of large macromolecular ensembles with *in situ* atomic force microscopy, in *From Solid-Liquid Interface to Nanostructure Engineering*, vol. 2. (ed. Lin, T., and DeYoreo, J. J.) Plenum/Kluwer Academic Publisher, New York, NY, pp. 201–238.

24. Brar, B., and Leonard, D. (1994) Spiral growth of GaSb on (001) GaAs using molecular beam epitaxy, *Appl. Phys. Lett.*, **66**, 463–465.

25. Maiwa, K., Plomp, M., van Enckevort, W. J. P., and Bennema, P. (1998) AFM observation of barium nitrate {111} and {100} faces: spiral growth and two-dimensional nucleation growth, *J. Cryst. Growth*, **186**, 214–223.

26. Van der Hoek, B., Van der Eerden, J. P., and Bennema, P. (1982) Thermodynamical stability conditions for the occurrence of hollow cores caused by stress of line and planar defects, *J. Cryst. Growth*, **56**, 621–632.

27. Kuznetsov, Yu. G., Malkin, A. J., Land, T. A., DeYoreo, J. J., Barba, A. P., Konnert, J., and McPherson, A. (1997) Molecular resolution imaging of macromolecular crystals by atomic force microscopy, *Biophys. J.*, **72**, 2357–2364.

28. Hollander, F. F. A., Plomp, M., van de Streek, C. J., van Enckevort, W. J. P. (2001) A two-dimensional Hartman-Perdok analysis of polymorphic fat surfaces observed with atomic force microscopy, *Surf. Sci.*, **471**, 101–113.

29. Kim, H.-S., Sherman, D., Johnson, F., and Aronson., A. I. (2004) Characterization of major *Bacillus anthracis* spore coat protein and its role in spore inactivation, *J. Bacteriol.*, **186**, 2413–2417.

30. Veis, A. (2003) Mineralization in organic matrix networks, in *Reviews in Mineralogy and Geochemistry*, vol. 54 (ed. Dove, De Yoreo, and Weiner) Mineralogical Society of America, Washington, DC, pp. 249–289.

31. Aizenberg, J., Lambert, G., Weiner, S., and Addadi, L. (2002) Factors involved in the formation of amorphous and crystalline calcium carbonate: a study of an ascidian skeleton, *J. Am. Chem. Soc.*, **124**, 32–39.

32. Belcher, A. M., Wu, X. N., Christensen, R. J., Hansma, P. K., Stucky, G. D., and Morse, D. E. (1996) Control of crystal phase switching and orientation by soluble mollusc-shell proteins, *Nature*, **381**, 56–58.

33. Yao, N., Epstein, A., and Akey, A. (2006) Crystal growth via spiral motion in abalone shell nacre, *J. Mater. Res.*, **21**, 1939–1946.

34. Wise, S. W., and deVilliers, J. (1971) Scanning electron microscopy of molluscan shell ultrastructures: screw dislocations in pelecypod nacre, *Trans. Am. Microsc. Soc.*, **90**, 376–380

35. Johnstone, K., Ellar, D. J., and Appleton, T. C. (1980) Location of metal ions in *Bacillus megaterium* sporesby high resolution electron probe X-ray microanalysis, *FEMS Microbiol. Lett.*, **7**, 97–101.

36. Stewart, M., Somlyo, A. P., Somlyo, A. V., Shuman, H., Lindsay, J. A., and Murrell, W. G. (1980) Distribution of calcium and other elements in cryosectioned *Bacillus cereus T* spores, determined by high-resolution scanning electron probe X-ray microanalysis, *J. Bacteriol.*, **143**, 481–491.

37. Ghosal, S., Fallon, S. J., Leighton, T. J., Wheeler, K. E., Kristo, M. J., Hutcheon, I. D., and Weber, P. K. (2008) Analysis of bacterial spores using nano-secondary ion mass spectrometry (NanoSIMS), *Analyt. Chem.*, **80**, 5986–5992.

38. Chernov, A. A. (1984) *Modern Crystallography: Vol III. Crystal Growth*, Springer-Verlag, Berlin.

39. Elhadj, S., Plomp, M., Velsko, S. V., and Malkin, A. J. (2010) In preparation.
40. Muller, D. J., and Dufrêne, Y. F. (2008) Atomic force microscopy as a multifunctional molecular toolbox in nanobiotechnology, *Nat. Nanotechnol.*, **3**, 261–269.
41. Dufrêne, Y. F. (2008) Towards nanomicrobiology using atomic force microscopy, *Nat. Rev. Microbiol.*, **6**, 674–680.
42. Robinson, J. M., Takizawa, T., and Vandre, D. (2000) Application of gold cluster compounds in immunochemistry and correlative microscopy: comparison with colloid gold, *J. Microsc.*, **199**, 163–179.
43. Bendayan, M. (2001) Worth its weight in gold, *Science*, **291**, 1363–1365.
44. Kaftan, D., Brumfeld, V., Nevo, R., Scherz, A., and Reich, Z. (2002) From chloroplasts to photosystems: in situ scanning force microscopy on intact thylakoid membranes, *EMBO J.*, **21**, 6146–6153.
45. Soman, P., Rice, Z., and Siedlecki, C. A. (2008) Immunological identification of fibrinogen in dual-component protein films by AFM imaging, *Micron*, **39**, 832–842.
46. Lin, H., Lal, R., and Clegg, D. O. (2000) Imaging and mapping heparin-binding sites on single fibronectin molecules with atomic force microscopy, *Biochemistry*, **39**, 3192–3196.
47. Putman, C. A. J., de Grooth, B. G., Hansma, P. K., and van Hulst, N. F. (1993) Immunogold labels-cell-surface markers in atomic force microscopy, *Ultramicroscopy*, **48**, 177–182.
48. Müller, D. J., Fotiadis, D., and Engel, A. (1998) Mapping flexible protein domains at subnanometer resolution with the atomic force microscope, *FEBS Lett.*, **430**, 105–111.
49. Moir, A. (2006) How do spores germinate? *J. Appl. Microbiol.*, **101**, 526–530.
50. Setlow, P. (2003) Spore germination, *Curr. Opin. Microbiol.*, **6**, 550–556.
51. Moir, A., Corfe, B. M., and Behravan, J. (2002) Spore germination, *Cell. Mol. Life Sci.*, **59**, 403–409.
52. Dobson, C. M. (2003) Protein folding and misfolding, *Nature*, **426**, 884–890.
53. Smith, J. F., Knowles, T. P. J., Dobson, C. M., MacPhee, C. E., and Welland, M. E. (2006) Characterization of the nanoscale properties of individual amyloid fibrils, *Proc. Natl. Acad. Sci. USA*, **103**, 15806–15811.
54. Grimbergen, R. F. P., Meeke, H., Bennema, P., Strom, C. S., and Vogels, L. J. P. (1998) On the prediction of crystal morphology. I. The Hartman-Perdok theory revisited, *Acta. Crystallogr.*, **A 54**, 491–500.
55. Meroueh, S. O., Bencze, K. Z., Hesek, D., Lee, M., Fisher, J. F., Stemmler, T. L., and Mobashery, S. (2006) Three-dimensional structure of the bacterial cell wall peptidoglycan, *Proc. Natl. Acad. Sci. USA*, **103**, 4404–4409.
56. Ward, J. B. (1973) Chain-length of glycans in bacterial cell-walls, *Biochem. J.*, **133**, 395–398.

57. Höltje, J. V., and Heidrich, C. (2001) Enzymology of elongation and constriction of the murein sacculus of *Escherichia coli*, *Biochimie*, **83**, 103–108.

58. Bettegowda C., Huang, X., Lin, J., Cheong, I., Kohli, M., Szabo, S. A., Zhang, X. S., Diaz, L. A., Velculescu, V. E., Parmigiani, G., Kinzler, K. W., Vogelstein, B., and Zhou, S. (2006) The genome and transcriptomes of the anti-tumor agent *Clostridium novyi-NT*, *Nat. Biotechnol.*, **24**, 1573–1580.

59. Asatiani, N. V., Abuladze, M. K., Kartvelishvili, T. M., Bakradze, N. G., Sapojnikova, N. A., Tsibakhashvili, N. Y., Tabatadze, L. V., Lejava, L. V., Asanishvili, L. L., and Holman, H. Y. (2004) Effect of chromium (VI) action on *Arthrobacter oxydans*, *Curr. Microbiol.*, **49**, 321–326.

60. Abuladze, M. K., Asatiani, N. V., Bakradze, N. G., Kartvelishvili, T. M., Holman, H. Y. N., Kalabegishvili, T. L., Mosulishvili, L. M., Rcheulishvili, A. N., Sapojnikova, N. A., and Tsibakhashvili, N. Y. (2002) Effect of chromium action on the protein composition of *Arthrobacter oxydans*, *Fresenius Environ. Bull.*, **11**, 562–567.

Chapter 5

UNDERSTANDING CELL SECRETION AND MEMBRANE FUSION PROCESSES ON THE NANOSCALE USING THE ATOMIC FORCE MICROSCOPE

Bhanu P. Jena
Department of Physiology, Wayne State University School of Medicine, 5245 Scott Hall, Detroit, MI 48201, USA
bjena@med.wayne.edu

5.1 ATOMIC FORCE MICROSCOPY: RESOLVING A MAJOR CONUNDRUM IN CELL SECRETION

Secretion is a fundamental cellular process as old as life itself and occurs in all living organisms, from the simple yeast to cells in humans. Secretion is responsible for a variety of physiological activities in living organisms, such as neurotransmission and the release of hormones and digestive enzymes. Secretory defects in cells are responsible for a host of debilitating diseases. Since the mid 1950s, it was believed that during cell secretion, secretory vesicles completely merge at the cell plasma membrane, resulting in the diffusion of intravesicular contents to the cell exterior and the compensatory retrieval of the excess membrane by endocytosis. In contrast, the observation of partially empty vesicles in cells following secretion could not be justified according to the aforementioned mechanism. Then in the 1960s, experimental data concerning neurotransmitter release mechanisms by Katz[1] and Folkow *et al.*[2] brilliantly hypothesized that limitation of the quantal packet may be set by the nerve membrane, in which case the size of the packet may

Life at the Nanoscale: Atomic Force Microscopy of Live Cells
Edited by Yves Dufrêne

www.panstanford.com

actually correspond to just a fraction of the vesicle content.[2-4] In the interim, a large body of work was published both for and against the complete merger of secretory vesicles at the cell plasma membrane during secretion, further deepening the controversy. The only definitive determination of the mechanism of cell secretion relied on the direct observation of the process at nanometre resolution in live cells. The conundrum was finally resolved using atomic force microscopy (AFM).[5] Isolated live pancreatic acinar cells in near-physiological buffer when imaged using AFM at high force (200–300 pN) demonstrate the size and shape of the secretory vesicles called zymogen granules or ZGs lying immediately below the apical plasma membrane of the cell (Fig. 5.1). Within 2.5 minutes of exposure to a physiological secretory stimulus (1 μM carbamylcholine), the majority of ZGs within cells swell (Fig. 5.1), followed by a decrease in ZG size, by which time secretion is complete (Fig. 5.1). These studies reveal for the first time in live cells the intracellular swelling of secretory vesicles following stimulation of secretion and their deflation following partial discharge of vesicular contents.[5] No loss of secretory vesicles is observed throughout the experiment, demonstrating transient fusion and partial discharge of vesicular contents during cell secretion.

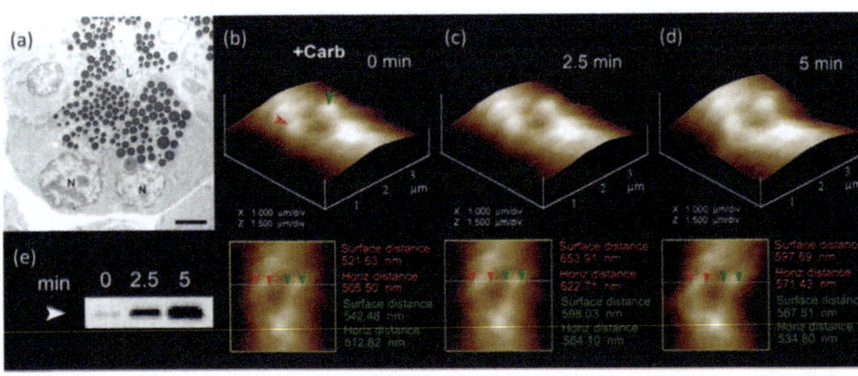

Figure 5.1. The swelling dynamics of ZGs in live pancreatic acinar cells. (a) Electron micrograph of pancreatic acinar cells showing the basolaterally located nucleus (N) and the apically located ZGs. The apical end of the cell faces the acinar lumen (L). Bar = 2.5 μm. (b–d) The apical ends of live pancreatic acinar cells were imaged by AFM, showing ZGs (red and green arrowheads) lying just below the apical plasma membrane. Exposure of the cell to a secretory stimulus using 1 μM carbamylcholine resulted in ZG swelling within 2.5 minutes, followed by a decrease in ZG size after 5 minutes. The decrease in the size of ZGs after 5 minutes is due to the release of secretory products such as α-amylase, as demonstrated by the immunoblot assay (e).[5]

The other major breakthrough in our understanding of cell secretion came with the discovery of a new cellular structure, the "porosome", using AFM.[6–13] In the past 12 years, the porosome has been determined to be the universal secretory machinery in cells. Porosomes are supramolecular lipoprotein structures at the cell plasma membrane, where membrane-bound secretory vesicles transiently dock and fuse to release intravesicular contents to the outside during cell secretion. The mouth of the porosome opening to the outside ranges in size from 150 nm in diameter in acinar cells of the exocrine pancreas to 12 nm in neurons, which dilates during cell secretion, returning to its resting size following the completion of the process. In the past decade, the composition of the porosome, its structure and dynamics at nanometre resolution and in real time, and its functional reconstitution into artificial lipid membrane, have all been elucidated. Since porosomes in exocrine and neuroendocrine cells measure 100–180 nm, and only 20–35% increase in porosome diameter is demonstrated following the docking and fusion of 0.2–1.2 μm in diameter secretory vesicles, it is concluded that secretory vesicles "transiently" dock and fuse at the base of the porosome complex to release their contents to the outside (Fig. 5.2).

5.2 DISCOVERY OF THE "POROSOME"

Porosomes were first discovered in acinar cells of the exocrine pancreas.[6] Exocrine pancreatic acinar cells are polarized secretory cells possessing an apical and a basolateral end. This well-characterized cell of the exocrine pancreas synthesizes digestive enzymes, which is stored within 0.2–1.2 μm diameter of apically located membranous sacs or secretory vesicles referred to as zymogen granules. Following a secretory stimulus, ZGs dock and fuse with the apical plasma membrane to release their contents to the outside. Contrary to neurons, where secretion of neurotransmitters occurs in the millisecond time regime, the pancreatic acinar cells secrete digestive enzymes over minutes following a secretory stimulus. As pancreatic acinar cells are slow secretory cells, they were ideal for investigation of the molecular steps involved in cell secretion. In the mid 1990s, AFM studies were undertaken on live pancreatic acinar cells to evaluate at high resolution the structure and dynamics of the apical plasma membrane in both resting and following stimulation of cell secretion. To our surprise, isolated live pancreatic acinar cells in physiological buffer, when imaged using AFM,[6] reveal new cellular structures. At the apical plasma membrane, a group of circular "pits" measuring 0.4–1.2 μm in diameter, containing smaller "depressions", were observed.

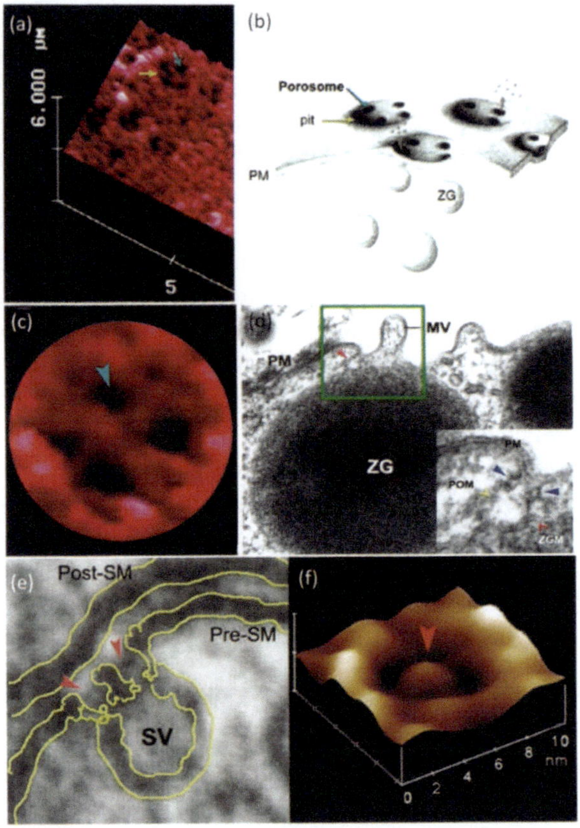

Figure 5.2. Porosomes or previously referred to as "depression" at the plasma membrane (PM) in pancreatic acinar cell and at the nerve terminal. (a) AFM micrograph depicting "pits" (yellow arrow) and "porosomes" within (blue arrow), at the apical PM in a live pancreatic acinar cell.[6] (b) To the right is a schematic drawing depicting porosomes at the cell PM, where membrane-bound secretory vesicles called zymogen granules (ZGs) dock and fuse to release intravesicular contents.[6] (c) A high-resolution AFM micrograph shows a single pit with four 100–180 nm porosomes within.[6] (d) An electron micrograph depicting a porosome (red arrowhead) close to a microvilli (MV) at the apical PM of a pancreatic acinar cell. Note the association of the porosome membrane (POM, yellow arrowhead) and the zymogen granule membrane (ZGM) (red arrow head) of a docked ZG (inset). Cross section of a circular complex at the mouth of the porosome is seen (blue arrow head).[11] (e) The bottom left panel shows an electron micrograph of a porosome (red arrowhead) at the nerve terminal, in association with a synaptic vesicle (SV) at the presynaptic membrane (Pre-SM). Notice a central plug at the neuronal porosome opening.[9] (f) The bottom right panel is an AFM micrograph of a neuronal porosome in physiological buffer, also showing the central plug (red arrowhead) at its opening.[9] It is believed that the central plug in neuronal porosomes may regulate its rapid close-open conformation during neurotransmitter release. The neuronal porosome is an order of magnitude smaller (10–15 nm) in comparison with porosome in the exocrine pancreas.

Each depression measures between 100 and 180 nm in diameter, and typically three to four depressions are found within a pit. The basolateral membranes in acinar cells are devoid of such structures. High-resolution AFM images of depressions in live acinar cells further reveal a cone-shaped morphology, and the depth of each cone measures 15–35 nm. Subsequent studies over the years demonstrate the presence of depressions in all secretory cells examined. Analogous to pancreatic acinar cells, examination of resting growth hormone (GH)-secreting cells of the pituitary[8] and chromaffin cells of the adrenal medulla[14] also reveals the presence of pits and depressions at the cell plasma membrane. The presence of depressions or porosomes in neurons, astrocytes, β-cells of the endocrine pancreas and mast cells has also been elucidated, demonstrating their universal presence.

Exposure of pancreatic acinar cells to a secretagogue (mastoparan) results in a time-dependent increase (25–45%) in both the diameter and relative depth of depressions. Studies demonstrate that depressions return to resting size on completion of cell secretion.[6,7] No demonstrable change in pit size is detected following stimulation of secretion.[6] Enlargement of depression diameter and an increase in its relative depth after exposure to secretagogue correlated with secretion. Aditionally, exposure of pancreatic acinar cells to cytochalasin B, a fungal toxin that inhibits actin polymerization and secretion, results in a 15–20% decrease in depression size and a consequent 50–60% loss in secretion.[6] Results from these studies suggest depressions to be the fusion pores in pancreatic acinar cells. Furthermore, these studies demonstrate the involvement of actin in regulation of both the structure and function of depressions. Similarly, depressions in resting GH cells measure 154 ± 4.5 nm (mean ± SE) in diameter, and following exposure to a secretagogue, there is a 40% increase in depression diameter (215 ± 4.6 nm; $p < .01$), with no appreciable change in pit size.[8] The enlargement of depression diameter during cell secretion and its subsequent decrease, accompanied by loss in secretion following exposure to actin depolymerizing agents,[6] also suggested them to be the secretory portal. A direct determination that depressions are indeed the portals via which secretory products are expelled from cells was unequivocally demonstrated using immuno-AFM studies (Fig. 5.3).[7] Localization at depressions of gold-conjugated antibody to secretory proteins finally provided the direct evidence that secretion occurs through depressions. ZGs contain the starch-digesting enzyme amylase. AFM micrographs of the specific localization of gold-tagged amylase-specific antibodies (Fig. 5.3) at depressions, following stimulation of cell secretion,[7,10] conclusively demonstrated depressions as the cellular secretory portal. Similarly, in somatotrophs of the pituitary gland, gold-tagged GH-specific antibody found to selectively localize at the depression openings following stimulation of secretion[8] established these sites too to be the secretory portal in these cells. Over the years, the term "fusion pore" has been loosely referred to plasma membrane dimples that originate following a secretory stimulus or to the continuity or channel established between opposing lipid membrane during membrane fusion. Therefore, for clarity, the term "porosome" was assigned to depressions.

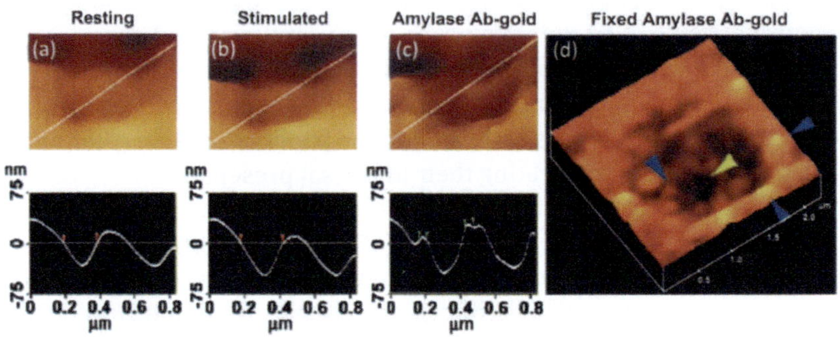

Figure 5.3. Porosomes dilate to allow expulsion of vesicular contents. (a and b) AFM micrographs and section analysis of a pit and two out of the four depressions or porosomes, showing enlargement of porosomes following stimulation of secretion. (c) Exposure of live cells to gold-conjugated amylase antibody (Ab) results in specific localization of gold to these secretory sites. Note the localization of amylase-specific immunogold at the edge of porosomes. (d) AFM micrograph of pits and porosomes with immunogold localization is also demonstrated in cells immunolabeled and then fixed. Blue arrowheads point to immunogold clusters and the yellow arrowhead points to a depression or porosome opening.[7]

The porosome structure, at the cytosolic compartment of the plasma membrane in the exocrine pancreas[10] and in neurons,[9] has also been determined at near-nanometre resolution in live tissue. To determine the morphology of porosomes at the cytosolic compartment of pancreatic acinar cells, isolated plasma membrane preparations in near-physiological buffered solution have been imaged at high resolution using AFM.[10] These studies reveal scattered circular disks measuring 0.5–1 µm in diameter, with inverted cup-shaped structures within.[10] The inverted cups at the cytosolic compartment of isolated pancreatic plasma membrane preparations range in height from 10 to 15 nm. On several occasions, ZGs ranging in size from 0.4 to 1 µm in diameter were observed in association with one or more of the inverted cups, suggesting the circular disks to be pits and the inverted cups to be porosomes. To further confirm that the cup-shaped structures are porosomes, where secretory vesicles dock and fuse, immuno-AFM studies were performed. Target membrane proteins SNAP-23[15] and syntaxin[16] (t-SNARE) and secretory vesicle-associated membrane protein v-SNARE or VAMP[17] are part of the conserved protein complex involved in fusion of opposing bilayers in the presence of calcium.[18-27] Since ZGs dock and fuse at the plasma membrane to release vesicular contents, it was hypothesized that if porosomes are the secretory sites, then plasma membrane-associated t-SNAREs should localize there. The t-SNARE protein SNAP-23 had previously been reported in pancreatic acinar cells.[28] A polyclonal monospecific SNAP-23 antibody recognizing a single 23 kDa protein in immunoblots of pancreatic plasma membrane fraction, when used in immuno-AFM studies, demonstrated selective localization to the base of the cup-shaped structures. These results demonstrate that the inverted cup-shaped structures in inside-out isolated pancreatic plasma membrane preparations are indeed porosomes, where secretory vesicles dock and fuse to release their contents during cell secretion.[10] The size and shape of the immunoisolated porosome complex have also been determined using both negative staining electron microscopy and AFM (Fig. 5.4).[9,11-13]

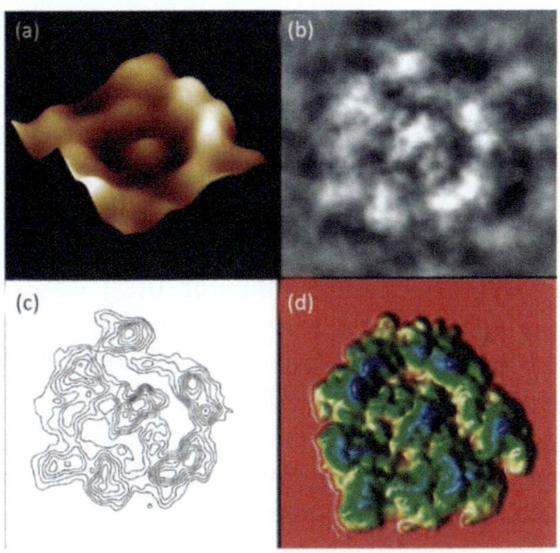

Figure 5.4. Nanoscale, three-dimensional contour map of protein assembly within the neuronal porosome complex. (a) Atomic force micrograph of an immunoisolated neuronal porosome, reconstituted in lipid membrane. Note the central plug of the porosome complex and the presence of approximately eight globular units arranged at the lip of the complex. (b) Negatively stained electron micrographs of isolated neuronal porosome protein complexes. Note the 10–12 nm complexes exhibiting a circular profile and having a central plug. Approximately 8–10 interconnected protein densities are observed at the rim of the structure, which are connected to a central element via spoke-like structures. (c) Electron density maps of negatively stained electron micrographs of isolated neuronal porosome protein complexes. (d) Three-dimensional topography of porosomes obtained from their corresponding electron density maps. The colors from yellow through green to blue correspond to the protein image density from lowest to the highest. The highest peak in each image represents 27 Å.[13]

The immunoisolated porosome complex has also been both structurally and functionally reconstituted into liposomes and lipid bilayer membranes.[9,11–13] Transmission electron micrographs of pancreatic porosomes reconstituted into liposomes exhibit a 150–200 nm cup-shaped basket-like morphology, similar to what is observed in its native state when co-isolated with ZGs. To test the functionality of the isolated porosome complex, purified porosomes obtained from exocrine pancreas or neurons were subjected to reconstitution in lipid membrane of the electrophysiological setup (EPC9) and challenged with isolated ZGs or synaptic vesicles. Electrical activity of the reconstituted membrane as well as the transport of vesicular contents from the *cis* to the *trans* compartments of the bilayer chambers was monitored. Results from these experiments demonstrate that the lipid membrane-reconstituted porosomes are indeed functional,[9,11] since in the presence of calcium, isolated secretory vesicles dock and fuse to transfer intravesicular contents from the *cis* to the *trans* compartment of the bilayer chamber. ZGs fused with the porosome-reconstituted bilayer as demonstrated by an increase in capacitance and conductance and a time-dependent transport of the ZG enzyme amylase from *cis* to the *trans* compartment of the bilayer chamber. Amylase is detected using immunoblot analysis of the buffer in the *cis* and *trans* chambers. As observed in immunoblot assays of isolated porosomes, chloride channel activity is present in the reconstituted porosome complex.[11] Furthermore, the chloride channel inhibitor DIDS was found to inhibit current activity through the porosome-reconstituted bilayer, demonstrating a requirement of the porosome-associated chloride channel activity in porosome function. Similarly, the structure and biochemical composition of the neuronal porosome, and the docking and fusion of synaptic vesicles at the neuronal porosome complex, have also been elucidated. In summary, these studies demonstrate. Porosomes to be permanent supramolecular lipoprotein structures at the cell plasma membrane, where membrane-bound secretory vesicles transiently dock and fuse to release intravesicular contents to the outside. Porosomes have therefore been designated as universal secretory machinery in cells.[29,30]

5.3 AFM: ELUCIDATING SNARE-INDUCED MEMBRANE FUSION IN CELLS

As outlined in the preceding section, in live cells, membrane fusion is mediated via a specialized set of proteins present in opposing bilayers.[15–27] Target membrane proteins, SNAP-25 and syntaxin (t-SNAREs) and secretory vesicle-associated protein (v-SNARE), are part of the conserved protein complex involved in fusion of opposing lipid membranes. The structure and arrangement of the membrane-associated full-length SNARE complex was first determined using AFM (Fig. 5.5).[20] Results from the study demonstrate that t-SNAREs and v-SNARE, when present in opposing bilayers, interact in a circular array to form supramolecular ring complexes each measuring a few nanometers. The size of the ring complex is directly proportional to the curvature of the opposing bilayers.[24] In the presence of calcium, the ring complex helps in establishing continuity between the opposing bilayers.[21-23] In contrast, in the absence of membrane, soluble v-SNARE and t-SNAREs fail to assemble in such specific and organized pattern and do not form such conducting channels. Once v-SNARE and t-SNAREs residing in opposing bilayers meet, the resulting SNARE complex overcome the repulsive forces between the opposing bilayers, bringing them closer to within a distance of 2.8–3 Å, allowing calcium bridging of the opposing phospholipids headgroups, leading to local dehydration and membrane fusion.[21-23]

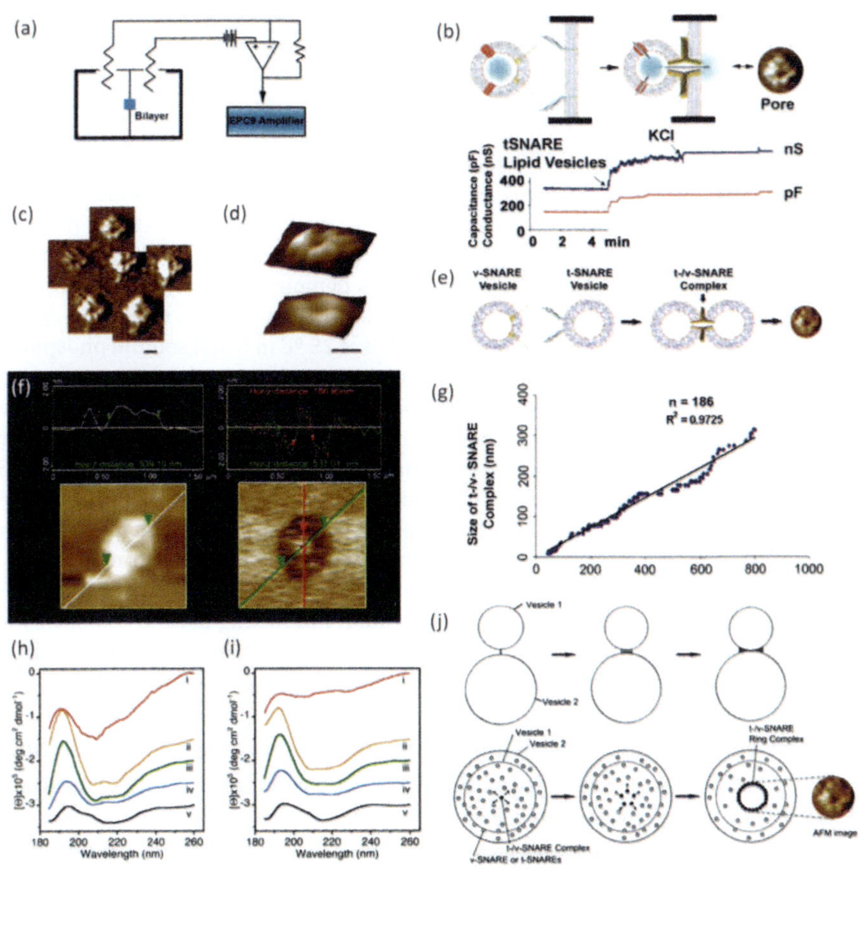

Figure 5.5. Membrane-directed assembly and the disassembly of SNAREs. Opposing bilayers containing t-SNARE and v-SNAREs, respectively, interact in a circular array to form conducting channels in the presence of calcium. (a) Schematic diagram of the bilayer-electrophysiology setup (EPC9). (b) Lipid vesicle containing nystatin channels (*red*) and membrane bilayer with SNAREs demonstrate significant changes in capacitance and conductance. When t-SNARE vesicles were added to a v-SNARE membrane support, the SNAREs in opposing bilayers arranged in a ring pattern, forming pores as shown in the AFM micrographs (c,d). t-/v-SNARE ring complex at low (c) and high resolution (d) is shown. Bar = 100 nm. A stepwise increase in capacitance and conductance (−60 mV holding potential) is demonstrated following docking and fusion of SNARE-reconstituted vesicles at the SNARE-reconstituted bilayer of the EPC9 electrophysiological set up (b). Docking and fusion of the vesicle at the bilayer membrane opens vesicle-associated nystatin channels and SNARE-induced pore formation, allowing conductance of ions from the *cis* to the *trans* side of the bilayer membrane (b). Further addition of KCl to induce gradient-driven fusion resulted in little or no further increase in conductance and capacitance, demonstrating that docked vesicles have already fused and that the membrane is intact (b). (e–g) The size of the t-/v-SNARE complex is directly proportional to the size of the SNARE-reconstituted vesicles. (e) Schematic diagram depicting the interaction of t-SNARE-reconstituted and v-SNARE-reconstituted liposomes. (f) AFM images of docked v-SNARE vesicle at t-SNARE-reconstituted membrane, before and after its dislodge using the AFM cantilever tip, exposing the t-/v-SNARE-ring complex at the center. (g) Note the high correlation coefficient between vesicle diameter and size of the SNARE complex. (h,i) CD data reflecting structural changes to SNAREs, both in suspension and in association with membrane. Structural changes, following the assembly and disassembly of the t-/v-SNARE complex, are further shown. (h) CD spectra of purified full-length SNARE proteins in suspension and (i) in membrane-associated; their assembly and (NSF–ATP)-induced disassembly is demonstrated. (i) v-SNARE; (ii) t-SNAREs; (iii) t-/v-SNARE complex; (iv) t-/v-SNARE + NSF; and (v) t-/v-SNARE + NSF + 2.5 mM ATP are shown. CD spectra were recorded at 25 ºC in 5 mM sodium phosphate buffer (pH 7.5), at a protein concentration of 10 μM. In each experiment, 30 scans were averaged per sample for enhanced signal to noise, and data were acquired on duplicate independent samples to ensure reproducibility. (j) Schematic diagram depicting the possible molecular mechanism of SNARE ring complex formation, when t-SNARE vesicles and V-SNARE vesicles meet. The process may occur because of a progressive recruitment of t-/v-SNARE pairs as the opposing vesicles are pulled toward each other, until a complete ring is established, preventing any further recruitment of t-/v-SNARE pairs to the complex. The top panel is a side view of two vesicles (one t-SNARE-reconstituted and the other v-SNARE reconstituted) interacting to form a single t-/v-SNARE complex, leading progressively (from left to right) to the formation of the ring complex. The lower panel is a top view of the two interacting vesicles.[30]

VAMP and syntaxin are both integral membrane proteins, with the soluble SNAP-25 associating with syntaxin. Hence, the key to our understanding of SNARE-induced membrane fusion requires determination of the atomic arrangement and interaction between membrane-associated v-SNARE and t-SNAREs. Ideally, the atomic coordinates of membrane-associated SNARE complex using X-ray crystallography would help elucidate the chemistry of SNARE-induced membrane fusion in cells. So far, such structural details at the atomic level of membrane-associated t-/v-SNARE complex have not been realized. This has been primarily due to solubility problems of membrane-associated SNAREs, compounded with the fact that v-SNARE and t-SNAREs need to reside in opposing membranes when they meet, to assemble in a physiologically relevant SNARE complex. The remaining option has been the use of nuclear magnetic resonance (NMR) spectroscopy. However, NMR spectroscopy too has been of little help, since the size of t-/v-SNARE ring complexes are beyond the maximum limit for NMR spectroscopy studies. Regardless of these setbacks, AFM force spectroscopy has provided for the first time at nanometre to sub-nanometre resolution an understanding of the structure, assembly and disassembly of membrane-associated t-/v-SNARE complexes in physiological buffer solution.[20-27] A bilayer electrophysiological setup allowed measurements of membrane conductance and capacitance during fusion of v-SNARE-reconstituted liposomes with t-SNARE-reconstituted membrane, and vice versa (Fig. 5.5a,b). Results from these studies demonstrated that t-SNAREs and v-SNARE when present in opposing membrane interact and assemble in a circular array, and in the presence of calcium, they form conducting channels.[20] The interaction of t-/v-SNARE proteins to form such a conducting channel is strictly dependent on the presence of t-SNAREs and v-SNARE in opposing bilayers. Addition of purified recombinant v-SNARE to a t-SNARE-reconstituted lipid membrane results in non-physiological interactions and without influence on the electrical properties of the membrane.[20] However, in the presence of calcium, when v-SNARE vesicles are added to t-SNARE-reconstituted membrane or vice versa, SNAREs assemble in a ring conformation. The resultant increase in membrane capacitance and conductance demonstrates the establishment of continuity between the opposing t-SNARE- and v-SNARE-reconstituted bilayers. These results confirm that t-SNARE and v-SNAREs are required to reside in opposing membranes, as they exist in the physiological state in cells, to allow appropriate t-/v-SNARE interactions that lead to membrane fusion in the presence of calcium. Studies using SNARE-reconstituted liposomes and bilayers[21,22] further demonstrate the following: (i) a low fusion rate (τ = 16 minutes) is obtained between t-SNARE- and v-SNARE-reconstituted liposomes in the absence of Ca^{2+} and (ii) exposure of t-/v-SNARE liposomes to Ca^{2+} drives vesicle fusion on a near-physiological relevant time-scale (τ ~10 seconds), demonstrating Ca^{2+} and SNAREs in combination to be the minimal fusion machinery in cells.[21,22] Native and synthetic vesicles exhibit a significant negative surface charge primarily owing to the polar phosphate head groups, generating a repulsive force that prevents the aggregation and fusion of opposing vesicles. In cells, SNAREs provide direction and specificity and bring opposing bilayers closer to within a distance of 2–3 Å,[21,22] enabling Ca^{2+} bridging and membrane fusion. The bound Ca^{2+} then leads to the expulsion of water between the bilayers at the bridging site, leading to lipid mixing and membrane fusion. Hence, SNAREs, besides bringing opposing bilayers closer, dictate the site and size of the fusion area during cell secretion. The size of the t-/v-SNARE complex is dictated by the curvature of the opposing membranes[24]; hence, smaller the vesicle, the smaller the t-/v-SNARE ring complex formed.

A unique set of chemical and physical properties of the Ca^{2+} ion makes it ideal for participating in the membrane fusion reaction. Calcium ion exists in its hydrated state within cells. The properties of hydrated calcium have been extensively studied using X-ray diffraction and neutron scattering, in combination with molecular dynamics simulations.[31–34] The molecular dynamic simulations include three-body corrections compared with *ab initio* quantum mechanics/molecular mechanics and molecular dynamics simulations. First-principles molecular dynamics has also been used to investigate the structural, vibrational and energetic properties of $[Ca(H_2O)_n]^{2+}$ clusters and the hydration shell of the calcium ion.[32] These studies demonstrate that hydrated calcium $[Ca(H_2O)_n]^{2+}$ has more than one shell around the Ca^{2+}, with the first hydration shell having six water molecules in an octahedral arrangement.[32] In studies using light scattering and X-ray diffraction of SNARE-reconstituted liposomes, it has been demonstrated that fusion proceeds only when Ca^{2+} ions are available between the t-SNARE- and v-SNARE-apposed proteoliposomes.[21,22] Mixing of t-SNARE and v-SNARE proteoliposomes in the absence of Ca^{2+} leads to a diffuse and asymmetric diffractogram in X-ray diffraction studies, a typical characteristic of short-range ordering in a liquid system.[33] In contrast, when t-SNARE and v-SNARE proteoliposomes in the presence of Ca^{2+} are mixed, it leads to a more structured diffractogram, with approximately a 12% increase in X-ray scattering intensity, suggesting an increase in the number of contacts between opposing bilayers, established presumably through calcium–phosphate bridges, as previously suggested.[21,22,34] The ordering effect of Ca^{2+} on inter-bilayer contacts observed in X-ray studies[21] is in good agreement with light, AFM and spectroscopic studies, suggesting close apposition of PO-lipid head groups in the presence of Ca^{2+}, followed by the formation of Ca^{2+}–PO bridges between the adjacent bilayers.[21,22,35] X-ray diffraction studies show that the effect of Ca^{2+} on bilayers orientation and inter-bilayer contacts is most prominent in the area of 3 Å, with additional appearance of a new peak at position 2.8 Å, both of which are within the ionic radius of Ca^{2+}.[21] These studies further suggest that the ionic radius of Ca^{2+} may make it an ideal player in the membrane fusion reaction. Hydrated calcium $[Ca(H_2O)_n]^{2+}$, however, with a hydration shell having six water molecules and measuring ~6 Å would be excluded from the t-/v-SNARE-apposed inter-bilayer space. Hence, calcium has to be present in the buffer solution when t-SNARE vesicles and v-SNARE vesicles meet. Indeed, studies demonstrate that if t-SNARE and v-SNARE vesicles are allowed to mix in a calcium-free buffer, there is no fusion following post addition of calcium.[22] How does calcium work? Calcium bridging of apposing bilayers may lead to the release of water from the hydrated Ca^{2+} ion, leading to bilayer destabilization and membrane fusion. Additionally, the binding of calcium to the phosphate head groups of the apposing bilayers may also displace the loosely coordinated water at the PO-lipid head groups, resulting in further dehydration, leading to destabilization of the lipid bilayer and membrane fusion. Recent studies in the laboratory,[23] using molecular dynamics simulations in the isobaric–isothermal ensemble to determine whether Ca^{2+} was capable of bridging opposing phospholipid head groups in the early stages of the membrane fusion process, indeed demonstrate this to be the case. Furthermore, the distance between the oxygen atoms of the opposing PO-lipid head groups bridged by calcium was found to be 2.92 Å, in agreement with the 2.8 Å distance previously determined using X-ray diffraction measurements. The hypothesis that there is loss of coordinated water both from the hydrated calcium ion and oxygen of the phospholipid head groups in opposing bilayers, following calcium bridging, is further demonstrated from the study.[23]

In the presence of ATP, the highly stable, membrane-directed and self-assembled t-/v-SNARE complex can be disassembled by a soluble ATPase, the *N*-ethylmaleimide-sensitive factor (NSF).[25-27] Careful examination of the partially disassembled t-/v-SNARE bundles within the complex using AFM demonstrates a left-handed super coiling of SNAREs. These results demonstrate that t-/v-SNARE disassembly requires the right-handed uncoiling of each SNARE bundle within the ring complex, demonstrating NSF to behave as a right-handed molecular motor.[26] Using circular dichroism (CD) spectroscopy, we reported[27] for the first time that both t-SNAREs and v-SNARE and their complexes in buffered suspension exhibit defined peaks at CD signals of 208 and 222 nm wavelengths, consistent with a higher degree of helical secondary structure. Surprisingly, when incorporated in lipid membrane, both SNAREs and their complexes exhibit reduced folding. Furthermore, these studies demonstrated that NSF, in the presence of ATP, disassembles the SNARE complex as reflected from the CD signals demonstrating elimination of α-helices within the structure. These results demonstrate that NSF–ATP is sufficient for the disassembly of the t-/v-SNARE complex. These studies[20-27] have provided a molecular understanding of SNARE-induced membrane fusion in cells.

5.4 CONCLUSION

In this chapter, the current understanding of the molecular machinery and mechanism of cell secretion and SNARE-induced membrane fusion is presented. Porosomes are specialized plasma membrane structures universally present in secretory cells, from exocrine and endocrine cells to neuroendocrine cells and neurons. Since porosomes in exocrine and neuroendocrine cells measure 100–180 nm, and only a 20–35% increase in porosome diameter is demonstrated following the docking and fusion of 0.2–1.2 μm in diameter secretory vesicles, it is concluded that secretory vesicles "transiently" dock and fuse at the base of the porosome complex to release their contents to the outside. Furthermore, isolated live cells in a near-physiological buffer when imaged using AFM demonstrate the size and shape of the secretory vesicles lying immediately below the apical plasma membrane of the cell. Following exposure to a secretory stimulus, secretory vesicles swell, followed by a decrease in vesicle size. No loss of secretory vesicles is observed following secretion, demonstrating transient fusion and partial discharge of vesicular contents during cell secretion. In agreement, "secretory granules are recaptured largely intact after stimulated exocytosis in cultured endocrine cells",[36] "single synaptic vesicles fusing transiently and successively without loss of identity",[37] "zymogen granule exocytosis is characterized by long fusion pore openings and preservation of vesicle lipid identity".[38] This is in contrast to the general belief that in mammalian cells, secretory vesicles completely merge at the cell plasma membrane, resulting in passive diffusion of vesicular contents to the cell exterior and the consequent retrieval of excess membrane by endocytosis at a later time. Additionally, a major logistical problem with complete merger of secretory vesicle membrane at the cell plasma membrane is the generation of partially empty vesicles following cell secretion observed in electron micrographs. It is fascinating how even single-cell organisms have developed such specialized secretory machinery, like the secretion apparatus of *Toxoplasma gondii*, the contractile vacuole in paramecium and the secretory structures in bacteria. Hence, it comes as no surprise that mammalian cells have evolved such highly specialized and sophisticated structure—the "porosome complex" for the precise and regulated release of secretory products during cell secretion. The discovery of the porosome, and an understanding of its structure and dynamics at nanometre resolution and in real time in live cells, its composition and its functional reconstitution in lipid membrane, and the molecular mechanism of SNARE-induced membrane fusion have greatly advanced our understanding of cell secretion. It is evident that the secretory process in cells is a well-coordinated, highly regulated and a finely tuned biomolecular orchestra. Clearly, these findings could not have advanced without AFM, and therefore this powerful tool has greatly contributed to a new understanding of the cell. AFM has enabled the determination of live cellular structure–function at sub-nanometer to angstrom resolution, in real time, contributing to the birth of the new field of "Nano Cell Biology". Future directions will involve an understanding of the protein distribution and their arrangement at atomic resolution in the porosome complex and a similar understanding of the structure of the t-/v-SNARE ring complex. Determination of the atomic structure of membrane-associated full-length SNAREs and their complexes, and of the neuronal porosome complex, is being further advanced using cryoelectron microscopy in the author's laboratory.

Acknowledgement

The author thanks the many students and collaborators who have participated in the various studies discussed in this article. Research in the author's laboratory was supported by the grants from the NIH, NSF and Wayne State University.

References

1. Katz, B. (1962) The transmission of impulses from nerve to muscle and the subcellular unit of synaptic action, *Proc. R. Soc. Lond. B Biol. Sci.*, **155**, 455–479.
2. Folkow, B., Häggendal, J., and Lisander, B. (1967) Extent of release and elimination of noradrenalin at peripheral adrenergic nerve terminal, *Acta Physiol. Scand. Suppl.*, **307**, 1–38.
3. Folkow, B., and Häggendal, J. (1970) Some aspects of the quantal release of the adrenergic transmitter, *Springer-Verlag Bayer Symp.*, **II**, 91–97.
4. Folkow, B., and Nilsson, H. (1997) Transmitter release at the adrenergic nerve endings: total exocytosis or fractional release? *News Physiol. Sci.*, **12**, 32–35.
5. Kelly, M., Cho, W. J., Jeremic, A., Abu-Hamdah, R., and Jena, B. P. (2004) Vesicle swelling regulates content expulsion during secretion, *Cell Biol. Int.*, **28**, 709–716.
6. Schneider, S. W., Sritharan, K. C., Geibel, J. P., Oberleithner, H., and Jena, B. P. (1997) Surface dynamics in living acinar cells imaged by atomic force microscopy: identification of plasma membrane structures involved in exocytosis, *Proc. Natl. Acad. Sci. USA*, **94**, 316–321.
7. Cho, S. J., Quinn, A. S., Stromer, M. H., Dash, S., Cho, J., Taatjes, D. J., and Jena, B. P. (2002) Structure and dynamics of the fusion pore in live cells, *Cell Biol. Int.*, **26**, 35–42.

8. Cho, S. J., Jeftinija, K., Glavaski, A., Jeftinija, S., Jena, B. P., and Anderson, L. L. (2002) Structure and dynamics of the fusion pores in live GH-secreting cells revealed using atomic force microscopy, *Endocrinology*, **143**, 1144–1148.
9. Cho, W. J., Jeremic, A., Rognlien, K. T., Zhvania, M. G., Lazrishvili, I., Tamar, B., and Jena, B. P. (2004) Structure, isolation, composition and reconstitution of the neuronal fusion pore, *Cell Biol. Int.*, **28**, 699–708.
10. Jena, B. P., Cho, S. J., Jeremic, A., Stromer, M. H., and Abu-Hamdah, R. (2003) Structure and composition of the fusion pore, *Biophys. J.*, **84**, 1–7.
11. Jeremic, A., Kelly, M., Cho, S. J., Stromer, M. H., and Jena, B. P. (2003) Reconstituted fusion pore, *Biophys. J.*, **85**, 2035–2043.
12. Cho, W. J., Jeremic, A., Jin, H., Ren, G., and Jena, B. P. (2007) Neuronal fusion pore assembly requires membrane cholesterol, *Cell Biol. Int.*, **31**, 1301–1308.
13. Cho, W. J., Ren, G., and Jena, B. P. (2008) EM 3D contour maps provide protein assembly at the nanoscale within the neuronal porosome complex, *J. Microsc.*, **232**, 106–111.
14. Cho, S. J., Wakade, A., Pappas, G. D., and Jena, B. P. (2002) New Structure Involved in transient membrane fusion and exocytosis, *Ann. N. Y. Acad. Sci.*, **971**, 254–256.
15. Oyler, G. A., Higgins, G. A., Hart, R. A., Battenberg, E., Billingsley, M., Bloom, F. E., and Wilson, M. C. (1989) The identification of a novel synaptosomal-associated protein, SNAP-25, differentially expressed by neuronal subpopulations, *J. Cell Biol.*, **109**, 3039–3052.
16. Bennett, M. K., Calakos, N., and Scheller, R. H. (1992) Syntaxin: a synaptic protein implicated in docking of synaptic vesicles at presynaptic active zones, *Science*, **257**, 255–259.
17. Trimble, W. S., Cowan, D. W., and Scheller, R. H. (1988) VAMP-1: a synaptic vesicle-associated integral membrane protein, *Proc. Natl. Acad. Sci. USA*, **85**, 4538–4542.
18. Malhotra, V., Orci, L., Glick, B. S., Block, M. R., and Rothman, J. E. (1988) Role of an N-ethylmaleimide-sensitive transport component in promoting fusion of transport vesicles with cisternae of the Golgi stack, *Cell*, **54**, 221–227.
19. Wilson, D. W., Whiteheart, S. W., Wiedmann, M., Brunner, M., and Rothman, J. E. (1992) A multisubunit particle implicated in membrane fusion, *J. Cell Biol.*, **117**, 531–538.
20. Cho, S. J., Kelly, M., Rognlien, K. T., Cho, J., Hörber, J. K., and Jena, B. P. (2002) SNAREs in opposing bilayers interact in a circular array to form conducting pores, *Biophys. J.*, **83**, 2522–2527.
21. Jeremic, A., Kelly, M., Cho, J. A., Cho, S.-J., Hörber, J. K., and Jena, B. P. (2004) Calcium drives fusion of SNARE-apposed bilayers, *Cell Biol. Int.*, **28**, 19–31.
22. Jeremic, A., Cho, W. J., and Jena, B. P. (2004) Membrane fusion: what may transpire at the atomic level, *J. Biol. Phys. Chem.*, **4**, 139–142.
23. Potoff, J. J., Issa, Z., Manke, Jr. C. W., and Jena, B. P. (2008) Ca^{2+}-dimethylphosphate complex formation: providing insight into Ca^{2+} mediated local dehydration and membrane fusion in cells, *Cell Biol. Int.*, **32**, 361–366.

24. Cho, W. J., Jeremic, A., and Jena, B. P. (2005) Size of supramolecular SNARE complex membrane-directed self-assembly, *J. Am. Chem. Soc.*, **127**, 10156–10157.

25. Jeremic, A., Quinn, A. S., Cho, W. J., Taatjes, D. J., and Jena, B. P. (2006) Energy-dependent disassembly of self-assembled SNARE complex: observation at nanometer resolution using atomic force microscopy, *J. Am. Chem. Soc.*, **128**, 26–27.

26. Cho, W. J., and Jena, B. P. (2007) N-ethymaleimide sensitive factor is a right-handed molecular motor, *J. Biomed. Nanotechnol.*, **3**, 209–211.

27. Cook, J. D., Cho, W. J., Stemmler, T. L., and Jena, B. P. (2008) Circular dichroism (CD) spectroscopy of the assembly and disassembly of SNAREs: the proteins involved in membrane fusion in cells, *Chem. Phys. Lett.*, **462**, 6–9.

28. Gaisano, H. Y., Sheu, L., Wong, P. P., Klip, A., and Trimble, W. S. (1997) SNAP-23 is located in the basolateral plasma membrane of rat pancreatic acinar cells, *FEBS Lett.*, **414**, 298–302.

29. Jena, B. P. (2008) Porosome: the universal molecular machinery for cell secretion, *Mol. Cells*, **26**, 517–529.

30. Jena, B. P. (2009) Porosome: the secretory portal in cells, *Biochemistry*, **49**, 4009–4018.

31. Chialvo, A. A., and Simonson, J. M. (2003) The structure of $CaCl_2$ aqueous solutions over a wide range of concentration. Interpretation of diffraction experiments via molecular simulation, *J. Chem. Phys.*, **119**, 8052–8061.

32. Bako, I., Hutter, J., and Palinkas, G. (2002) Car-Parrinello molecular dynamics simulation of the hydrated calcium ion, *J. Chem. Phys.*, **117**, 9838–9843.

33. McIntosh, T. J. (2000) Short-range interactions between lipid bilayers measured by X-ray diffraction, *Curr. Opin. Struct. Biol.*, **10**, 481–485.

34. Portis, A., Newton, C., Pangborn, W., and Papahadjopoulos, D. (1979) Studies on the mechanism of membrane fusion: evidence for an intermembrane Ca^{2+} phospholipid complex, synergism with Mg^{2+}, and inhibition by spectrin, *Biochemistry*, **18**, 780–790.

35. Laroche, G., Dufourc, E. J., Dufoureq, J., and Pezolet, M. (1991) Structure and dynamics of dimyristoylphosphatidic acid/calcium complex by 2H NMR, infrared, spectroscopies and small-angle x-ray diffraction, *Biochemistry*, **30**, 3105–3114.

36. Taraska, J. W., Perrais, D., Ohara-Imaizumi, M., Nagamatsu, S., and Almers, W. (2003) Secretory granules are recaptured largely intact after stimulated exocytosis in cultured endocrine cells, *Proc. Natl. Acad. Sci. USA*, **100**, 2070–2075.

37. Aravanis, A. M., Pyle, J. L., and Tsien, R. W. (2003) Single synaptic vesicles fusing transiently and successively without loss of identity, *Nature*, **423**, 643–647.

38. Thorn, P., Fogarty, K. E., and Parker, I. (2004) Zymogen granule exocytosis is characterized by long fusion pore openings and preservation of vesicle lipid identity, *Proc. Natl. Acad. Sci. USA*, **101**, 6774–6779.

Chapter 6

NANOPHYSIOLOGY OF CELLS, CHANNELS AND NUCLEAR PORES

Hermann Schillers, Hans Oberleithner and Victor Shahin
Institute of Physiology II, Medical Department, University of Münster,
Robert-Koch Street 27b, D-48149 Münster, Germany
shahin@uni-muenster.de

6.1 PLASMA MEMBRANE

6.1.1 Plasma Membrane and Channels

The plasma membrane separates the cell interior from the extracellular space by using a lipid bilayer. This lipid bilayer accommodates diverse membrane proteins, including integral membrane proteins such as receptors, ion channels and transporters, as well as certain antigens that are peripherally associated with the membrane. Because of their important roles in cell growth, differentiation and cell–cell signalling, the structures of the plasma membrane and the proteins associated with it have attracted wide attention and have been extensively investigated.

Several techniques are available to investigate the heterogeneity of cell membranes, but they show limitations in terms of resolution or artificial conditions. For biochemical approaches, membranes are usually fractionalized, and therefore the arrangement of proteins and membrane domains is hardly observable at the scale of a cell. Atomic force microscopy (AFM) is a surface probe that visualizes protein structures at nanometre range in native membranes without using fixatives. This allows protein counting and protein height measurements essential for the determination of individual molecular masses and protein distribution on the cell surface.

Life at the Nanoscale: Atomic Force Microscopy of Live Cells
Edited by Yves Dufrêne

www.panstanford.com

A single-molecule approach provides considerable advantages, as it removes the data-averaging drawback inherent in biochemical techniques that record measurements over large ensembles of molecules. Hence, AFM is a valuable tool to study membranes and membrane proteins down to single-molecule level.

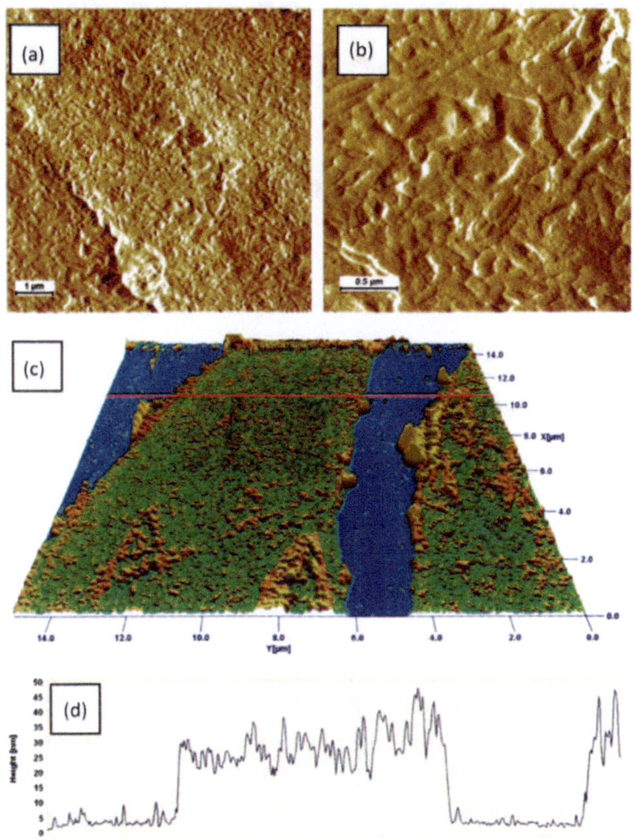

Figure 6.1. (a) AFM images of extracellular (a, b) and intracellular (c) faces of a 16HBE14o$^-$ cell membrane. The extracellular apical surface shows a high density of microvilli (a) with heights up to 1 µm. Even at higher resolution, protein structures on microvilli are hardly detectable (b). Contrarily the cytosolic face of isolated membranes is rather flat. Figure 6.1c shows a colour-coded view of a 64 µm^2 scan area containing large plasma membrane fragments attached to the poly-L-lysine-coated glass surface. Poly-L-lysine-coated glass is shown in "blue", the lipid bilayer membrane is shown in "turquoise" and the membrane proteins are shown in "brown". The red line in Fig. 6.1c corresponds to the profile line in 6.1d. The section line shows the poly-L-lysine-coated glass surface and the plasma membrane with a high density of protein structures protruding from the inner surface of the plasma membrane with heights up to 40 nm.

The ideal sample for high-resolution AFM imaging is hard and flat. Hardness reduces vertical and lateral movement of sample structures as well as a flat sample minimizes tip convolution artefacts. The plasma membrane of eucaryotic cells is generally anything but flat. The curvature of a cell is formed by lamellopodia, cell body and cell nucleus. The membrane shows major structures like membrane ruffles, microvilli and cilia and also submembranous structures like the cytoskeleton. A huge variety of proteins are heterogeneously distributed within the membrane, and many membrane proteins are equipped with highly branched sugars forming the glycocalyx. The glycocalyx, a network of polysaccharides that protrudes up to 100 nm from cellular surfaces, limits the tip access to the membrane surface, thus reducing the resolution. Therefore, we isolated cell membranes on a solid support in such a way that the intracellular face of the membrane is accessible for the AFM tip ("inside-out" orientation). An example of images obtained for human lung epithelial cells (cell line 16HBE14o$^-$) is shown in Fig. 6.1.

The extracellular face of the cell membrane shows a dense distribution of microvilli maintaining the mucocilliary clearance. These microvilli are up to 1 μm in length with diameters of several hundred nanometres (Fig. 6.1a). This microvilli layer generates an extreme roughness of the surface, which causes a serious tip convolution and therefore a reduction of resolution. Protein structures are hardly detectable even in smaller scan areas with increased resolution (Fig. 6.1b). Contrarily the intracellular faces of theses membranes are rather flat with height differences below 50 nm (Fig. 6.1c,d). Protein structures are clearly detectable, enabling AFM studies on single-molecule level.

Together, isolated, "inside-out" oriented membranes offer several advantages: (i) the membrane is flat because no curvature is imposed by underlying structures (e.g. nucleus, cytoskeleton), (ii) the membrane is hard because it lies on a hard support instead of a soft cytosol, (iii) the intracellular face of the membrane can be imaged by AFM, which means that no glycocalyx disturbs high-resolution imaging. Furthermore, the majority of membrane proteins are located intracellularly and therefore accessible by the AFM tip in an inside-out configuration.

6.1.2 The Cystic Fibrosis Transmembrane Conductance Regulator

A membrane protein of clinical importance is the cystic fibrosis transmembrane conductance regulator (CFTR).[1] CFTR is a plasma membrane cyclic AMP-activated Cl^- channel that is expressed in several functionally diverse tissues, including the kidney, pancreas, intestine, heart, vas deferens, sweat duct and lung. It is a protein of the ATP-binding cassette transporter superfamily, known to play a crucial role in maintaining the salt and water balance on the epithelium. Stimulating CFTR by cAMP increases channel activity and increases the total number of CFTR channels in the membrane, which is achieved by the insertion and removal of CFTR channels from the plasma membrane. Mutations in CFTR affect the number of channels in the plasma membrane, channel activity and the intracellular trafficking of CFTR. A mutation in the gene encoding for CFTR results in cystic fibrosis (CF), a very common lethal genetic disease.

One of the goals of our AFM studies on CFTR in isolated cell membranes was to quantify this protein in its native environment and to elucidate membrane trafficking of CFTR.

The most predominant mutation, ΔF508, results in a defective protein trafficking, which manifests in organ pathology.[2] To perform its task, CFTR has to be correctly incorporated into the cell membrane in a sufficient number. The issue regarding the number of CFTR within the cellular membrane is gaining increasing interest for developing ΔF508-CFTR-rescuing strategies and gene therapies for CF. Although a wealth of information has been gathered using different quantification approaches, the conclusions obtained so far regarding the CFTR number were indirectly drawn. Identification of CFTR within the cell membrane at single-molecule level makes it feasible to visualize the distribution and organization of CFTR proteins within the cell membrane of healthy individuals and CF patients.

6.1.3 Visualisation and Quantification of Plasma Membrane Dynamics

We used *Xenopus l.* oocytes as expression system for human CFTR. We examined the expression of CFTR after injection of CFTR-cRNA with voltage-clamp experiments and isolated the membranes of oocytes exhibiting cAMP-inducible currents in voltage-clamp analysis. This experimental procedure ensures that the plasma membrane investigated by AFM contains functional CFTR. We used the AFM to image the cytoplasmic surface of native plasma membranes of CFTR-expressing *Xenopus l.* oocytes before and after cAMP stimulation.

AFM revealed large patches of inside-out oriented plasma membrane and areas without membrane. Edges of membrane were used to determine total height of plasma membrane and its protruding structures. Figure 6.2 shows a 3D colour-coded view of a 9 μm^2 scan area containing plasma membrane fragments attached to the poly-L-lysine-coated glass surface. Poly-L-lysine-coated glass is shown in "blue", the lipid bilayer membrane is shown in "turquoise" and the membrane proteins are shown in "brown". Membrane fragmentation occurs frequently because of the preparation method we used. The broken line in the upper part of Fig. 6.2 corresponds to the profile line in the lower part showing the lipid bilayer with a height of about 5 nm. Proteins appear with different shapes and protrude from the inner surface of the plasma membrane with heights up to 20 nm.

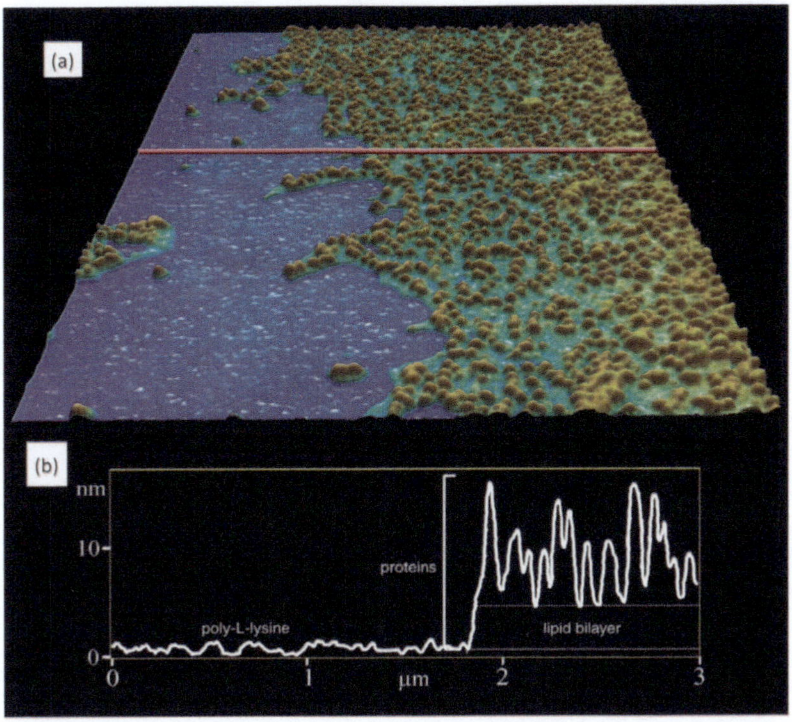

Figure 6.2. (a) Colour-coded view of a 9 µm^2 AFM scan area containing plasma membrane fragments attached to the poly-L-lysine-coated glass surface. Poly-L-lysine-coated glass is shown in "blue", the lipid bilayer membrane is shown in "turquoise" and the membrane proteins are shown in "brown". (b) The red line in Fig. 6.2a corresponds to the profile line in 6.2b. The section line shows three height levels: (1) the poly-L-lysine-coated glass surface, (2) the lipid bilayer with a height of about 5 nm and (3) proteins protruding from the inner surface of the plasma membrane with a height up to 20 nm.[3]

The main difference between cAMP-stimulated and non-stimulated oocyte membrane is the protein density. Quantification of protein distribution is shown in Fig. 6.3. Molecular volumes were estimated from protein heights measured by AFM. Molecular weights were then calculated from the respective volume measurements.

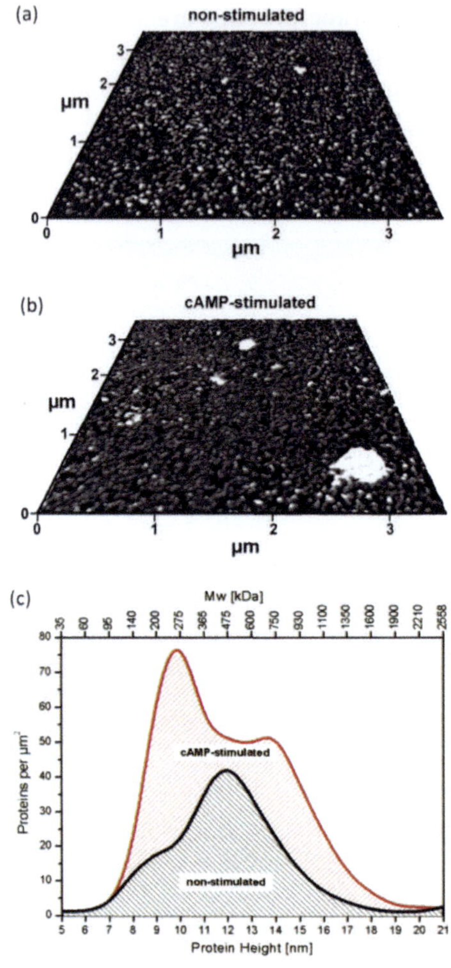

Figure 6.3. (a) The membrane patch shown in Fig. 6.3a was isolated before cAMP stimulation. (b) The membrane shown in Fig. 6.3b was isolated during cAMP stimulation. The proteins exhibit heights from 6 nm to 20 nm. The large white spots are yet unidentified intracellular structures like yolk proteins. (c) Protein distribution of CFTR-expressing plasma membrane. The hatched areas represent the respective height distributions of stimulated and non-stimulated oocytes.[4]

CFTR-expressing oocytes show an average protein height of 12 nm, corresponding to a molecular mass of 475 kDa (Fig. 6.3c, black hatched area). Stimulation with cAMP dramatically changes protein distribution. In membranes of CFTR-expressing oocytes, the protein density increases in response to IBMX from 200 to 400 proteins per μm^2, with an average protein height of 11.8 nm, corresponding to a molecular mass of 464 kDa. Protein distribution shows two peak values, at 9 nm and at 14 nm, corresponding to molecular masses of 275 kDa and 750 kDa, respectively (Fig. 6.3c, red hatched area). The data obtained in CFTR-expressing oocytes indicate that the protein covered area increases in response to cAMP by about 110%. This observation strongly suggests protein insertion into the plasma membrane. CFTR-expressing oocytes exhibited upon cAMP stimulation two new peaks at 275 kDa and 750 kDa. Since both peaks do not appear in CFTR-negative oocytes in response to cAMP stimulation, we conclude that the two peaks are caused by CFTR. Considering the molecular mass of 180 kDa for a CFTR monomer, the peak at 275 kDa and 750 kDa could be multimeric CFTR or CFTR forming clusters with other proteins. Together, upon stimulation with cAMP, CFTR is inserted into the plasma membrane, indicated by a shift in protein density and protein distribution. Insertion of CFTR in the plasma membrane leads to the formation of clusters, heteromeric structures composed of CFTR and other proteins with yet unknown stoichiometry.

These data show that the dynamics of plasma membrane protein distribution could be visualized and quantified with AFM.

6.1.4 Quantification of CFTR in Human Red Blood Cells

CFTR is distributed in various cell types, and it is also shown for red blood cells (RBCs).[5-8] Interestingly, CF patients do not show hematological disorders; therefore, the meaning of CFTR on RBCs is unclear. We performed a quantification study of the CFTR copies in RBC membranes at single-molecule level and compared the difference between healthy donors and CF patients with the homozygous ΔF508 mutation. For this purpose, two different AFM techniques were used: (1) immunostaining with quantum dot (Qdot)-labelled antibodies and (2) topography and recognition imaging.

The membrane isolation approach was used not only to achieve high resolution but also to have the intracellular portion of CFTR freely accessible for antibodies. RBCs are non-adherent cells, and therefore we glued them onto poly-L-lysine-coated glass. These attached RBCs were sheared open with a jet stream of isotonic phosphate buffered saline (with 0.2 mM EGTA).[9] The quality of membrane preparations was assessed, with AFM revealing large areas of freely accessible intracellular plasma membrane surfaces (Fig. 6.4). Membranes appear as flat round structures, with protrusions up to 25 nm in height. A high density of erythrocytes during preparation causes an overlapping of membrane edges, resulting in multilayered membrane areas clearly visible in the AFM images (Fig. 6.4a–c).

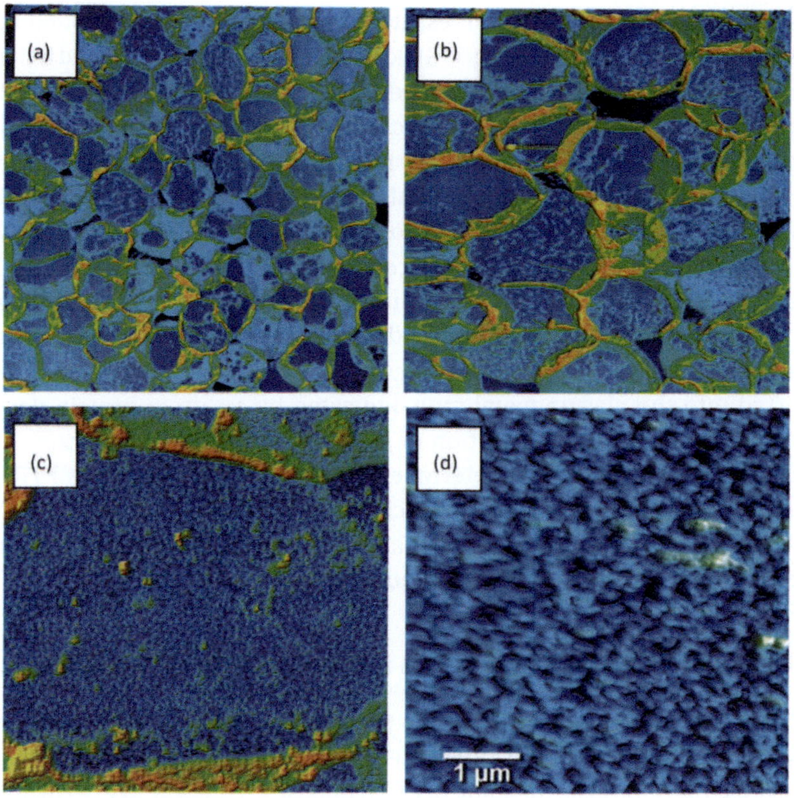

Figure 6.4. Series of AFM images showing inside-out oriented isolated RBC membranes. The size of scanned areas is (a) 80 × 80 µm, (b) 40 × 40 µm, (c) 10 × 10 µm and (d) 5 × 5 µm. Images are colour-coded, poly-L-lysine-coated glass is shown in "dark blue" and the lipid bilayer membrane with proteins is shown in "light blue". Overlapping membranes are shown in "turquoise", and multiple overlaps are shown in "brown".

6.1.5 Qdot-Labelled Antibodies

Monoclonal antibody against the C-terminus of CFTR in combination with secondary Qdot-labelled antibodies were used for immunostaining.[10] Because of their properties as particles with defined size and superior fluorescence, the Qdots were used as excellent AFM and fluorescence markers for CFTR localization. Fluorescence microscopy of the site-specific Qdot-labelled CFTR showed clearly the presence of CFTR on human RBC membranes (Fig. 6.5a,c). The weak non-specific autofluorescence of the membranes, due to glutaraldehyde fixation, enabled us to visualize each RBC membrane. However, the membranes from CF patients (Fig. 6.5b,d) showed drastically reduced fluorescent events. This result indicates that the ΔF508-CFTR is misprocessed also in RBC, where only a small amount of CFTR reaches the cell surface. Within the diffraction limitation of fluorescence microscopy, two closely positioned Qdots (e.g. ~200 nm distance) would be detected as one bright spot. Therefore, to achieve single-molecule detection, our next step was to apply the high-resolution AFM to the Qdot-labelled membranes.

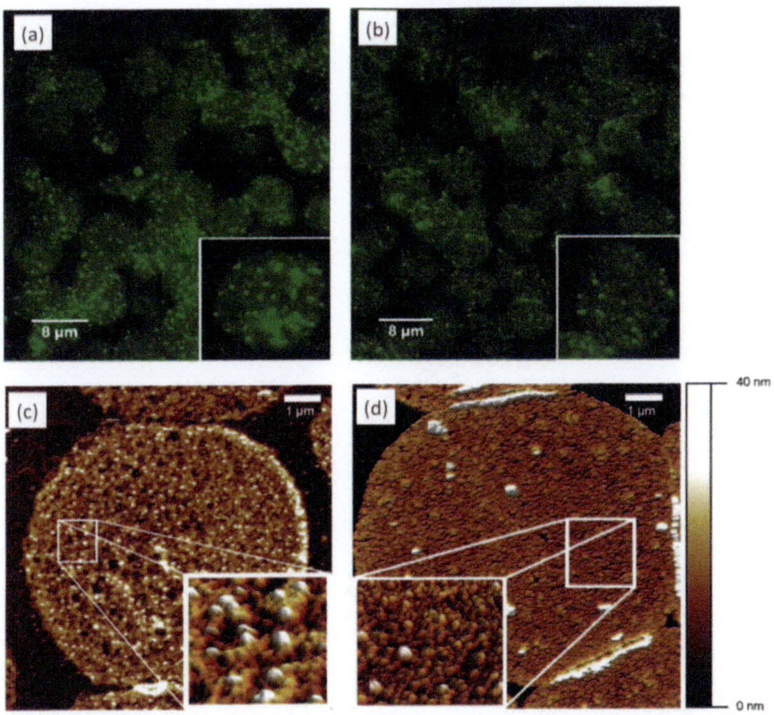

Figure 6.5. Immunostaining of CFTR in isolated RBC membrane patches with Qdot-labelled antibodies. The upper panel represents fluorescence images of non-CF (a) and CF (b) RBC membrane patches. Each inset shows a single membrane with clearly distinguishable bright fluorescence events. The lower panel shows AFM images of non-CF (c) and CF (d) RBC membrane patches. High-resolution scans, shown in (c) and (d), identify the Qdot as high structures (~15 nm, colour code: white) with specific shapes. It is evident that the number of Qdot-labelled CFTR molecules is much higher in non-CF RBC than in CF RBC.[10]

The crystalline nature of the Qdots, however, allowed single-molecule detection with AFM. In the AFM images, Qdots appear as structures with uniform height and shape (Fig. 6.5c,d). Since the AFM provided the required single-molecule resolution for further detailed quantification of Qdot-labelled CFTR proteins at the RBC membranes, we used the images taken with AFM.

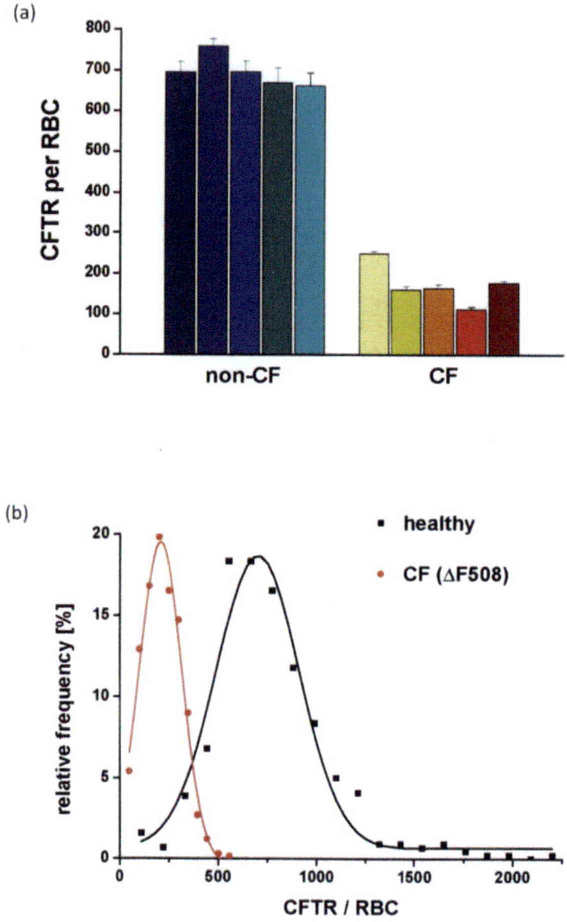

Figure 6.6. Histogram of CFTR distribution on human red blood cells. Single-molecule counting of Qdot-labelled CFTR molecules on RBC from five non-CF donors and five CF patients (100–120 RBC membrane patches from each individual) reveals a Gaussian distribution of CFTR within the RBC population. The histograms show peak values of 642 for non-CF-RBC (black curve) and 204 for CF patients (red curve), respectively. The isolated membrane patches represent approximately 40% of the RBC membrane, and the results were extrapolated to the total RBC surface area of 130 μm^2.[10]

Quantification of Qdots on non-CF RBC membranes revealed a mean value of about 650 CFTR molecules per RBC. In contrast, in CF patients, we found only about 200 CFTR molecules per RBC (Fig. 6.6). Assuming that each Qdot represents a single CFTR molecule, we could determine a CFTR density of ~5 CFTR per μm^2 for non-CF RBC and of ~1.6 CFTR per μm^2 for RBC from CF patients. The observed CFTR density of non-CF RBC is in good correlation to electrical measurements in Calu-3 cells, a human airway epithelial cell line.[11]

6.1.6 Topography and Recognition Imaging of Human RBCs

TREC[12,13] allows mapping of topographical details of a specimen and simultaneous investigation of the distribution of proteins on the surface (see also Chapter 7). To provide a specific recognition, an antibody is covalently bound to the scanning AFM tip. While oscillating over the surface, the AFM tip approaches the surface that contains cognate antigen, and an antibody–antigen bond is formed. During a subsequent retraction of the tip, the bond will cause a measurable tension. Since topographical features affect only the lower part of the oscillation, the latter is used for the piezo feedback loop. In contrast, the molecular recognition influences only the upper part of the oscillation, which is separated in an electronic circuit for localizing the recognition events (Fig. 6.7). As a result, topographical and the simultaneous recorded recognition images provide structural and chemical information of the investigated surface.

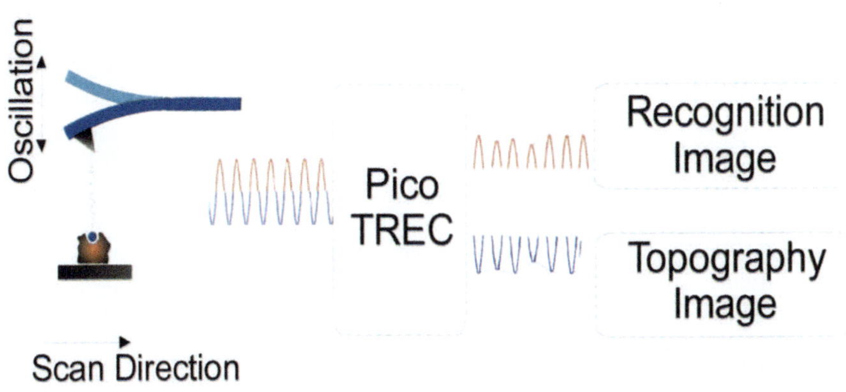

Figure 6.7. TREC working principle. A ligand functionalized AFM tip is oscillated over the sample surface. The lower part of the amplitude is used for driving the AFM feedback loop, resulting in the topography image, whereas the upper part is affected by molecular recognition, yielding a simultaneously acquired recognition image.[14]

First, the topographical images of both non-CF (Fig. 6.8a) and CF (Fig. 6.8d) erythrocyte membranes revealed similar structures protruding out of the membranes with 10–12 nm in height, representing the membrane proteins. The structures were comparable with the topography of membrane proteins obtained with standard AFM. Visualization at single-molecule level was achieved without compromising its topographic imaging performance, despite the fact that the tip was carrying a tethered antibody. Second, the simultaneously acquired recognition images (Fig. 6.8b,e) showed dark spots, corresponding to interactions between the antibody on the tip and membrane proteins. These binding sites can be assigned to particular topographical structures, allowing the identification of CFTR among the abundance of different proteins present in the membrane. The most prominent observation was that CF samples (Fig. 6.8e) revealed clearly fewer recognition spots compared with non-CF samples (Fig. 6.8b). The specificity of the antibody-CFTR recognition process was successfully proven in a control experiment where the antibody-CFTR interaction was blocked. The block resulted in almost complete abolishment of the recognition spots, confirming clearly that the recognition events arise from the specific interaction of anti-CFTR antibody on the tip with the CFTR on the surface (Fig. 6.8c,f).

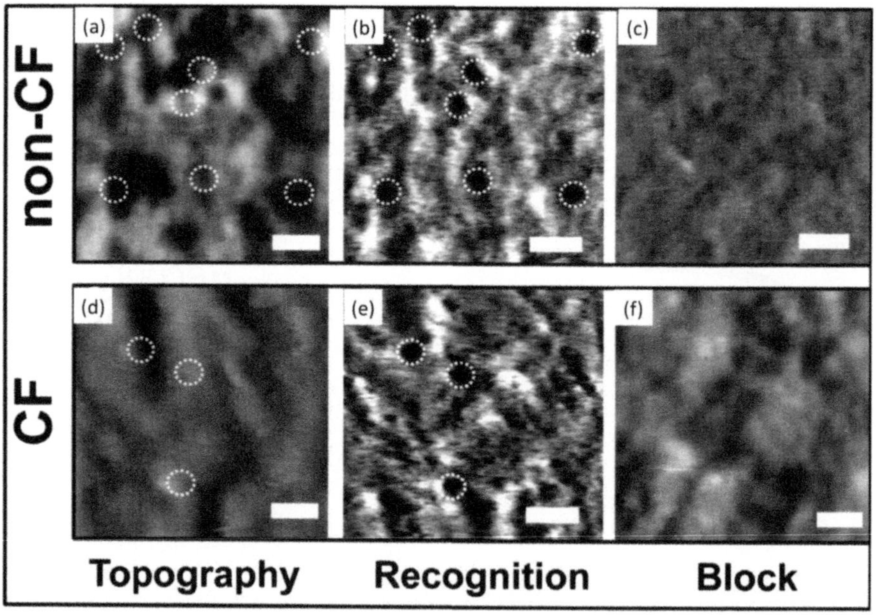

Figure 6.8. Topography and recognition images of isolated erythrocyte membranes. TREC imaging topography of a non-CF (a) and of a CF (d) erythrocyte membrane. Dark spots in the recognition images b and e represent the specific interaction sites between the modified tip (i.e. anti-CFTR antibody tip) and CFTR, corresponding to the same areas as shown in a and d. The CF membrane (e) clearly reveals fewer recognition events compared with the non-CF membrane (b). Blocking the membrane of non-CF (c) and CF (f) erythrocytes with free anti-CFTR antibody results in the disappearance of the recognition signals (block efficiency > 90%), confirming the specificity of recognition. Scale bar is 200 nm, *z* scale 80 nm.[14]

Quantification of the recognition events revealed values of six and two CFTR molecules per μm^2 for non-CF and CF erythrocytes, respectively. Extrapolating the results to the total erythrocyte surface area of 130 μm^2 results in about 800 CFTR/erythrocyte for non-CF and about 250 CFTR/erythrocyte for CF samples. These values are slightly higher but still in good correlation to the observation made with Qdot-labelled antibodies to quantify CFTR on erythrocyte membranes. Qdots are several nanometres in size, and therefore they could cause sterical hindering when individual CFTR molecules are located in close vicinity. With the TREC technique, sterical hindering does not occur, possibly explaining the slightly higher number of CF recognition sites. Clearly, the TREC has an advantage over the labelling method where direct visualization of the molecule is not possible since Qdots lay on or very close to the target molecule.

In conclusion, these studies show that AFM experiments on isolated plasma membranes not only allow quantification and localization of membrane proteins but also provide insight in their dynamics at a single-molecule level.

6.2 ENDOTHELIAL CELLS

6.2.1 Mechanodynamics of Vascular Endothelial Cells

In physics, the term stiffness is clearly defined: "Stiffness is a measure of the resistance offered by an elastic body to deformation". Importing this term into cellular physiology stiffness indicates a force (Newton) necessary to compress a cell for a certain distance (metre). Application of force happens to most tissues in real life, particularly to vascular endothelium. Hemodynamic forces, born by the beating heart, generate shear stress at the endothelial surface. It is inevitable that the apical cell surfaces undergo reversible deformations. This mechanical stimulus triggers the activation of the endothelial nitric oxide synthase (eNOS) and the release of NO. The latter diffuses to the adjacent vascular smooth muscle cells, leading to vasodilation. This regulatory mechanism distributes the blood in the organism according to the metabolic demands and, at the same time, maintains systemic blood pressure within physiological limits. Therefore, it is obvious that the same shear force should cause a stiff (less deformable) cell to release less NO as compared with a soft (more deformable) cell. This leads to the conclusion that endothelial mechanical stiffness is a key parameter in the control of local blood supply and arterial blood pressure.

6.2.2 AFM Used as a Mechanical Nanosensor

The "tool of choice" for quantitatively measuring stiffness (given in N/m) of living adherent endothelial cells is an AFM. In principle, the AFM is used as a mechanical tool, i.e. the AFM tip is pressed against the cell so that the membrane is indented. This distorts the AFM cantilever which serves as a soft spring. The cantilever deflection, measured by a laser beam reflected from the gold-coated cantilever surface, permits force–distance curves of single cells. The slope of such curves is directly related to the force (expressed in Newton) necessary to indent the cell for a given distance (expressed in metre). At least two different slopes (Fig. 6.9) can be identified depending on the depth of indentation. The initial rather flat slope (indentation depth: up to several 100 nanometres) reflects the soft plasma membrane stiffness including the cortical cytoskeleton (cell shell), while the late rather steep slope reflects the stiffness of the more rigid cell centre.

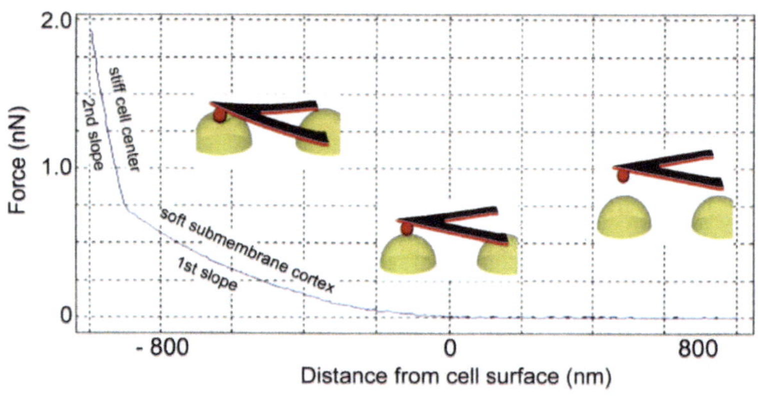

Figure 6.9. Indentation technique using atomic force microscopy. Indentation curve with two different slopes.

The so-called force–distance curves can be obtained on single living cells. Important parameters for reliable measurements are (i) how fast (indentation velocity), (ii) how deep (indentation depth which is related to the loading force) and (iii) how often (indentation frequency) force curves are being obtained in a single cell. Endothelial cells tolerate such measurements for hours when no more than 12 indentations per minute are performed, indentation velocity is not exceeding 1 µm per second and indentation depth is not beyond 20% of the cell height. Another important technical improvement is the use of spheres mounted to the AFM tip. The spherical tips (sphere diameter = 1 µm) gently interact with the cell surface, which results in "low-noise" force curves.[15]

In the past, the term "cell stiffness" has been used as a global description of a mechanical cell property. More recently, cell stiffness could be split into at least two components, one describing the stiffness of the plasma membrane including the spiderweb-like submembranous actin network (cell shell), the other describing the stiffness of the bulk cytoplasm.[16] By a further reduction of the indentation velocity and indentation depth, a third stiffness component can be separated from the other two, which is located above the cell membrane and most likely relates to the glycocalyx. It is a very soft layer (several 100 nm thick, stiffness is about 0.2 pN/nm) and neglected pending further experiments.

6.2.3 Sodium: "Stiffener" of Vascular Endothelial Cells

Hypertension, stroke, coronary heart disease and kidney fibrosis are related to high sodium intake as shown in many studies.[17] Although the deleterious action of high sodium intake is obvious, the underlying mechanisms how salt (NaCl) acts at the organ, tissue and cellular levels are still unclear. High sodium causes fibrosis in kidney and heart[18] and supports inflammatory processes.[19] When dietary salt intake exceeds renal excretion capacity, sodium is stored in the space between cells bound to extracellular organic material.[20] A close look at the plasma sodium concentration shows that there is a small but significant rise in sodium concentration (3 to 4 mM) when dietary salt intake is high.[21,22] Hence, it was postulated that changes in plasma sodium could directly affect blood pressure.

Stimulated by this work, it was tested whether endothelial cells directly respond to small changes in extracellular sodium. Indeed, cells stiffen within minutes when extracellular sodium is elevated. This mechanical response happens only when aldosterone, a sodium-saving steroid hormone, is present in the culture media. Surprisingly, endothelial cells are highly sensitive to sodium in a narrow physiological range (Fig. 6.10). Small increments of extracellular sodium between 140 and 145 mM increase cell stiffness by more than 20%, indicating a relevant physiological role in the control of endothelial function.

Cells exposed to low sodium are more deformable by shear force than the same cells exposed to high sodium as demonstrated in the two AFM images of Fig. 6.11. Endothelial cells scanned at constant force and constant frequency in a solution of 135 mM sodium (=low sodium) are visibly flattened by the scanning AFM tip. In contrast, in medium containing 150 mM sodium (=high sodium), the same endothelial cells resist the "pressure" of the scanning AFM tip and remain prominent.[23]

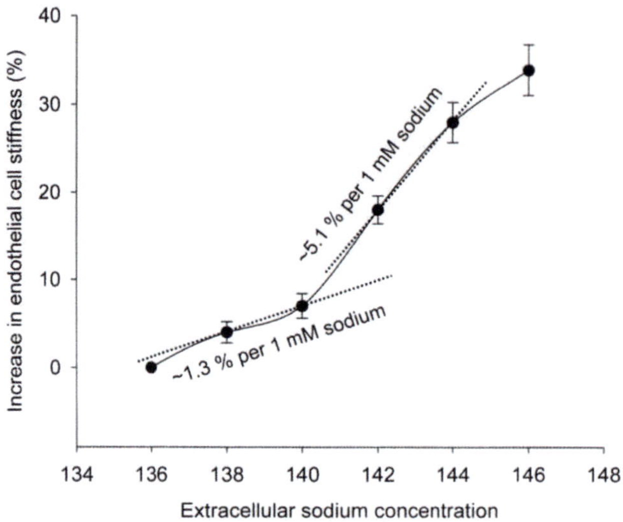

Figure 6.10. Endothelial cell stiffness depends upon extracellular sodium concentration. The dotted lines represent the estimated slopes of the relation "stiffness change over sodium change".

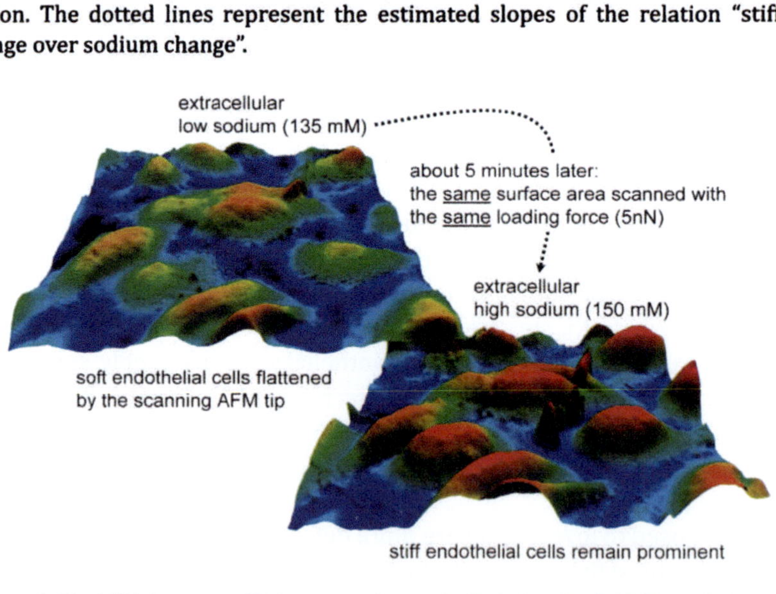

Figure 6.11. AFM images of living vascular endothelial cells (height scale is colour coded). An endothelial cell monolayer was scanned with constant force (5 nN) and constant frequency (1 Hz) at two different sodium concentrations in the bath. At low extracellular sodium concentration (135 mM), cells are flattened by the force of the AFM stylus applied to the endothelial cells. At high sodium (150 mM), cells resist the applied forces and thus appear prominent.

6.2.4 Potassium: "Softener" of Vascular Endothelial Cells

In contrast to natural food, processed food products are rich in sodium and, at the same time, poor in potassium. There is no question that a high-potassium, low-sodium diet exerts beneficial effects on the cardiovascular system[24,25] and even improves mood states, for example, depression, tension and vigor.[26] Potassium deficiency is difficult to detect since 98% are hidden inside the cells. Nevertheless, plasma potassium is maintained in narrow limits, between 4 and 5 mM, and a subtle indicator for any disturbances of potassium homeostasis.

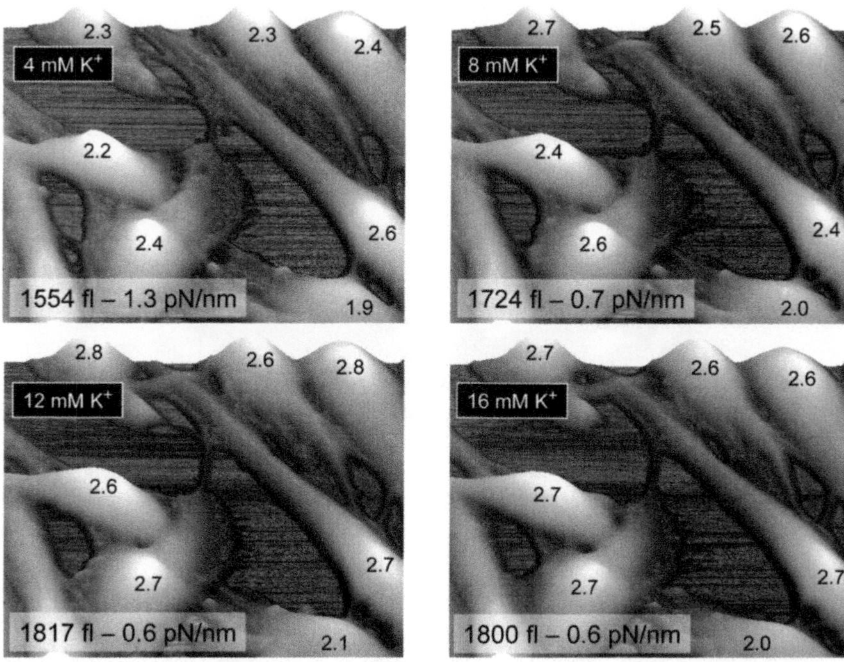

Figure 6.12. AFM imaging of living vascular endothelial cells, exposed for 5 minutes to increasing concentrations of extracellular potassium. Paired experiment showing the same cells at different conditions. Numbers on cells indicate the respective cell heights (in μm). Numbers at the left lower corners of the images refer to the average cell volume (given in femtolitre) and to the average mechanical cell stiffness (given in pico-Newton/nm). Cells swell and soften in response to increasing potassium.

Recently, the question was addressed whether extracellular potassium could directly alter the function of endothelial cells. This is indeed the case. AFM imaging of living endothelial cells reveals that cell height and cell volume increase when extracellular potassium is stepwise increased. At the same time, cells soften (Fig. 6.12). Elevated extracellular ("systemic") potassium concentrations (>5.5 mM) usually measured in the blood plasma frequently occur as a severe symptom associated with kidney disease. However, "local" potassium concentrations up to 15 mM are absolutely normal in the interstitium of muscle during physical exercise[27] and during increased neuronal activity in brain.[28] As a functional consequence of swelling and softening, vascular endothelial cells undergo more pronounced shear-stress-mediated (reversible) deformations which result in enhanced NO formation.

6.2.5 The "Solation-Gelation" Hypothesis

Endothelial cells are subjected to large changes in cell shape (e.g. during dilation/constriction of blood vessels, particularly with each contraction of the heart) and can adjust best to such alterations if the deformability (physical compliance) of the cells is high.

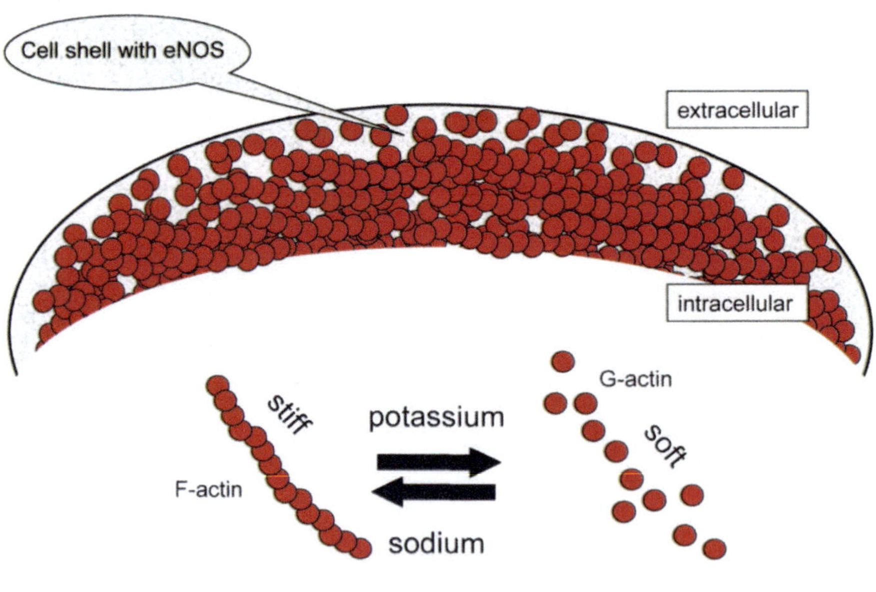

Figure 6.13. Concept of how sodium and potassium control the dynamic cortical zone (cell shell).

At least two linear slopes have been described in the indentation curves, the first tends to be flat while the second is steeper (see Fig. 6.9). The first flat slope indicates a low stiffness and is limited to the submembranous cortex of the cell (cell shell). Obviously, there is a fluidic layer beneath the plasma membrane, which is highly dynamic in terms of thickness and viscosity. A cellular model describes this concept mainly based on AFM measurements (Fig. 6.13). The cortical cytoskeleton of vascular endothelial cells is highly dynamic, and the state of polymerization of cortical actin determines the structure and mechanical properties of this layer.[29,30] Monomeric globular actin (G-actin) can rapidly polymerize into filamentous actin (F-actin), which causes a rapid change in local viscosity. The switch from F-actin to G-actin by using the polymerization inhibitor cytochalasin D is associated with solation of the cortex.[31] An increase in extracellular potassium mimics this response, indicating that potassium per se softens the cortical actin cytoskeleton by changing F-actin to G-actin. G-actin is known to colocalize with the endothelial eNOS and to increase eNOS activity.[32,33] This could explain the activation of eNOS by high potassium.

Sodium is possibly a functional antagonist in this system. Sodium influx increases the viscosity of the submembranous layer by stiffening the cytoskeleton. When sodium is in the high physiological range, F-actin dominates over monomeric actin. This explains the sodium-induced increase in cell stiffness. When potassium is elevated, actin filaments disaggregate into actin monomers, and the endothelial cell softens. Both F-actin and G-actin are negatively charged molecules, and the interaction with Na^+ and K^+ will finally depend upon local concentrations and specific affinities of the respective ions. It has to be kept in mind that this scenario is supposed to happen in a quite restricted cytosolic space, directly underneath the plasma membrane, most likely at the caveolae.[34] Since this cytosolic submembranous zone (cell shell) is only a few hundred nanometres thick, about 90% of the cell body remains unchallenged.

Taken together, "local" mechanodynamics, i.e. the mechanical properties underneath the plasma membrane, determines the function of vascular endothelial cells.

6.3 NUCLEAR PORES

6.3.1 Apoptosis: Physiological Relevance of Apoptosis

For every cell, there is a time to live and a time to die, and cell death can be executed by various injury types or by suicide. Unlike cell death by injury, the process of cell death by suicide is highly orderly and is often referred to as programmed cell death or apoptosis. Apoptosis is the regulated elimination of cells that occurs naturally during the course of development, as well as in many pathological circumstances that require cell death for the benefit of the organism.

In the adult organism, the number of cells is kept relatively constant through cell death and division. Cells must be replaced when they malfunction or become diseased, but proliferation must be offset by cell death.[35] This control mechanism is part of the homeostasis required by living organisms to maintain their internal states within certain limits. Loss of cells by injury, for instance trauma, is undesired. By contrast, apoptosis generally confers key advantages during the life cycle of a multicellular organism. Apoptosis occurs during the development of multicellular organisms and goes on throughout the adult life. For example, the differentiation of fingers and toes in developing human embryo occurs because cells between fingers commit suicide and the consequence is that the digits are separate. A severely damaged cell commits suicide to prevent damage from being spread on to surrounding cells. Apoptosis is thus involved in fundamental processes of life, like embryonic development, tissue homeostasis or immune defence. Defects in apoptosis cause or contribute to developmental malformation, cancer and degenerative disorders.

6.3.2 The Process of Apoptosis

In contrast to the diversity of stimuli generating apoptosis, signalling and execution mechanisms are strongly conserved.[36] As seen in Fig. 6.14, the execution of apoptosis is mainly driven by caspases, a family of cysteine proteases. Activation of caspases, in turn, occurs as a consequence of cytochrome c release from mitochondria.[37] The apoptotic program can also be initiated artificially by delivering a load of exogenous cytochrome c into the cytosol.[38] Hallmarks of apoptosis are numerous. They comprise cell shrinkage, plasma membrane blebbing, nuclear and DNA fragmentation and the formation of apoptotic bodies.[39]

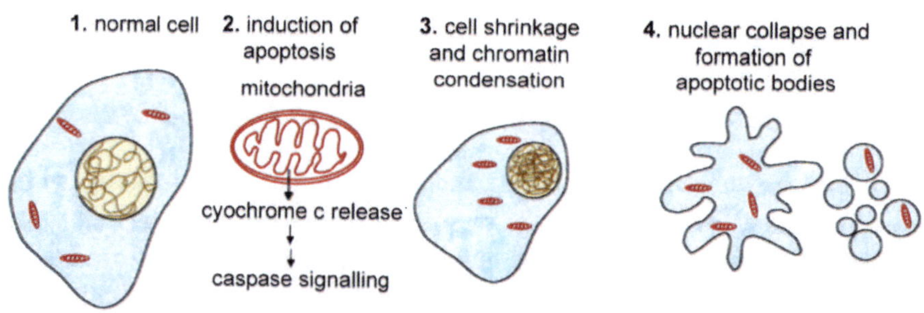

Figure 6.14. Schematic of cell destruction during apoptosis.

6.3.3 The Nuclear Envelope Is a Key Target of Apoptosis

Cell destruction during apoptosis proceeds in a strategic manner. Thereby, critical cellular components, including the cell nucleus, are sequentially targeted and dismantled. Nuclear dismantling, in turn, requires key changes in structure and mechanics of the nuclear envelope, which separates the cytosol from the nucleus (Fig. 6.15a). The nuclear envelope shields the nuclear DNA, mediates the pivotal nucleocytoplasmic exchange of material through nuclear pore complexes (NPCs) (Fig. 6.15b), is involved in regulation of gene expression[40] and confers essential structural stability to the cell nucleus through the underlying nuclear lamina (Fig. 6.15a). The nuclear envelope is therefore one of the major cellular targets of apoptosis.

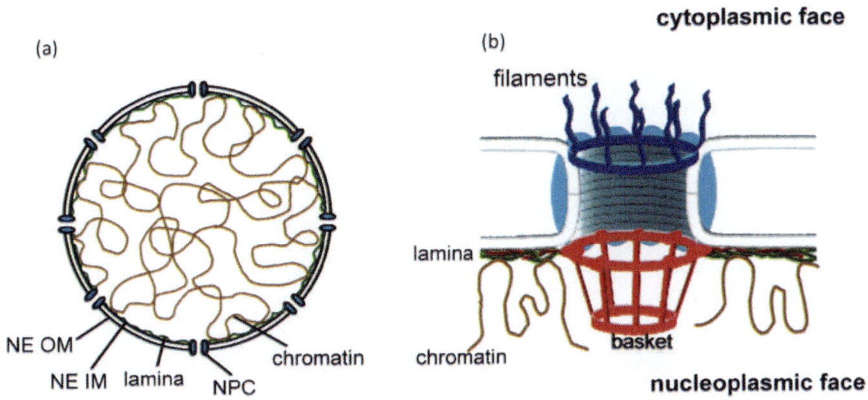

Figure 6.15. The nuclear envelope. (a) and (b) are schematics of the cell nucleus and the nuclear pore complex (NPC), respectively. NE IM and NE OM stand for nuclear envelope inner and outer membranes, respectively.

6.3.4 AFM Unravels the Fate of the Nuclear Envelope During Apoptosis

Nanoscale investigation of structure and mechanics of the nuclear envelope in the normal state but also during apoptosis has remained an unfulfilled wish because of the lack of an appropriate approach. The development of AFM,[41] a powerful emerging approach capable of simultaneous structural and mechanical investigations at the nanoscale and in fluid, has made this wish come true.[42] Using AFM structural and mechanical properties of the nuclear envelope can be investigated under various physiological conditions including apoptosis.[42–44]

For this purpose, *Xenopus l.* oocytes can be committed to apoptosis by microinjection of cytochrome c into the cytosol of oocytes. Figure 6.16 depicts an experimental approach for investigating, with AFM, the structure and mechanics of the nuclear envelope following induction of apoptosis in oocytes.

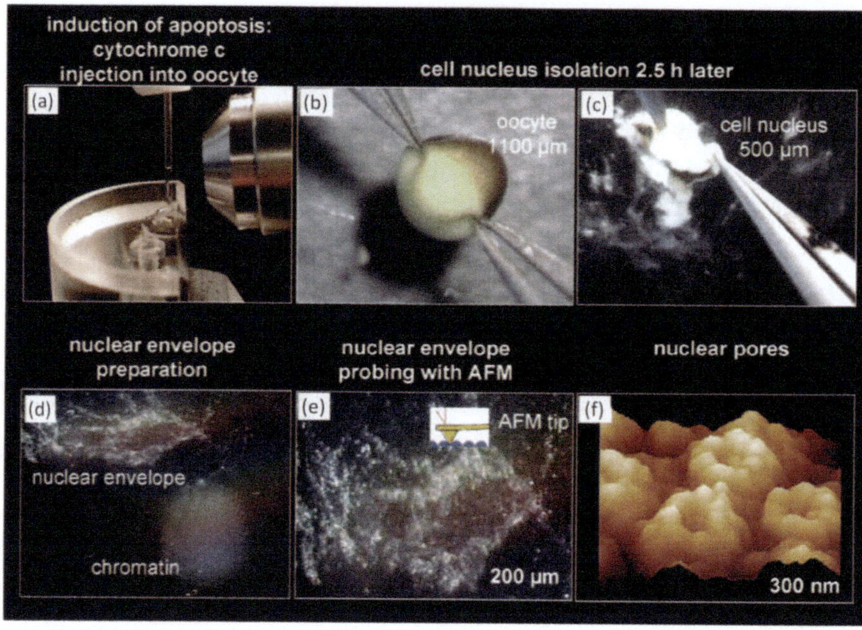

Figure 6.16. Experimental approach for the AFM investigation of the structure and mechanics of the nuclear envelope following induction of apoptosis in oocytes. (a) Induction of apoptosis in *Xenopus l.* oocytes. (b, c) Isolation of the cell nucleus 2.5 hours after injection. (d) Preparation of the nuclear envelope. (e) Application of AFM to structurally and mechanically investigate the nuclear envelope in fluid at the nanoscale. (f) Individual nuclear pores visualised with AFM.

6.3.4.1 Disfiguration and softening of the doomed nuclear envelope upon degradation of its prominent structural and functional features, the nuclear basket and the nuclear lamina

As shown in Fig. 6.17, both the NPC basket and the nuclear lamina degrade during apoptosis, and the consequences of degradation to both the nuclear envelope and the cell nucleus are severe. The NPC basket is indispensable for the nucleocytoplasmic cross-talk. It mediates export of ribonucleoproteins and other molecules from the nucleus to the cytosol, and this cross-talk is consequently impaired following NPC basket degradation. Nuclear lamina

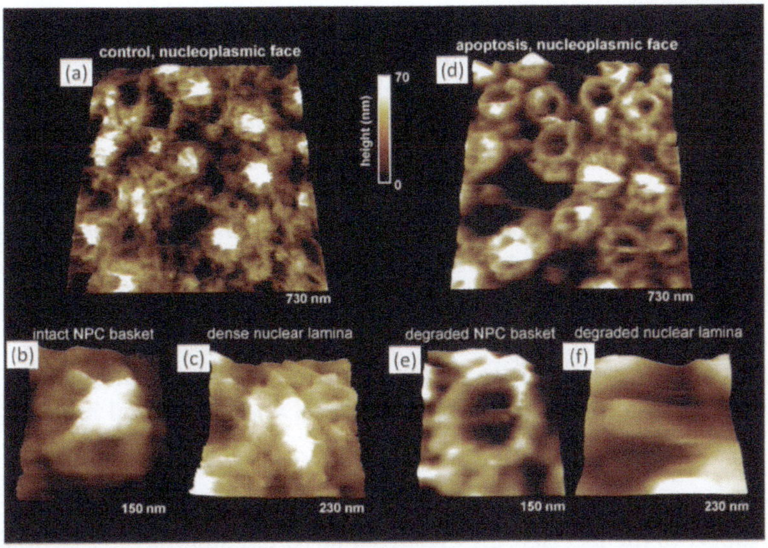

Figure 6.17. AFM images of the nucleoplasmic faces of control (left) versus apoptotic (right) nuclear envelopes of *Xenopus l.* oocyte.[42]

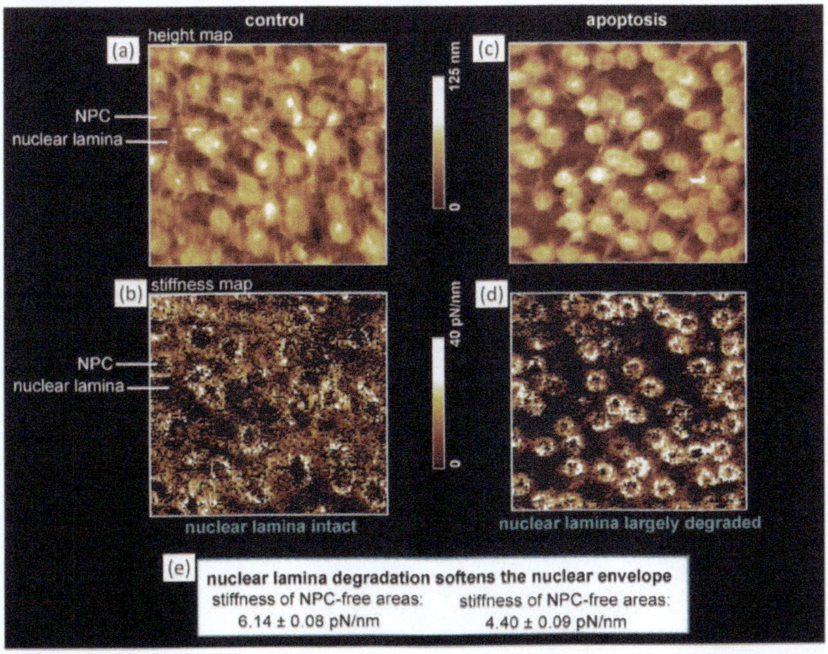

Figure 6.18. AFM-based nano-structural and indentation investigations of the nucleoplasmic faces of control (left) versus apoptotic (right) nuclear envelopes of *Xenopus l.* oocyte.[42]

degradation causes further serious damage to both the nuclear envelope and the cell nucleus. The nuclear lamina confers crucial structural and mechanical stability to the whole nucleus and is directly involved in the regulation of gene expression. Mutations in the nuclear lamina are known to lead to severe diseases (laminopathies).[45] All in all, degradation of the NPC basket and the nuclear lamina disrupts the essential cross-talk between the chromatin and the nuclear envelope as well as between the nucleo- and cytoplasma. Nuclear lamina degradation also leads to nuclear envelope softening, which ultimately destabilises the cell nucleus (Fig. 6.18).

6.3.4.2 The cytoplasmic face of the doomed nuclear envelope is deprived of its prominent structural and functional features, the NPC filaments

As seen in Fig. 6.19, the cytoplasmic filaments of NPC degrade during apoptosis. With respect to the fact that the cytolplasmic NPC filaments are essential for import of proteins from the cytosol into the nucleus, the consequence of their degradation is impaired nucleocytoplasmic cross-talk and a loss of nuclear import selectivity. This in turn promotes the nuclear access of generally excluded cytosolic apoptotic factors.

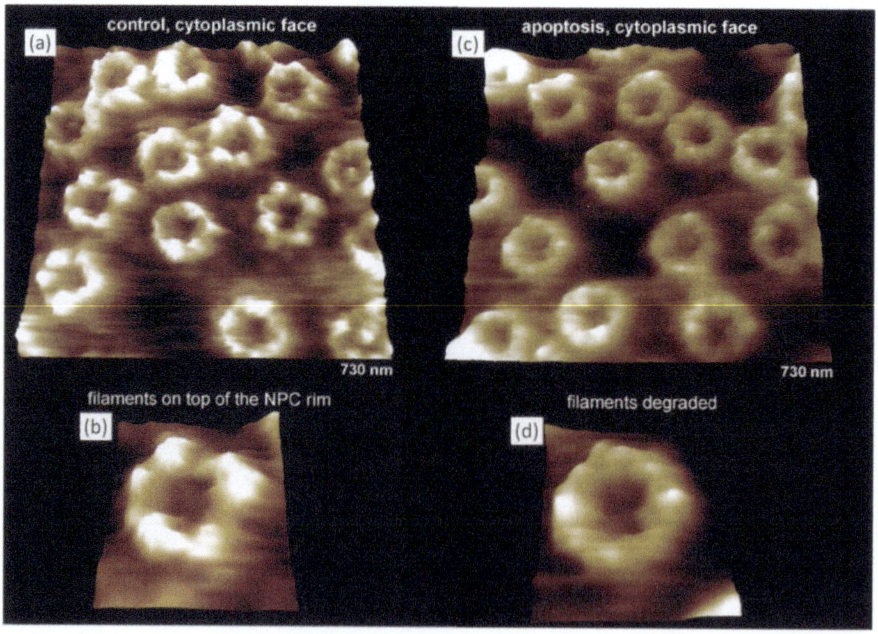

Figure 6.19. AFM images of the cytoplasmic faces of control (left) versus apoptotic (right) nuclear envelopes of *Xenopus l.* oocyte.[42]

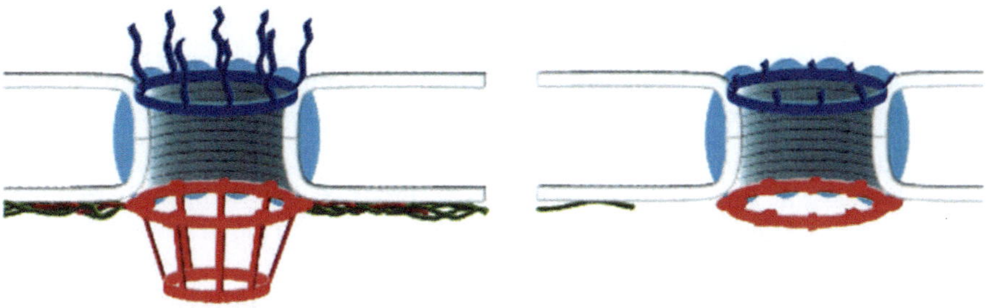

Figure 6.20. Schematic model of nuclear pore structural disruption during apoptosis.

All in all, apoptosis requires a remodelling of structure and mechanics of both nuclear envelope faces to bring about nuclear collapse (Fig. 6.20). Degradation of the cytoplasmic NPC filaments as well as the nuclear basket deprives the NPC of its transport selectivity and thus leads to disruption of selective nucleocytoplasmic cross-talk. This in turn promotes the exchange of apoptotic factors between the cytosol and the nucleus. Simultaneous degradation of the nuclear lamina cuts off the cross-talk between the chromatin and the nuclear envelope and leads to a destabilisation of the cell nucleus, ultimately promoting nuclear collapse.

References

1. Guggino, W. B., and Stanton, B. A. (2006) New insights into cystic fibrosis: molecular switches that regulate CFTR, *Nat. Rev. Mol. Cell Biol.*, **7**, 426–436.
2. Wine, J. J. (2003) Rules of conduct for the cystic fibrosis anion channel, *Nat. Med.*, **9**, 827–828.
3. Schillers, H., Danker, T., Schnittler, H. J., Lang, F., and Oberleithner, H. (2000) Plasma membrane plasticity of *Xenopus laevis* oocyte imaged with atomic force microscopy, *Cell Physiol. Biochem.*, **10**, 99–107.
4. Schillers, H., Danker, T., Madeja, M., and Oberleithner, H. (2001) Plasma membrane protein clusters appear in CFTR-expressing *Xenopus laevis* oocytes after cAMP stimulation, *J. Membr. Biol.*, **180**, 205–212.
5. Sprague, R. S., Ellsworth, M. L., Stephenson, A. H., Kleinhenz, M. E., and Lonigro, A. J. (1998) Deformation-induced ATP release from red blood cells requires CFTR activity, *Am. J. Physiol.*, **275**, 1726–1732.
6. Sterling, K. M., Jr., Shah, S., Kim, R. J., Johnston, N. I., Salikhova, A. Y., and Abraham, E. H. (2004) Cystic fibrosis transmembrane conductance regulator in human and mouse red blood cell membranes and its interaction with ecto-apyrase, *J. Cell Biochem.*, **91**, 1174–1182.

7. Verloo, P., Kocken, C. H., Van der, W. A., Tilly, B. C., Hogema, B. M., Sinaasappel, M., Thomas, A. W., and De Jonge, H. R. (2004) Plasmodium falciparum-activated chloride channels are defective in erythrocytes from cystic fibrosis patients, *J. Biol. Chem.*, **279**, 10316–10322.
8. Stumpf, A., Wenners-Epping, K., Walte, M., Lange, T., Koch, H. G., Haberle, J., Dubbers, A., Falk, S., Kiesel, L., Nikova, D., Bruns, R., Bertram, H., Oberleithner, H., and Schillers, H. (2006) Physiological concept for a blood based CFTR test, *Cell Physiol. Biochem.*, **17**, 29–36.
9. Swihart, A. H., Mikrut, J. M., Ketterson, J. B., and Macdonald, R. C. (2001) Atomic force microscopy of the erythrocyte membrane skeleton, *J. Microsc.*, **204**, 212–225.
10. Lange, T., Jungmann, P., Haberle, J., Falk, S., Duebbers, A., Bruns, R., Ebner, A., Hinterdorfer, P., Oberleithner, H., and Schillers, H. (2006) Reduced number of CFTR molecules in erythrocyte plasma membrane of cystic fibrosis patients, *Mol. Membr. Biol.*, **23**, 317–323.
11. Haws, C., Finkbeiner, W. E., Widdicombe, J. H., and Wine, J. J. (1994) CFTR in Calu-3 human airway cells: channel properties and role in cAMP-activated Cl-conductance, *Am. J. Physiol.*, **266**, 502–512.
12. Stroh, C., Wang, H., Bash, R., Ashcroft, B., Nelson, J., Gruber, H., Lohr, D., Lindsay, S. M., and Hinterdorfer, P. (2004) Single-molecule recognition imaging microscopy, *Proc. Natl. Acad. Sci. USA*, **101**, 12503–12507.
13. Ebner, A., Kienberger, F., Kada, G., Stroh, C. M., Geretschlager, M., Kamruzzahan, A. S., Wildling, L., Johnson, W. T., Ashcroft, B., Nelson, J., Lindsay, S. M., Gruber, H. J., and Hinterdorfer, P. (2005) Localization of single avidin-biotin interactions using simultaneous topography and molecular recognition imaging, *Chemphyschem*, **6**, 897–900.
14. Ebner, A., Nikova, D., Lange, T., Haeberle, J., Falk, S., Duebbers, A., Bruns, R., Oberleithner, H., and Schillers, H. (2008) Determination of CFTR densities in erythrocyte plasma membranes using recognition imaging, *Nanotechnology*, **19**, 384017.
15. Carl, P., and Schillers, H. (2008) Elasticity measurement of living cells with an atomic force microscope: data acquisition and processing, *Pflugers Arch.*, **457**, 551–559.
16. Iyer, S., Gaikwad, R. M., Subba-Rao, V., Woodworth, C. D., and Sokolov, I. (2009) Atomic force microscopy detects differences in the surface brush of normal and cancerous cells, *Nat. Nanotechnol.*, **4**, 389–393.
17. Adrogue, H. J., and Madias, N. E. (2007) Sodium and potassium in the pathogenesis of hypertension, *N. Engl. J. Med.*, **356**, 1966–1978.
18. Yu, H. C., Burrell, L. M., Black, M. J., Wu, L. L., Dilley, R. J., Cooper, M. E., and Johnston, C. I. (1998) Salt induces myocardial and renal fibrosis in normotensive and hypertensive rats, *Circulation*, **98**, 2621–2628.
19. Sanders, P. W. (2009) Vascular consequences of dietary salt intake, *Am. J. Physiol. Renal Physiol.*, **297**, 237–243.

20. Titze, J., Lang, R., Ilies, C., Schwind, K. H., Kirsch, K. A., Dietsch, P., Luft, F. C., and Hilgers, K. F. (2003) Osmotically inactive skin Na+ storage in rats, *Am. J. Physiol. Renal Physiol.*, **285**, 1108–1117.

21. Adams, J. M., Bardgett, M. E., and Stocker, S. D. (2009) Ventral lamina terminalis mediates enhanced cardiovascular responses of rostral ventrolateral medulla neurons during increased dietary salt, *Hypertension*, **54**, 308–314.

22. He, F. J., Markandu, N. D., Sagnella, G. A., de Wardener, H. E., and MacGregor, G. A. (2005) Plasma sodium: ignored and underestimated, *Hypertension*, **45**, 98–102.

23. Oberleithner, H., Riethmuller, C., Schillers, H., MacGregor, G. A., de Wardener, H. E., and Hausberg, M. (2007) Plasma sodium stiffens vascular endothelium and reduces nitric oxide release, *Proc. Natl. Acad. Sci USA*, **104**, 16281–16286.

24. He, F. J., de Wardener, H. E., and MacGregor, G. A. (2007) Salt intake and cardiovascular mortality, *Am. J. Med.*, **120**, e5.

25. Haddy, F. J., Vanhoutte, P. M., and Feletou, M. (2006) Role of potassium in regulating blood flow and blood pressure, *Am. J. Physiol. Regul. Integr. Comp. Physiol.*, **290**, 546–552.

26. Torres, S. J., Nowson, C. A., and Worsley, A. (2008) Dietary electrolytes are related to mood, *Br. J. Nutr.*, **100**, 1038–1045.

27. Mohr, M., Nordsborg, N., Nielsen, J. J., Pedersen, L. D., Fischer, C., Krustrup, P., and Bangsbo, J. (2004) Potassium kinetics in human muscle interstitium during repeated intense exercise in relation to fatigue, *Pflugers Arch.*, **448**, 452–456.

28. Kofuji, P., and Newman, E. A. (2004) Potassium buffering in the central nervous system, *Neuroscience*, **129**, 1045–1056.

29. Pesen, D., and Hoh, J. H. (2005) Micromechanical architecture of the endothelial cell cortex, *Biophys. J.*, **88**, 670–679.

30. Kasas, S., Wang, X., Hirling, H., Marsault, R., Huni, B., Yersin, A., Regazzi, R., Grenningloh, G., Riederer, B., Forro, L., Dietler, G., and Catsicas, S. (2005) Superficial and deep changes of cellular mechanical properties following cytoskeleton disassembly, *Cell Motil. Cytoskeleton*, **62**, 124–132.

31. Oberleithner, H., Callies, C., Kusche-Vihrog, K., Schillers, H., Shahin, V., Riethmuller, C., MacGregor, G. A., and de Wardener, H. E. (2009) Potassium softens vascular endothelium and increases nitric oxide release, *Proc. Natl. Acad. Sci USA*, **106**, 2829–2834.

32. Searles, C. D., Ide, L., Davis, M. E., Cai, H., and Weber, M. (2004) Actin cytoskeleton organization and posttranscriptional regulation of endothelial nitric oxide synthase during cell growth, *Circ. Res.*, **95**, 488–495.

33. Su, Y., Edwards-Bennett, S., Bubb, M. R., and Block, E. R. (2003) Regulation of endothelial nitric oxide synthase by the actin cytoskeleton, *Am. J. Physiol. Cell Physiol.*, **284**, 1542–1549.

34. Rizzo, V., McIntosh, D. P., Oh, P., and Schnitzer, J. E. (1998) In situ flow activates endothelial nitric oxide synthase in luminal caveolae of endothelium with rapid caveolin dissociation and calmodulin association, *J. Biol. Chem.*, **273**, 34724–34729.

35. Thompson, C. B. (1995) Apoptosis in the pathogenesis and treatment of disease, *Science*, **267**, 1456–1462.

36. Ashkenazi, A., and Dixit, V. M. (1998) Death receptors: signaling and modulation, *Science*, **281**, 1305–1308.

37. Taylor, R. C., Cullen, S. P., and Martin, S. J. (2008) Apoptosis: controlled demolition at the cellular level, *Nat. Rev. Mol. Cell Biol.*, **9**, 231–241.

38. Buendia, B., Courvalin, J. C., and Collas, P. (2001) Dynamics of the nuclear envelope at mitosis and during apoptosis, *Cell Mol. Life Sci.*, **58**, 1781–1789.

39. Wyllie, A. H., Kerr, J. F., and Currie, A. R. (1980) Cell death: the significance of apoptosis, *Int. Rev. Cytol.*, **68**, 251–306.

40. Stewart, C. L., Roux, K. J., and Burke, B. (2007) Blurring the boundary: the nuclear envelope extends its reach, *Science*, **318**, 1408–1412.

41. Binnig, G., Quate, C. F., and Gerber, C. (1986) Atomic force microscope, *Phys. Rev. Lett.*, **56**, 930–933.

42. Kramer, A., Liashkovich, I., Oberleithner, H., Ludwig, S., Mazur, I., and Shahin, V. (2008) Apoptosis leads to a degradation of vital components of active nuclear transport and a dissociation of the nuclear lamina, *Proc. Natl. Acad. Sci. USA*, **105**, 11236–11241.

43. Shahin, V., Ludwig, Y., Schafer, C., Nikova, D., and Oberleithner, H. (2005) Glucocorticoids remodel nuclear envelope structure and permeability, *J. Cell Sci.*, **118**, 2881–2889.

44. Shahin, V., Hafezi, W., Oberleithner, H., Ludwig, Y., Windoffer, B., Schillers, H., and Kuhn, J. E. (2006) The genome of HSV-1 translocates through the nuclear pore as a condensed rod-like structure. *J. Cell Sci.*, **119**, 23–30.

45. Gruenbaum, Y., Margalit, A., Goldman, R. D., Shumaker, D. K., and Wilson, K. L. (2005) The nuclear lamina comes of age, *Nat. Rev. Mol. Cell Biol.*, **6**, 21–31.

Chapter 7

TOPOGRAPHY AND RECOGNITION IMAGING OF CELLS

Lilia Chtcheglova, Linda Wildling and Peter Hinterdorfer
University of Linz, Altenbergerstrasse 69, A-4040 Linz, Austria
peter.hinterdorfer@jku.at

7.1 INTRODUCTION

Determining the distribution of specific binding sites on biological samples with high spatial accuracy (in the order of several nanometres) is an important challenge in many fields of biological science.[1] TREC (for "simultaneous topography and recognition imaging") is a recently developed atomic force microscopy (AFM) imaging technique, which has become an indispensable tool for high-resolution receptor mapping. So far, this method has been successfully applied to model protein systems, such as avidin–biotin,[2,3] to histones within remodelled chromatin structures,[4] to protein lattices[5] and to isolated red blood cell membranes.[6]

The TREC technique was also applied to cells, and this chapter gives an overview of the most recent TREC applications for cellular systems. High-resolution AFM imaging is combined with single-molecule interaction measurements.

7.2 AFM TIP CHEMISTRY VIA POLYETHYLENE GLYCOL LINKERS

Both molecular recognition force spectroscopy and TREC measurements require the AFM tip to be transformed into a biospecific molecular sensor by attaching a ligand onto the tip. One of the most elegant ways is to anchor a few

Life at the Nanoscale: Atomic Force Microscopy of Live Cells
Edited by Yves Dufrêne

www.panstanford.com

ligands onto the AFM tip via a long, flexible tether, such as polyethylene glycol (PEG) chains.[7] The immobilization of the sensor molecule via the flexible linker gives the ligand the freedom to adopt the correct orientation, and this indeed increases the chances of receptor detection on the surface.

The attachment of ligands onto AFM tips via PEG chains is usually performed in three steps as illustrated in Fig. 7.1. Firstly, amino ($-NH_2$) groups are produced on the tip either by the esterification of the superficial silicon oxide layer with ethanolamine hydrochloride in dimethylsulfoxide[8] (Fig. 7.1, step Ia) or gas phase silanization with 3-aminopropyltriethoxysilane similar to the procedure described by Lyubchenko and co-workers[9] (Fig. 7.1, step Ib). It has proved critical to use methods that do not significantly increase roughness and/or stickiness of the tip surface. In the second step, heterobifunctional PEG chains are attached by one end to the amino group

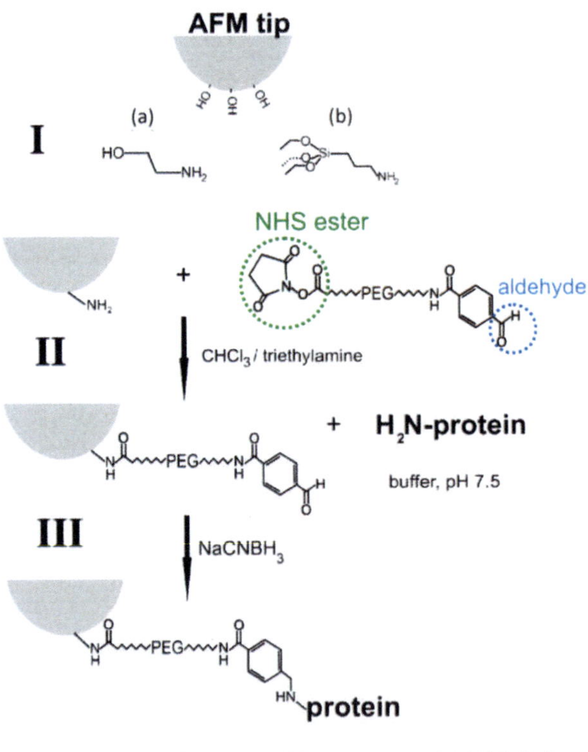

Figure 7.1. AFM tip functionalization with proteins via PEG linkers. (I) Amino-functionalization of silicon nitride tips either via (a) esterification with ethanolamine or (b) silanization with 3-aminopropyltriethoxysilane (APTES) from the gas phase. (II) Use of heterobifunctional NHS-PEG-aldehyde linker for flexible attachment of underivatized protein onto the AFM tip. (III) The C=N double bond is usually fixed by a reaction with sodium cyanoborohydride ($NaCNBH_3$).

on the tip (Fig. 7.1, step II). This is always done by amide bond formation, for which, all PEG linkers possess an activated carboxy (-COOH) group, in the form of an *N*-hydroxysuccinimide ester (NHS ester). The PEG solution is normally adjusted to ensure low density of cross-linkers on the Si_3N_4 tip surface, and therefore single-molecule detection by the tip is enabled. In the last step, a ligand molecule is coupled to another free functional end of the PEG linker as shown in Fig. 7.1, step III. One of the most suitable PEG linkers used is an aldehyde linker[10] (abbreviated as NHS-PEG-aldehyde), which can link underivatized antibodies and other proteins via their lysine residues, of which 80–90 are found per antibody molecule. Finally, functionalized tips can be stored in PBS at 4 °C for several weeks until use.

7.3 OPERATING PRINCIPLES OF TOPOGRAPHY AND RECOGNITION IMAGING

In contrast to common recognition imaging based on force spectroscopy a recently developed AFM imaging technique termed simultaneous topography and recognition imaging (named TREC) overcomes some of the limitations regarding lateral resolution and imaging speed by using dynamic force microscopy with a functionalized sensor tip that is oscillated during scanning across the surface.

The operating principle of TREC is based on MAC (magnetic alternating current) mode AFM,[11] where a magnetically coated cantilever is oscillated through an alternating magnetic field. The tip functionalized with a ligand molecule via a short (~8–10 nm) flexible PEG linker (tip functionalization procedure is described earlier) is oscillated close to its resonance frequency while scanning over the surface. When such a tip-tethered ligand binds to its receptor on the sample surface (i.e., when specific molecular recognition occurs), the PEG linker will be stretched during upward movement of the cantilever. The resulting loss in energy will in turn cause the top peaks of the oscillations to be lowered. The ligand–receptor-binding events thus become visible because of a reduction in the oscillation amplitude, as a result of specific recognition during the lateral scan. In contrast to "normal" MAC mode imaging, TREC uses the lower part of the oscillation to drive a feedback loop for obtaining the topography image, whereas the upper part of the oscillation is used for the generation of the recognition image. Moreover, using half-amplitude feedback allows accurate determination of the surface topography.[12] To provide more details, the time-resolved deflection signal of the oscillating cantilever is low-pass filtered to remove the thermal noise and the DC (direct current) is offset levelled and amplified before splitting into the lower (U_{down}) and upper (U_{up}) parts of the oscillations. The signal passes a trigger threshold on each path, and the lower peak of each oscillation period is determined by means of sample and hold analysis. Subsequent peaks form a staircase function, which is then filtered and fed into the AFM controller, where U_{down} drives the feedback loop to record the topography image and U_{up} provides the information to establish the corresponding recognition image. Moreover, the utilization of cantilevers with low Q factor (~1 in liquid) in combination with a proper chosen driving frequency and amplitude regime enables that both types of information are unrelated.[4,12] Generally, the ideal amplitude regime for the observation of recognition events differs from one functionalized cantilever to another. It depends on the length of the linker molecule, on the exact location of the linker molecule on the tip apex and on the size of the attached molecule. It typically lies in the range of 10–20 nm.

To summarize, the topography and recognition images can be simultaneously and independently obtained using a specially designed electronic circuit (PicoTREC, Agilent Technologies, Chandler, Arizona), which splits the cantilever oscillation amplitude into the lower and upper parts (with respect to the cantilever resting position) and contains specific information about topography and recognition, respectively (Fig. 7.2).

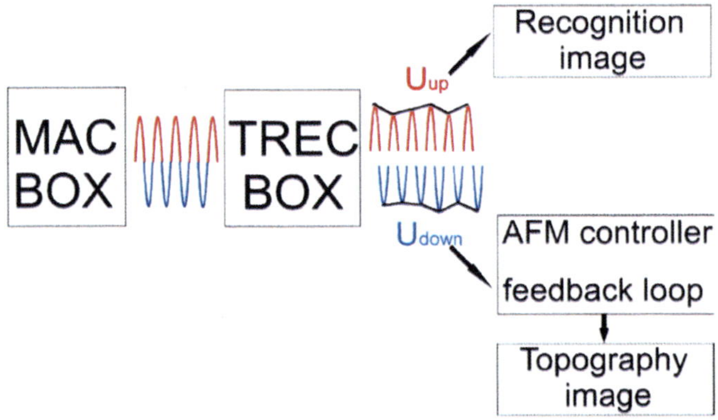

Figure 7.2. Schematic of TREC functioning. The raw cantilever deflection signal is fed into the TREC box, where the maxima (U_{up}) and the minima (U_{down}) of each oscillation period are used for the recognition and the topography image, respectively.

7.4 APPLICATIONS OF TREC TO CELLS

Mapping receptor-binding sites on cellular surfaces is a challenging task in molecular cell biology. This information can usually be obtained from the extensive exploitation of common optical techniques such as immunostaining (or immunocytochemistry) or even by somewhat sophisticated techniques such as single-molecule optical microscopy,[13] near-field scanning optical microscopy (for more details see Chapter 9)[14,15] or stimulated emission depletion microscopy.[16] The lateral resolution in these studies ranged from a few tens of nanometres[13–16] to ~200 nm because of the diffraction phenomenon also known as Abbe limits. In addition, no information about topography can be attained. At present, AFM offers an exceptional solution for obtaining topography images with nanoscale resolution and single molecular interaction forces on different biological specimens such as proteins, DNA, membranes, cells, etc., under physiological or near-physiological conditions and without the need for scrupulous sample preparation or labelling.[17] Hence, the spatial nano-mapping of molecular recognition sites can be obtained by performing AFM adhesion force mapping using the force–volume technique, which represents the molecular recognition imaging using force spectroscopy[18–21] (for more details see Chapter 12). On the other hand, dynamic recognition mapping (TREC) is faster and enables better lateral resolution than adhesion force mapping.[1,2,4] Because of the continuous progress in the technical aspects of the AFM and "smart" tip functionalization procedures, the investigations of receptor–ligand interactions on living cells at the single-molecule level have become achievable. Because cells represent systems of more complex composition, organization and processing in space and time than proteins, the next goal is the application of TREC to cellular membranes that contain different functional domains enriched in sphingolipids, cholesterol and specific transmembrane proteins.[22]

7.4.1 Nano-Mapping of Vascular Endothelial-Cadherin on Endothelial Cells

The first TREC studies on cells were conducted on microvascular endothelial cells from mouse myocardium (MyEnd) to locally identify vascular endothelial (VE)-cadherin binding sites and correlate their position with membrane topographical features (Fig. 7.3).[23] VE-cadherin belongs to the widespread and functionally important family of calcium-dependent cell adhesion molecules, cadherins (this name arises from the approximate contraction of "Calcium dependent *ADHER*ent prote*IN*"), which are single-pass transmembrane glycoproteins known to be crucial for calcium-dependent, homotypic (or homophilic) cell–cell adhesion[24] and are also essential for the morphogenesis of tissues and the maintenance of tissue function. In the case of vascular endothelium, the adhesion between cells has to be strong enough to resist the hydrodynamic forces created by blood flow (shear stress of up to 10 Pa) or blood pressure (wall distension). VE-cadherin is strictly located at intercellular junctions of essentially all types of endothelium. This molecule not only regulates adhesive intercellular endothelial junctions (e.g., adherent junctions[25] known to be primarily responsible for mechanical linkage between cells), in which VE-cadherins are clustered and linked through their cytoplasmic domain to the actin-based cytoskeleton,[26,27] but also plays an essential role in the remodelling, gating and maturation of vascular vessels.[28,29] VE-cadherin belongs to the classical type II cadherin subgroup and shares the common structure with other classical cadherins. It consists of extracellular ectodomain (EC) containing five similar repeated subdomains (EC1–EC5), a single-pass transmembrane domain and a highly conserved cytoplasmic segment, through which cadherins are connected inside the cell to a cluster of catenins and thus linked to the actin microfilaments (Fig. 7.3a). This cytoskeletal anchorage is thought to be important for strengthening the cadherin-mediated adhesion.[27] Homophilic cell–cell adhesion is mediated by the cadherin extracellular domains,[30] which enable association in parallel lateral *cis*-dimers in physiological Ca^{2+} concentration (~1.8 mM)[31–36] as schematically represented in Fig. 7.3b. The parallel *cis*-dimer is thought to be the basic structural functional unit for promoting the homophilic bond between cells,[31,33,36,37] and these cadherin dimers are assumed to contain one or two binding sites[31,33,34,38] to form *trans*-interacting antiparallel tetramers or adhesion dimers[34] (Fig. 7.3c).

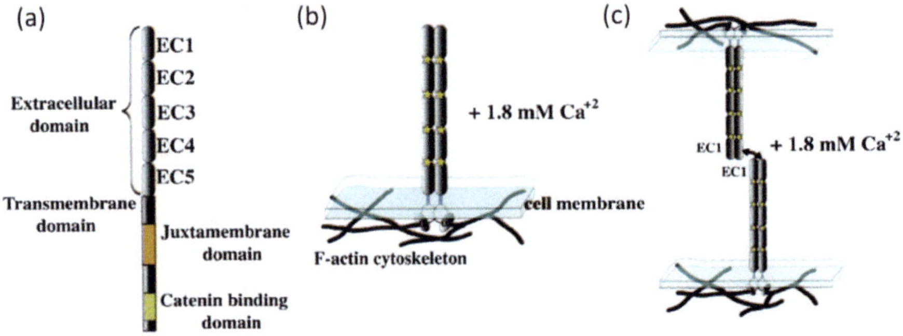

Figure 7.3. (a) Schematic of VE-cadherin domain organization. As with other cadherins, VE-cadherin is characterized by the presence of five sequence repeats of ~110 amino acids, which form folded Greek-key topology extracellular (EC) domains. The connections between successive domains are rigidified by conserved Ca^{2+}-binding sites representing the most significant feature of the repeat sequences. The cytoplasmic domain of VE-cadherin includes the "juxtamembrane region" that binds p120-catenin ($p120^{ctn}$) and the "catenin binding domain" that interacts with *β*-catenin and plakoglobin. (b) In the presence of extracellular calcium (1.8 mM), the rigid cadherin extracellular domains (shown as grey rods) enable association in functional *cis*-dimers. The calcium-binding sites between extracellular domains are shown as yellow stars. (c) These active cadherin *cis*-dimers promote a homophilic bond between adjacent cells by forming a *trans*-adhesion dimer.

To overcome issues associated with cell elasticity and lateral diffusion of receptors, a fixation procedure can be applied similar to immunochemistry experiences. The fixation procedure usually makes the soft biological objects stiffer, and as a consequence, it generally gives access to high lateral resolution in AFM images as was observed with proteins (GroES).[39] However, the common fixation of cells in buffer solution at room temperature causes the smoothing of the cell surface with the loss of most filamentous features, which were seen in AFM pictures of living cells.[40] The nucleus also most probably becomes visible because of the membrane collapse during dehydration caused by the fixation procedure. When the unpurified solution of glutaraldehyde is used, the undesirable formation of globular large features on the cell surface (e.g., polymers of glutaraldehyde) can also be detected. A method has been found to gently fix the cells with a solution of glutaraldehyde containing monomers (EM grade) similar to the procedure described by Oberleithner and co-workers.[41] The prepared stock solution of glutaraldehyde (~200 µL, 5% in Hank's balanced salt solution [HBSS]) was added and gently mixed with the culture medium (~2 mL), and the cells were then incubated at 37 °C for 1–2 hours. Such a method is likely to prevent unexpected osmotic and temperature changes in the cell culture medium. As a result, the cell volume[41] athe filamentous structures of cytoskeleton (Fig. 7.4a) are mostly preserved, which makes further AFM investigations possible at a subcellular level.

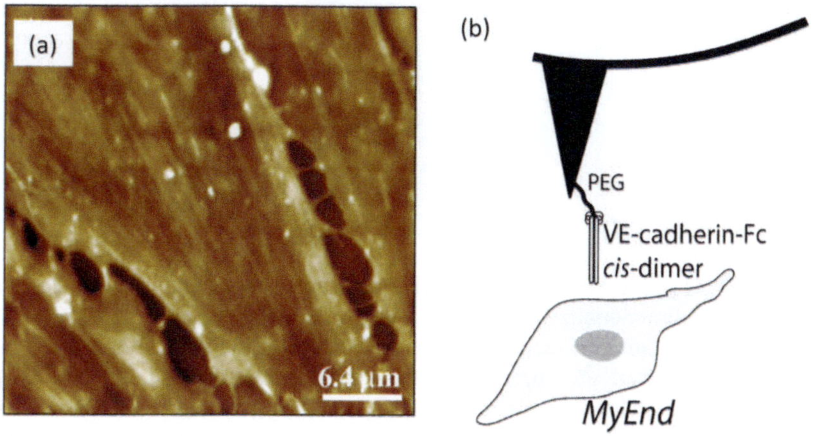

Figure 7.4. (a) AFM topography image of gently fixed MyEnd cells. Colour scale (dark brown to white) is 0–400 nm. (b) Schematic of dynamic recognition imaging to visualize VE-cadherin on MyEnd cell surface.

AFM functional imaging was performed with magnetically coated AFM tips that were decorated with a recombinant VE-cadherin-Fc *cis*-dimer via PEG linker (for more details, refer to section 7.2.) (Fig. 7.4b). Since VE-cadherin is cell specific and located at intercellular junctions,[25,42] TREC images were collected on the contact region between adjacent cells in calcium buffer solution (i.e., HBSS containing 1.8 mM Ca^{2+}) at ambient temperature. The topography of a scanned cell surface area shows a complex picture of linear and branched filaments, likely representing filaments of the peripheral actin belt, with some globular features (Fig. 7.5a). The oscillation amplitude was accurately adjusted to obtain the proper recognition with high efficiencies and repeatability (>90%). Accordingly, a recognition signal corresponds to the amplitude reduction due to the specific VE-cadherin *trans*-interaction (seen as dark red spots in recognition image). These spots reflect microdomains

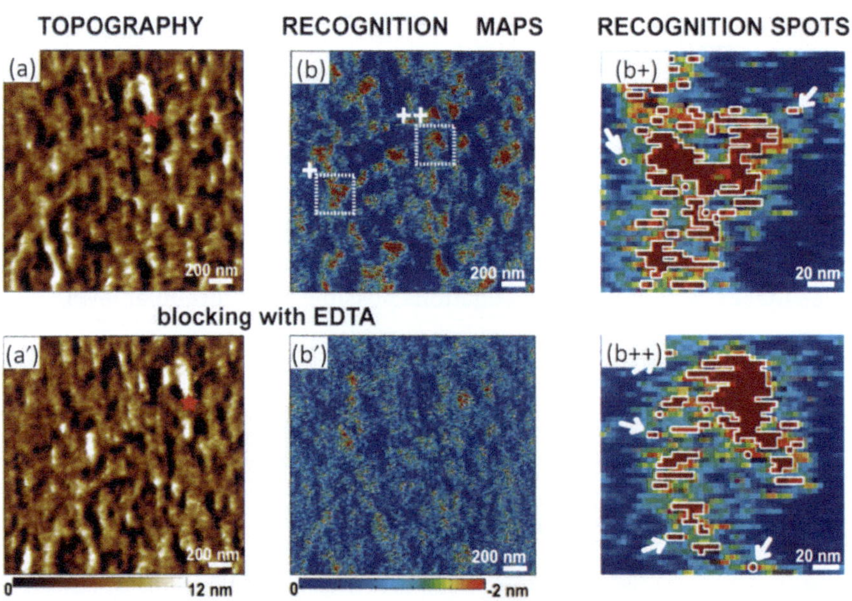

Figure 7.5. Mapping VE-cadherin on the vascular endothelial cell surface with VE-cadherin-Fc-functionalized tip. (a, a′) Topography images simultaneously recorded with recognition maps b and b′, respectively. Red stars indicate the AFM scanner lateral drift of ~5 nm/min. (b) Recognition image of VE-cadherin domains representing an amplitude reduction due to a specific binding between VE-cadherin on the AFM tip and VE-cadherin molecules on the cell surface. (b′) The recognition clusters practically disappeared in Ca^{2+}-free conditions, since the active VE-cadherin *cis*-dimers on the AFM tip dissociated in inactive monomers, thereby abolishing specific VE-cadherin *trans*-interaction. After blocking with 5 mM EDTA, topography (a′) remains unchanged, indicating that the blocking does not affect membrane topography. (b+, b++) Examples of recognition spots taken from b. Recognition areas are depicted by threshold analysis (threshold = −1.7 nm) and bordered by white lines. Single VE-cadherin *cis*-dimers are clearly seen (arrows).[23]

with dimensions from ~10 to ~100 nm, which were non-uniformly distributed on the cellular surface (Fig. 7.5b). The recognition efficiency was high and remained so during subsequent rescans. The specificity of binding was confirmed by the addition of 5 mM EDTA (Ca^{2+} free conditions) leading to the disappearance of almost all binding events in the recognition image (Fig. 7.5b'), whereas no change in the topography image was detected (Fig. 7.5a'). Figures 7.5b^{+} and b^{++} illustrate a closer look at the recognition spots. "Hot" spots consisting of one to two large domains (50–80 nm) could clearly be seen surrounded by smaller domains (10–20 nm) or even single-molecule spots (typically 1–4 pixels long, 1 pixel ~4 nm) by taking into account the size of the VE-cadherin *cis*-dimer (diameter 3 nm) and the free orientation of PEG linker leading to the specific binding event before/after the binding site position. More than 600 single specific events were recognized, and around 6000 active *cis*-dimers were calculated over the scanned area (4 μm^2).

The receptor-binding sites can properly be assigned to the topographical features for heterogeneous biological samples such as chromatin.[4] Figure 7.6a illustrates the superimposition of the recognition map onto the corresponding topographical image. This procedure allows revealing the locations of receptors in the topographical image with high lateral resolution and high efficiency. Interestingly, only a few VE-cadherin domains were found directly on the top of filaments, whereas most domains were located near and between filaments. The last observation indicates that at this stage of cell maturation (day 1 or 2 after seeding), the clustering of VE-cadherin is incomplete.

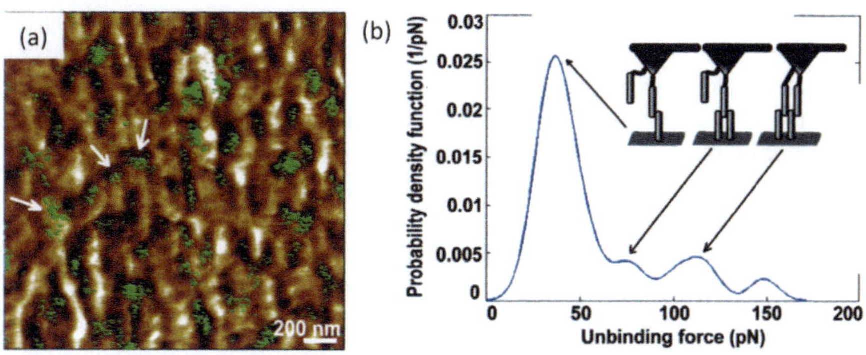

Figure 7.6. (a) Overlay of recognition map of VE-cadherin (in green) onto the corresponding topography image. A few VE-cadherin domains are situated directly on the top of filaments (arrows). Colour scale (dark brown to white) is 0–12 nm. (b) Force distribution recorded on gently fixed MyEnd surface with VE-cadherin-Fc-coated tip in Ca^{2+}-rich conditions.[23]

In addition, standard single-molecule force spectroscopy measurements were conducted on the same scanned surface area with the AFM tip functionalized with VE-cadherin-Fc. Force curves were accumulated ($n \sim 500$) before and after the blocking experiment. The force distribution of cadherin–cadherin dissociation illustrates multiple force peaks of one-, two- and threefold binding with a force quantum of ~40 pN (Fig. 7.6b). This characteristic force fingerprint is very similar to an isolated VE-cadherin system.[35] The specific unbinding events were abolished in free Ca^{2+} conditions (addition of 5 mM EDTA) accompanied by a reduction in binding probability (from ~30% to ~1%). Therefore, force spectroscopy data explicitly confirm that specific domains contain active VE-cadherin *cis*-dimers.

7.4.2 Localization of Ergtoxin-1 Receptors on the Voltage-Sensing Domain of hERG K^+ Channel

TREC and single-molecule force spectroscopy have been recently introduced as a novel way to investigate the properties of voltage-gated channels in cells.[43] Usually, the information about the structure and function of different voltage-gated channels in living cells was gained from patch-clamp investigations. The single-molecule AFM techniques have been exploited to detect a new receptor site(s) for ergtoxin-1 (ErgTx1) in the voltage-sensing domain of the human ether-à-go-go-related gene (hERG) potassium (K^+) channel,[44] with the aim of expanding an understanding about the microscopic mechanism of the hERG K^+ channel blockade with ErgTx1. hERG K^+ channel plays an important role in the heart,[44] peripheral sympathetic ganglia, brain and tumour cells. hERG channels are largely involved in myocardial repolarization[45,46] and are associated with both the inherited and the acquired (drug induced) long QT syndromes that may be responsible for fatal cardiac arrhythmias. ErgTx1 belongs to scorpion toxins,[43] which are K^+ channel blockers, and which binds to the hERG channel with 1:1 stoichiometry and high affinity (Kd ~ 10 nM). Peptide toxins usually block the pore of the channel, either directly by occupying the selective filter or by binding to an electrostatic ring surrounding the pore. Previously, it has been identified that ErgTx1 binds to the outer vestibule of the hERG channel.[47] Nevertheless, a characteristic feature of the action of ErgTx1 on hERG is an incomplete block of macroscopic current events at concentrations orders of magnitude higher than the K_d value. Such results suggest that ErgTx1 is a gating modifier rather than a pore blocker.[48,49] In addition, it binds near the pore and cannot fully occlude the permeation pathway.[50,51] The binding site for ErgTx1 on hERG is thought to be formed, at least in part, by the extracellular linker between S5 transmembrane helix and the pore helix (S5P linker),[48] which is critically involved in voltage-dependent inactivation in hERG.[52]

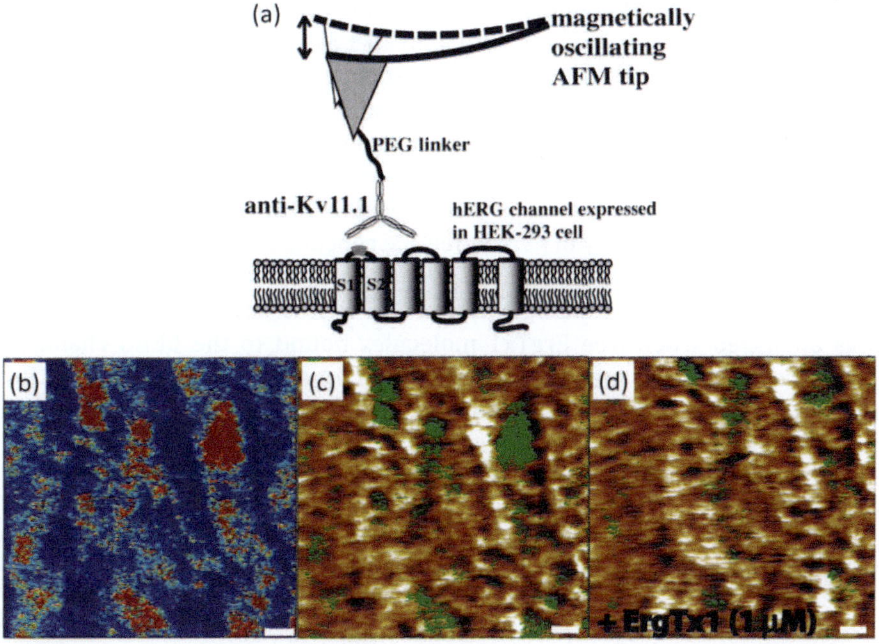

Figure 7.7. Nano-mapping of hERG K⁺ channels on hERG HEK-293 cell surface. (a) Schematic representation of recognition imaging to detect hERG K⁺ channels (here binding sites of extracellular epitope [*shown in light grey*] situated between S1 and S2 domains of hERG subunit). (b) Recognition map obtained with anti-Kv11.1-coated tip. (c) Superimposition of recognition map (*in green*) onto the corresponding topography image. (d) Recognition clusters disappeared only in part in the presence of high concentrations of ErgTx1 (~1 μM), whereas no visible effect was obtained at lower concentrations of ErgTx1 (~400 nM) (data not shown). Scale bars on all images are 170 nm.[43]

Therefore, AFM functional dynamic imaging (TREC) has been applied to test the presence of extracellular binding sites of hERG K^+ channels on gently fixed HEK-293 cells expressing hERG channels. Measurements were started by scanning the whole cell surface with subsequent zooming into small areas of ~4 μm^2. TREC images were acquired with magnetically coated AFM tips (MAC tips) which were functionalized with an antibody anti-Kv11.1 (against epitope tags present on the hERG subunits) via PEG linker as previously mentioned (Fig. 7.7a). All images were taken in HBSS (1.8 mM Ca^{2+}) at 25 °C. The oscillation amplitude was adjusted to be less than the extended PEG linker to provide the proper recognition image with high efficiencies and repeatability (>90%). Accordingly, the recognition map represents an amplitude reduction due to specific binding between anti-Kv11.1 on the tip and epitope tags on the cell surface ("dark" spots in Fig. 7.7b). Figure 7.7c illustrates non-uniform distribution of microdomains (in green) on the cellular surface with domain sizes from ~30 up to ~350 nm, with a mean ± SD of 99 ± 81 nm (n = 25) on the long domain axis. During several subsequent rescans, recognition maps of hERG channels remained unchanged. Next, ErgTx1 was very slowly (~50 μL/min) injected in the fluid cell while scanning the same sample. After the first and second injection of ErgTx1 (concentration of ~400 nM), no visual changes in the recognition maps were observed. However, the recognition clusters disappeared, only in part, when the concentration of ErgTx1 reached 1 μM (Fig. 7.7d), whereas no change in the topography image was observed (Fig. 7.7d). The specific binding between anti-Kv11.1 and the cellular surface was abolished when free ErgTx1 molecules bound to the hERG channels and thus blocked the antibody access to interact with epitope tags on hERG subunits. The topography of a scanned cell surface area showed a complex picture of linear and branched filamentous structures with some globular features. Most domains were found to be located near and between filaments (Fig. 7.7b). TREC results suggest that ErgTx1 does not only interact with the extracellular surface of the pore domain (S5–S6) but might also interact with the voltage-sensing domains (S1–S4) of the hERG K^+ channel.

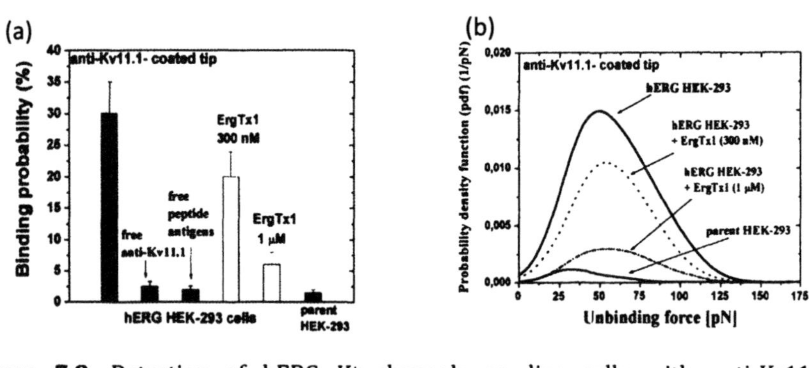

Figure 7.8. Detection of hERG K^+ channels on live cells with anti-Kv11.1-functionalized AFM tip. (a) Quantitative comparison of binding probabilities obtained on living HEK-293 cells expressing hERG K^+ channels in the absence (*left light gray*) and presence of either free anti-Kv11.1 antibodies (*middle light gray*) or free antigen peptides (*right light gray*); binding probably on parent HEK-293 cells is shown in *black*. Consecutive injection of ErgTx-1 (300 nM, 1 μM) reduces the binding probability (*white*). Values are mean ± SEM, n = 2000–4000. (b) Force distributions (pdf) observed in the absence of ErgTx1 (*solid blue line*) and in the presence of ErgTx1 (*dot* [300 nM] and *short dashed-dot lines* [1 μM]). Areas under the curves are scaled to the corresponding binding probabilities.

To extend TREC measurements, AFM force–distance cycles with a tip carrying an epitope-specific antibody (anti-Kv11.1) were collected on living and gently fixed hERG HEK-293 cells. Both studies conducted on living and fixed cells lead to similar results (force distributions and binding probabilities). The data obtained with living cells are presented in Fig. 7.8. The anti-K_v11.1 (hERG)-extracellular antibody is known to bind to the voltage-sensor domain (S1–S2 region) of HERG K^+ channel (Fig. 7.7a). The specific binding of the antibody to the extracellular part of hERG channel was characterized by a unique unbinding force. To confirm the specificity of this binding, blocking experiments were carried out by injecting either free antibodies or free peptide antigen. In both cases, almost no unbinding events were observed. Binding probabilities (probability to record an unbinding event in force–distance cycles) from several experiments were also calculated (Fig. 7.8a). The binding probability of ~30% was calculated for the interaction between anti-Kv11.1-extracellular antibody and hERG HEK-293 cells. When free anti-Kv11.1 antibodies or free peptide antigens were present in solution, the binding probability drastically decreased to the level of ~2% (Fig. 7.8a). By constructing an empirical probability density function of the unbinding forces (Fig. 7.8b), the maximum of the distribution was found to be 45 ± 9 pN. Another indicator of the specificity, a very low binding probability (~1.5%) with a force peak of ~25 pN (Fig. 7.8b), was found for the parent HEK-293 cells not expressing hERG K^+ channels. These results illustrate that the extracellular part of hERG K^+ channel expressed in living cells can be specifically detected at the molecular level by using epitope-specific antibodies.

The possible effects of ErgTx1 on antibody binding were further investigated. Force curves were accumulated before and after ErgTx1 multiple injections in the same scan area with the same functionalized tip. In the presence of ErgTx1 at different concentrations, the peak force for force distributions (Fig. 7.8b) remains at the same position, whereas the binding probability between antibody and living hERG HEK 293 cells dramatically decreased following multiple ErgTx1 injections (Fig. 7.8). These results provide support about a possible new binding site of ErgTx1 in the voltage-sensor domain of hERG K^+ channel.

Thus, it has been demonstrated that the combination of dynamic molecular recognition imaging (TREC) with single-molecule force spectroscopy is a suitable method to obtain information about the structure and function of hERG K^+ channels. Both techniques exploit AFM tips with a very low surface density of ligands (~400 molecules per μm^2) and thus allow the detection of single molecular events. Functionalization of AFM tips with anti-Kv11.1 (hERG)-extracellular antibody enabled them to detect binding sites of hERG on the cell surface expressed hERG channels. The main outcome of this study reveals that the voltage-sensing domain (S1–S4) of hERG K^+ channel might be one of the binding sites of ErgTx1.

In summary, this chapter illustrates the great potential of TREC for the investigation and localization of membrane proteins on cell surfaces with several piconewton force resolution and a few nanometre positional accuracy. In the future, the technique should be applicable to a wide variety of cell types, including not only animal cells but also plant cells and microorganisms.

References

1. Hinterdorfer, P., and Y. F. Dufrêne, Y. F. (2006) Detection and localization of single molecular recognition events using atomic force microscopy, *Nat. Methods*, **3**, 347–355.
2. Stroh, C. M., Ebner, A., Geretschläger, M., Freudenthaler G., Kienberger, F., Kamruzzahan, A. S. M., Smith-Gill, S. J., Gruber, H. J., and Hinterdorfer, P. (2004) Simultaneous topography and recognition imaging using force microscopy, *Biophys. J.*, **87**, 1981–1990.
3. Ebner, A., Kienberger, F., Kada, G., Stroh, C. M., Geretschläger, M., Kamruzzahan, A. S. M., Wildling, L., Johnson, W. T., Ashcroft, B., Nelson, J., Lindsay, S. M., Gruber, H. J., and Hinterdorfer, P. (2005) Localization of single avidin-biotin interactions using simultaneous topography and molecular recognition imaging, *ChemPhysChem*, **6**, 897–900.
4. Stroh, C., Wang, H., Bash, R., Ashcroft, B., Nelson, J., Gruber, H., Lohr, D., Lindsay, S. M., and Hinterdorfer, P. (2004) Single-molecule recognition imaging microscopy, *Proc. Natl. Acad. Sci. USA.*, **101**, 12503–12507.
5. Tang, J., Ebner, A., Badelt-Lichtblau, H., Völlenkle, C., Rankl, C., Kraxberger, B., Leitner, M., Wildling, L., Gruber, H. J., Sleytr, U. B., Ilk, N., and Hinterdorfer, P. (2008) Recognition imaging and highly ordered molecular templating of bacterial S-layer nanoarrays containing affinity-tags, *Nano Lett.*, **8**, 4312–4319.
6. Ebner, A., Nikova, D., Lange, T., Haberle, J., Falk, S., Dubbers, A., Bruns, R., Hinterdorfer, P., Oberleithner, H., and Schillers, H. (2008) Determination of CFTR densities in erythrocyte plasma membranes using recognition imaging, *Nanotechnology*, **19**, 384017–384022.
7. Hinterdorfer, P., Baumgartner, W., Gruber, H. J., Schilcher, K., and Schindler., H. (1996) Detection and localization of individual antibody-antigen recognition events by atomic force microscopy, *Proc. Natl. Acad. Sci. USA*, **93**, 3477–3481.
8. Riener, C. K., Stroh, C. M., Ebner, A., Klampfl, C., Gall, A. A., Romanin, C., Lyubchenko, Y. L., Hinterdorfer, P., and Gruber, H. J. (2003) Simple test for single molecule recognition force microscopy, *Anal. Chim. Acta*, **479**, 59–75.
9. Lyubchenko, Y. L., Blankenship, R. E., Gall, A. A., Lindsay, S. M., Thiemann, O., Simpson, L., and Shlyakhtenko, L. S. (1996) Atomic force microscopy of DNA, nucleoproteins and cellular complexes: the use of functionalized substrates, *Scanning Microsc. Suppl.*, **10**, 97–109.
10. Bonanni, B., Kamruzzahan, A. S. M., Bizzarri, A. R., Rankl, C., Gruber, H. J., Hinterdorfer, P., and Cannistraro, S. (2005) Single molecule recognition between

cytochrome C 551 and gold-immobilized azurin by force spectroscopy, *Biophys. J.*, **89**, 2783–2791.

11. Han, W., Lindsay, S. M., and Jing, T. (1996) A magnetically driven oscillating probe microscope for operation in liquids, *Appl. Phys. Lett.*, **69**, 4111–4113.
12. Preiner, J., Ebner, A., Chtcheglova, L., Zhu, R., and Hinterdorfer, P. (2009) Simultaneous topography and recognition imaging: physical aspects and optimal imaging conditions, *Nanotechnology*, **20**, 215103–215121.
13. Schmidt, Th., Schütz, G. J., Baumgartner, W., Gruber, H. J., and Schindler, H. (1996) Imaging of single molecule diffusion, *Proc. Natl. Acad. Sci. USA*, **93**, 2926–2929.
14. Koopman, M., Cambi, A., de Bakker, B. I., Joosten, B., Figdor, C. F., van Hulst, N. F., and Garcia-Parajo, M. F. (2004) Near-field scanning optical microscopy in liquid for high resolution single molecule detection on dendritic cells, *FEBS Lett.*, **573**, 6–10.
15. Ianoul, A., Street, M., Grant, D., Pezacki, J., Taylor, R. S., and Johnston, L. J. (2004) Near-field scanning fluorescence microscopy study of ion channel clusters in cardiac myocyte membranes, *Biophys. J.*, **87**, 3525–3535.
16. Willig, K. I., Rizzoli, S. O., Westphal, V., Jahn, R., and Hell, S. W. (2006) STED microscopy reveals that synaptotagmin remains clustered after synaptic vesicle exocytosis, *Nature*, **440**, 935–939.
17. Hörber, J. K. H., and Miles, M. J. (2003) Scanning probe evolution in biology, *Science*, **302**, 1002–1005.
18. Grandbois, M., Dettmann, W., Benoit, M., and Gaub, H. E. (2000) Affinity imaging of red blood cells using an atomic force microscope, *J. Histochem. Cytochem.*, **48**, 719–724.
19. Lehenkari, P. P., Charras, G. T., Nykänen, A., and Horton, M. A. (2000) Adapting atomic force microscopy for cell biology, *Ultramicroscopy*, **82**, 289–295.
20. Almqvist, N., Bhatia, R., Primbs, G., Desai, N., Banerjee, S., and Lal, R. (2004) Elasticity and adhesion force mapping reveals real-time clustering of growth factor receptors and associated changes in local cellular rheological properties, *Biophys. J.*, **86**, 1753–1762.
21. Gilbert, Y., Deghorain, M., Wang, L., Xu, B., Pollheimer, P. D., Gruber, H. J., Errington, J., Hallet, B., Haulot, X., Verbelen, C., Hols, P., and Dufrêne, Y. F. (2007) Single-molecule force spectroscopy and imaging of the vancomycin/D-Ala-D-Ala interaction, *Nano Lett.*, **7**, 796–801.
22. Simons, K., and Ikonen, E. (1997) Functional rafts in cell membranes, *Nature*, **387**, 569–572.
23. Chtcheglova, L. A., Waschke, J., Wildling, L., Drenckhahn, D., and Hinterdorfer, P. (2007) Nano-scale dynamic recognition imaging on vascular endothelial cells, *Biophys. J.*, **93**, L11–L13.
24. Vincent, P. A., Xiao, K., Buckley, K. M., and Kowalczyk, A. P. (2004) VE-cadherin: adhesion at arm's length, *Am. J. Physiol. Cell Physiol.*, **286**, C987–C997.

25. Lampugnani, M. G., and Dejana, E. (1997) Interendothelial junctions: structure, signaling and functional roles, *Curr. Opin. Cell Biol.*, **9**, 674–682.

26. Hirano, S., Nose, A., Hatta, K., Kawakami, A., and Takeichi, M. (1987) Calcium-dependent cell-cell adhesion molecules (cadherins): subclass specificities and possible involvement of actin bundles, *J. Cell Biol.*, **105**, 2501–2510.

27. Yap, A. S., Brieher, W. M., and Gumbiner, B. M. (1997) Molecular and functional analysis of cadherin-based adherent junctions, *Annu. Rev. Cell Dev. Biol.*, **13**, 119–146.

28. Allport, J. R., Muller, W. A., and Luscinskas, F. W. (2000) Monocytes induce reversible focal changes in vascular endothelial cadherin complex during transendothelial migration under flow, *J. Cell Biol.*, **148**, 203–216.

29. Bibert, S., Jaquinod, M., Concord, E., Ebel, C., Hewat, E., Vanbelle, C., Legrand, P., Weidenhaupt, M., Vernet, T., and Gulino-Debrac, D. (2002) Synergy between extracellular modules of vascular endothelial cadherin promotes homotypic hexameric interactions, *J. Biol. Chem.*, **277**, 12790–12801.

30. Shan, W.-S., Tanaka, H., Philips, G. R., Arndt, K., Yoshida, M., Colman, D. R., and Shapiro, L. (2000) Functional *cis*-heterodimers of N- and R-Cadherins, *J. Cell Biol.*, **148**, 579–590.

31. Shapiro, L., Fannon, A. M., Kwong, P. D., Thompson, A., Lehmann, M. S., Grübel, G., Legrand, J.-F., Als-Nielsen, J., Colman, D. R., and Hendrickson, W. A. (1995) Structural basis of cell-cell adhesion by cadherins, *Nature*, **374**, 327–337.

32. Nagar, B., Overduin, M., Ikura, M., and Rini, J. M. (1996) Structural basis of calcium-induced E-cadherin rigidification and dimerization, *Nature*, **380**, 360–364.

33. Takeda, H., Shimoyama, Y., Nagafuchi, A., and Hirohashi, S. (1999) E-cadherin functions as a *cis*-dimer at the cell-cell adhesive interface in vivo, *Nat. Struct. Biol.*, **6**, 310–312.

34. Koch, A. W., Bozic, D., Pertz, O., and Engel, J. (1999) Homophilic adhesion by cadherins, *Curr. Opin. Struct. Biol.*, **9**, 275–281.

35. Baumgarther, W., Hinterdorfer, P., Ness, W., Raab, A., Vestweber, D., Schindler, H., and Drenckhahn, D. (2000) Cadherin interaction probed by atomic force microscopy, *Proc. Natl. Acad. Sci. USA*, **97**, 4005–4010.

36. Brieher, W. M., Yap, A. S., and Gumbiner, B. M. (1996) Lateral dimerization is required for the homophilic binding activity of C-cadherin, *J. Cell Biol.*, **135**, 487–496.

37. Chappuis-Flament, S., Wong, E., Hicks, L. D., Kay, C. M., and Gumbiner, B. M. (2001) Multiple cadherin extracellular repeats mediate homophilic binding and adhesion, *J. Cell Biol.*, **154**, 231–243.

38. Yap, A. S., Niessen, C. M., and Gumbiner, B. M. (1998) The juxtamembrane region of the cadherin cytoplasmic tail supports lateral clustering, adhesive strengthening, and interaction with $p120^{ctn}$, *J. Cell Biol.*, **141**, 779–789.

39. Mou, J., Czajkowsky, D. M., Sheng, S. J., Ho, R., and Shao, Z. (1996) High resolution surface structure of E. coli GroES oligomer by atomic force microscopy, *FEBS Lett.*, **381**, 161–164.

40. Pesen, D., and Hoh, J. H. (2005) Micromechanical architecture of the endothelial cell cortex, *Biophys. J.*, **88**, 670–679.

41. Oberleithner, H., Schneider, S. W., Albermann, L., Hillebrand, U., Ludwig, T., Riethmüller, C., Shahin, V., Schäfer, C., and Schillers, H. (2003) Endothelial cell swelling by aldosterone, *J. Membr. Biol.*,, **196**, 163–172.

42. Baumgartner, W., Schütz, G. J., Wiegand, J., Golenhofen, N., and Drenckhahn, D. (2003) Cadherin function probed by laser tweezer and single molecule fluorescence in vascular endothelial cells, *J. Cell Sci.*, **116**, 1001–1011.

43. Chtcheglova, L. A., Atalar, F., Ozbek, U., Wildling, L., Ebner, A., and Hinterdorfer, P. (2008) Localization of ergtoxin-1 receptors on the voltage sensing domain of hERG K^+ channel by AFM recognition imaging, *Pflugers Arch.*, **456**, 247–254.

44. Gurrola, G. B., Rosati, B., Rocchetti, M., Pimienta, G., Zaza, A., Arcangeli, A., Olivotto, M., Possani, L. D., and Wanke, E. (1999) A toxin to nervous, cardiac, and endocrine ERG K^+ channels isolated from *Centruroides noxius* scorpion venom, *FASEB. J.*, **13**, 953–962.

45. Trudeau, M. C., Warmke, J. W., Ganetzky, B., and Robertson, G. A. (1995) HERG, a human inward rectifier in the voltage-gated potassium channel family, *Science*, **269**, 92–95.

46. Sanguinetti, M. C., Jiang, C., Curran, M. E., and Keating, M. T. (1995) A mechanistic link between an inherited and an acquired cardiac arrhythmia: HERG encodes the I_{Kr} potassium channel, *Cell*, **81**, 299–307.

47. Pardo-Lopez, L., Garcia-Valdes, L., Gurrola, G. B., Robertson, G. A., and Possani, L. D. (2002) Mapping the receptor site for ergtoxin, a specific blocker of ERG channels, *FEBS Lett.*, **510**, 45–49.

48. Pardo-Lopez, L., Zhang, M., Liu, J., Jiang, M., Possani, L. D., and Tseng, G. N. (2002) Mapping the binding site of a human ether-a-go-go related gene-specific peptide toxin (ErgTx) to the channel's outer vestibule, *J. Biol. Chem.*, **277**, 16403–16411.

49. Torres, A. M., Bansal, P., Alewood, P. F., Bursill, J. A., Kuchel, P. W., and Vandenberg, J. I. (2003) Solution structure of CnErg1 (Ergtoxin), a HERG specific scorpion toxin, *FEBS. Lett.*, **539**, 138–142.

50. Rodriguez de la Vega, R. C., Merino, E., Becerril, B., and Possani, L. D. (2003) Novel interactions between K^+ channels and scorpion toxins, *Trends Pharmacol. Sci.*, **24**, 222–227.

51. Xu, C. Q., Zhu, S. Y., Chi, C. W., and Tytgat, J. (2003) Turret and pore block of K^+ channels: what is the difference? *Trends Pharmacol. Sci.*, **24**, 446–449.

52. Clarke, C. E., Hill, A. P., Zhao, J., Kondo, M., Subbiah, R. N., Campbell, T. J., and Vandenberg, J. I. (2006) Effect of S5P alpha-helix charge mutants on inactivation of hERG K^+ channels, *J. Physiol.*, **573**, 291–304.

Chapter 8

HIGH-SPEED ATOMIC FORCE MICROSCOPY FOR DYNAMIC BIOLOGICAL IMAGING

Takayuki Uchihashi and Toshio Ando

Department of Physics, Kanazawa University, Kakuma-machi, Kanazawa 920-1192, Japan, and Core Research for Evolutional Science and Technology (CREST) of the Japan Science and Technology Agency, Sanban-cho, Chiyoda-ku, Tokyo 102-0075, Japan

tuchi@kenroku.kanazawa-u.ac.jp

8.1 INTRODUCTION

Proteins are inherently dynamic molecules that undergo structural changes and interactions with other molecules over a wide timescale range, from nanoseconds to milliseconds or longer.[1] Protein motions play an important biological role in the assembly into protein complexes, ligand binding and enzymatic reactions. Therefore, understanding the dynamic behaviour of a protein is a requisite for gaining insight into biological processes. Experimental determination of protein structures has been made using X-ray crystallography and nuclear magnetic resonance. However, dynamic changes in protein molecules usually occur spontaneously and unsynchronously and thus are difficult to detect using these ensemble-average methods.

Recent advances in single-molecule fluorescence microscopy have allowed us to determine the localization of individual protein molecules with high accuracy. This enables the precise measurement of translational or rotational motions of individual fluorophores attached to biological molecules and, in some cases, the measurement of the association and dissociation reactions of biological molecules. Single-molecule fluorescence resonance energy transfer measurement is a powerful approach to analyzing intramolecular and intermolecular interaction dynamics in proteins. Thus, the "directness"

Life at the Nanoscale: Atomic Force Microscopy of Live Cells
Edited by Yves Dufrêne

www.panstanford.com

of our understanding of dynamic processes played by biological molecules is enhanced. However, this directness is not sufficient. These single-molecule fluorescence techniques observe protein molecules indirectly, and therefore, we still need to fill the gap between the recorded fluorescence images and the actual behaviour of the labelled biological molecules. To further enhance the directness, we need techniques that allow us to directly observe biological molecules with nanometre spatial and millisecond temporal resolution.

The atomic force microscope (AFM) is capable of directly visualizing unstained biological samples in liquids at nanometre resolution.[2] Since the invention, biologists have hoped that its unique capability would allow us to observe the dynamic behaviour of protein molecules at work. However, the imaging speed was limited to several tens of seconds per frame, and hence, it could not trace the fast dynamic processes progressing within a sample. Over the past decade, various efforts have been directed towards increasing the imaging rate of AFM.[3–8] The most advanced high-speed AFM can now capture images at 30–60 ms/frame over a scan range of ~250 nm with ~100 scan lines.[5–7] Importantly, the tip–sample interaction force has been greatly reduced without sacrificing the imaging rate, so that weak dynamic interactions between biological macromolecules are not significantly disturbed.

In this chapter, first we briefly review the limiting factors of imaging speed, and key techniques for high-speed imaging. For details of the instrumentation, readers may refer to a comprehensive review.[8] Then, we demonstrate some examples of successful imaging of protein, focusing on dynamics in two-dimensional (2D) protein crystals. In the last section, we describe the potential of high-speed AFM for cell imaging.

8.2 HIGH-SPEED IMAGING TECHNIQUES

High-speed AFM for biological samples in solutions is based on the tapping mode[9,10] in which the AFM tip is vertically oscillated and periodically brought into contact to a sample surface during scanning. The tip oscillation reduces the lateral force between tip and sample and thus minimizes damage and/or deformation of biological molecules. The vertical tip force acting on a sample is controlled by a PID (proportional-integral-derivative) feedback controller so that the oscillation amplitude of the cantilever is kept constant. Precise and fast feedback control is highly required for fast and low-invasive imaging. In this section, we simply describe the quantitative relationship between the feedback bandwidth and the various factors involved in AFM devices and the scanning conditions.[6] Then, the elemental techniques in the AFM for fast imaging are described.

8.2.1 Feedback Bandwidth and Imaging Rate

Supposing that an image is taken in time t for a scan range $W \times W$ with scan lines N, the scan velocity V_s in the x-direction is simply given by $V_s = 2WN/t$. For $W = 240$ nm, $N = 100$ and $t = 30$ ms, V_s becomes 1.6 mm/s. Here, assuming that the sample surface has a sinusoidal shape with a periodicity λ in the x-direction, the sample stage should move in the z-direction with a frequency of $f = V_s/\lambda$ to keep the tip–sample distance constant. When $\lambda = 10$ nm and $V_s =$ 1.6 mm/s, f becomes 160 kHz. The feedback bandwidth f_B should be equal to f or higher and thus can be expressed as

$$f_B \geq 2WN/\lambda t \tag{8.1}$$

Equation (8.1) gives the relationship between the image acquisition time t and the feedback bandwidth f_B. Because of the chasing-after nature of feedback control, sample topography is always traced with a phase delay, θ, which is given by $\sim 2 \times 2\pi f \Delta\tau$, where $\Delta\tau$ is the open-loop time delay (the sum of time delays of devices contained in the feedback loop). The main delays in tapping-mode AFM are the reading time of the cantilever oscillation amplitude, the cantilever response time, the z-scanner response time, the integral time of error signals in the feedback controller and the parachuting time. Here, "parachuting" means that the cantilever tip completely detaches from the sample surface at a steep down-hill region of the sample and thereafter takes time until it lands on the surface again. It takes at least a time of $1/(2f_c)$ to measure the amplitude of a cantilever that is oscillating at its resonant frequency f_c. The response time of second-order resonant systems such as cantilevers and piezoactuators is expressed as $Q/\pi f_0$, where Q and f_0 are the quality factor and the resonant frequency, respectively. The feedback bandwidth is usually defined by the feedback frequency that results in a phase delay of $\pi/4$. With this definition, we obtain $f_B = 1/(16\Delta\tau)$.

8.2.2 Key Devices For High-speed AFM

8.2.2.1 Cantilever

Cantilevers for fast and low-invasive imaging should have a high resonant frequency and a small spring constant. Regarding the feedback bandwidth, it is most important that the amplitude detection time and the cantilever's response time decrease in inverse proportion to the resonant frequency. To realize both, i.e., a small spring constant and a high resonant frequency, the size of cantilevers must be reduced. The small cantilevers most recently developed are made of silicon nitride and coated with gold of thickness ~ 20 nm. They have dimensions of length ~ 6 μm, width ~ 2 μm and thickness ~ 90 nm, which results in resonant frequencies of ~3.0 MHz in air and ~1.2 MHz in water, a spring constant of ~0.2 N/m and Q ~ 2.5 in water. The small cantilevers with a sharp tip are not commercially available at present. We therefore use electron beam deposition to grow an amorphous carbon tip on the original tip,[11] which can be sharpened by a plasma etching in argon gas.

8.2.2.2 Optical beam deflection detector

To focus an incident laser beam onto a small cantilever, a lens with a high numerical aperture (resulting in a short working distance) has to be used. An objective lens with a long working distance of 8 mm is used; a laser beam reflected back from the rear side of a cantilever is collected and collimated using the same objective lens as that used for focusing the incident laser beam onto the cantilever.[3] The focused spot is 3–4 μm in diameter. The incident and reflected beams can be separated using a quarter wavelength plate and a polarization splitter.

8.2.2.3 Amplitude detection

Conventional rms-to-dc converters use a rectifier circuit and a low-pass filter, which requires at least several oscillation cycles to output an accurate rms value. To detect the cantilever oscillation amplitude at the periodicity of a half oscillation cycle, we developed a peak-hold method; the peak and bottom voltages are captured and then their difference is output as the amplitude.[3] This amplitude detector is the fastest detector, and the phase delay has a minimum value of π, resulting in a bandwidth of $f_c/4$.

8.2.2.4 High-speed scanner

The scanner is the device most difficult to optimize for high-speed scanning. High-speed scanning of mechanical devices with macroscopic dimensions tends to produce unwanted vibrations. Several conditions are required to establish a high-speed scanner: (a) high resonant frequencies, (b) a small number of resonant peaks in a narrow frequency range, (c) sufficient maximum displacements, (d) small crosstalk between the three-dimensional (3D) axes, (e) low quality factors. We employ flexure stages made of blade springs for the x- and y-scanners. The flexure stages are made by monolithic processing to minimize the number of resonant peaks.[3] The maximum displacements of the x- and y-scanners at 100 V are 1 and 3 μm, respectively.

The *x*-piezoactuator is held at both ends with flexures, so that its centre of mass is hardly displaced and, consequently, no large mechanical excitation is produced. The *z*-piezoactuator (maximum displacement, 2 μm at 100 V; self-resonant frequency, 400 kHz) is held only at the four side-rims parallel to the displacement direction. The *z*-piezoactuator can be displaced almost freely in both counter directions, and consequently, impulsive forces are barely exerted on the holder. This holding method has an additional advantage in that the resonant frequency is not lowered by holding, although the maximum displacement decreases by half. The *x*-scanner is actively damped either by the previously developed *Q*-control technique[5] or by feed-forward control using inverse compensation.[12] The *z*-scanner is also actively damped by the *Q*-control technique. The *z*-scanner bandwidth f_s is extended to ~500 kHz, and the quality factor Q_s is reduced to ~0.5. Therefore, its response time τ_s ($=Q_s/\pi f_s$) is ~0.32 μs.

8.2.2.5 Dynamic PID control

The force reduction is quite important for biological AFM imaging. A shallower amplitude set point can reduce the tapping force but promotes "parachuting" during which the error signal is saturated and therefore the parachuting time is prolonged with increasing set-point amplitude, resulting in a decrease in the feedback bandwidth. The feedback gain cannot be increased to shorten the parachuting time, as a larger gain induces an overshoot at up-hill regions of the sample, resulting in the instability of the feedback operation. To solve this problem, a novel PID controller named "dynamic PID controller" was developed.[6] It can automatically change the feedback gain depending on the oscillation amplitude. Namely, the feedback gain is increased when the error signal exceeds a threshold level, which shortens the parachuting time or avoids parachuting. The dynamic PID controller can avoid parachuting in fact even when the set-point amplitude is increased up to 90% of the free oscillation amplitude.

8.3 HIGH-SPEED AFM IMAGING OF PROTEIN SAMPLES

High-speed AFM is not completely established yet as a tool for routinely observing biomolecular processes, although the performance of high-speed AFM has been markedly improved in the last 3–4 years. At present, it is important to examine whether we can really image biological processes that have been expected or known to occur. Further, high-speed AFM has not yet been applied to observe cellular structures because of some technological reasons described in the last section. Yet it would be valuable to demonstrate the potential of high-speed AFM for observing dynamic events of intermolecular interactions of proteins which take place in cell membrane fractions.

The current view of cell membrane structure derives from the fluid mosaic model in which proteins are considered to diffuse freely within a fluid lipid bilayer.[13] The first direct evidence for protein diffusion within cell membranes was provided by hybrid cell experiments.[14] Since then, various techniques including fluorescence recovery after photobleaching microscopy[15] and single particle tracking microscopy[16] have provided a more detailed understanding of the mobile nature of proteins in biological membranes. In particular, it has been shown that proteins in native membranes may not diffuse freely but are in fact confined to specific domains. Cells use several confining mechanisms such as anchoring to the cytoskeleton through hetero-bifunctional proteins,[17] diffusion barriers formed by the accumulation of proteins anchored to cytoskeleton meshes[18] or self-assembly into large 2D crystalline patches. Despite these advances, an understanding of membrane dynamics at the nanoscale remains a major challenge primarily because of the lack of measurement techniques allowing simultaneous spatial and temporal observation of single molecules within native membranes.

In this section, we introduce the capability of high-speed AFM for observing intermolecular interactions, lateral organization and rotational dynamics in 2D protein crystals.

8.3.1 Defect Diffusion in Streptavidin 2D Crystals

For protein crystal formation, the protein–protein association energy must be in an appropriate range. However, the association energy at each contact point had not been assessed experimentally. Here, we show that high-speed AFM imaging can enable its estimation using streptavidin as a model sample.[19]

Streptavidin is a protein that consists of four identical subunits: each specifically binds to one biotin.[20] It is easily crystallized in a 2D form on biotinylated lipid layers, which is considered to be an ideal model system to investigate 2D crystals grown on lipid layers. On the biotinylated lipid layers, two biotin-binding sites are occupied by the biotin moiety of a lipid layer, while the other two are exposed to an aqueous environment and therefore are free from biotin as depicted in Fig. 8.1a.

2D crystals of streptavidin were formed on a supported lipid bilayer (SLB) as follows. The lipid composition used was dioleoylphosphatidylcholine (DOPC), dioleoylphosphatidylserine (DOPS) and 1,2-dioleoyl-*sn*-glycero-3-phosphoethanolamine-*N*-(cap biotinyl) (biotin-cap-DOPE) (7 : 2 : 1, weight ratio). Dried lipid films were obtained by mixing lipids dissolved in chloroform followed by evaporating the solvent with argon. The lipid films were further dried in a desiccator by aspirating for more than 30 minutes. To obtain multilamellar vesicles (MLVs), the dried lipid films were resuspended in a buffer (10 mM HEPES-NaOH, 150 mM NaCl, 2 mM $CaCl_2$ [pH 7.4]) by vortexing. Small unilamellar vesicles (SUVs) were produced from the MLV suspension by sonications with a tip-sonicator for a few seconds.

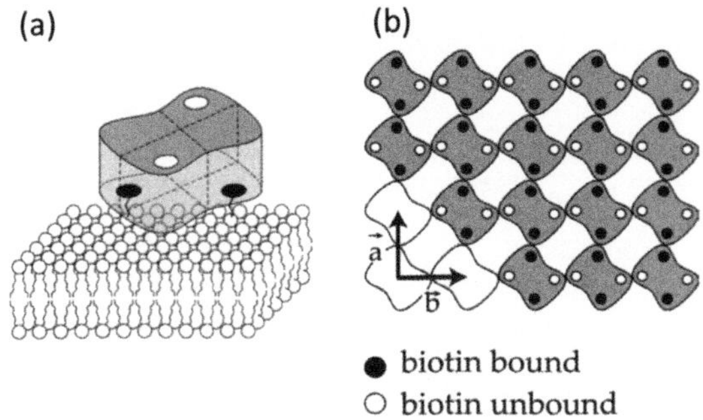

Figure 8.1. (a) Schematic of a streptavidin molecule on a biotinylated lipid bilayer. (b) Schematic of streptavidin arrays in a *C*222 crystal.

SLBs were prepared by depositing 0.1 mg/ml SUVs onto a freshly cleaved mica surface and incubated for 30 minutes in a chamber with saturated humidity at room temperature. After that, the excess lipids were washed out with the buffer. 2D crystallization of streptavidin on biotin-containing SLBs was performed by injecting streptavidin dissolved in an appropriate buffer at a final concentration of 0.1 mg/ml and incubating for 2 hours in a chamber with saturated humidity at room temperature. The buffer used for streptavidin crystallization has the same composition as the one used for the SLB formation. Then, excess streptavidin molecules were washed out with the buffer. As shown in Fig. 8.1b, in the *C*222 crystal, the intermolecular contacts between biotin-bound subunits are contiguously aligned along one crystal axis (*a*-axis), while the contacts between biotin-unbound subunits are contiguously aligned along the other axis (*b*-axis).

Monovacancy defects in the streptavidin 2D crystals were produced by increasing the tapping force onto the sample from the oscillating tip. Then, diffusion of point defects in the crystals was observed. Figure 8.2a shows images of the streptavidin 2D crystal and monovacancy defects therein, which are clipped from successively captured high-speed AFM images. In Fig. 8.2b, the trajectories of two monovacancy defects are shown. The mobility of the monovacancy defects was obviously anisotropic with respect to thew two axes of the crystalline lattice. These defects have larger mobility along the *b*-axis than along the *a*-axis.

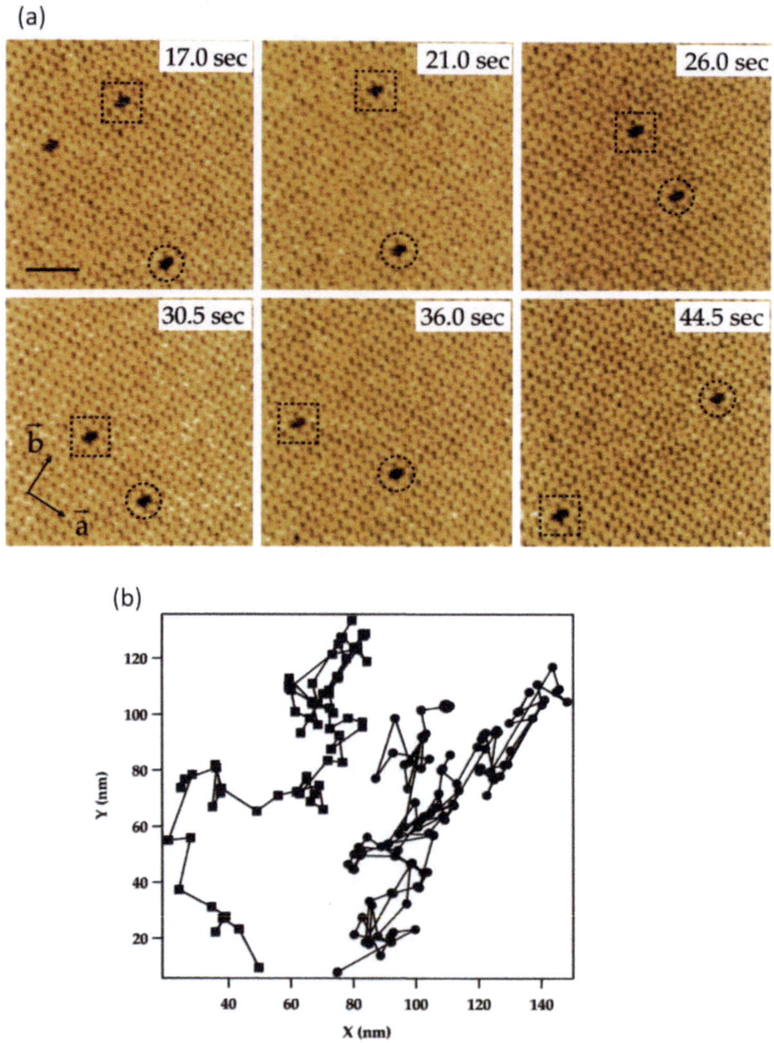

Figure 8.2. Migration of monovacancy defects in streptavidin 2D crystal. (a) High-speed AFM images of streptavidin 2D crystal and monovacancy defects therein. The monovacancy defects are enclosed by dashed squares and circles. The directions of the lattice vectors of the crystal are also indicated. Successive images were obtained at an imaging rate of 0.5 s/frame with a scan area of 150 × 150 nm^2. (b) Trajectories of individual monovacancy defects. Closed squares and circles correspond to defects indicated by open squares and circles shown in (a), respectively.

The mobility of monovacancy defects along each axis of the *C*222 crystal was quantified by measuring the mean-square displacements (MSDs) at various intervals (see Fig. 8.3). From the linear increase in MSDs with time, the diffusion rate constants of migrating monovacancy defects were determined to be D_a = 20.5 nm^2/s along the *a*-axis, which includes rows of contiguous biotin-bound subunits, and D_b = 48.8 nm^2/s along the *b*-axis, which includes rows of contiguous biotin-unbound subunits.

The linear relationship between MSDs and time of this migration indicates that the migration of monovacancy defects occurs by a random walk. The one-dimensional diffusion constant D is expressed as $D = \delta^2/2\tau$, where δ is the step length and τ is the time for each step (stepping time).[21] In the *C*222 crystal of streptavidin, the step length δ is 5.9 nm in both axes because the minimum step length corresponds to the lattice constant. Therefore, the stepping time τ for the movements along each axis can be estimated to be τ_a = 0.85 seconds and τ_b = 0.36 seconds for the *a*- and *b*-axis, respectively.

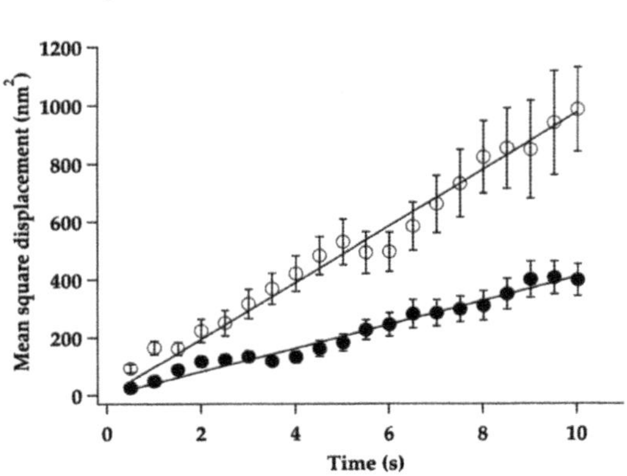

Figure 8.3. Plot of mean-square displacements (MSDs) of monovacancy defects against time. The MSDs as a function of time was measured from 94 trajectories. Error bars indicate standard error. The MSDs of defects along the *a*- and *b*-axes in the *C*222 crystal are compared. Data fitted to a linear function yielded diffusion constants D_a = 20.5 nm^2/s and D_b = 48.8 nm^2/s for the directions along the *a*- and *b*-axes, respectively. Closed circle, MSDs with the *a*-axis that includes rows of contiguous biotin-bound subunits; open circle, MSDs with the *b*-axis that includes rows of contiguous biotin-unbound subunits.

This anisotropy in lateral mobility (i.e., $D_b > D_a$) arises from a free energy difference between the biotin-bound subunit–subunit interaction and biotin-unbound subunit–subunit interaction. When a streptavidin molecule adjacent to a monovacancy defect moves to the defect site along the *a*-axis, two intermolecular bonds between biotin-unbound subunits ("u–u bond") and one intermolecular bond between biotin-bound subunits ("b–b bond") are broken. On the other hand, when it moves to the defect site along the *b*-axis, one u–u bond and two b–b bonds are broken (see Fig. 8.1). Therefore, the difference in the activation energies E_a and E_b for the step movement of a monovacancy defect along the respective *a*- and *b*-axes simply corresponds to the difference between the free energy changes G_{u-u} and G_{b-b} produced by the formation of the respective u–u bond and b–b bond, namely, $E_b - E_a = G_{u-u} - G_{b-b}$. Therefore, the observed relationship $D_b > D_a$ indicates $G_{u-u} < G_{b-b}$; namely, the u–u bond affinity is higher than the b–b bond. The ratio of the two diffusion rate constants (D_b/D_a) can be expressed by

$$D_b/D_a = \exp[-(E_b - E_a)/(k_B T)] \tag{8.2}$$

where k_B is Boltzmann constant and T is the absolute temperature. Thus, from the observed value of $D_b/D_a \sim 2.4$, the free energy difference $G_{u-u} - G_{b-b}$ is estimated to be approximately $-0.88\ k_BT$ ($T \sim 300$ K), which corresponds to –0.52 kcal/mol.

8.3.2 Crystal Dynamics of Purple Membrane

The purple membrane (PM) exists in the plasma membrane of *Halobacterium halobium*, and its constituent protein, bacteriorhodopsin (bR), functions as a light-driven proton pump. In the PM, bR monomers are associated to form a trimeric structure, and the trimers are arranged in a hexagonal lattice.[22] However, several aspects in the crystal formation remain open; for example, (i) trimer–trimer interaction sites and (ii) the existence of preformed trimers in the fluidic non-crystal region. In the 2D crystal of bR and any crystals in general, they are in dynamic equilibrium with the constituents at the interface between the crystal and the liquid. Here, we visualized dynamic events at the interface in the PM to provide information on the crystal formation and intermolecular interactions.[23]

The PM adsorbed on a mica surface in a buffer solution (10 mM Tris-HCl [pH 8.0] and 300 mM KCl) exhibits flat, roundly shaped patches (Fig. 8.4a). A 2D crystal lattice of bR is formed over the inner region surrounded by a dotted line in Fig. 8.4a, whereas in the peripheral outer region, there are no bR crystals. Figure 8.4b shows a magnified image of an edge region of the PM captured at 1 s/frame. There is a distinct border between the crystal and non-crystal areas. We found that the border shape fluctuates with time, indicating that the border region of the crystal is unstable and seems to be in dynamic equilibrium with bR molecules in the non-crystal area. In fact, spike noises were frequently observed in the non-crystal area and very likely to be produced by moving bR molecules which are too fast to be clearly captured at the imaging rate used (1 s/frame).

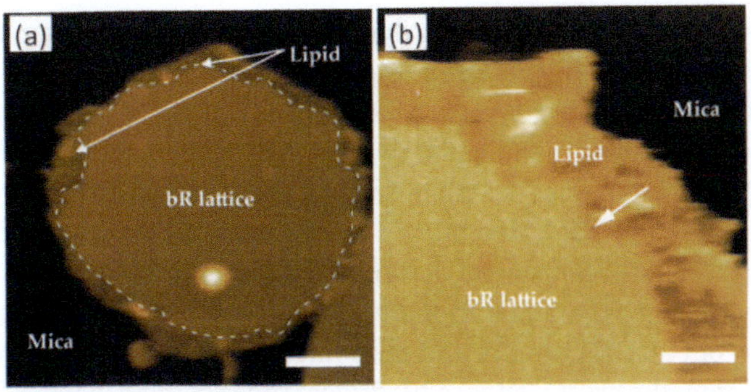

Figure 8.4. AFM images of purple membranes adsorbed onto a mica surface. (a) Low magnification image indicating that the purple membrane patch consists of a crystal area (encircled with a dotted line) and a non-crystal area (the periphery of the crystal area). (b) A magnified image of the edge region of the membrane patch captured at 1 frame/s. Scale bars: (a) 80 nm (b) 20 nm.

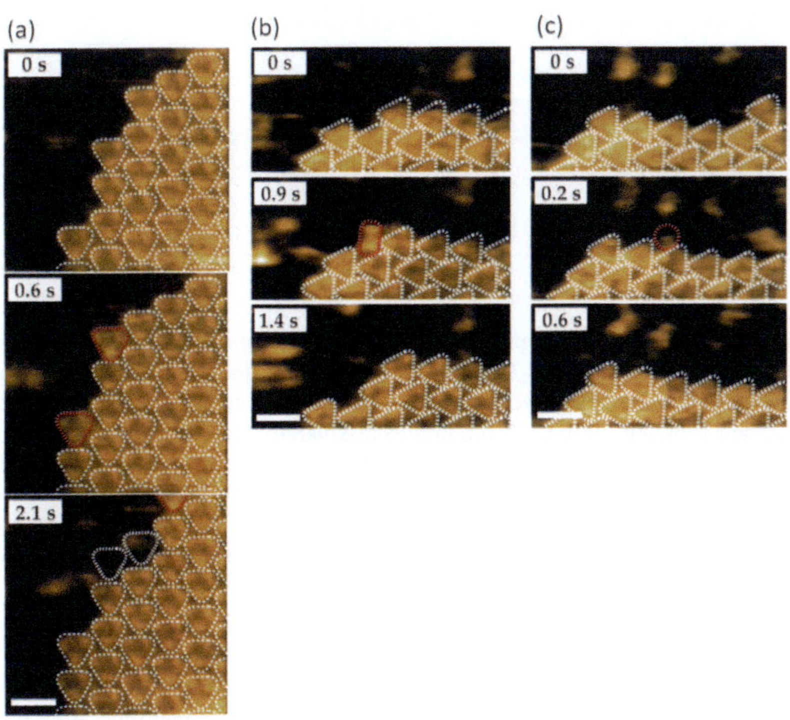

Figure 8.5. Time-lapse high-magnification AFM images of purple membranes on the borders between the crystal and non-crystal areas. The bR molecules encircled by the red dotted line indicate newly bound bR trimer (a), dimer (b) and monomer (c). The white triangles indicate the previously bound trimers. Scale bars: 5 nm (a), 10 nm (b, c). Imaging rate: (a) 0.3 s/frame, (b), (c) 0.1 s/frame.

To investigate the details of the dynamic structural changes in the crystal edge of PM, higher magnification images were acquired for the boundary region. Figure 8.5a shows typical AFM images taken at 0.3 s/frame. The bR trimers in the crystal are indicated by the thinlined triangles. At 0.6 seconds (Fig. 8.5a), two bR trimers (red triangles) have newly bound to the crystal edge. At 2.1 seconds (Fig. 8.5a), two bR trimers (white triangles) have dissociated and another trimer (red triangle) has bound to the crystal edge. One of the two dissociated trimers remained in the crystal area for ~0.9 seconds. Not only bR trimers but also bR dimers and monomers were observed to bind to and dissociate from the crystal edge. A bR dimer (red rectangle in Fig. 8.5b) stayed bound to the edge for ~0.5 seconds, whereas a bR monomer (red circle in Fig. 8.5c) remained bound to the edge for ~0.4 seconds. These residence times for monomers and dimers are shorter than those of the trimers. According to the analysis for 239 observed binding events, the binding of trimeric bR occurred predominantly (82%), whereas binding of dimeric bR (6.7%) was only about half that of monomeric bR (11.3%).

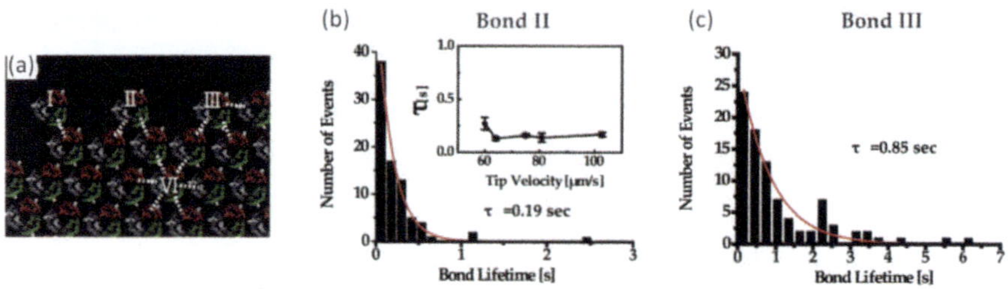

Figure 8.6. (a) Schematic representing the binding manner of a purple membrane trimer at the crystal edge (I, II, III) and in the crystal interior (VI). The Roman numbers indicate the number of interaction bonds (dotted lines) containing W12 residue. (b) Histogram showing the type II binding events versus lifetime. This histogram was fitted with a single-exponential function (red line). The inset shows the average lifetime as a function of tip velocity. The lifetime was ~0.2 seconds irrespective of the tip velocity. Error bars indicate the standard deviation for the nonlinear least-square curve fitting. (c) Histogram showing the type III binding events versus lifetime. The red line indicates the best fit with a single-exponential function

To estimate the inter-trimer interaction energy, we analyzed the residence time of newly bound bR trimers at the crystal edge and its dependence on the number of interaction sites. For our analysis, we assumed that within the 2D bR crystal, a trimer can interact with the surrounding trimers through six sites as indicated by the dotted lines in "VI" of Fig. 8.6a. Inter-trimer interactions around W12 residues participate in lattice formation.[24,25] Following the same model, the number of interaction sites at the crystal edge is reduced, depending on the binding position, as indicated by "I", "II" and "III" in Fig. 8.6a. Successive AFM images as exemplified in Fig. 8.5 showed many binding and dissociation events in which bR trimers bound to different sites at the border between the crystal and non-crystal areas. These events can be classified into types "I", "II" and "III" depending on the number of interaction sites involved. Type II binding events are predominant (~74%), whereas type I (~6%) and type III (~20%) events are less frequent. The lifetime of the type I bonds was too short to obtain clear images of the corresponding event, preventing reliable statistics.

Figure 8.6b shows a histogram of the lifetime of type II bonds which was measured using AFM images taken at 0.1 s/frame (tip velocity, 75 μm/s). This histogram could be well fitted by a single exponential (correlation coefficient, $r = 0.9$), from which the average lifetime τ_2 was estimated to be 0.19 ± 0.01 seconds. To ensure that the observed dissociation events are not significantly affected by the AFM tip during scanning, we examined the dependence of the average lifetime on the tip velocity while a constant vertical force was maintained. The inset in Fig. 8.6b shows the average lifetime as a function of tip velocity and indicates that the average lifetime is about 0.17 ± 0.06 seconds, irrespective of tip velocity. Thus, we conclude that the tip–sample interaction does not significantly affect the natural association and dissociation kinetics of the bR trimer. Figure 8.6c shows a histogram of the type III bond lifetime, from which the average lifetime τ_3 was estimated to be 0.85 ± 0.08 seconds. The longer lifetime of type III bonds compared with type II obviously arises from a relationship of $E_3 < E_2 < 0$, where E_2 and E_3 are the association energies responsible for type II and type III interactions, respectively. The average lifetime ratio, τ_2/τ_3, is given by

$$\tau_2/\tau_3 = \exp[(E_3 - E_2)/k_BT] \qquad (8.3)$$

Because the type II interaction contains two elementary bonds, whereas the type III interaction contains three, the energy difference $E_3 - E_2$ corresponds to the association energy of the single elementary bond. From the ratio $\tau_2/\tau_3 = 0.22$ and Eq. (8.3), this elementary association energy is estimated to be about $-1.5\ k_BT$, which corresponds to −0.9 kcal/mol at 300 K. This value is approximately consistent with that estimated by differential scanning calorimetry.[26,27]

8.3.3 Crystal Dynamics of Annexin V

Annexin V is a soluble protein, belonging to a protein family that binds to negatively charged phospholipids, in particular DOPS, in the presence of calcium ions. It undergoes 2D crystallization on lipid monolayers.[28] The property of annexin V self-organization has been proposed to be functionally relevant in its biological function.[29] The structure of its soluble form has been solved by X-ray crystallography and that of the membrane-bound form was investigated extensively by electron crystallography and AFM.[30] The fundamental oligomeric state of annexin V is a trimer and trimers assemble into two common crystal forms with p3 or p6 symmetry. In this section, we briefly demonstrate some dynamic events, such as crystal growth, dynamic equilibrium between the 2D crystal and the liquid phase, observed in annexin V crystals with p6 symmetry.

As lipids, we here used DOPC, DOPS and DOPE (5 : 2 : 3 w/w). The lipid bilayer supported on a mica surface was prepared by the same method described in section 8.3.1. Two-dimensional crystallization of annexin V on the bilayer was performed by injecting an annexinV solution into the bilayer sample during AFM imaging. The buffer used for the observation was 50 mM Tris-HCl pH 8.0, 5 mM KCl, 2.5 mM $MgCl_2$, 3 mM $CaCl_2$.

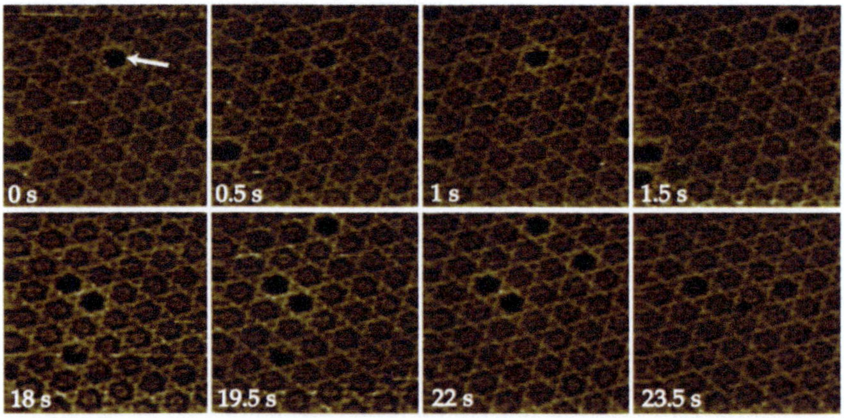

Figure 8.7. Binding and dissociation dynamics of annexin V trimers on the 2D crystal with p6 symmetry. For example, the hole in the honeycomb lattice indicated by an arrow in the 0-second image is filled by a trimer diffusing on the crystal surface at 0.5 seconds and then dissociate in the next 0.5 seconds. Successive images were obtained at an imaging rate of 0.5 s/frame with a scan area of 150×150 nm^2.

The annexin V crystal with p6 symmetry exhibits a honeycomb structure. The "holes" of the honeycomb structure tend to be occupied with a relatively mobile trimer, which undergoes a more relaxed interaction with its surrounding cage than a molecule forming part of the honeycomb lattice. Because of this mobility, the central trimer, therefore, shows a less sharply defined density in the EM images of the crystal.[31,32] Figure 8.7 shows successive AFM images obtained at an imaging rate of 0.5 s/frame. The hole indicated by the arrow in Fig. 8.7 (0 second) is occupied by a trimer in the next frame. This trimer is weakly bound and then dissociates soon at 1 second. The hole is filled again at 1.5 seconds but the trimer is bound more stably at this time. The dissociation and association of centre trimers occur at several places in the crystal (images between 18 and 23.5 seconds). This observation also indicates that unbound trimers exist on the crystal surface and are rapidly diffusing on it.

High-speed AFM imaging also revealed rotational diffusion of a centre trimer weakly bound to the surrounding cage. Figure 8.8 shows the images captured at 0.2 s/frame. The centre trimer encircled in Fig. 8.8 (0 second) rotates counterclockwise with a 60° step. In the cage surrounded with six trimers in the lattice, the central trimer can assume two stable positions with identical association energy. This rotational motion indicates the association energy to be in the order of ~1 k_BT.

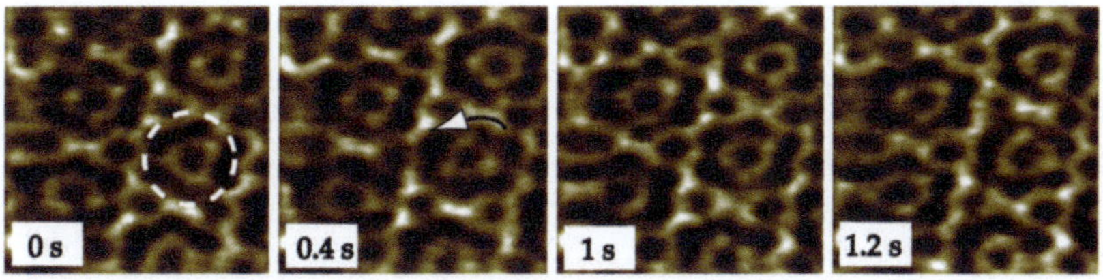

Figure 8.8. Rotational diffusion of the annexin V trimer trapped in a lattice cage. Successive images were obtained at an imaging rate of 0.2 s/frame with a scan area of 50×50 nm^2.

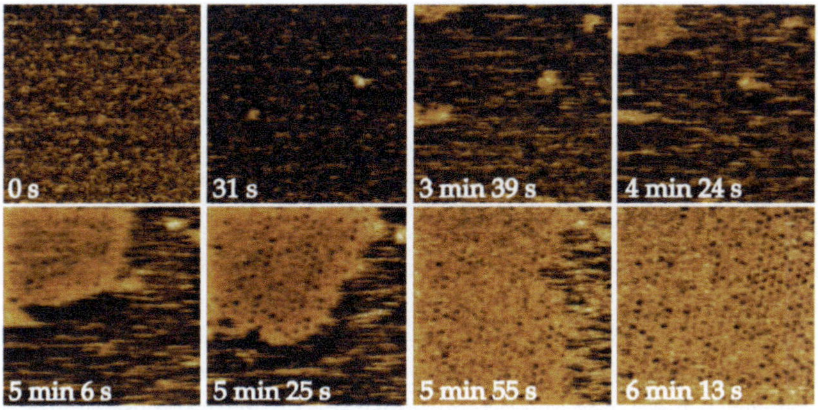

Figure 8.9. Crystal growth of annexin V. At 0 second, the image shows only the lipid surface and a large noise induced by diffusing molecules on the surface. $CaCl_2$ solution was injected at 7 seconds. Successive images were obtained at an imaging rate of 1 s/frame with a scan area of 400×400 nm^2.

Figure 8.9 shows successive AFM images captured during the crystal growth of annexin V. In this measurement, the buffer solution at the initial condition does not contain $CaCl_2$. During scanning, $CaCl_2$ was injected into the sample chamber to give a final concentration of 3 mM. At 0 second, the lipid surface is primarily observed because of the absence of Ca^+ ions. This image shows large noise which is caused by annexin V molecules rapidly diffusing on the surface. A $CaCl_2$ solution was injected at 7 seconds. Soon after the injection, small particles appear (see the image at 31 seconds). This is probably the first stage of the assembly in which three molecules cluster together in an almost irreversible manner to form a trimer. Surface-diffusing annexin molecules come into contact with the surface-bound trimer and gradually increase the cluster size with time. However, in the captured images, crystallization progresses more predominantly from the top left in the images. The precursor protein clusters observed at the first stage are incorporated into large crystals during progression of the crystallization. Eventually, the lipid surface is completely covered by the crystal in a few minutes.

8.4 FUTURE PROSPECTS: TOWARDS DYNAMIC IMAGING OF LIVE CELLS

Current high-speed AFMs can film dynamic processes played by purified protein molecules. The video images of molecular processes provide insight into their functional mechanisms in a much more straightforward manner than other techniques. However, at present, high-speed AFM cannot be applied to observe dynamics on cell membranes. To change this situation, we have to overcome some technical difficulties. In this section, we discuss whether and how we can achieve such a new generation of high-speed AFM.

8.4.1 Lower Interaction Force and Non-Contact Imaging

Since cell membranes are suspended and hence extremely soft, achieving small tip–sample interaction force is essential for their stable and high-resolution imaging. Also, membrane proteins that are not anchored to cytoskeletons or not clustered diffuse very fast within the membranes. High-speed AFM requires much higher imaging speed. Generally, to reduce cantilever stiffness, one must compromise the resonant frequency and *vice versa*. The most advanced small cantilevers deem to have almost achieved the ultimate goal of balancing these two mechanical quantities. Therefore, reduction of the interaction force by using softer and smaller cantilevers seems impossible. The ultimate minimization of the tip–sample interaction force is attained by non-contact imaging. True non-contact-AFM (nc-AFM) has only been realized in a vacuum environment by utilizing a cantilever with a significantly large quality factor in vacuum.[33] If high-speed nc-AFM is realized in liquid conditions, we can use stiffer cantilevers with much higher resonant frequencies, which will promise markedly higher imaging rates.

The non-contact imaging capability in liquids has already been achieved by ion-conductance scanning probe microscopy (ICSPM).[34] Owing to progress in fabrication techniques for producing very sharp glass capillaries with a small pore at the apex, the spatial resolution of ICSPM has reached a few nanometre.[35] Immobile protein molecules with ~14 nm in size on living cell membranes have been successfully imaged.[36] However, it seems difficult to increase the imaging rate of ICSPM; the bandwidth of ion-conductance detection cannot be easily increased, because the ionic current through the small pore of the capillary electrode is very low.

Although not for high-speed nc-AFM, control algorithms to reconcile a large quality factor of the cantilever with high-speed imaging have been proposed.[37,38] The position and velocity of the oscillating cantilever are continuously monitored (or discretely monitored with small time-bins). From these measured quantities, an estimator calculates the tip–sample interaction force of each tapping cycle. A model-based predictor uses the estimated force to control the tip–sample distance in the next tapping cycle. Experiments with conventional AFMs implemented with the new controllers demonstrated regulation of the tip–sample interaction force at each tapping cycle, irrespective of the time delay of the cantilever's response. However, to apply this method to a real high-speed AFM, extremely fast digitization and calculations are required.

8.4.2 High-Speed AFM Combined with Optical Microscopy

The size of a cell is generally over a few tens of micrometers in width and a few micrometers in height. On the other hand, the extension ranges of the high-speed scanner we normally use are approximately 3 μm × 1 μm × 2 μm in *x*- × *y*- × *z*-directions, which are too small to be used for imaging a cell. In practice, such a small imaging area makes it difficult to find cells to be imaged.

One of the solutions is combining a high-speed scanner with a conventional low-speed scanner for wide area imaging. Another solution is combining high-speed AFM with an optical microscope. Since optical microscopy and high-speed AFM have advantages and disadvantages over each other, combining these techniques into a single instrument would therefore be useful. From the optical image covering a wide area of the sample, we can quickly find a much narrower area to be scanned by AFM. Further, fluorescence microscopy provides complimentary information to high-speed AFM images, such as the identification of proteins observed by high-speed AFM, the simultaneous recording of topographic changes in protein molecules and optical signals for chemical reactions such as ATP hydrolysis.

In current high-speed AFM, a raster scanning is carried out by moving the sample stage relatively to the fixed cantilever. In this design, the sample stage should be very small so that the resonant frequency of the z-scanner is not lowered. For simultaneous optical and AFM imaging, a stand-alone AFM, in which the cantilever is scanned relative to the fixed sample, has to be adapted to ensure optical transparency of the sample stage. Micro-electro-mechanical fabrication techniques, which have been employed to produce self-sensing and/or self-actuation cantilevers[39] and sensor-combined scanners,[40] could be the key to the realization of a combined system as well as to the significant enhancement of high-speed AFM performance.

8.4.3 High-Speed AFM for Intracellular Imaging

Recently, it has been reported that AFM can have a capability of subsurface imaging.[41] This new modal AFM is called scanning near-field ultrasound holography (SNFUH) and has been successfully used for intracellular imaging under ambient conditions.[41] In its application, a high-frequency acoustic wave is launched from under the sample stage and propagates through the sample. Materials embedded in the sample with different elastic moduli modulate the phase and amplitude of the propagating acoustic wave. These modulations affect the nonlinear acoustic interference that occurs at the cantilever tip excited with another high-frequency acoustic wave with different frequency. The interference produces a wave with a frequency corresponding to their frequency difference. By adjusting the frequency difference to the cantilever resonant frequency, the cantilever is effectively oscillated by the nonlinear acoustic interference. SNFUH has no resolution in the z-direction. However, using multiple images obtained from different launching angles of the ultrasonic wave, it is probably possible to reconstitute a 3D image. Combining SNFUH with high-speed scanning techniques will enable the high-resolution 3D imaging of various intracellular processes in live cells which take place spontaneously or as a result of their responses to extracellular stimuli.

8.5 SUMMARY

We have described various studies on the instrumentation and imaging of biomolecules carried out in the last decade. The direct and real-time observation of dynamic biomolecular processes is straightforward and can give deep insights into their functional mechanisms. Therefore, this new microscopy will markedly change our style of considering biological questions. Nevertheless, there are presently only few setups of high-speed bio-AFM that can capture dynamic biomolecular processes at 10–30 frames/s, and consequently, the user population is limited. Besides, to our knowledge, only two manufacturers are producing small cantilevers for high-speed bio-AFM. We hope that this current situation will be quickly improved by manufacturers.

In the near future, high-speed AFM will be actively used to observe a wide range of dynamic processes that occur on isolated proteins, protein assemblies and protein–DNA complexes. More complex systems including live cells and organisms will become targets of high-speed AFM after some technical advances described earlier are successfully overcome. The in vivo and in vitro visualization of various processes at the molecular level will become possible including the responses of membrane receptors to stimuli, nuclear envelope formation and disassembly, chromosome replication and segregation processes, phagocytosis, protein synthesis in the endoplasmic reticulum and the targeting processes of synthesized proteins through the Golgi apparatus. Thus, high-speed AFM-based visualization techniques have great potential to bring about breakthroughs not only in biochemistry and biophysics but also in cell biology, physiology and pharmaceutical and medical sciences. To open up such unprecedented fields, steady efforts have to be carried out towards expanding the capability of high-speed AFM and related techniques.

Acknowledgements

We thank D. Yamamoto, N. Kodera, M. Shibata, H. Yamashita and all previous students for their dedicated studies for developing high-speed AFM. This work was partially supported by the Japan Science, Technology Agency (JST; the CREST program and a Grant-in-Aid for Development of Systems, Technology for Advanced Measurement and Analysis and Strategic International Cooperative Program), the Japan Society for the Promotion of Science (JSPS; a Grant-in-Aid for Basic Research (S), Grant-in-Aid for Science Research on Priority Areas; innovative nanoscience of supramolecular motor proteins working in biomembranes), industrial technology research grant program in '04 from New Energy and Industrial Technology Development Organization (NEDO).

References

1. Henzler-Wildman, K., and Kern, D. (2007) Dynamic personalities of proteins, *Nature*, **450**, 964–972.
2. Müller, D. J., Fotiadis, D., Scheuring, S., Müller, S. A., and Engel, A. (1999) Electrostatically balanced subnanometer imaging of biological specimens by atomic force microscope, *Biophys. J.*, **76**, 1101–1111.
3. Ando, T., Kodera, N., Takai, E., Maruyama, D., Saito, K., and Toda, A. (2001) A high-speed atomic force microscope for studying biological macromolecules, *Proc. Natl. Acad. Sci. USA*, **98**, 12468–12472.
4. Ando, T., Kodera, N., Maruyama, D., Takai, E., Saito, K., and Toda, A. (2002) A high-speed atomic force microscope for studying biological macromolecules in action, *Jpn. J. Appl. Phys.*, **41**, 4851–4856.
5. Kodera, N., Yamashita, Y., and Ando, T. (2005) Active damping of the scanner for high-speed atomic force microscopy, *Rev. Sci. Instrum.*, **76**, 053708 (5 pp.).
6. Kodera, N., Sakashita, M., and Ando, T. (2006) Dynamic proportional-integral-differential controller for high-speed atomic force microscopy, *Rev. Sci. Instrum.*, **77**, 083704 (7 pp.).
7. Yamashita, H., Uchihashi, T., Kodera, N., Miyagi, A., Yamamoto, D., and Ando, T. (2007) Tip-sample distance control using photo-thermal actuation of a small cantilever for high-speed atomic force microscopy, *Rev. Sci. Instrum.*, **78**, 083702 (5 pp.).
8. Ando, T., Uchihashi, T., and Fukuma, T. (2008) High-speed atomic force microscopy for nano-visualization of dynamic biomolecular processes, *Prog. Surf. Sci.*, **83**, 337–437.
9. Zhong, Q., Inniss, D., Kjoller, K., and Elings, V. B. (1993) Fractured polymer/silica fiber surface studied by tapping mode atomic force microscopy, *Surf. Sci. Lett.*, **290**, L688–L692.
10. Hansma, P. K., Cleveland, J. P., Radmacher, M., Walters, D. A., Hillner, P. E., Bezanilla, M., et al. (1994) Tapping mode atomic force microscopy in liquids, *Appl. Phys. Lett.*, **64**, 1738–1740.
11. Wendel, M., Lorenz, H., and Kotthaus, J. P. (1995) Sharpened electron beam deposited tips for high resolution atomic force microscope lithography and imaging, *Appl. Phys. Lett.*, **67**, 3732–3734.
12. Schitter, G., and Stemmer, A. (2004) Identification and open-loop tracking control of a piezoelectric tube scanner for high-speed scanning-probe microscopy, *IEEE Trans Control Systems Technol.*, **12**, 449–454.
13. Singer, S. J., and Nicolson, G. L. (1972) The fluid mosaic model of the structure of cell membranes, *Science*, **175**, 720–731.
14. Frye, L. D., and Edidin, M. (1970) The rapid intermixing of cell surface antigens after formation of mouse-human heterokaryons, *J. Cell Sci.*, **7**, 319–335.
15. Swaminathan, R., Hoang, C. P., and Verkman, A. S. (1997) Photobleaching recovery and anisotropy decay of green fluorescent protein GFP-S65T in

solution and cells: cytoplasmic viscosity probed by green fluorescent protein translational and rotational diffusion, *Biophys. J.*, **72**, 1900–1907.

16. Kusumi, A., Sako, Y., and Yamamoto, M. (1993) Confined lateral diffusion of membrane receptors as studied by single particle tracking (nanovid microscopy). Effects of calcium-induced differentiation in cultured epithelial cells, *Biophys. J.*, **65**, 2021–2040.
17. Byers, T. J., and Branton, D. (1985) Visualization of the protein associations in the erythrocyte membrane skeleton, *Proc. Natl. Acad. Sci. USA*, **82**, 6153–6157.
18. Halenda, R. M., Primakoff, P., and Myles, D. G. (1987) Actin filaments, localized to the region of the developing acrosome during early stages, are lost during later stages of guinea pig spermiogenesis, *Biol. Reprod.*, **36**, 491–499.
19. Yamamoto, D., Uchihashi, T., Kodera, N., and Ando, T. (2008) Anisotropic diffusion of point defects in two-dimensional crystal of streptavidin observed by high-speed atomic force microscopy, *Nanotechnology*, **19**, 384009 (9 pp.).
20. Weber, P. C., Ohlendorf, D. H., Wendoloski, J. J., and Salemme, F. R. (1989) Structural origins of high-affinity biotin binding to streptavidin, *Science*, **243**, 85–88.
21. Berg, H. C. (1993) *Random Walks in Biology*, Princeton University Press, Princeton, New Jersey.
22. Henderson, R., Baldwin, J. M., Ceska, T. A., Zemlin, F., Beckmann, E., and Downing, K. H. (1990) Model for the structure of bacteriorhodopsin based on high-resolution electron cryo-microscopy, *J. Mol. Biol.*, **213**, 899–929.
23. Yamashita, H., Voïtchovsky, K., Uchihashi, T., Contera, S. A., Ryan, J. F., and Ando, T. (2009) Dynamics of bacteriorhodopsin 2D crystal observed by high-speed atomic force microscopy, *J. Struct. Biol.*, **167**, 153–158.
24. Weik, M., Patzelt, H., Zaccai, G., and Oesterhelt, D. (1998) Localization of glycolipids in membranes by in vivo labeling and neutron diffraction, *Mol. Cell*, **1**, 411–419.
25. Sapra, K. T., Besir, H., Oesterhelt, D., and Müller, D. J. (2006) Characterizing molecular interactions in different bacteriorhodopsin assemblies by single-molecule force spectroscopy, *J. Mol. Biol.*, **355**, 640–650.
26. Jackson, M. B., and Sturtevant, J. M. (1978) Phase transitions of the purple membranes of Halobacterium halobium, *Biochemistry*, **17**, 911–915.
27. Koltover, I., Raedler, J. O., Salditt, T., Rothschild, K. J., and Safinya, C. R. (1999) Phase behavior and interactions of the membrane-protein bacteriorhodopsin, *Phys. Rev. Lett.*, **82**, 3184–3187.
28. Richter, R. P., Lai Kee Him, J., Tessier, B., Tessier, C., and Brisson, A. R. (2005) On the kinetics of adsorption and two-dimensional self-assembly of annexin A5 on supported lipid bilayers, *Biophys. J.*, **89**, 3372–3385.
29. Oling, F., Sopkova-de Oliveira Santos, J., Govorukhina, N., Mazères-Dubut, C., Bergsma-Schutter, W., Oostergetel, G., Keegstra, W., Lambert, O., Lewit-Bentley, A., and Brisson, A. (2000) Structure of membrane-bound annexin A5 trimers: a hybrid cryo-EM-X-ray crystallography study, *J. Mol. Biol.*, **304**, 561–573.

30. Reviakine, I., Bergsma-Schutter, A., Morozov, A. N., and Brisson, A. (2001) Two-dimensional crystallization of annexin A5 on phospholipid bilayers and monolayers: a solid-solid phase transition between crystal forms, *Langmuir*, **17**, 1680–1686.

31. Reviakine, I., Bergsma-Schutter, W., Mazeres-Dobut, C., Govorukhina, N., and Brisson, A. (2000) Surface topography of the p3 and p6 annexin V crystal forms determined by atomic force microscopy, *J. Struct. Biol.*, **131**, 234–239.

32. Mosser, G., Ravanat, C., Freyssinet, J.-M., and Brisson, A. (1991) Sub-domain structure of lipid-bound annexin-V resolved by electron image analysis, *J. Mol. Biol.*, **217**, 241–245.

33. Giessibl, F. J., Hembacher, S., Bielefeldt, H., and Mannhart, J. (2000) Subatomic features on the silicon (111)-(7 × 7) surface observed by atomic force microscopy, *Science*, **289**, 422–425.

34. Hansma, P. K., Drake, B., Marti, O., Gould, S. A., and Prater, C. B. (1989) The scanning ion-conductance microscope, *Science*, **243**, 641–643.

35. Ying, L., Bruckbauer, A., Zhou, D., Gorelik, J., Shevchuk, A., Lab, M., Korchev, Y., and Klenerman, D. (2005) The scanned nanopipette: a new tool for high resolution bioimaging and controlled deposition of biomolecules, *Phys. Chem. Chem. Phys.*, **7**, 2859–2866.

36. Shevchuk, A. I., Frolenkov, G. I., Sanchez, D., James, P. S., Freedman, N., Lab, M. J., Jones, R., Klenerman, D., and Korchev, Y. E. (2006) Imaging proteins in membranes of living cells by high-resolution scanning ion conductance microscopy, *Angew. Chem. Int. Ed. Engl.*, **45**, 2212–2216.

37. Sahoo, D. R., Sebastian, A., and Salapaka, M. V. (2003) Transient-signal-based sample-detection in atomic force microscopy, *Appl. Phys. Lett.*, **26**, 5521–5523.

38. Jeong, Y., Jayanth, G. R., Jhiang, S. M., and Menq, C. H. (2006) Direct tip-sample interaction force control for the dynamic mode atomic force microscopy, *Appl. Phys. Lett.*, **88**, 204102 (3 pp.).

39. Barrett, R. C., and Quate, C. F. (1991) High-speed, large-scale imaging with the atomic force microscope, *J. Vac. Sci. Technol.*, **B 9**, 302–306.

40. Degertekin, F. L., Onaran, A. G., Balantekin, M., Lee, W., Hall, N. A., and Quate, C. F. (2005) Sensor for direct measurement of interaction forces in probe microscopy, *Appl. Phys. Lett.*, **87**, 213109 (3 pp.).

41. Shekhawat, G. S., and Dravid, V. P. (2005) Nanoscale imaging of buried structures via scanning near-field ultrasound holography, *Science*, **310**, 89–92.

Chapter 9

NEAR-FIELD SCANNING OPTICAL MICROSCOPY OF BIOLOGICAL MEMBRANES

Thomas S. van Zanten[a] **and Maria F. Garcia-Parajo**[a,b]
[a] *Single Molecule BioNanophotonics group, IBEC-Institute for Bioengineering of Catalonia and CIBER-bbn, Baldiri Reixac 15-21, 08028 Barcelona, Spain*
[b] *ICREA-Institució Catalana de Recerca i Estudis Avançats, 08010 Barcelona, Spain*
mgarcia@pcb.ub.es

9.1 A VIEW ON CELL MEMBRANE COMPARTMENTALIZATION

One of the most fascinating but also controversial fields in cell biology concerns the organization of the cellular plasma membrane. In fact, the view of the cell membrane as a two-dimensional homogeneous structure has changed radically in recent years by demonstrations of lateral heterogeneities, patches and the existence of protein domains in the membrane.[1-3] The general consensus points to a direct relation between the lateral organization of proteins and lipids and their specific cellular function.[4-7] Similarly, a large body of evidence indicates that the size of many of these membrane domains is in the range of 30 to 800 nm.[6,8] However, other workers in the field have seriously questioned the existence of some membrane domains in living cells, in particular those known as membrane "rafts".[9] Part of the controversy regarding the existence of membrane domains lays in their physical size, being smaller than the diffraction limit of light, and thus not resolvable by classical optical means. Moreover, there is increasing evidence that the assembly and disassembly of such complexes are rather dynamic and thus difficult to visualize using standard optical microscopy settings.[10] Finally, biochemical and biophysical approaches aimed at the study of protein domains have lead

Life at the Nanoscale: Atomic Force Microscopy of Live Cells
Edited by Yves Dufrêne

www.panstanford.com

in many cases to contradictory results.[11] There is therefore a need for new high-resolution methodologies capable of directly imaging domains within the plasma membrane of intact cells.

Fluorescence microscopy has become one of the most prominent and versatile research tools used in modern cell biology and in principle ideal to investigate cell membrane organization in living cells.[12] The reasons for it are essentially twofold. First, light-based microscopy allows the study of living specimens in their native environment in a non-invasive manner. Additionally, fluorescence microscopy offers chemical specificity by exploiting polarization, lifetime and spectral contrast.[13] Furthermore, progress in detector technology has recently pushed fluorescence microscopy to its ultimate level of sensitivity: the detection of individual molecules.[14–16] Second, enormous progress on the development of specific and highly efficient fluorescent probes for exogenous labelling has been achieved. In parallel to external antibody labelling, the advent of green fluorescent protein (GFP) technology has revolutionized live cell imaging because an autofluorescent molecule can be genetically encoded as a fusion with the c-DNA of interest.[17] Indeed, the spectral variants of GFP and the unrelated red fluorescent protein (DsRed) make it possible to perform nowadays multicolour imaging in living cells.[17,18]

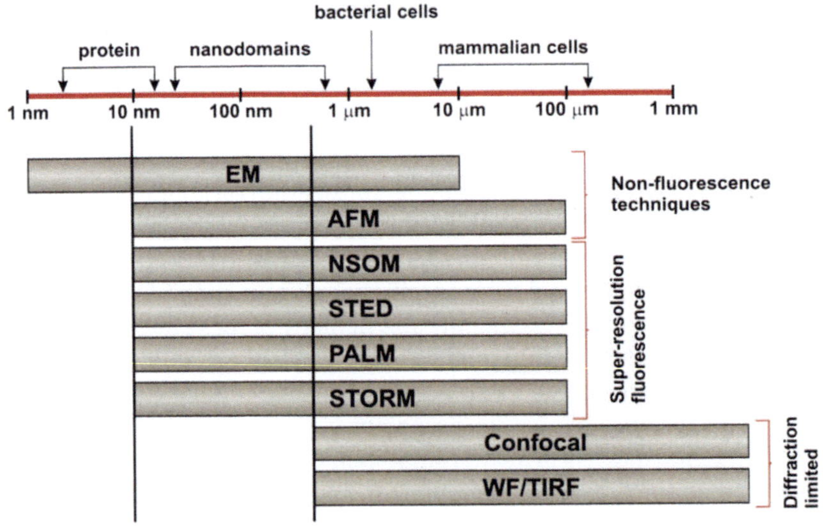

Figure 9.1. Comparison of spatial resolution techniques for biological imaging. WF: wide-field microscopy; TIRF: total internal reflection fluorescence microscopy; STED: stimulated emission depletion; PALM: photoactivated localization microscopy; STORM: stochastic optical reconstruction microscopy; EM: electron microscopy; AFM: atomic force microscopy. STED, PALM and STORM belong to far-field super-resolution techniques while NSOM is a near-field super-resolution technique.

In the last few years, a number of fluorescent-based techniques have been applied to study the organization of the cellular plasma membrane. In particular, confocal, wide-field and total internal reflection microscopy can resolve structures on the cell membrane and track proteins and other biomolecules in living cells (Fig. 9.1). However, a major drawback of standard light microscopy is the fundamental limit of the attainable spatial resolution, which is dictated by the laws of diffraction. This diffraction limit originates from the fact that it is impossible to focus light to a spot smaller than half its wavelength. In practice, this means that the maximal resolution in optical microscopy is ~250–300 nm. Since a large body of evidence indicates that dynamic cell-signalling events start by oligomerization and interaction of individual proteins (i.e., on the molecular scale), the need for imaging techniques that have a higher resolution is growing.

Traditionally, high-resolution cell biology has been the arena of electron microscopy (Fig. 9.1), which offers superb resolution but lacks the aforementioned advantages of fluorescence microscopy. The advent of scanning probe microscopy (Fig. 9.1), and especially atomic force microscopy (AFM), in which an atomically sharp probe attached to a cantilever is scanned over the surface of interest, has made nanometre resolution also attainable on living cells.[19,20] However, although AFM produces a high-resolution topographical image of the sample, it lacks biochemical specificity. Hence, although individual molecules can be seen, their identities cannot be defined. This seriously limits the usefulness of AFM for high-resolution imaging on cells. A promising way around the problem relies on specific labelling of the AFM probe with biomolecules (e.g., with antibodies or ligands). This introduces a contrast mechanism based on specific interactions between the probe and a certain type of molecules in the specimen.[21] More recently, molecular recognition imaging using AFM and biofunctionalized probes has been successfully implemented by the Hinterdorfer group (see Chapter 7).[22] Although extremely sensitive, the experimental approach is, so far, restricted to a single type of interaction being probed. The combination of scanning probe microscopy with an optical contrast mechanism, affording spatial super-resolution imaging and spectroscopy, biochemical specificity and versatility, and ultra-fast time response, is the domain of near-field scanning optical microscopy (NSOM) and the main topic of this chapter.

As a side note, it is worth mentioning that in recent years, several new far-field super-resolution imaging techniques have also broken the diffraction limit of light, producing fluorescence images in the nanometre range, not only laterally but also in three dimensions (Fig. 9.1). In short, these techniques take advantage of specific photophysical properties of fluorescence probes in conjunction with tailored ways of illumination to either achieve direct[23] or reconstructed[24–26] imaging at the nanoscale. For instance, in stimulated emission depletion microscopy, the resolution is enhanced by reversible saturable transitions of the fluorescent probes,[23] while in photoactivatable localization microscopy[24,25] and stochastic optical reconstruction microscopy,[26] the ascertainable localization accuracy (rather than resolution) depends strongly on the total number of detected photons. Several recent scientific contributions have highlighted so far the advantages and current limitations in terms of spatial and temporal resolution of these emerging techniques, as well as current challenges on fluorescence probe technology.[27,28] The reader is referred to these contributions for further inside on super-resolution far-field optical microscopy.

9.2 NEAR-FIELD SCANNING OPTICAL MICROSCOPY

A different concept that breaks the diffraction limit of light providing optical super-resolution at the nanometre scale is NSOM. In NSOM, as in the case of AFM, a sharp probe physically scans the sample surface (Fig. 9.2a) generating a topographic imaging of the sample under study. However, in contrast to AFM, NSOM is capable to simultaneously generate optical images. A typical NSOM configuration is shown in Fig. 9.2a. The practical feasibility of this kind of NSOM was first demonstrated by Pohl *et al.*, immediately following the advent of scanning probe microscopy and in fact before the introduction of the AFM.[29] The most generally applied near-field optical probe consists of a small aperture, typically 20–120 nm in diameter (i.e., much smaller than the wavelength of the excitation light), at the end of a metal-coated tapered optical fibre (Fig. 9.2b). The probe funnels the incident light wave to dimensions that are substantially below the diffraction limit. This results in a light source that has the size of the aperture. However, in contrast to common light sources such as lightbulbs and lasers, the light emitted by the probe is predominantly composed of evanescent waves rather than propagating waves. The intensity of the evanescent light decays exponentially and to insignificant levels ~100 nm away from the aperture. Effectively, the probe can excite fluorophores only within a layer of <100 nm from the probe—that is, in the "near-field" region (inset Fig. 9.2a). The sample fluorescence can subsequently be collected by conventional optics and transformed into an optical image of the sample surface in which the resolution is now primarily dictated by the aperture dimensions rather than by the wavelength of the light.

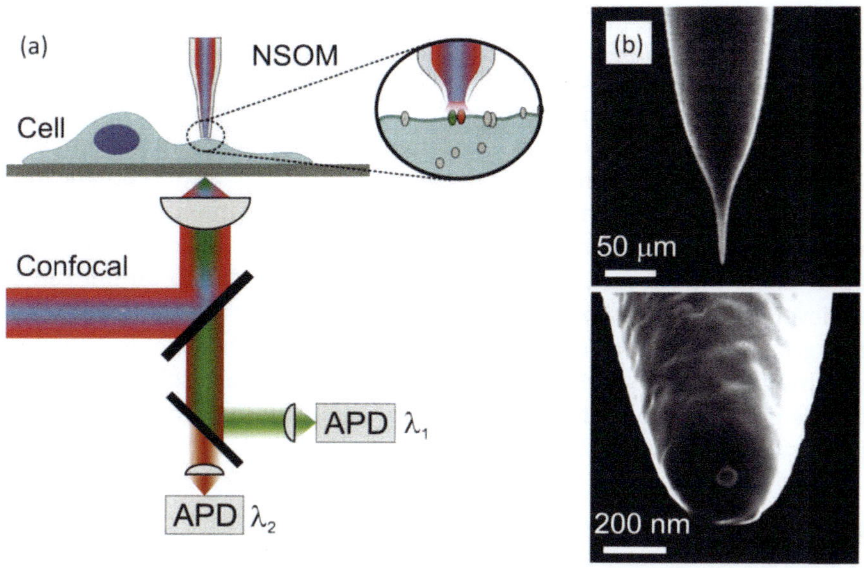

Figure 9.2. (a) Schematic layout of a combined confocal/near-field optical set-up. Laser light is focused onto the sample using a high NA objective (confocal excitation) or alternatively by the use of a subwavelength aperture probe. Fluorescence is collected by a conventional inverted microscope. Dual-channel optical detection allows wavelength and/or polarization discrimination. The inset illustrates the principle of surface-specific excitation where only fluorophores close to the aperture end (red dots) are efficiently excited, in contrast to those outside the near-field region (gray dots). The optical near-field generated at the aperture has a significant intensity at distances < 100 nm away from the aperture, selectively exciting fluorophores that are in close proximity to the cell surface. (b) Tapered NSOM fibre (above) together with a SEM image of the probe end (below). Schematic adapted with permission from Ref. 67. © (2009) National Academy of Sciences, USA.

9.2.1 Different NSOM Configurations

The most commonly used NSOM configuration for biological applications is based on aperture-type fibre probes as described earlier, although other types of approaches have also been implemented. For instance, instead of using the probe to illuminate the sample, one can employ far-field optics to illuminate the sample and use the probe to collect the evanescent field in close proximity to the sample surface. Although perfectly suitable for some photonic applications, its use in fluorescence imaging is less appropriate since far-field illumination translates in unnecessary sample photobleaching. A different experimental strategy to NSOM is based on the use of metallic tips, known in the literature as apertureless NSOM[30] when the tip is used as passive scatterer, or tip-enhanced NSOM when the metallic tip is excited to enhance the electromagnetic field at the end of the tip apex.[31] In both cases, the sample is illuminated in the far field and a metal probe is placed in the tight focus of the illumination beam. The local interaction with the sample surface is subsequently detected as a modulation in the scattered far field. Extreme sensitivity is required to observe the weakly scattered light from the nanometre-sized tip in the presence of the light scattered by the sample. When combined with fluorescence, and the tip is properly excited with radial fields along the tip axis, optical resolutions in the order to 30 nm can be achieved.[32-34] This method is however accompanied by a large fluorescence background generated from far-field illumination of the sample, therefore requiring modulation techniques to recover the high-resolution signal.[35] On the positive side of the balance, this method is free from the associated practical difficulties of fabricating circular apertures.

9.2.2 Fabrication of NSOM Probes

The most crucial component of aperture-type NSOM is the fabrication of the actual probe. Many different concepts for aperture probes were explored during the past 15 years, each of them with distinct advantages.[36] Commonly, a fibre is pulled to an apex of nanometre dimensions and coated with aluminium to confine the light inside the tapered region. Aluminium is commonly preferred to other opaque materials because of its very small penetration depth, which implies a high reflectivity. However, probes that combine all necessary demands for NSOM have only scarcely been produced. Generally, the evaporated aluminium coating has a grainy structure, resulting in pinholes and an irregularly shaped aperture with asymmetric polarization behaviour. Moreover, the grains increase the distance between aperture and sample, causing reduction of resolution and loss of local excitation intensity. Also, the damage threshold of the coating generally limits the probe brightness to <10 nW in the far field. In a refined approach, we, and others, have fabricated high-definition aperture probes, combining superior polarization characteristics and high throughput, by making use of the focused ion beam (FIB) technique, which is capable of polishing on a nanometre scale.[37] In the FIB apparatus, a beam of Ga ions, collimated to 7 nm, is used to remove a very thin slice of material from the aluminium-coated probe end. The resulting "FIB probe" has a flat-end face with a roughness below 7 nm and a well-defined circular aperture. Figure 9.3 shows a series of apertures probe after FIB milling, as imaged in the FIB apparatus at low beam dose. We managed to fabricate apertures as small as 20 nm. The polarization extinction ratio exceeds 100:1 for all polarization directions, with brightness up to 1 mW for 70 to 90 nm aperture probes.

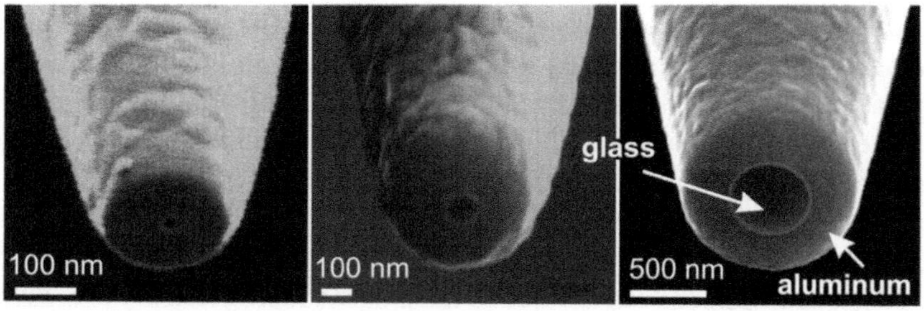

Figure 9.3. Examples of aperture probes of different diameters after FIB treatment. Left: 35 nm aperture; middle: 95 nm aperture; right: 530 nm aperture.

9.2.3 Shear-Force Feedback to Control Probe–Sample Distance

Since the near-field intensity exiting the subwavelength aperture probe decays exponentially with distance from the probe, for efficient excitation it is essential to have accurate control of the probe–sample distance during scanning. Several different techniques have been implemented so far to monitor the vertical position of the probe tip. First NSOMs relied on electron-tunnelling feedback, later extended to photon tunnelling in the photon scanning tunnelling microscopy. Today, the majority of NSOMs utilize two distinct feedback methods, which have analogous sensitivity and performance and are similar to noncontact AFM. These techniques are called shear-force feedback and tapping-mode feedback. This latter method implies the use of bent tips, which are usually characterized by lower optical throughput, technical difficulties in the fabrication and higher mechanical vibrations.

To date, the most commonly employed NSOM configuration relies on shear-force feedback based on the use of quartz tuning forks.[38] In this approach, the NSOM tip is glued onto one of the arms of the tuning fork. The tuning fork-probe system is oscillated at its resonance frequency in a lateral vibrational mode (with a <1 nm amplitude). When in proximity to the sample, shear forces dampen this motion and induce measurable changes in the oscillation amplitude and phase. An electronic feedback system, controlling the probe–sample distance directly through the piezo-electric scan stage, is subsequently used to maintain a constant oscillation amplitude/phase during scanning. In this way, a constant probe–sample distance of <10 nm is realized. The feedback signal itself, as in AFM, is used to generate a topographic map of the sample surface with comparable resolution and sensitivity as tapping-mode AFM (Fig. 9.4a). Of course, unique to NSOM is the fact that a corresponding fluorescence map is simultaneously generated.

One of the major obstacles that have restricted the use of NSOM in cell biology has been related to its difficulty to operate in liquid conditions, a crucial step towards live cell imaging. Successful control of the tip–sample distance has been routinely achieved in air by using tuning forks as sensing elements and driven at resonance, as explained earlier. However, this approach systematically failed once the tuning fork was immersed in a liquid. Our group has demonstrated that, in aqueous environments, sensitivity of the surface topography can be regained by keeping the tuning fork dry in a "diving bell" enclosure just above the probe.[39,40] Using this system, we have been able to measure the topology of intact cell membranes without compromising sensitivity or resolution (Fig. 9.4b). Alternatively, Höppener and colleagues used the tuning fork with the tip placed perpendicular to the prongs of the fork and protruding about ~2 mm below the fork. The configuration works thus as "tapping mode" with the tip immersed in solution and the tuning fork kept dry above the liquid.[41] An alternative method for position control is based on ion conductance. The method relies on the use of sharp micropipettes. As the probe approaches the sample, ion conduction is partially blocked and the change in conductivity is used as a measure of the tip–sample distance.[42] This mechanism has been coupled to NSOM to obtain images of living cells.[42]

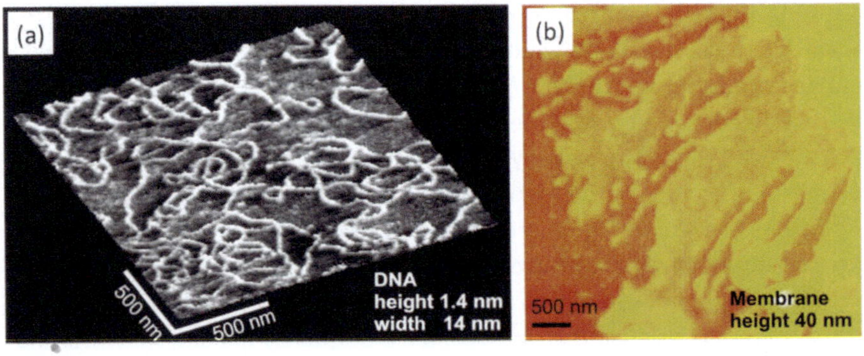

Figure 9.4. (a) Shear-force image of plasmid DNA deposited on a mica surface. (b) Shear-force image of an intact monocyte cell membrane measured in liquid conditions using the "diving bell" concept.

9.2.4 Excitation and Detection Paths in Fluorescence NSOM

For biological applications, the most widely used configuration is an aperture-type NSOM working in fluorescence, incorporated into an inverted optical microscope, with near-field excitation and far-field detection[43,44] (see Fig. 9.2a). This scheme preserves most of the conventional imaging modes (confocal microscopy for instance), which remain available in combination with the near-field approach. Light that is emitted by the aperture locally excites fluorescent markers attached to the biological molecules under investigation (proteins and/or lipids). The emitted fluorescence emerging from the imaging zone must be collected with the highest possible efficiency. For this purpose, high numerical aperture (oil immersion) microscope objectives are usually employed. The collected light is directed to sensitive detectors, such as avalanche photodiodes or photo-multiplier tubes, via suitable dichroic mirrors for spectral splitting or through a polarizing beam splitter cube for polarization detection. Filters are also commonly used to select the spectral regions of interest removing unwanted spectral components. In this sense, inverted optical microscopes are an advantageous solution for light collection, redistribution and filtering.

9.3 APPLICATION OF NSOM TO MODEL AND CELL MEMBRANES

9.3.1 Model Membranes Inspected by NSOM

Model membranes have been used for a long time to investigate the segregation behaviour of lipids and different proteins in predetermined lipid mixtures, while reducing the complexity of the cell membrane. The typical binary or ternary lipid mixtures used to mimic the lipid composition of cell membranes indeed phase-segregate into liquid-condensed (LC) and liquid-expanded (LE) phases. By transferring monolayers of a lipid mixture on a substrate using standard Langmuir–Blodgett techniques, Hwang *et al.* used NSOM in dry and buffer conditions to reveal previously unresolved features of around 50 nm.[45,46] When a higher pressure was used to form the monolayer, the domains of the LC phase appeared to decrease in size, and an increasingly complex fine web structure of the LE phase emerged.[45,46] Cholesterol addition, typically enriching the LC phase, resulted in the formation of elongated thin LC domains. From these morphology changes, it was concluded that cholesterol reduced the line tension between the domains in regions of LC/LE coexistence. Likewise, the addition of the ganglioside GM1, again a LC constituent, affected the monolayer morphology significantly. Moreover, GM1 induced a more pronounced segregation between the LC and LE phases. These results suggested the formation of genuine distinct domains, thus favouring the occurrence of a lipid raft type of phenomenon on model membranes. The lipids typically enriching the LC phase are significantly more saturated than lipids constituting the LE phase. Thus, when all lipids pack in their subsequent phase, the LE phase will be lower in height. Indeed, by specifically labelling the LE phase, a perfect correlation was found between topographical and fluorescence signals.[47] To extend these findings, Hollars and Dunn used tapping-mode feedback NSOM in air to additionally obtain compliance information of the lipid monolayer.[48] Because the carbohydrate chains of the lipids from the LC phase are highly saturated, they pack in an ordered fashion as compared with the lipids from the LE phase. As expected, the LC phase was found less compliant than the LE phase.[48] As such Hollars and Dunn demonstrated the strength of NSOM as compared with fluorescence or scanning probe techniques on their own.

In a more recent work supporting the formation of lipid rafts on model membranes, small amounts of labelled GM1 revealed that GM1 is not homogeneously distributed throughout the LC phase. Instead, they were seen to constitute their own 100–200 nm sized domains.[49] In fact, upon closer examination, the labelled GM1 distribution appeared to be more complex. To better characterize the GM1 behaviour, GM1 lipids were labelled with Bodipy. This fluorophore displays a redshift in the emission spectra when present in higher concentrations because of excimer formation, thus being able to probe the local lipid density.[50] Because of the strong tendency of GM1 to partition in gel or liquid-ordered phases, high-concentration GM1 was found in the LC phase, showing the redshifted emission, even while using low deposition pressures.[51] Nevertheless, a rather large fraction of single Bodipy-GM1 was still found randomly distributed in the LE phase. Upon increasing the deposition pressure towards expected cell membrane pressures, the LC domain phases became smaller, and the labelled GM1 appeared to preferentially partition into the LC phase.[51]

The use of NSOM to investigate monolayers has also been extended towards bilayers[52] and protein containing lipid layers in dry or buffer conditions.[41,53-55] The addition of proteins to such lipid phase-segregated model systems will be an important step in understanding how lipid-based interaction can influence protein distribution. Subsequently, monitoring the dynamics would then provide a more complete spatio-temporal map of proteins and lipids in a lipid bilayer. Work in this direction has been performed using AFM in combination with fluorescence correlation spectroscopy (FCS).[56] The recent proof-of-principle indication that dynamical studies can be also performed with NSOM[57] opens up an exciting field that combines high-resolution imaging with ultrafast dynamics. Indeed, the advantage of performing FCS on confined volumes has been recently demonstrated on living cells.[58,59] The incorporation of this approach in NSOM would also provide, in addition to surface sensitivity, topography and resolution, temporal information.

9.3.2 Cell Membrane Compartmentalization Inspected by NSOM

Regarding cell membrane quantitative imaging, NSOM has been mainly used to investigate the degree of clustering of different receptors on the cell membrane. Since live cell imaging still remains a technological challenge for NSOM, most of the work reported so far has been performed in dry conditions and with cells being subject to several treatments before inspection: dry air, several steps of methanol and then dried or paraformaldehyde. The latter is advantageous when working in liquid conditions, since it chemically fixes the cells, without altering cell membrane morphology (<2%) and therefore prevents mobility of membrane components during imaging. In some cases, the association of multiple components has also been investigated using dual-colour NSOM. In the context of receptor clustering, our group has used NSOM to image pathogen recognition receptors with high spatial resolution on cells of the immune system, providing insight into the mechanisms exploited by the cell to ensure high performance of these receptors[40,60] (Fig. 9.5). By labelling the pathogen recognition receptor DC-SIGN with a specific monoclonal antibody, we found that as much as 80% of DC-SIGN is clustered on the cell membrane of immature dendritic cells, imaged either in dry or buffer conditions.[40,60] These domains were randomly distributed over the plasma membrane with a size distribution centered at ~185 nm. Interestingly, we discovered a remarkable heterogeneity of the DC-SIGN packing density within the clusters. This suggests that the large spread in DC-SIGN density per cluster likely serves to maximize the chances of DC-SIGN binding to a large variety of viruses and pathogens having different binding affinities.[60] Indeed, the organization of DC-SIGN in nanodomains appeared crucial for efficient binding and internalization of pathogens.[7]

Recently, Chen *et al.* used NSOM in dry conditions in combination with quantum dots to label the T cell receptor (TCR) of T cells in live animals before and after cell stimulation.[61] In the resting state, the TCR complexes were found monomerically organized on the T cell membrane. Upon T cell stimulation, the TCR complexes reorganized and formed 270–390 nm sized domains. Interestingly, these small-sized domains were not only formed but also sustained for days. Additional experiments showed that although unstimulated cells could produce an immune response, stimulated cells produced significant higher levels of cytokines.[61] By means of these high-resolution NSOM experiments, it was shown that the TCR reorganization plays a significant role in antigen recognition and cytokine production.

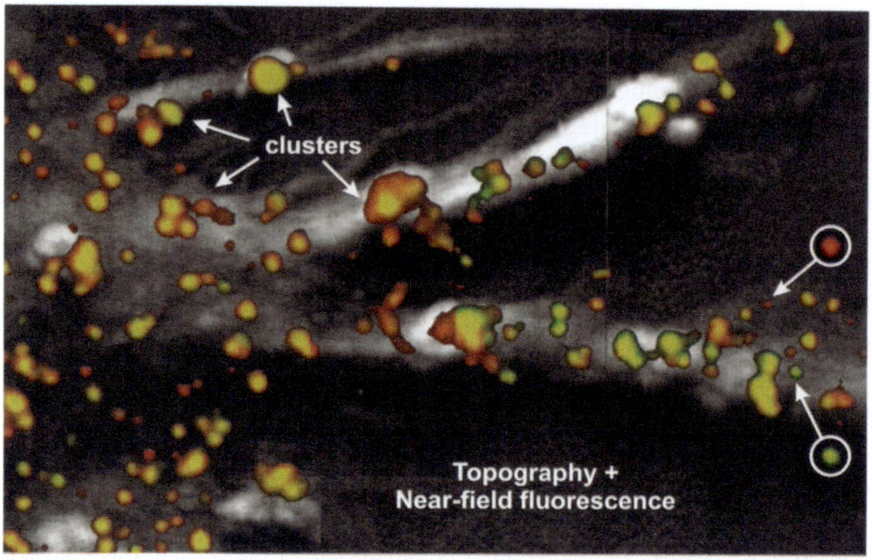

Figure 9.5. Combined topography (gray) and NSOM image (colour) of the pathogen recognition receptor DC-SIGN expressed on immature dendritic cells. Spots have different size and intensity reflecting the nanocluster organization of DC-SIGN. Image adapted with permission from Ref. 60. © (2007) Wiley-VHC.

In the case of members of the epidermal growth factor (EGF) receptor tyrosine kinase family, clustering is thought to have an adverse effect. Some EGFs, like the erbB2 receptor, are found to be over-expressed in breast cancerous cells. It is thought that this over-expression leads to cluster formation causing the highly oncogenic activation of very potent kinase activity. Indeed, by applying NSOM in air, the clustering behaviour of EGF receptors was found to be associated with the activation state of the cell.[62] Additionally, it was found that EGF cluster sizes increased if the quiescent cells were treated with EGF activators to the same extend as cells over-expressing these EGFs.[62] Since activation of the EGF signalling pathways requires extensive interaction between individual members of the EGF family, it is likely that concentrating one of these EGF receptors in clusters increases the likelihood of co-clustering of other EGF members. This co-clustering would then subsequently increase the EGF signalling efficiency. In other words, a higher local concentration will decrease the lag time for direct inter-receptor contact.

Cell-signalling events commonly involve a multitude of spatially segregated proteins and lipids. As such, standard confocal microscopy studies in biology usually involve multiple colours corresponding to multiple specifically labelled proteins. However, inherent to all lens-based techniques are chromatic aberrations that cause multiple wavelengths to never perfectly overlap. In contrast, NSOM guarantees a perfect overlay between multiple excitation wavelengths, an essential requirement to resolve the true nanoscale landscape of cell membranes. Already in 1997, Enderle *et al.* used for the first time dual-colour NSOM to directly measure the association of a host protein (protein4.1) and parasite proteins (MESA and PfHRP1) in malaria (*Plasmodium falciparum*)-infected dried erythrocytes.[63] As the parasitic proteins interact with the host proteins, 100 nm sized knob-like topographical features appear on the membrane of the host cell. To investigate the direct interaction of host and parasite proteins, the proteins were specifically labelled and subsequently imaged with NSOM. As expected, the fluorescence from the two labelled parasitic proteins and the labelled host protein were found on the knob-like structures. The high-resolution of NSOM however demonstrated that host and parasite proteins did not physically colocalized in the same compartments.[63]

The increased co-localization of individual components on the cell membrane has been actually demonstrated on two members of the interleukin family by combining dual-colour excitation and single-molecule detection NSOM on dried T cells.[64] IL2R and IL15R did not interact if their organization was monomeric. However, in their clustered form, both receptors were found to co-localize significantly, suggesting that clustering of both receptors takes place in the same nanocompartments.[64] Interestingly, IL2R and IL15R clusters were found to have a constant packing density albeit forming domains of different sizes.[64] Although the receptors were found to pack at different densities, the linear increase in the number of receptors with domain size suggested a general *building block* type of assembly for these receptors[64] as opposed to the heterogeneous packing exhibited by DC-SIGN.[60]

Ianoul *et al.* have also used dual-colour NSOM to investigate the association of β-adrenergic receptors (βAR) and caveolae of the surface of dried cardiac myocytes.[65] The study showed that ~15–20% β_2ARs colocalize in caveolae. The lack of complete colocalization of β_2AR with the caveolae suggested that the diverse functional properties of the β_2AR could arise from its association with multiprotein complexes of different compositions that may not be caveolar in nature. Interestingly, the fraction of β_2ARs not colocalizing with caveolae appeared proximal to it, indicating β_2AR complexes are pre-assembled in, or near caveolae.[65] More conventionally used techniques such as fluorescence resonance energy transfer are unable to report on such a proximity effect at spatial scales >10 nm. On the other extreme, diffraction limited techniques such as confocal microscopy will not be able to reveal a lack of co-localization if multiple components are located at distances <300 nm. As such, NSOM is capable of bridging the gap between 10 and 300 nm, providing valuable information at these important spatial scales.

NSOM in dry conditions has also been used to spatially relate topographical features to two different lipid species.[66] Both GM1 and GM3 were seen to cluster in 40–360 nm domains that distributed randomly on the plasma membrane of epithelial cells. However, upon closer examination, it appeared that the GM3 clusters were localized on the peaks of microvillus-like structures.[66] In contrast, the majority of the GM1 lipid clusters were found in the valleys or slopes of these topographical protrusions.[66] These results highlight the importance of correlating topography and optical information uniquely afforded by NSOM. Along these lines, it is worthy to mention that several groups have also implemented AFM in combination with confocal microscopy to correlate topography with fluorescence information, albeit at lower optical resolution (diffraction-limited). On the other hand, a combination of AFM and confocal FCS can also provide complementary information on the dynamics of different nano-environments on membranes and correlate it with topographic information as afforded by AFM.[56]

More recently, our group has applied dual-colour NSOM in physiological conditions together with detailed statistical analysis to follow the spatial nano-scale organization of the integrin receptor LFA-1 and its association with membrane "rafts" along different stages of integrin activation.[67] Rafts have been implicated in regulation of integrin-mediated cell adhesion, although the underlying mechanism has remained elusive. We used single-molecule NSOM with localization accuracy of ~3 nm, to capture the spatio-functional relationship between the integrin LFA-1 and raft components (GPI-APs) on immune cells. Our experiments showed that in resting cells, LFA-1 organizes in small nanoclusters of about 80 nm, while GPI-APs organized largely as monomers. Interestingly, a 20% subpopulation of GPI-APs formed small oligomers on the cell surface and concentrated in regions smaller than 250 nm, suggesting a hierarchical pre-arrangement of GPI-APs on resting monocytes. Dual-colour NSOM demonstrated that integrin nanoclusters are spatially different but reside proximal to GPI-AP nanodomains, forming hotspots on the cell surface (Fig. 9.6). Ligand-mediated integrin activation resulted in an interconversion from monomers to nanodomains of GPI-APs and the generation of nascent adhesion sites where integrin and GPI-APs colocalized at the nanoscale. Cholesterol depletion significantly affected the reciprocal distribution pattern of LFA-1 and GPI-APs in the resting state, and LFA-1 adhesion to its ligand. As such, our data demonstrated the existence of nanoplatforms as essential intermediates in nascent cell adhesion.[67] Since raft association with a variety of membrane proteins other than LFA-1 has been documented, we proposed that hotspots regions enriched with raft components and functional receptors may constitute a prototype of nanoscale inter-receptor assembly and correspond to a generic mechanism to offer cells with privileged areas for rapid cellular function and responses to the outside world.

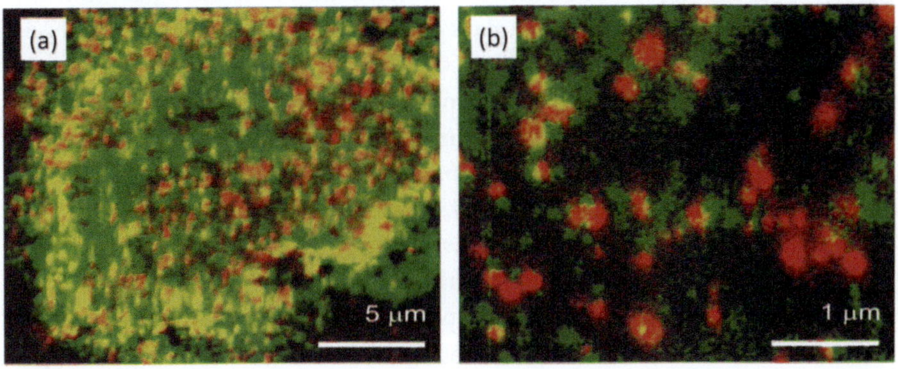

Figure 9.6. (a) Dual-colour confocal image of the integrin receptor LFA-1 (red) and GPI-anchored proteins (green) on the cell membrane of monocytes. The large extent of yellow patches on the image resulting from the lack of sufficient spatial resolution suggests colocalization of LFA-1 and GPIs. (b) Dual-colour super-resolution NSOM demonstrates that in the resting state, LFA-1 and GPIs do not colocalize at the nanometre scale and reside in different nanocompartments of the cell membrane. Image adapted with permission from Ref. 67. © (2009) National Academy of Sciences, USA.

9.4 FUTURE PROSPECTS IN NSOM

The examples summarized in this chapter clearly illustrate the potential of NSOM as a quantitative microscopy tool for biological imaging. Nevertheless, the technical complexities associated with NSOM have limited so far its widespread use in the biological community. In turn, far-field super-resolution approaches are gaining increasing importance in the last two to three years. From the technical point-of-view, one of the major challenges of NSOM is the fabrication of bright, robust and truly nanometre-sized probes required for high resolution. In aperture-type of illumination NSOM, only a small fraction (10^{-4} to 10^{-6}) of the light coupled to the fibre is emitted through the aperture, resulting in low transmission. Together with the finite skin depth of the metal, the practical resolution is thus constrained to ~50 nm. Fortunately, recent developments in the field of nanophotonics and specifically nanoplasmonics are triggering a renewed interest in the NSOM community. This is because optical antennas in combination with plasmonics promise super-resolution at the nanometre scale accompanied by a great degree of electric field enhancement and thus brighter local illumination sources.

The main idea of optical antennas is to localize and enhance the optical radiation to a nanometric region, similar to electromagnetic antennas, which convert propagating radiation into a confined zone. In the biological context, gold nanoparticles attached to glass tips have been exploited as nano-antennas[33] and used to image single Ca^{2+} channels on erythrocyte plasma membranes at 50 nm optical resolution.[34] Unfortunately, the method relies on far-field illumination to excite the antenna, therefore adding a significant background contribution to the antenna response and requiring modulation techniques to reduce the background.[35] A different excitation scheme that suppresses background illumination was first proposed by Frey *et al.*[68] and more recently refined by Taminiau *et al.*[69] In these tip-on-aperture antennas, the local illumination properties of aperture-type NSOM are used to drive the antenna to resonance. Using this configuration, single-molecule detection with 30 nm resolution and virtually no background has been recently demonstrated.[69] Although nanoscale imaging of biological samples should be one of the most promising applications of this approach,[70] its use in intact cell membranes in physiological conditions has not been explored until very recently.

Our group has recently demonstrated the potential of optical antennas for nanobioimaging of individual receptors and nanodomains on intact cells of the immune system.[71] The probe-based monopole optical antennas were fabricated by carving of the antenna on the tip apex of conventional NSOM probes at the glass–metal interface using (Ga+)-FIB milling.[71] The geometry, i.e., length, width and radius, of the curvature of the antennas can be carefully controlled during FIB to maximize their response in liquid conditions. In our case, the dimensions of the fabricated antennas varied from 50 to 60 nm in width, ~20 nm of radius of curvature and lengths between 90 and 135 nm (Fig. 9.7). These probes were then used under appropriate excitation antenna conditions to image individual antibodies in liquid conditions with an unprecedented resolution of 26 ± 4 nm and virtually no surrounding background. On intact cell membranes in physiological conditions, the obtained resolution is currently 30 ± 6 nm. Importantly, the method allowed us to distinguish individual proteins from nanodomains and to quantify the degree of clustering by directly measuring physical size and intensity of individual fluorescent spots.[71] Improved antenna geometries by carefully reducing the width and adjusting the length to optimum resonance in liquid conditions should lead to true live cell imaging below 10 nm resolution with position accuracy in the sub-nanometric range.

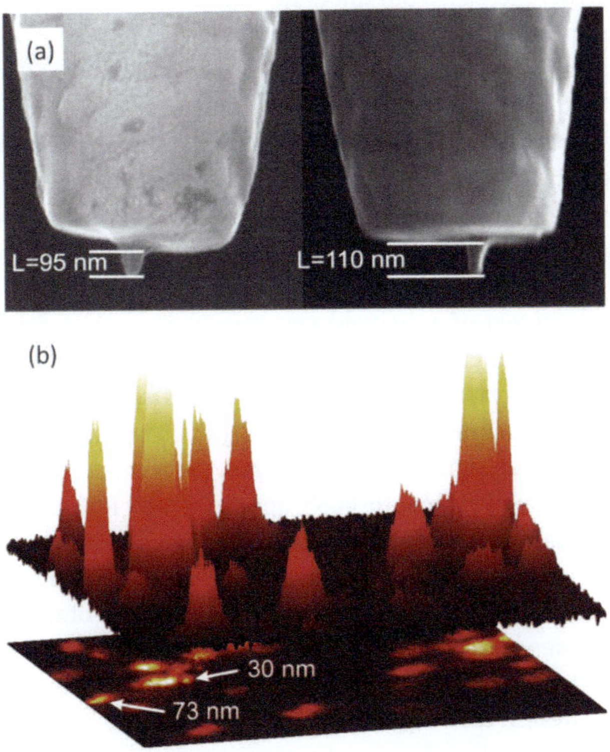

Figure 9.7. (a) Different antenna probes fabricated using FIB milling. (b) Super-resolution image of LFA-1 on monocytes in liquid conditions obtained with an antenna probe. Image adapted with permission from Ref. 71. © (2010) Wiley-VHC.

Aside from these developments in nanophotonics to increase further the resolution of NSOM, there is also growing interest in incorporating into NSOM other capabilities, which have been recently reported for AFM. In particular, it will be extremely appealing to implement the molecular recognition technique (TREC) in NSOMs. TREC, which has been introduced by the Hinterdorfer group a few years ago,[22] involves the use of chemically functionalized tips to identify and localize specific molecules on the cell membrane during scanning.[72] When combined with NSOM, one could in principle obtain simultaneous topography, molecular recognition and optical imaging at the nanometre scale and in one single measurement.

9.5 CONCLUSIONS

The past few years have witnessed tremendous technical advances in super-resolution optical microscopy using both far- and near-field methods. This has in turn further increased our understanding on the compartmentalization of the cell membrane and its implications in cellular function and diseases. However, a significant number of questions are still open and awaiting for techniques that combine high spatial and temporal resolution in one and the same instrument. Far-field super-resolution methods have already demonstrated the possibility of following the dynamics of slowly moving receptors on the cell membrane on small fields of views or in combination with a FCS approach. Further developments of probes and instrumentation will certainly lead to improvement of these techniques.

Within the context of near-field super-resolution, first demonstrations of NSOM measurements on living cells have been reported although high-resolution dynamics on the membrane of living cells is yet to be demonstrated. Obviously, if the scanning speed is not significantly faster than protein diffusion, the optical signal will be blurred. Nevertheless, the promising demonstration of subwavelength-FCS[57] opens the way for probing dynamics at relevant spatial scales potentially revealing the driving mechanisms for nanodomain formation and evolution during cell activation. Additionally, multicolour cross-correlation should indicate if certain proteins are diffusing in identical or separate domains. The combination of capabilities that is offered by NSOM makes the technique a worthy and essential asset in the spectra of biophysical techniques available nowadays.

References

1. Yechiel, E., and Edidin, M. (1987) Micrometer-scale domains in fibroblast plasma membranes, *J. Cell Biol.*, **105**, 755–760.
2. Jacobson, K., Sheets, E. D., and Simson, R. (1995) Revisiting the fluid mosaic model of membranes, *Science*, **268**, 1441–1442.
3. Kusumi, A., and Sako, Y. (1996) Cell surface organization by the membrane skeleton, *Curr. Opin. Cell Biol.*, **8**, 566–574.
4. Simons, K., and Ikonen, E. (1997) Functional rafts in cell membranes, *Nature*, **387**, 569–572.
5. Simons, K., and Toomre, D. (2000) Lipid rafts and signal transduction, *Nat. Rev. Mol. Cell Biol.*, **1**, 31–39.
6. Vereb, G., Szollosi, J., Matko, J., Nagy, P., Farkas, T., Vigh, L., Matyus, L., Waldmann, T. A., and Damjanovich, S. (2003) Dynamic, yet structured: the cell membrane

three decades after the Singer–Nicolson model, *Proc. Nat. Acad. Sci. USA*, **100**, 8053–8057.

7. Cambi, A., de Lange, F., van Maarseveen, N. M., Nijhuis, M., Joosten, B., van Dijk, E. M. H. P., de Bakker, B. I., Fransen, J. A. M., Bovee-Geurts, P. H. M., van Leeuwen, F. N., van Hulst, N. F., and Figdor, C. G. (2004) Microdomains of the C-type lectin DC-SIGN are portals for virus entry into dendritic cells, *J. Cell Biol.*, **164**, 145–155.
8. Sharma, P., Varma, R., Sarasij, R. C., Ira, Gousset, K., Krishnamoorthy, G., Rao, M., and Mayor, S. (2004) Nanoscale organization of multiple GPI-anchored proteins in living cell membranes, *Cell*, **116**, 577–589.
9. Shaw, A. S. (2006) Lipid rafts: now you see them, now you don't, *Nat. Immunol.*, **7**, 1139–1142.
10. Kusumi, A., Nakada, C., Ritchie, K., Murase, K., Suzuki, K., Murakoshi, H., Kasai, R. S., Kondo, J., and Fujiwara, T. (2005) Paradigm shift of the plasma membrane concept from the two-dimensional continuum fluid to the partitioned fluid: high-speed single-molecule tracking of membrane molecules, *Annu. Rev. Biophys. Biomol. Struct.*, **34**, 351–378.
11. Jacobson, K., Mouritsen, O. G., and Anderson, R. G. W. (2007) Lipid rafts: at a crossroad between cell biology and physics, *Nat. Cell Biol.*, **9**, 7–14.
12. Stephens, D. J., and Allan, V. J. (2003) Light microscopy techniques for live cell imaging, *Science*, **300**, 82–86.
13. Michalet, X., Kapanidis, A. N., Laurence, T., Pinaud, F., Doose, S., Pfughoefft, M., and Weiss, S. (2003) The power and prospects of fluorescence microscopies and spectroscopies, *Annu. Rev. Biophys. Biomol. Struct.*, **32**, 161–182.
14. Betzig, E., and Chichester, R. J. (1993) Single molecules observed by near-field scanning optical microscopy, *Science*, **262**, 1422–1425.
15. Weiss, S. (1999) Fluorescence spectroscopy of single biomolecules, *Science*, **283**, 1676–1683
16. Moerner, W. E. (2002) A dozen years of single molecule spectroscopy in physics, chemistry and biology, *J. Phys. Chem. B.*, **106**, 910–927.
17. Tsien, R. Y. (1998) The green fluorescent protein, *Annu. Rev. Biochem.*, **67**, 509–544.
18. Lippincott-Schwartz, J., and Patterson, G. H. (2003) Development and use of fluorescence protein markers in living cells, *Science*, **300**, 87–91.
19. Hansma, P. K., Cleveland, J. P., Radmacher, M., Walters, D. A., Hillner, P. E., Bezanilla, M., Fritz, M., Vie, D., Hansma, H. G., and Prater, C. B. (1994) Tapping mode atomic force microscopy in liquids, *Appl. Phys. Lett.*, **64**, 1738–1740.
20. Putman, C. A., van der Werf, K. O., de Groot, B. G., van Hulst, N. F., and Greve, J. (1994) Tapping mode atomic force microscopy in liquid, *Appl. Phys. Lett.*, **64**, 2454–2456.
21. Willemsen, O. H., Snel, M. M., Cambi, A., Gerve, J., de Groot, B. G., and Figdor, C. G. (2000) Biomolecular interactions measured by atomic force microscopy, *Biophys. J.*, **79**, 3267–3281.

22. Stroh, C., Wang, H., Bash, R., Ashcroft, B., Nelson, J., Gruber, H., Lohr, D., Lindsay, S. M., and Hinterdorfer, P. (2004) Single-molecule recognition imaging microscopy, *Proc. Natl. Acad. Sci. USA*, **101**, 12503–12507.
23. Hell, S. W. (2007) Far-field optical nanoscopy, *Science*, **316**, 1153–1158.
24. Betzig, E., Patterson, G. H., Sougrat, R., Lindwasser, O. W., Olenych, S., Bonifacino, J. S., Davidson, M. W., Lippincott-Schwartz, J., and Hess, H. F. (2006) Imaging intracellular fluorescent proteins at nanometer resolution, *Science*, **313**, 1642–1645.
25. Hess, S. T., Girirajan, T. P. K., and Mason, M. D. (2006) Ultra-high resolution imaging by fluorescence photoactivation localization microscopy, *Biophys. J.*, **91**, 4258–4272.
26. Rust, M. J., Bates, M., and Zhuang, X. (2006) Sub-diffraction-limit imaging by stochastic optical reconstruction microscopy (STORM), *Nat. Methods*, **3**, 793–795.
27. Lippincott-Schwartz, J., and Manley, S. (2009) Putting super-resolution fluorescence microscopy to work, *Nat. Methods*, **6**, 21–23.
28. Fernandez-Suarez, M., and Ting, Y. A. (2008) Fluorescence probes for super-resolution imaging in living cells, *Nat. Rev. Mol. Cell Biol.*, **9**, 929–943.
29. Pohl, D. W., Denk, W., and Lanz, M. (1984) Optical stethoscopy: image recording with resolution $\lambda/20$, *Appl. Phys. Lett.*, **44**, 651–653.
30. Zenhausern, F., Martin, Y., and Wickramasinghe, H. K. (1995) Scanning interferometric apertureless microscopy: optical imaging at 10 Angstrom resolution, *Science*, **269**, 1083–1085.
31. Novotny, L., and Stranick S. J. (2006) Near-field optical microscopy and spectroscopy with pointed probes, *Annu. Rev. Phys. Chem.*, **57**, 303–331.
32. Sanchez, E. J., Novotny, L., and Sunney-Xie, X. (1999) Near-field fluorescence microscopy based on two-photon excitation with metal tips, *Phys. Rev. Lett.*, **82**, 4014–4017.
33. Anger, P., Bharadwaj, P., and Novotny, L. (2006) Enhancement and quenching of single-molecule fluorescence, *Phys. Rev. Lett.*, **96**, 1130021–1130024.
34. Höppener, C., and Novotny, L. (2008) Antenna-based optical imaging of single Ca^{2+} transmembrane proteins in liquids, *Nano Lett.*, **8**, 642–646.
35. Höppener, C., Beams, R., and Novotny, L. (2009) Background suppression in near-field optical imaging, *Nano Lett.*, **9**, 903–908.
36. Paesler, M. A., and Moyer, P. J. (1996) *Near-Field Optics: Theory, Instrumentation and Applications*, Wiley, New York, USA.
37. Veerman, J. A., Otter, A. M., Kuipers, L., and van Hulst, N. F. (1998) High definition aperture probes for near-field optical microscopy fabricated by focused ion beam milling, *Appl. Phys. Lett.*, **72**, 3115–3117.
38. Karrai K., and Grober, R. D. (1995) Piezoelectric tip-sample distance control for near field optical microscopes, *Appl. Phys. Lett.*, **66**, 1842–1844.
39. Koopman, M., de Bakker, B. I., Garcia-Parajo, M. F., and van Hulst, N. F. (2003) Shear force imaging of soft samples in liquid using a diving bell concept, *Appl. Phys. Lett.*, **83**, 5083–5085.

40. Koopman, M., Cambi, A., de Bakker, B. I., Joosten, B., Figdor, C. G., van Hulst, N. F., and Garcia-Parajo, M. F. (2004) Near-field scanning optical microscopy in liquid for high resolution single molecule detection on dendritic cells, *FEBS Lett.*, **573**, 6–10.

41. Höppener, C., Siebrasse, J. P., Peters, R., Kubitscheck, U., and Naber, A. (2005) High-resolution near-field optical imaging of single nuclear pore complexes under physiological conditions, *Biophys. J.*, **88**, 3681–3688.

42. Korchev, Y. E., Raval, M., Lab, M. J., Gorelik, J., Edwards, C. R. W., Rayment, T., and Klenerman, D. (2000) Hybrid scanning ion conductance and scanning near-field optical microscopy for the study of living cells, *Biophys. J.*, **78**, 2675–2679.

43. de Lange, F., Cambi, A., Huijbens, R., de Bakker, B. I., Rensen, W., Garcia-Parajo, M. F., van Hulst, N. F., and Figdor, C. G. (2001) Cell biology beyond the diffraction limit: near-field scanning optical microscopy, *J. Cell Sci.*, **114**, 4153–4160.

44. Garcia-Parajo, M. F., de Bakker, B. I., Koopman, M., Cambi, A., de Lange, F., Figdor, C. G., and van Hulst, N. F. (2005) Near-field fluorescence microscopy: an optical nanotool to study protein organization at the cell membrane, *Nanobiotechnology*, **1**, 113–120.

45. Hwang, J., Tamm, L. K., Bohm, C., Ramalingam, T. S., Betzig, E., and Edidin, M. (1995) Nanoscale complexity of phospholipid monolayers investigated by near-field scanning optical microscopy, *Science*, **270**, 610–614.

46. Hwang, J., Gheber, L. A., Margolis, L., and Edidin, M. (1998) Domains in cell plasma membranes investigated by near-field scanning optical microscopy, *Biophys. J.*, **74**, 2184–2190.

47. Hollars, C. W., and Dunn, R. C. (1998) Submicron structure in l-a-dipalmitoylphosphatidyl-choline monolayers and bilayers probed with confocal, atomic force, and near-field microscopy, *Biophys. J.*, **75**, 342–353.

48. Hollars, C. W., and Dunn, R. C. (1997) Submicron fluorescence, topography, and compliance measurements of phase-separated lipid monolayers using tapping-mode near-field scanning optical microscopy, *J. Phys. Chem. B*, **101**, 6313–6317.

49. Burgos, P., Yuan, C., Viriot, M.-L., and Johnston, L. J. (2003) Two-color near-field fluorescence microscopy studies of microdomains ("Rafts") in model membranes, *Langmuir*, **19**, 8002–8009.

50. Dahim, M., Mizuno, N., Li, X. M., Momsen, W. E., Momsen, M. M., and Brockman, H. L. (2002) Physical and photophysical characterization of a BODIPY phosphatidylcholine as a membrane probe, *Biophys. J.*, **83**, 1511–1524.

51. Coban, O., Burger, M., Laliberte, M., Ianoul, A., and Johnston, L. J. (2007) Ganglioside partitioning and aggregation in phase-separated monolayers characterized by bodipy GM1 monomer/dimer emission, *Langmuir*, **23**, 6704–6711.

52. Ianoul, A., Burgos, P., Lu, Z., Taylor, R. S., and Johnston, L. J. (2003) Phase separation in supported phospholipid bilayers visualized by near-field scanning optical microscopy in aqueous solution, *Langmuir*, **19**, 9246–9254.

53. Flanders, B. N., and Dunn, R. C. (2002) A near-field microscopy study of submicron domain structure in a model lung surfactant monolayer, *Ultramicroscopy*, **91**, 245–251.

54. Sibug-Aga, R., and Dunn, R.C. (2004) High-resolution studies of lung surfactant collapse, *Photochem. Photobiol.*, **80**, 471–476.

55. Murray, J., Cuccia, L., Ianoul, A., Cheetham, J., and Johnston, L. J. (2004) Imaging the selective binding of synapsin to anionic membrane domains, *Chembiochem*, **5**, 1489–1494.

56. Chiantia, S., Kahya, N., Ries, J., and Schwille, P. (2006) Effects of ceramide on liquid-ordered domains investigated by simultaneous AFM and FCS, *Biophys. J.*, **90**, 4500–4508.

57. Vobornik, D., Banks, D. S., Lu, Z., Fradin, C., Taylor, R., and Johnston, L. J. (2008) Fluorescence correlation spectroscopy with sub-diffraction-limited resolution using near-field optical probes, *Appl. Phys. Lett.*, **93**, 163904–163906.

58. Wenger, J., Conchonaud, F., Dintinger, J., Wawrezinieck, L., Ebbesen, T. W., Rigneault, H., Marguet, D., and Lenne, P. F. (2007) Diffusion analysis within single nanometric apertures reveals the ultrafine cell membrane organization, *Biophys. J.*, **92**, 913–919.

59. Eggeling, C., Ringemann, C., Medda, R., Schwarzmann, G., Sandhoff, K., Polyakova, S., Belov, V. N., Hein, B., von Middendorff, C., Schonle, A., and Hell, S. W. (2009) Direct observation of the nanoscale dynamics of membrane lipids in a living cell, *Nature*, **457**, 1159–1162.

60. de Bakker, B. I., de Lange, F., Cambi, A., Korterik, J. P., van Dijk, E. M. H. P., van Hulst, N. F., Figdor, C. G., and Garcia-Parajo, M. F. (2007) Nanoscale organization of the pathogen receptor DC-SIGN mapped by single-molecule high-resolution fluorescence microscopy, *Chemphyschem*, **8**, 1473–1480.

61. Chen, Y., Shao, L., Ali, Z., Cai, J., and Chen, Z. W. (2008) NSOM/QD-based nanoscale immunofluorescence imaging of antigen-specific T-cell receptor responses during an in vivo clonal Vg2V*d*2 T-cell expansion, *Blood*, **111**, 4220–4232.

62. Nagy, P., Jenei, A., Kirsch, A. K., Sazollosi, J., Damjanovich, S., and Jovin, T. M. (1999) Activation-dependent clustering of the erbB2 receptor tyrosine kinase detected by scanning near-field optical microscopy, *J. Cell Sci.*, **112**, 1733–1741.

63. Enderle, T., Ha, T., Ogletree, D. F., Chemla, D. S., Magowan, C., and Weiss, S. (1997) Membrane specific mapping and colocalization of malarial and host skeletal proteins in the *Plasmodium falciparum* infected erythrocyte by dual-color near-field scanning optical microscopy, *Proc. Natl. Acad. Sci. USA*, **94**, 520–525.

64. de Bakker, B. I., Bodnar, A., van Dijk, E. M. H. P., Vámosi, G., Damjanovich, S., Waldmann, T. A., van Hulst, N. F., Jenei, A., and Garcia-Parajo, M. F. (2008) Nanometer-scale organization of the alpha subunits of the receptors for IL2 and IL15 in human T lymphoma cells, *J. Cell Sci.*, **121**, 627–633.

65. Ianoul, A., Grant, D. D., Rouleau, Y., Bani-Yaghoub, M., Johnston, L. J., and Pezacki, J. P. (2005) Imaging nanometer domains of β-adrenergic receptor complexes on the surface of cardiac myocytes, *Nat. Chem. Biol.*, **1**, 196–202.

66. Chen, Y., Qin, J., and Chen, Z. (2008) Fluorescence-topographic NSOM directly visualizes peak-valley polarities of GM1/GM3 rafts in cell membrane fluctuations, *J. Lipid Res.*, **49**, 2268–2275.

67. van Zanten, T. S., Cambi, A., Koopman, M., Joosten, B., Figdor, C. G., and Garcia-Parajo, M. F. (2009) Hotspots of GPI-anchored proteins and integrin nanoclusters function as nucleation sites for cell adhesion, *Proc. Natl. Acad. Sci. USA*, **106**, 18557–18562.

68. Frey, H. G., Witt, S., Felderer, K., and Guckenberger, R. (2004) High-resolution imaging of single fluorescent molecules with the optical near-field of a metal tip, *Phys. Rev. Lett.*, **93**, 200801–200804.

69. Taminiau, T. H., Moerland, R. J., Segerink, F. B., Kuipers, L., and van Hulst, N. F. (2007) $\lambda/4$ resonance of an optical monopole antenna probed by single molecule fluorescence, *Nano Lett.*, **7**, 28–33.

70. Garcia-Parajo, M. F. (2008) Optical antennas focus in on biology, *Nat. Photonics*, **2**, 201–203.

71. van Zanten, T. S., Lopez, M. J., and Garcia-Parajo, M. F. (2010) Imaging individual proteins and nanodomains on intact cell membranes with a probe-based optical antenna, *Small*, **6**, 270–275.

72. Chtcheglova, L. A., Waschke, J., Wildling, L., Drenckhahn, D., and Hinterdorfer, P. (2007) Nano-scale dynamic recognition imaging on vascular endothelial cells, *Biophys. J.*, **93**, L11–L13.

Chapter 10

QUANTIFYING CELL ADHESION USING SINGLE-CELL FORCE SPECTROSCOPY

Anna Taubenberger, Jens Friedrichs and Daniel J. Müller
Biotechnological Center, University of Technology Dresden, Tatzberg 47-51, 01307 Dresden, Germany, and Biosystems Science and Engineering, ETH Zürich, Mattenstr. 26, 4058 Basel, Switzerland
daniel.mueller@bsse.ethz.ch

10.1 ON THE IMPORTANCE OF CELL ADHESION

Adhesive interactions of cells with their surrounding regulate cell growth, differentiation, migration and survival in multicellular organisms and are therefore essential to tissue homeostasis.[1-7] Adhesive interactions are mediated by different types of cell adhesion molecules (CAMs), mainly cadherins, integrins, selectins and adhesion molecules of the immunoglobulin family. CAMs are transmembrane proteins, composed of an intracellular domain that interacts with cytoplasmic proteins including the cytoskeleton and an extracellular domain that specifically binds to adhesion partners.[8] CAMs mediate homotypic (cadherins) and heterotypic (selectins, integrins) interactions between cells as well as interactions between cells and extracellular matrix (ECM) proteins (mainly integrins). Since adhesive interactions of CAMs are essential to cell physiology as well as pathology, their basic binding and regulation mechanisms are of great interest. This chapter presents an overview of methods that allow detection and quantification of cell adhesion, with an emphasis on atomic force microscopy (AFM)-based single-cell force spectroscopy (SCFS).

Life at the Nanoscale: Atomic Force Microscopy of Live Cells
Edited by Yves Dufrêne

www.panstanford.com

10.2 METHODS TO MEASURE CELL ADHESION

Different assays have been developed to qualitatively and quantitatively study cell adhesion. Usually, these assays probe the ability of cells to remain attached to an adhesive substrate when exposed to a certain detachment force. This adhesive substrate can be another cell, a surface or an organic matrix. Adhesion assays can be classified into *bulk* or *single-cell* assays.

10.2.1 Bulk Assays

The most commonly used bulk assay to study cell adhesion is the washing assay. In this assay, cells are seeded onto an adhesive substrate of interest, allowed to adhere for a given time and rinsed with physiological buffer. Thereby, non- or weakly attached cells are dislodged from the substrate and the fraction of attached cells is determined.[9–12] Washing assays are hardly reproducible since the applied shear forces are unknown, unevenly distributed and difficult to control. Moreover, washing assays can hardly provide quantitative information about cell adhesion strengths or energies. Reproducible results that qualitatively describe cell adhesion have been obtained using parallel plate chamber setups, spinning disc devices[13] or centrifugation assays.[14,15] However, these techniques also have limitations since the resistance of cells to detachment by hydrodynamic or centrifugal forces depends not only on the number, distribution and strength of the formed adhesion bonds, but also on the spread area and surface topography of the cells. Thus, with these assays, the adhesive strength of cells can only be estimated.

On the other hand, using bulk assays, statistically relevant data can be easily obtained as large numbers of cells are tested in each experiment. However, these assays only analyze the average behaviour of large cellular populations. Therefore, they are rather constricted in determining potential differences in the adhesion of individual cells. Adhesive subpopulations might result from different functional states of individual cells. Such valuable information can be easily missed in bulk assays.

10.2.2 Single-Cell Assays

For a more quantitative approach, techniques that measure the adhesion of single cells are needed. Compared with bulk assays, SCFS assays are usually time-consuming, since only a single cell is analyzed in each experiment. However, a clear advantage of single-cell approaches over bulk assays is that adhesive subpopulations of cells can be identified. Most SCFS techniques allow characterizing cell adhesion down to the single-molecule level, thereby providing detailed insights into regulation mechanisms of adhesion receptors.

SCFS techniques include micropipettes, laser tweezers, magnetic tweezers and AFM.

Several *micropipette techniques*, including the step pressure technique[16] and the biomembrane force-probe (BFP),[17-20] have been developed that operate both at the cellular and molecular level. These methods were applied to study single-molecule interactions, membrane tether formation from single cells and overall cell adhesion.[21] However, whereas the step pressure technique[16] is characterised by a low force resolution (≈100 pN), the BFP is applicable over a rather limited range of forces (≈0.1 to 1000 pN).[16,17,22] Disadvantageously, this force range does not allow to monitor the formation of higher-ordered adhesive structures.[60]

Optical tweezers have been employed to measure the interaction of cells with functionalized microspheres.[23,24] Alternatively, whole cells have been trapped in laser beams, and their adhesion to functionalized substrates has been probed. The confined force detection range of optical tweezers (0.1–100 pN)[25] limits their applicability mainly to single-molecule studies. Moreover, the high laser intensity at the focus of the beam can cause local temperature increase, which may damage cells.[25]

Magnetic tweezers have also been applied to measure cell–substrate interactions. In these experiments, a substrate-coated magnetic microsphere is brought into contact with a cell and detached by generating a magnetic field.[25] Alternatively, a bead-coupled cell can be probed on an adhesive substrate.[26] Similar to optical tweezers, the force detection range of the two latter magnetic tweezer setups (0.01–100 pN)[25] limits their applicability to single-molecule studies. Recently, a novel magnetic tweezer setup was introduced that allows applying forces of up to 100 nN to magnetic beads.[27] This later setup designed to measure rather high forces is sufficient to study the rupture of complex cell adhesion sites. However, it shows a rather poor force sensitivity that hardly enables to detect adhesive forces of single CAMs. To sum up, the SCFS techniques described are restricted either to the analysis of single-molecule interactions or to the detection of overall cell adhesion at lower force resolution. In the following, we will introduce the principles and advantages of AFM-based SCFS to characterize the adhesion of single cells to molecular resolution.

10.3 AFM-BASED SCFS

AFM-based SCFS is the most versatile among the mentioned SCFS techniques since it allows the largest range of forces, from ≈5 pN to about 100 nN, to be measured.[28] In this setup, a cell is attached to an AFM cantilever, and the adhesive strength to a protein-coated surface (Fig. 10.1a) or another cell (Fig. 10.1c) is quantified.[29-31] Alternatively, the cantilever is functionalized with a protein of interest, and adhesion is probed to an immobilized cell (Fig. 10.1b).[32] When using a functionalized AFM tip and reduced contact times and contact forces between tip and cell surface, the binding probability of the probing tip with the cell surface is rather low. In this case, single binding events dominate, and the assay may be rather related to single-molecule force spectroscopy approaches such as discussed in Chapters 11, 12 and 15. In the majority of cell–surface interaction studies, the first setup (Fig. 10.1a) has been applied and will therefore be detailed.

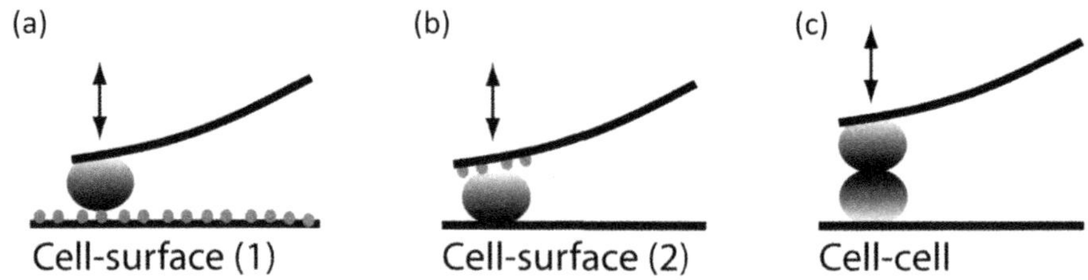

Figure 10.1. SCFS setups to measure cellular interactions with adhesive substrates. (a) A single cell is immobilized to a tip-less AFM cantilever, and the adhesion of the cell to a substrate is probed. (b) A cell, attached to a supporting surface, is probed with a ligand-coated cantilever. (c) To quantify cell–cell adhesion, a cell immobilized to a supporting surface is probed with another cell attached to a cantilever.

10.3.1 Converting a Living Cell into a Probe

To attach a living cell to the cantilever, the cantilever surface has to be functionalized with an adhesive substrate. For the immobilization of eukaryotic cells, concanavalin A, a lectin that binds mannose residues of glycoproteins on the cell surface,[33] is frequently used.[30,34–39] For certain cell types, e.g., T cells, the use of concanavalin A may be problematic since it can lead to cell activation.[40] Alternatively, wheat germ agglutinin,[41] ECM proteins,[42,43] polyphenolic proteins extracted from marine mussels[44,45] or antibodies[42] can be used to attach different cell types to the AFM cantilever. In other studies, cells were biotinylated and attached to a streptavidin-modified cantilever,[46] or cells were directly grown on the cantilever.[31]

To attach a cell to a functionalized cantilever, suspended cells are added into a temperature-controlled fluid chamber. Cantilever and cell are visualized by light microscopy and positioned relative to each other. Then, the cantilever is lowered onto a single cell, gently pressed on it and withdrawn to capture the cell (Fig. 10.2).

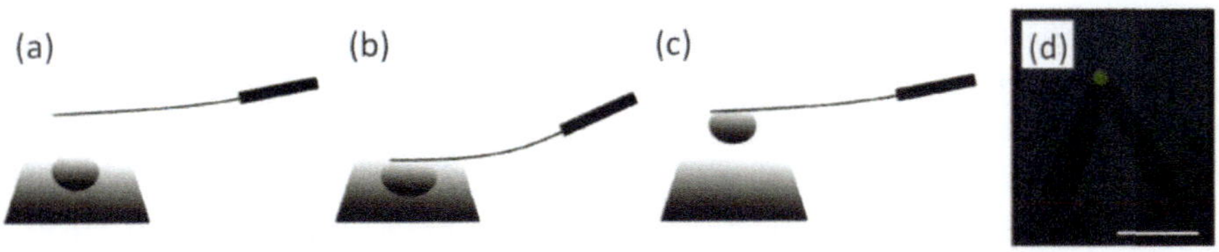

Figure 10.2. Attaching a living cell to the AFM cantilever. (a) The functionalized cantilever is positioned above a cell and gently pressed onto the cell. (b) During contact, adhesive interactions are established between the cell and functionalized cantilever. (c) Thereafter, the cell–cantilever couple is separated from the support. During the next minutes, firm attachment between cell and cantilever is established. (d) A green-fluorescent fibroblast (vinculin-GFP) immobilized on an AFM cantilever. The picture is an overlay of images recorded by phase contrast and epi-fluorescence microscopy. The scale bar corresponds to 100 μm.

10.3.2 Probing Adhesive Interactions of the Cell with the Substrate

To quantify adhesive interactions of the immobilized cell with a given substrate, the cantilever is approached to the substrate until a preset contact

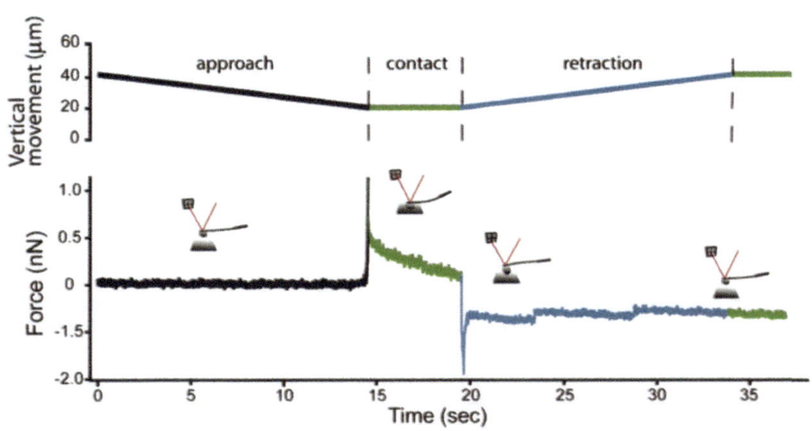

Figure 10.3. Monitoring adhesive forces of a single cell. The vertical approach of the cantilever-attached cell (above) and the force acting on the cantilever (below) versus time during an F-D cycle is depicted. The cell is approached onto a substrate until a given force set-point is reached (black). In the constant height-mode, the vertical position of the cell is kept constant during contact (green). Because of the viscous properties of the cell, the force acting on the cantilever decays rapidly within the first seconds of contact. During retraction (blue), the cantilever bends downwards because of the adhesion established between the cell and substrate. The steps of the retraction F-D curve reflect rupture events of the adhesive interactions that have been formed between cell and substrate. The baseline force level is reached when all connections between cell and substrate have been ruptured.

force is reached. After a defined contact time, the cell is withdrawn at constant velocity. Bonds between cell and substrate sequentially break until cell and substrate are completely separated (Fig. 10.3). During the approach and retract process, the force acting on the cantilever, which is proportional to the cantilever deflection, is recorded in a force–distance (F-D) curve. Parameters influencing the experimental results, such as contact force and time, contact condition (constant height, constant force) and pulling velocity can be precisely controlled.

10.3.3 Interpretation of F-D Curves

When interpreting F-D curves, it must be considered that they contain information about the established interactions between cell and substrate and about mechanical properties of the cell.[28,35,47] Often F-D curves (Fig. 10.4) show characteristic complex interaction patterns. Approach F-D curves (Fig. 10.4, black) are characterized by a steep force increase occurring as soon as the cell is in contact with the substrate. The slope of the F-D curve in this contact region (dashed ellipse E) can be used to extract elastic properties of the cell. The retraction F-D curve (Fig. 10.4, blue) is typically characterized by the maximum force required to separate the cell from the substrate referred to as the detachment force (F_D). F_D is followed by step-like events, which are either preceded (jumps) or not preceded (tethers) by a ramp-like increase in force.

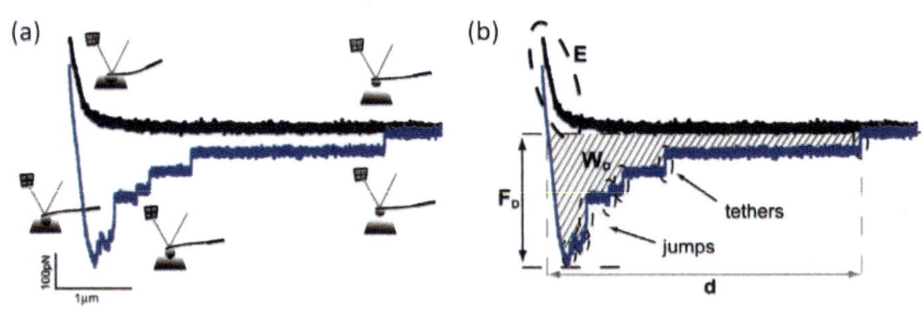

Figure 10.4. Information extracted from an F-D curve. (a) Schematic representation of the F-D cycle and recorded F-D curves. The approach curve is shown in black, retraction curve in blue. (b) Information that can be obtained from the F-D curves: F_D maximal force required to detach the cell from the substrate, E elastic properties, W_D adhesion work, d distance required for complete separation of cell from the substrate, discrete force steps (jumps and tethers).

10.3.3.1 Detachment force F_D

F_D is the maximal force required to separate the cell from the substrate (Fig. 10.4b). But how can detachment force be interpreted? By light microscopy, it is usually observed that a circular area approximates the contact zone between cell and substrate. During the initial detachment phase, bonds in the outer contact zone are predominantly stressed. The cell is stretched until a maximal force is reached. Upon bond failure, the contact zone shrinks. Assuming a homogenous distribution of receptors over the contact zone, more bonds per radial section will have formed at the periphery of the contact zone than in the inner region. Consequently, a maximal force is detected before the bonds at the periphery begin to rupture. Subsequently, the force decreases quickly since the applied force load is shared by fewer receptors in the inner

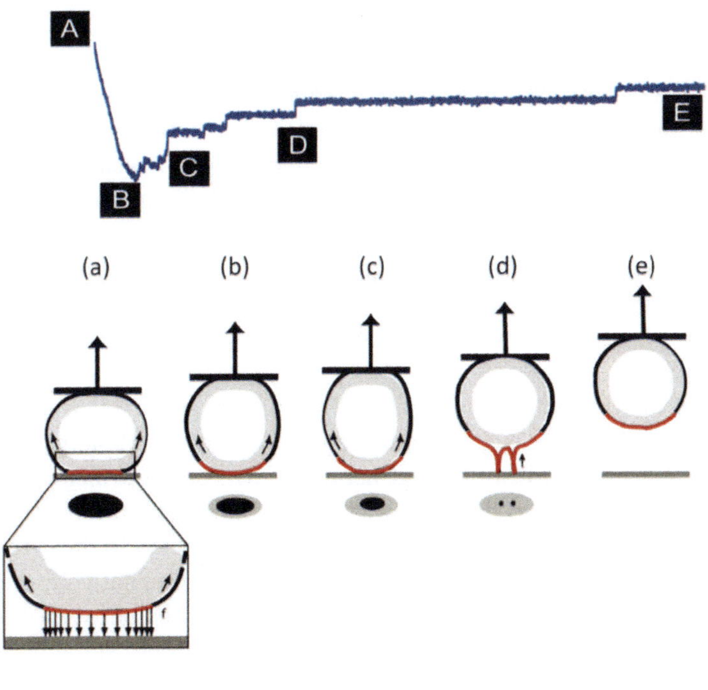

Figure 10.5. Schematic representation of the cell detachment process. The detachment process of a cell can be separated into different phases. (a) The cell is in contact with the substrate. In the contact zone (red) adhesive interactions are established. (b,c) During cell detachment, the established interactions (specific and non-specific) bonds rupture and the contact zone shrinks. When the cell body is separated from the substrate, membrane tethers (nanotubes) link cell and substrate (d) until the cell is fully detached from the substrate (e).

contact zone, and the probability that these bonds can resist large rupture forces decreases (Fig. 10.5).[48] Since the total number of receptors and their binding strengths contribute to F_D, it is most commonly used to quantify the overall cell adhesion. This overall cell adhesion is defined as the sum of all adhesive interactions established between cell and substrate.

10.3.3.2 Analyzing discrete force steps

F-D curves usually display small discrete force steps (Fig. 10.4b) that can be distinguished into jumps and tethers (Fig. 10.4). A non-linear force loading typically precedes jump-events, whereas a force plateau is detected prior to tether-events (Fig. 10.6). Force gradient prior rupture (>0 for jumps, ≈0 for tethers) along with the distance at which force jumps occur can be used to distinguish jump- and tether-events. Separating jump- and tether-events is necessary, since they contribute to different detachment scenarios (Fig. 10.6). *Jump-events* can be interpreted as the unbinding of single or few receptors from the substrate. The non-linear force loading prior to rupture suggests that the probed receptors are connected to the actin cytoskeleton[49] (Fig. 10.6a). To give an example, integrins often localize to specialized complexes involving assemblies of cytoskeletal linker and signalling proteins. Stretching these membrane–cytoskeleton linkers leads to a non-linear force increase prior bond rupture.[49] The magnitude of the force step reflects the stochastic survival of this ligand–receptor bond under an increasing force load.[48,49] The ensemble of jump-events can provide information on the affinity and avidity of receptors.

Tether-events can be found at pulling distances up to several tens of micrometers after the major rupture peak. The force plateau preceding the force step originated from the extraction of membrane tethers (membrane nanotubes) from the cell membrane (Fig. 10.6b). Membrane tethers are formed when receptors that are not or weakly connected to the cytoskeleton are pulled from the cell membrane (Fig. 10.6b).[49] Alternatively, membrane tethers can form by "unspecific" interactions established between membrane and AFM tip. Specific blocking experiments may be suitable to show unambiguously by which of the two binding events membranes tethers are formed.[50]

The force plateau measured upon extraction of a membrane tether shows that the force required to extract the tether is constant over long extraction lengths. The membrane tether restoring force depends on the extraction velocity and does not reflect the strength of the receptor–ligand bond anchoring the membrane tether at its tip.[49,51] Thus, native cell membranes establish force-clamped membrane tethers that can be employed to measure the lifetime of their tethering receptor–ligand or other bond.[50] Extracting membrane tethers AFM can be used to characterize cell membrane properties such as its anchoring to the cytoskeleton or its viscosity.[52,53] Changes in the receptor-cortex anchoring strength are revealed by comparing the number of bond ruptures (jumps) to tether-events.[39] With weaker anchoring of membrane and cytoskeleton, the probability of pulling tethers rises whereas the force required to extract tethers decreases.[50–52]

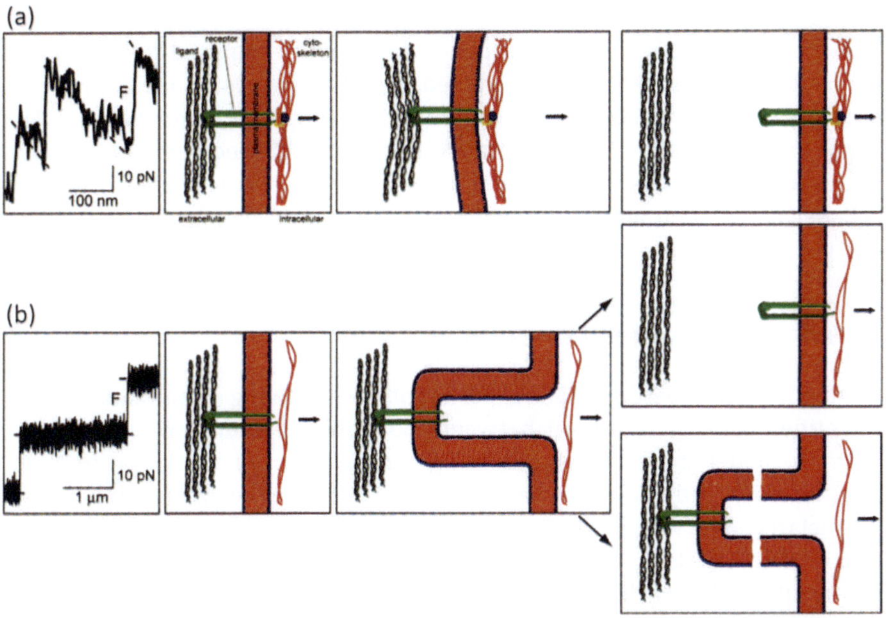

Figure 10.6. Scenarios causing jump- and tether-events. (a) Exemplary jump-events extracted from an F-D curve (left). A receptor anchored to the cytoskeleton binds to a ligand. Upon cantilever retraction, the receptor–membrane–cytoskeleton linker is stretched and the force acting on the cantilever increases. Upon rupture of the receptor–ligand bond, the force on the cantilever rapidly decreases. (b) Exemplary tether-events extracted from an F-D curve (left). A receptor that is not or weakly anchored to the cytoskeleton is extracted from the cell body at the tip of a membrane tether. The force acting on the ligand of the cantilever remains constant during membrane tether extraction. When the receptor–ligand bond breaks (upper sketch), the membrane tether fails (sketch below) or the receptor is pulled out of the membrane, the force on the cantilever decreases staircase-like.

10.3.3.3 Separation distance

The separation distance, *d* (Fig. 10.4b) is the distance at which all linkages between cell and substrate have been ruptured. This length is highly influenced by membrane tethers.

10.3.3.4 Elastic properties of the cell

The slope of the approach F-D curve in the contact region (Fig. 10.4b, dashed ellipse E) is influenced by the elastic properties of the cell. Organized actin filaments of the cell cortex can substantially influence the elastic properties of most cell types.

10.3.3.5 Detachment work

The detachment work, W_D, corresponds to the work required to detach the cell from the substrate. W_D is extracted from F-D curves by measuring the area enclosed by retraction curve and baseline (Fig. 10.4; hatched area). W_D is not only determined by the overall adhesion established between cell and substrate, but is also substantially influenced by the elastic properties of the cell. Because of their length, often tens of micrometres, membrane nanotubes significantly contribute to the W_D.

10.3.4 AFM-Based SCFS—State of the Art

Pioneering AFM-based SCFS experiments with living mammalian cells quantified cell–cell adhesion between trophoblasts and uterine epithelial cells to model interactions occurring during embryo implantation.[31] In another early study, Lehenkari *et al.* analysed adhesion between osteoblasts/osteoclasts and different RGD-containing ligands.[32] In other studies, endothelial cell–leukocyte adhesion[54,55] has been investigated, and the contribution of integrin- and selectin-mediated interactions to this adhesion was demonstrated using blocking antibodies. These experiments have provided insights into the mechanisms underlying adhesive interactions between leukocytes and endothelium that are crucial to initiate the process of transmigration during inflammatory response. Adhesive interactions between endothelial cells and leukocytes involve LFA (lymphocyte function-associated molecule)-1–ICAM (intercellular adhesion molecule)[33,56] and $\alpha_4\beta_1$-integrin–VCAM (vascular cell adhesion molecule).[57] Interactions between these two proteins were explored at the single-molecule level. Thereby, bond-specific parameters such as bond dissociation rates could be determined. Recently, homophilic JAM-A (junctional adhesion molecule) interactions were characterized, and a role of LFA-1 in their regulation was shown.[58] Homophilic and heterophilic N-, E- and VE-cadherin interactions were characterized at the single-molecule level using SCFS.[46,59] Furthermore, adhesive interactions of cell surface integrins with nanopatterned RGD-peptide-coated substrates were quantified revealing insights into the spatial organization of RGD ligands.[60]

For several years, we have established SCFS as a tool to quantify cell adhesion. An important improvement of the experimental setup was the development of a commercially available AFM that features an enhanced pulling range (>100 μm), precision (ΔDistance ≈ 0.1nm) and sensitivity (ΔForce ≈ 5 pN) such as required to conduct cell–cell adhesion measurements.[61] This optimized setup further allowed probing adhesion over a broad range of detachment forces, from single-molecule interactions to high forces exerted by more complex adhesion sites.[62] Moreover, by combining SCFS with advanced optical microscopy techniques, a better control of the experiment was provided. Using this setup, adhesion of gastrulating zebrafish cells to fibronectin-coated surfaces could be quantified and allowed characterizing the role of Wnt11 in modulating integrin-mediated adhesion.[30] In a similar setup, the impact of Wnt11 on intercellular adhesion of gastrulating zebrafish cells was studied. Wnt11 was found to modulate E-cadherin-mediated adhesion via a Rab5-dependent mechanism.[63] Furthermore, it was determined whether the adhesion of germ layers cells contribute to the gastrulation of zebrafish embryos.[38] Fundamental mechanisms underlying cellular sorting during gastrulation could be experimentally verified, and the contribution of cell adhesion and cell cortex tension to cell sorting could be deciphered. In another study, SCFS was applied to characterize integrin $\alpha_2\beta_1$-mediated adhesion to nanopatterned collagen type I matrices. Because of the high force resolution of the setup, it could be resolved that integrin receptors cooperatively assemble to higher-order adhesion structures to enhance cell adhesion.[62] In two further studies, the contribution of galectins to the adhesion of epithelial Madin-Darby canine kidney (MDCK) cells to ECM proteins was quantified. It was found that early adhesion of MDCK cells to laminin-111 was integrin-independent and mediated by carbohydrate interactions and galectins. When adhering to collagen type I and IV, MDCK cells frequently entered an enhanced adhesion state which was characterized by significantly increased detachment forces. Although MDCK cell adhesion was mediated by integrins, adhesion enhancement was observed for those cells in which a certain member of the galectin family, galectin-3, has been depleted. It was proposed that galectin-3 influences integrin-mediated adhesion complex formation.[36,37] In another example, the tyrosine kinase BCR/ABL, a hallmark of chronic myeloid leukaemia, was found to increase the concentration of integrin β_1-subunits on the cell surface and to enhance adhesion of leukemic cells to ECM proteins that have been secreted by bone marrow stromal cells.[34]

10.3.5 Strengths and Limitations of AFM-Based SCFS

With the described AFM-based SCFS setup, various cell adhesion experiments can be performed under near-physiological conditions. AFM-based SCFS allows to measure adhesive forces that range from a few piconewtons up to several hundred nanonewtons. Thus, interactions mediated by single CAMs[28,29,35,46,64] or adhesive interactions established by larger adhesive complexes can be detected in the cellular context.[41,60] Although other SCFS methods can provide a better force resolution (optical tweezer, BFP), AFM-based SCFS is more versatile in terms of the detectable range of adhesive forces. This advantage makes it possible to address a broad range of biological questions. Another advantage over other SCFS assays is the high precision with which the cell can be temporally and spatially manipulated. AFM-based SCFS can be easily combined with optical techniques such as total internal reflection fluorescence microscopy, confocal microscopy, fluorescent microscopy and conventional transmission light microscopy. AFMs specifically developed to perform SCFS are commercially available and relatively easy to use. In the future, establishing cellular assays and standardized experimental protocols for SCFS together with automated data analysis tools will enable newcomers and professionals to explore the molecular mechanisms of cell adhesion.

References

1. Adams, J. C., and Watt F. M. (1993) Regulation of development and differentiation by the extracellular matrix, *Development*, **117**, 1183–1198.
2. Bissell, M. J., Hall, H. G., and Parry, G. (1982) How does the extracellular matrix direct gene expression? *J. Theor. Biol.*, **99**, 31-68.
3. Cukierman, E., Pankov, R., and Yamada, K. M. (2002) Cell interactions with three-dimensional matrices, *Curr. Opin. Cell Biol.*, **14**, 633-639.
4. Damsky, C., Sutherland, A., and Fisher, S. (1993) Extracellular matrix 5: adhesive interactions in early mammalian embryogenesis, implantation, and placentation, *FASEB J.*, **7**, 1320-1329.
5. Damsky, C. H., and Ilic, D. (2002) Integrin signalling: it's where the action is, *Curr. Opin. Cell Biol.*, **14**, 594-602.
6. Juliano, R. L., and Haskill, S. (1993) Signal transduction from the extracellular matrix, *J. Cell Biol.*, **120**, 577-585.
7. Meighan, C. M., and Schwarzbauer, J. E. (2008) Temporal and spatial regulation of integrins during development, *Curr. Opin. Cell Biol.*, **20**, 520-524.
8. Bell, G. I. (1978) Models for specific adhesion of cells to cells, *Science*, **200**, 618-627.
9. Connors, W. L., and Heino, J. (2005) A duplexed microsphere-based cellular adhesion assay, *Anal. Biochem.*, **337**, 246-255.

10. Garcia, A. J., and Gallant, N. D. (2003) Stick and grip: measurement systems and quantitative analyses of integrin-mediated cell adhesion strength, *Cell Biochem. Biophys.*, **39**, 61-73.

11. Klebe, R. J. (1974) Isolation of a collagen-dependent cell attachment factor, *Nature*, **250**, 248-251.

12. Klebe, R. J., Hall, J. R., Rosenberger, P., and Dickey, W. D. (1977) Cell attachment to collagen: the ionic requirements, *Exp. Cell Res.*, **110**, 419-425.

13. Kaplanski, G., Farnarier, C., Tissot, O., *et al.* (1993) Granulocyte-endothelium initial adhesion. Analysis of transient binding events mediated by E-selectin in a laminar shear flow, *Biophys. J.*, **64**, 1922-1933.

14. Lotz, M. M., Burdsal, C. A., Erickson, H. P., and Mcclay, D. R. (1989) Cell adhesion to fibronectin and tenascin: quantitative measurements of initial binding and subsequent strengthening response, *J. Cell Biol.*, **109**, 1795-1805.

15. Mcclay, D. R., Wessel, G. M., and Marchase, R. B. (1981) Intercellular recognition: quantitation of initial binding events, *Proc. Natl. Acad. Sci. USA*, **78**, 4975-4979.

16. Sung, K. L., Sung, L. A., Crimmins, M., Burakoff, S. J., and Chien, S. (1986) Determination of junction avidity of cytolytic T cell and target cell, *Science*, **234**, 1405-1408.

17. Evans, E., Ritchie, K., and Merkel R. (1995) Sensitive force technique to probe molecular adhesion and structural linkages at biological interfaces, *Biophys. J.*, **68**, 2580-2587.

18. Evans, E. A., Waugh, R., and Melnik, L. (1976) Elastic area compressibility modulus of red cell membrane, *Biophys. J.*, **16**, 585-595.

19. Simson, D. A., Ziemann, F., Strigl, M., and Merkel, R. (1998) Micropipet-based pico force transducer: in depth analysis and experimental verification, *Biophys. J.*, **74**, 2080-2088.

20. Zarnitsyna, V. I., Huang, J., Zhang, F., Chien, Y. H., Leckband, D., and Zhu, C. (2007) Memory in receptor-ligand-mediated cell adhesion, *Proc. Natl. Acad. Sci. USA*, **104**, 18037-18042.

21. Shao, J. Y., Xu, G., and Guo, P. (2004) Quantifying cell-adhesion strength with micropipette manipulation: principle and application, *Front. Biosci.*, **9**, 2183-2191.

22. Leckband, D., and Israelachvili, J. (2001) Intermolecular forces in biology, *Q. Rev. Biophys.*, **34**, 105-267.

23. Andersson, M., Madgavkar, A., Stjerndahl, M., *et al.* (2007) Using optical tweezers for measuring the interaction forces between human bone cells and implant surfaces: system design and force calibration, *Rev. Sci. Instrum.*, **78**, 074302.

24. Thoumine, O., Kocian, P., Kottelat, A., and Meister, J. J. (2000) Short-term binding of fibroblasts to fibronectin: optical tweezers experiments and probabilistic analysis, *Eur. Biophys. J.*, **29**, 398-408.

25. Neuman, K. C., and Nagy, A. (2008) Single-molecule force spectroscopy: optical tweezers, magnetic tweezers and atomic force microscopy, *Nat. Methods*, **5**, 491-505.

26. Walter, N., Selhuber, C., Kessler, H., and Spatz, J. P. (2006) Cellular unbinding forces of initial adhesion processes on nanopatterned surfaces probed with magnetic tweezers, *Nano Lett.*, **6**, 398-402.

27. Kollmannsberger, P., and Fabry, B. (2007) High-force magnetic tweezers with force feedback for biological applications, *Rev. Sci. Instrum.*, **78**, 114301.

28. Helenius, J., Heisenberg, C. P., Gaub, H. E., and Muller, D. J. (2008) Single-cell force spectroscopy, *J. Cell Sci.*, **121**, 1785-1791.

29. Li, F., Redick, S. D., Erickson, H. P., and Moy, V. T. (2003) Force measurements of the alpha5beta1 integrin-fibronectin interaction, *Biophys. J.*, **84**, 1252-1262.

30. Puech, P. H., Taubenberger, A., Ulrich, F., Krieg, M., Muller, D. J., and Heisenberg, C. P. (2005) Measuring cell adhesion forces of primary gastrulating cells from zebrafish using atomic force microscopy, *J. Cell. Sci.*, **118**, 4199-4206.

31. Thie, M., Rospel, R., Dettmann, W., Benoit, M., Ludwig, M., Gaub, H. E., and Denker, H. W. (1998) Interactions between trophoblast and uterine epithelium: monitoring of adhesive forces, *Hum. Reprod.*, **13**, 3211-3219.

32. Lehenkari, P. P., and Horton, M. A. (1999) Single integrin molecule adhesion forces in intact cells measured by atomic force microscopy, *Biochem. Biophys. Res. Commun.*, **259**, 645-650.

33. Wojcikiewicz, E. P., Zhang, X., Chen, A., and Moy, V. T. (2003) Contributions of molecular binding events and cellular compliance to the modulation of leukocyte adhesion, *J. Cell Sci.*, **116**, 2531-2539.

34. Fierro, F. A., Taubenberger, A., Puech, P. H., Ehninger, G., Bornhauser, M., Muller, D. J., and Illmer, T. (2008) BCR/ABL expression of myeloid progenitors increases beta1-integrin mediated adhesion to stromal cells, *J. Mol. Biol.*, **377**, 1082-1093.

35. Franz, C. M., Taubenberger, A., Puech, P. H., and Muller, D. J. (2007) Studying integrin-mediated cell adhesion at the single-molecule level using AFM force spectroscopy, *Sci. STKE*, **2007**, pl5.

36. Friedrichs, J., Manninen, A., Muller, D. J., and Helenius, J. (2008) Galectin-3 regulates integrin alpha2beta1-mediated adhesion to collagen-I and -IV, *J. Biol. Chem.*, **283**, 32264-32272.

37. Friedrichs, J., Torkko, J. M., Helenius, J., *et al.* (2007) Contributions of galectin-3 and -9 to epithelial cell adhesion analyzed by single cell force spectroscopy, *J. Biol. Chem.*, **282**, 29375-29383.

38. Krieg, M., Arboleda-Estudillo, Y., Puech, P. H., Kafer, J., Graner, F., Muller, D. J., and Heisenberg, C. P. (2008) Tensile forces govern germ-layer organization in zebrafish, *Nat. Cell Biol.*, **10**, 429-436.

39. Tulla, M., Helenius, J., Jokinen, J., Taubenberger, A., Muller, D. J., and Heino, J. (2008) TPA primes alpha2beta1 integrins for cell adhesion, *FEBS Lett.*, **582**, 3520-3524.

40. Orme, I. M., and Shand, F. L. (1982) Concanavalin A-induced alteration of surface marker expression on murine T cells, *Int. J. Immunopharmacol.*, **4**, 137-142.

41. Benoit, M. (2002) Cell adhesion measured by force spectroscopy on living cells, *Methods Cell Biol.*, **68**, 91-114.

42. Nugaeva, N., Gfeller, K. Y., Backmann, N., Lang, H. P., Duggelin, M., and Hegner, M. (2005) Micromechanical cantilever array sensors for selective fungal immobilization and fast growth detection, *Biosens. Bioelectron.*, **21**, 849-856.

43. Saif, M. T., Sager, C. R., and Coyer, S. (2003) Functionalized biomicroelectromechanical systems sensors for force response study at local adhesion sites of single living cells on substrates, *Ann Biomed. Eng.*, **31**, 950-961.

44. Waite, J. H. (1983) Evidence for a repeating 3,4-dihydroxyphenylalanine- and hydroxyproline-containing decapeptide in the adhesive protein of the mussel, Mytilus edulis L, *J. Biol. Chem.*, **258**, 2911-2915.

45. Waite, J. H., and Tanzer, M. L. (1981) Polyphenolic substance of Mytilus edulis: novel adhesive containing L-dopa and hydroxyproline, *Science*, **212**, 1038-1040.

46. Panorchan, P., Thompson, M. S., Davis, K. J., Tseng, Y., Konstantopoulos, K., and Wirtz D. (2006) Single-molecule analysis of cadherin-mediated cell-cell adhesion, *J. Cell. Sci.*, **119**, 66-74.

47. Radmacher, M. (2002) Measuring the elastic properties of living cells by the atomic force microscope, *Methods Cell Biol.*, **68**, 67-90.

48. Evans, E., and Ritchie, K. (1997) Dynamic strength of molecular adhesion bonds, *Biophys. J.*, **72**, 1541-1555.

49. Evans, E. A., and Calderwood, D. A. (2007) Forces and bond dynamics in cell adhesion, *Science*, **316**, 1148-1153.

50. Krieg, M., Helenius, J., Heisenberg, C. P., and Muller, D. J. (2008) A bond for a lifetime: employing membrane nanotubes from living cells to determine receptor-ligand kinetics, *Angew. Chem. Int. Ed. Engl.*, **47**, 9775-9777.

51. Sheetz, M. P. (2001) Cell control by membrane-cytoskeleton adhesion, *Nat. Rev. Mol. Cell Biol.*, **2**, 392-396.

52. Sun, M., Graham, J. S., Hegedus, B., Marga, F., Zhang, Y., Forgacs, G., and Grandbois, M. (2005) Multiple membrane tethers probed by atomic force microscopy, *Biophys. J.*, **89**, 4320-4329.

53. Sun, M., Northup, N., Marga, F., Huber, T., Byfield, F. J., Levitan, I., and Forgacs, G. (2007) The effect of cellular cholesterol on membrane-cytoskeleton adhesion, *J. Cell. Sci.*, **120**, 2223-2231.

54. Zhang, X., Chen, A., De Leon, D., Li, H., Noiri, E., Moy, V. T., and Goligorsky M. S. (2004) Atomic force microscopy measurement of leukocyte-endothelial interaction, *Am. J. Physiol. Heart Circ. Physiol.*, **286**, H359-H367.

55. Zhang, X., Wojcikiewicz, E. P., and Moy, V. T. (2006) Dynamic adhesion of T lymphocytes to endothelial cells revealed by atomic force microscopy, *Exp. Biol. Med. (Maywood)*, **231**, 1306-1312.

56. Zhang, X., Wojcikiewicz, E., and Moy, V. T. (2002) Force spectroscopy of the leukocyte function-associated antigen-1/intercellular adhesion molecule-1 interaction, *Biophys. J.*, **83**, 2270-2279.

57. Zhang, X., Craig, S. E., Kirby, H., Humphries, M. J., and Moy, V. T. (2004) Molecular basis for the dynamic strength of the integrin alpha4beta1/VCAM-1 interaction, *Biophys. J.*, **87**, 3470-3478.

58. Wojcikiewicz, E., Koenen, R. R., Fraemohs, L., Minkiewicz, J., Azad, H., Weber, C., and Moy, V. T. (2009) LFA-1 binding destabilizes the JAM-A homophilic interaction during leukocyte transmigration, *Biophys. J.*, **96**, 285-293.

59. Panorchan, P., George, J. P., and Wirtz, D. (2006) Probing intercellular interactions between vascular endothelial cadherin pairs at single-molecule resolution and in living cells, *J. Mol. Biol.*, **358**, 665-674.

60. Selhuber-Unkel, C., Lopez-Garcia, M., Kessler, H., and Spatz, J. P. (2008) Cooperativity in adhesion cluster formation during initial cell adhesion, *Biophys. J.*, **95**, 5424-5431.

61. Puech, P. H., Poole, K., Knebel, D., and Muller, D. J. (2006) A new technical approach to quantify cell adhesion forces by AFM, *Ultramicroscopy*, **106**, 637-644.

62. Taubenberger, A., Cisneros, D. A., Friedrichs, J., Puech, P. H., Muller, D. J., and Franz, C. M. (2007) Revealing early steps of alpha2beta1 integrin-mediated adhesion to collagen type I by using single-cell force spectroscopy, *Mol. Biol. Cell.*, **18**, 1634-1644.

63. Ulrich, F., Krieg, M., Schotz, E. M., *et al.* (2005) Wnt11 functions in gastrulation by controlling cell cohesion through Rab5c and E-cadherin, *Dev. Cell*, **9**, 555-564.

64. Benoit, M., Gabriel, D., Gerisch, G., and Gaub H. E. (2000) Discrete interactions in cell adhesion measured by single-molecule force spectroscopy, *Nat. Cell Biol.*, **2**, 313-317.

Chapter 11

PROBING CELLULAR ADHESION AT THE SINGLE-MOLECULE LEVEL

Félix Rico,[a] Xiaohui Zhang[b] and Vincent T. Moy[c]

[a] *Physico Chimie Curie, UMR168 CNRS, Institut Curie, 11 Rue Pierre et Marie Curie, 75231 Paris Cedex 5, France*
felix.rico@curie.fr

[b] *Bioengineering Program & Department of Mechanical Engineering and Mechanics, Lehigh University, 19 Memorial Drive West, Bethlehem, PA 18015, USA*
xiz310@lehigh.ed

[c] *Department of Physiology and Biophysics, University of Miami School of Medicine, 1600 NW 10th Ave, Miami, FL 33136 USA*
vmoy@med.miami.edu

11.1 INTRODUCTION

Cell adhesion is involved in the formation and the functional and structural integrity of multicellular organisms. Cell adhesion molecules (CAMs) connect cells to each other and to the extracellular matrix mediating intercellular communication and providing mechanical stability.[1,2] Moreover, CAMs are known actors of mechanotransduction, or the conversion of mechanical stimuli into biochemical response, such as gene expression. Among many other examples, firm adhesion is present in cell–cell and cell–matrix contacts in tissues, which are regulated by active assembly and disruption of receptor and ligand bonds during development and growth. A paradigmatic example of dynamic adhesion is found during the leukocyte adhesion cascade, in which different types of bonds are formed and disrupted at precise rates as the cell passes through different activating steps.[3] In both cases, cell

Life at the Nanoscale: Atomic Force Microscopy of Live Cells
Edited by Yves Dufrêne

www.panstanford.com

adhesion complexes are continuously subjected to mechanical stresses. It is thus important to study cell adhesion under applied force, which is the most physiologically relevant scenario. The relatively recent development of adequate tools to manipulate and apply forces at the nanoscale has allowed us to probe receptor–ligand interactions one molecule at a time on living cells. The present chapter will try to briefly introduce the molecular players involved in cell adhesion and to describe the particular application of the atomic force microscope (AFM) to study the behaviour of these peculiar proteins under force.

The principal actor in cellular adhesion is the CAM. There is a wide variety of CAMs that have different functions depending on the type and state of the cell, and the biophysical and biochemical microenvironment in which they exist. Figure 11.1 shows the four major families of CAMs. *Selectins* are transmembrane multidomain glycoproteins that mediate heterotypic cell–cell adhesion. Selectins bind specific carbohydrate structures through calcium-dependent interactions via their lectin domains. The cytoplasmic domains can be linked to the actin cytoskeleton.[1,2] Members of the *immunoglobulin superfamily* (IgSF) are multidomain proteins that mediate homo- and heterotypic cell–cell adhesion. Members of the IgSF can present a varying number of the Ig fold, a compact and rigid structure with one or more disulfide bonds, from which one or more are binding domains. These molecules are present in cells from the nervous system, the vascular endothelium and blood cells and are commonly involved in cell recognition. Some members of the IgSF are linked to the actin cytoskeleton via other proteins, e.g. ezrin. Both homo- and heterotypic interactions of some IgSF members have been probed with AFM.[4–6] *Cadherins* are transmembrane multidomain glycoproteins that primarily mediate calcium-dependent homotypic cell–cell adhesion. The five repeats can interact with any of the repeats of an opposing cadherin, leading to multiple possibilities of cadherin–cadherin interaction. Specificity of tissues is in great part mediated by cadherins. Structural stability is accomplished through the formation of structures such as desmosomes and adherens junction, in which cadherins can link to intermediate filaments and the actin cytoskeleton, respectively. The adhesion strength of various types of cadherins has been studied at both multiple and single-molecule levels on purified proteins and on living cells.[7–9] *Integrins* are heterodimers and mediate both cell–cell and cell–extracellular matrix heterotypic binding. Integrins are composed of an α and a β chain, containing a large extracellular domain, a single membrane spanning region and a short cytoplasmic domain. Humans have 19 α and 8 β integrin subunits that combine to form more than

25 different integrin receptors.[1,2] Many integrins require the presence of divalent cations for binding, such as Mg^{2+} and Mn^{2+}. Some force measurements on integrins are described in section 11.4.

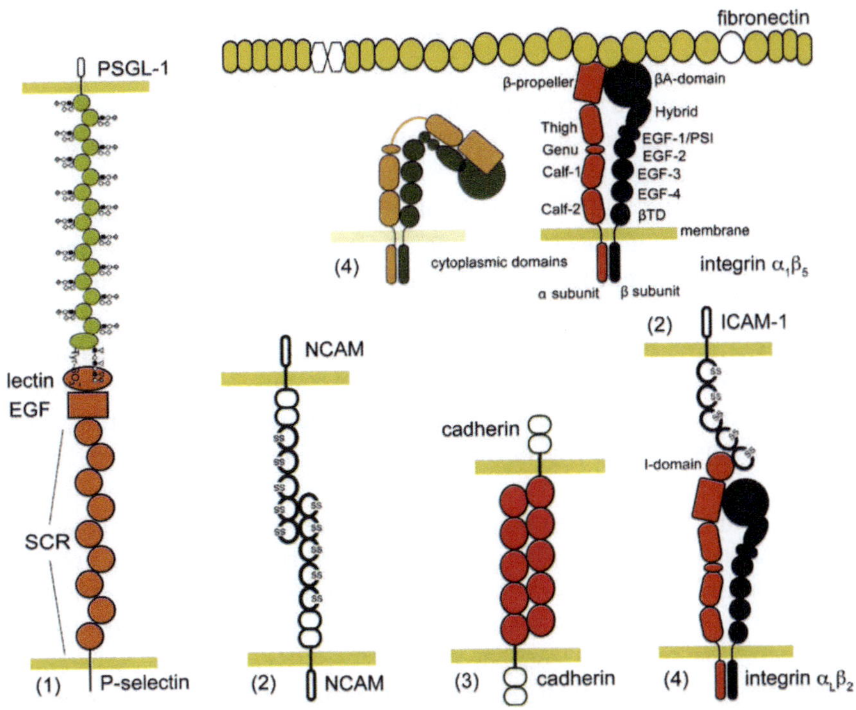

Figure 11.1. Major families of cell adhesion molecules with ligands. (1) *Selectins.* P-selectins are expressed in activated platelets and endothelial cells and their major ligand is P-selectin glycoprotein ligand-1 (section 11.4 describes published force measurements on selectins). (2) *Immunoglobulin superfamily (IgSF).* NCAM is expressed in neural cells and mediate cell–cell homotypic adhesion. Other members of the IgSF are intercellular adhesion molecules (ICAMs), which are known ligands of some integrins, such as $\alpha_L\beta_2$ (depicted). (3) *Cadherins.* Cadherins' subtypes are mainly found in specific tissues, e.g. VE-cadherin in the vascular endothelium, N-cadherin in neuronal cells or E-cadherin in the epithelium.[2] (4) *Integrins* ($\alpha_1\beta_5$/fibronectin, top; $\alpha_L\beta_2$/ICAM-1, bottom). Certain leukocytic intregrins allow firm adhesion and migration to the vascular endothelium, such as $\alpha_L\beta_2$ that binds to ICAM-1 (depicted). Integrins expressed in tissues, such as $\alpha_5\beta_1$, mediate firm adhesion to extracellular matrix proteins, including fibronectin and collagen, by forming focal adhesion complexes, in which integrins cluster and link their cytoplasmic domains to the actin cytoskeleton via other proteins, such as talin and α-actinin. Integrins mediate inside-out and outside-in communication and are primary candidates for cell mechanotransduction. Major conformational changes have been observed on integrins (shaded cartoon).

The biological bonds between CAMs are noncovalent bonds that generate strong, specific interaction between two molecules. Specificity is accomplished by the complementary geometry of the interacting molecules that favours force cooperation.[10] Unlike covalent bonds, which are strong and can be broken only by expending a significant amount of energy, the weaker receptor–ligand interactions that can rupture by spending one single ATP molecule are mediated by a combination of noncovalent interactions (van der Waals, electrostatic, etc.) and are normally described by an intrinsic lifetime or affinity constant.[10]

In tissues, cells adhere to each other and to the extracellular matrix forming adhesion complexes, such as focal adhesions, adherens junctions and gap junctions. Different types of proteins are responsible for the formation of these complexes, including, among many other, integrins, cadherins and connexins.[1,2] Even if the binding process is dynamic in nature, integrins and cadherins, for example, have long effective lifetimes, as they are known to form clusters, which also allow for the rapid rebinding of a protein, when occasionally dissociates. This permanent contact enables cells to form tissues creating barriers that restrict the passage of fluids, macromolecules and other cells. In addition, these types of adhesion complexes provide structural integrity and specificity to different tissues. Structural integrity is provided by the connection of CAMs to the cytoskeleton through complex structures composed by various proteins.[1,2] For example, focal adhesions are formed by the adhesion molecule integrin, which tethers the cell to the extracellular matrix, while the cytoplasmic region binds to talin and other proteins that connect it to the actin cytoskeleton. Specificity is accomplished by expressing particular types of CAMs that recognize specific ligands. In the case of the cadherin superfamily, which mainly mediates cell–cell adhesion, there are different subtypes that are characteristic of specific tissues. Most tissues are constantly subjected to mechanical stress. For example, lung and cardiac tissues are cyclically stretched because of breathing and heart beating, respectively. This involves that cell adhesion complexes are subjected to mechanical force. Then, their adhesion strength has to be strong enough to support these cyclic forces.

The leukocyte adhesion cascade provides the most relevant example of dynamic adhesion. Leukocytes (white blood cells) travel with the bloodstream without firmly adhering to the vascular tissue but patrolling it in search of signals of injury or inflammation.[3] Under pathological conditions, the cell passes through states of rolling, firm adhesion and transmigration. These steps involve different types of adhesion molecules with varying binding affinities. Rolling is mainly mediated by selectins and some types of integrins that are weakly linked to the cytoskeleton and allow the formation of long membrane tethers. On the other hand, firm adhesion and transmigration of leukocytes is mainly mediated by integrins, which have the interesting ability to change their conformation and affinity state. It has been proposed that under pathological conditions, leukocytes become activated and integrins change from a bend conformation with low binding affinity to an extended conformation with high binding affinity (Fig. 11.1, shaded cartoon).[11,12] These changes in conformation and binding affinity allow the cell to arrest and firmly adhere to the endothelium. During this process, the blood flow exerts dragging forces on leukocytes that, on the one hand, allow selectins to dissociate at a certain rate necessary for rolling, while, on the other hand, prevent integrins from firmly adhere unless they change their affinity state.

The adhesive capacity of cells depends not only on the adhesion strength of individual CAMs, but also on other factors including the mechanical properties of the cell, its activation state, the biochemical microenvironment and the distribution and expression level of receptors and ligands.[1,2,13–15] However, to better understand cell adhesion, it is important to determine the mechanism by which individual CAMs adhere and the effect that force has on it. The most straightforward approach to study the adhesion strength of single receptor–ligand interactions is by trying to answer the question: *What is the force required for breaking a bond?* This apparently simple question has fundamental drawbacks both experimentally and theoretically. In the following sections, we will try to describe the AFM as a model tool to answer this question, the available approaches that have been applied and the theoretical framework available to time to describe receptor–ligand interactions.

11.2 NANOTECHNOLOGY TO STUDY CELL ADHESION

The study of molecules at the single-molecule level was not possible until the development of adequate tools that provide positioning with nanometre (nm) resolution and force application in the picoNewton (pN) to nanoNewton (nN) range. During the last decades, many tools have been developed in this direction, including magnetic and optical tweezers, the biomembrane force probe or the AFM.[16] The AFM is probably the most widely used technique given its versatility and applicability to many different fields, from pure material sciences to cellular biology or surface chemistry. Indeed, the AFM has been used for both imaging and manipulating biological systems at the individual molecule level. In this section, we will describe its principle of operation and its applicability to the measurement of binding strength of receptor–ligand complexes.

11.2.1 The AFM

Being originally designed as an imaging tool to characterize the topography of electrically nonconductive materials, the AFM was rapidly applied to measure surface forces. Its application to the characterization of biological samples was accomplished by improvement of the method to detect the cantilever deflection, which allowed the operation under fluid conditions. The AFM was thereafter extensively used to characterize nonspecific forces in air and liquid as well as the mechanical properties of samples.

A schematic diagram of an AFM is shown in Fig. 11.2a. The most important part of the AFM is the cantilever and its tip, the latter being responsible of making contact with the sample, and the former of applying pushing and pulling forces. The cantilever tip is moved relative to the sample with subnanometre resolution by means of piezoelectric elements. An optical system, composed of a light source and a segmented photodiode, allows us to monitor the deflection, by focusing the laser beam on the backside of the cantilever and detecting the reflected light with the photodiode. For small deflections (d), the cantilever responds like a spring of constant k, and the force (F) is obtained following Hooke's law, $F = kd$. When working on living cells, it is very helpful to visualize the cells using optical microscopy to enable us to position the AFM tip on the precise location of the cell. The coupling

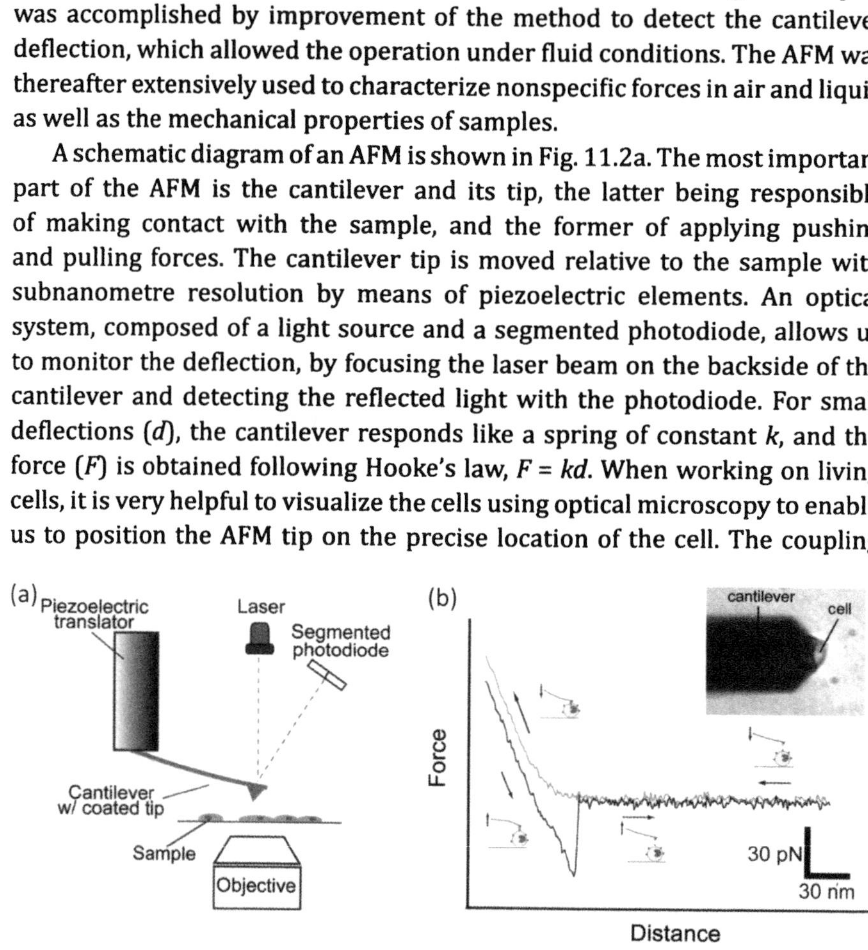

Figure 11.2. (a) Schematic diagram showing the major components of an atomic force microscope coupled to an inverted optical microscope to allow observation of biological samples. (b) Representative example of a force–distance curve (approach in gray) showing a single-molecule rupture event during retraction (black line) of the interaction between an integrin expressed on the surface of a monocytic cell and a ligand immobilized on the tip. The diagrams represent the position of the cantilever and substrate during the force curve cycle. The optical micrograph shows a THP-1 cell immobilized on a poly-L-lysine-coated dish and an AFM cantilever (Biolever, Olympus) during the force measurements (the cantilever width is ~30 μm).

of the AFM to an inverted optical microscope facilitates this task. Transmitted light microscopy, specially phase contrast or Normarsky microscopies, allows us to visualize cells and even organelles within cells and to easily position the AFM tip on the desired location. In addition, fluorescence microscopy is also suitable for visualization of fluorescently tagged molecules immobilized on the substrate or expressed on cells. Figure 11.2b shows an optical micrograph showing an AFM cantilever and a monocytic cell immobilized on the substrate.

Contact mode imaging consists of scanning the sample with the tip in the horizontal plane by applying a constant compressing force. The vertical piezo continuously corrects its position to account for the changes in topography of the sample and then keep the applied force constant. This continuous correction leads to the topographic image of the sample surface.

11.2.2 Force Spectroscopy

In the force spectroscopy mode or, simply, force mode, the cantilever is moved in the vertical direction in approaching and withdrawing cycles (Fig. 11.2b). The cantilever starts away from the sample and the piezo approaches the tip making contact until a set force is reached, then it withdraws away to the initial position. During approach and retraction, the vertical displacement (z) and the cantilever deflection is monitored generating what is known as a force–distance (F-z) curve, or just a force curve (Fig. 11.2b). Force measurements require routinely calibrating the sensitivity of the optical detection system (optical lever sensitivity, OLS) and the spring constant of the cantilever before each session. The OLS is calculated from the slope in a force–distance curve obtained on a hard substrate (such as glass). When pressing on a hard substrate the deflection of the cantilever is the same as the displacement of the piezo, thus we can transform the voltage signal detected by the photodiode into deflection of the cantilever. To calibrate the spring constant of the cantilever, different methods exist. The most widely used is probably the thermal fluctuations method.[17] It consists of measuring the mean square displacement $\langle d^2 \rangle$ of the cantilever due to thermal fluctuations. Assuming the cantilever response to be linear and with a single degree of freedom, the equipartition theorem can be applied to equate the average elastic potential energy $\langle E \rangle = k\langle d^2 \rangle/2$ to the thermal energy $k_B T/2$, k_B being the Boltzmann constant and T, the absolute temperature. Doing so, we can estimate the spring constant as $k = k_B T/\langle d^2 \rangle$. Another common calibration method is the Sader method, which takes into account the geometry and material properties of the cantilever.[18] The uncertainty in force due to systematic errors in the calibration of the system has been estimated to be

10–20%.[19] Force curve measurements have been extensively used to probe the viscoelastic properties of cells, the mechanical unfolding of proteins and the disruption forces of receptor–ligand bonds.[20,21] There are various types of force curves that will be described in the following sections.

11.3 SINGLE-MOLECULE MEASUREMENTS OF CELL ADHESION

As mentioned before, cellular adhesion is ultimately mediated by individual receptor–ligand interactions that are subjected to mechanical stress. Thus, it is relevant to understand the forced dissociation behaviour of individual receptor–ligand complexes on living cells. The AFM is particularly suitable for probing adhesion interactions one molecule at a time, given its high spatial resolution and fast time response. To that end, the AFM is used in force spectroscopy mode to measure the interaction forces between two adhering surfaces. The biomolecules of interest have to be immobilized on opposing surfaces (AFM tip and substrate or cell, Fig. 11.3) that will be brought into contact and allowed to interact. Then, the bond will be forcibly dissociated by pulling from receptor and ligand. Figure 11.4 shows four possible configurations to probe cell adhesion interactions with the AFM at the single-molecule level together with a schematic example of a resulting force curve. The various approaches have advantages and disadvantages and the final application will determine the best configuration to be used. In this section, we will describe the experimental set-up for single-molecule force measurements of cell adhesion using the AFM, including the necessary coating of cantilevers and substrates. The second part of this section will describe different force spectroscopy approaches, such as dynamic force spectroscopy (DFS) and force clamping techniques, which can be used to characterize biological bonds. In the last section we will introduce the theoretical framework to interpret force spectroscopy measurements.

11.3.1 Experimental Set-Up

Force spectroscopy measurements can be carried out on purified proteins immobilized on the tip and substrate, on proteins expressed on the surface of living cells or a combination of both. The use of a living cell has the main advantage of having the protein in its native environment, which means the protein is fully functional. Some CAMs on the cell membrane are dynamically linked to the cytoskeleton and the pulling response of a protein with or without link to the cytoskeleton varies importantly. In some cases, e.g. when studying the interaction of proteins not expressed on the cell membrane, the use of purified protein is almost required. The chemical modification of tip or substrate surfaces is, in any case, a crucial first step.

11.3.1.1 Surface coating

To immobilize biomolecules to study receptor–ligand interactions with the AFM, molecules can be physisorbed, covalently attached or linked via specific tags. The best approach will depend on the nature of the protein and the material the surface is made of. AFM tips are normally made of silicon or silicon nitride although they can come with different metal coatings, such as gold or aluminium. Gold-coated tips have the advantage of being biocompatible and relatively inert, providing low unspecific adhesion. An available technique to functionalize gold surfaces is the use of alkanethiols with a specific functional group to form a self-assembled monolayer to which covalently attach the protein of interest. Figure 11.3a shows a protocol to covalently attach biomolecules to gold-coated AFM tips using this approach. By covalently attaching the molecule, the probability of disrupting it from the tip is minimal. An important drawback of using amine groups as linking sites is that the orientation of the molecule is not controlled. Linking the molecule using specific tags that recognize known epitopes of the molecule can solve this disadvantage (Fig. 11.3b). Some groups have used antibodies as linkers.[22,23] However, it is important to first ensure that the specific bond is stronger than the probed one.[22] The histidine tag, widely used in protein purification, has been shown to be strong enough to support the binding forces of some receptor–ligand interactions.[24] The use of long linkers reduces nonspecific events between the tip and the substrate by increasing the distance between the two surfaces. In addition, knowing the exact length of the linker allows us to discriminate between specific and nonspecific interactions, and between single or multiple events by analyzing the force-deformation response prior to rupture, which is well described for some linkers by the freely jointed chain model.[25,26] In addition, long linkers provide more mobility to the molecule, allowing the molecule to orient properly during binding. Another coating method involves the reconstitution of proteins into lipid bilayers that will be then immobilized on the surface of the tip or substrate. This method is particularly convenient when using integral membrane proteins that may not be stable in solution.[23,27] However, in that case, it is important to ensure that the molecule is not removed from the lipid bilayer when pulled by the AFM tip.

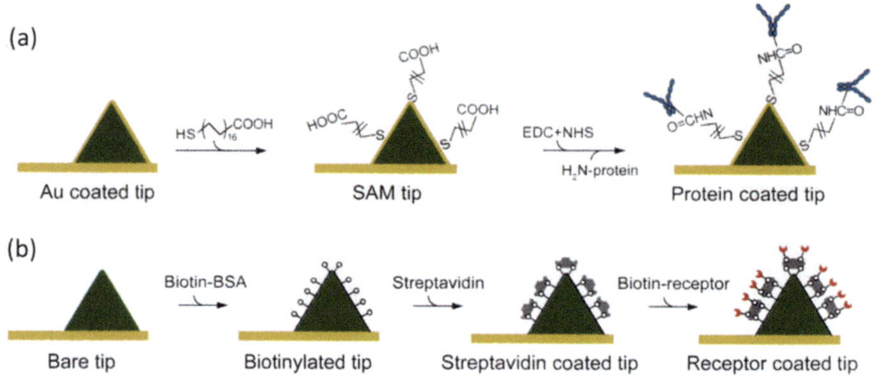

Figure 11.3. Tip surface chemistry modification. (a) Protein crosslinking to gold-coated cantilevers using alkanethiols. Cantilevers are incubated overnight in ACID16 (Nanothinks, Sigma), which contains a thiol group (-HS), a short linker and a functional carboxyl group. After rinsing with ethanol, cantilevers are incubated for 15 minutes in a solution containing EDC (1-ethyl-3-(3-dimethlaminopropyl)carbodiimide) and NHS (N-hydroxysuccinimide) to activate the free carboxyl groups. After rinsing with PBS, cantilevers are incubated in the protein solution overnight at 4°C, rinsed again before measurements and blocked with BSA prior to measurements. (b) Cantilever coating for cell immobilization using a biotinylated receptor. Biotinylated bovine serum albumin (biotin-BSA) is first adsorbed to the AFM tip by overnight incubation at 4°C. After rinsing with PBS, tips are incubated with soluble streptavidin (50 μg/ml) for 1 hour at room temperature and then rinsed again with PBS. In the last step, the biotinylated receptor (biotin receptor, such as biotin-concanavalin-A) is coupled to the free streptavidin sites through incubation for 1 hour at room temperature. Cantilevers are rinsed again with PBS before measurements. This method can also be used to attach proteins to the tip to measure receptor–ligand interactions, after assuring that streptavidin–biotin interaction is strong enough.

To immobilize a cell on the AFM cantilever, we also need to functionalize the surface to make it attractive to the cell. In the case of immune cells, special care should be taken on not using coatings that may activate signalling cascades, since this will affect the binding affinity of some receptors, such as integrins. A very simple method to functionalize the cantilever is the use of poly-L-lysine that would provide a positively charged surface to which cells would adhere unspecifically. Another widely used method is the use of concanavalin-A that recognizes sugars present in the cell's glycocalyx. The most common protocol to coat silicon nitride cantilevers with a biotinylated receptor is shown in Fig. 11.3b. After modifying the cantilever with the desired chemistry, suspended cells sitting on the sample surface are picked up by gently pressing on them and waiting a few seconds to allow contacts to form. The cell can then be probed against a ligand-functionalized substrate.

The opposing surface, or substrate, also needs to be functionalized to link the molecule of interest or suspended cells. Similar strategies to those described for AFM tips can be used. If the biomolecule of interest is relatively large (>50 kDa), physisorption can be used to immobilize it to the substrate by incubation during several hours.[5,6,28–30] For smaller molecules, partial denaturing due to physisorption could compromise their binding capacity, thus, an alternative method such as the one described for tip coating is suggested. Suspended cells can be easily immobilized on Petri dishes coated with poly-L-lysine at low concentration, allowing them to maintain their round shape. Monolayers of adherent cells, such as endothelial or epithelial cells, growing on cell culture dishes or coverslips can be directly used for AFM force measurements.[31,32]

11.3.1.2 Receptor–ligand configurations

Different configurations can be adopted to probe cell adhesion at the single-molecule level. Figure 11.4 shows the four main configurations commonly used. Measurements will be similar in all four, although the experimental conditions will slightly change. The use of purified proteins allows us to have a controlled and clean set-up in which only the biomolecules of interest are present. As mentioned before, molecules can be immobilized on the AFM tip and on the substrate using various methods (Fig. 11.3). An advantage of using purified protein is that it provides optimal conditions for single-molecule measurements. Protein can be diluted to very low concentrations and the relatively small area of contact between the AFM tip and the substrate reduces the probability of bond formation. Low binding probability (<30%) ensures that most of the events are mediated by single bonds. AFM cantilever tips come in different sizes and shapes. Unsharpened pyramidal tips have been extensively used, because the blunted apex of radius 20–50 nm provides a relatively bigger area than sharpened tips and wear is less pronounced.[5,25,33–36] Even if the use of purified proteins is very convenient and can be used as a first approach if the protein is available, purification of some proteins is not always possible and, more importantly, purification may alter the native conformation and adhesive properties of the protein, as it is known for some integrins.[37] Thus, when possible, it is recommended to work with proteins expressed on living cells.

Suspended cells, such as blood cells, can be immobilized on a coated AFM cantilever to carry out force measurements (Fig. 11.3).[5,38] For that purpose, it is recommended to use tipless cantilevers or cantilevers with tips of height smaller than the cells themselves, to avoid any interaction between the AFM tip and the substrate that will prevent the cell to contact it (see Fig. 11.4b). As

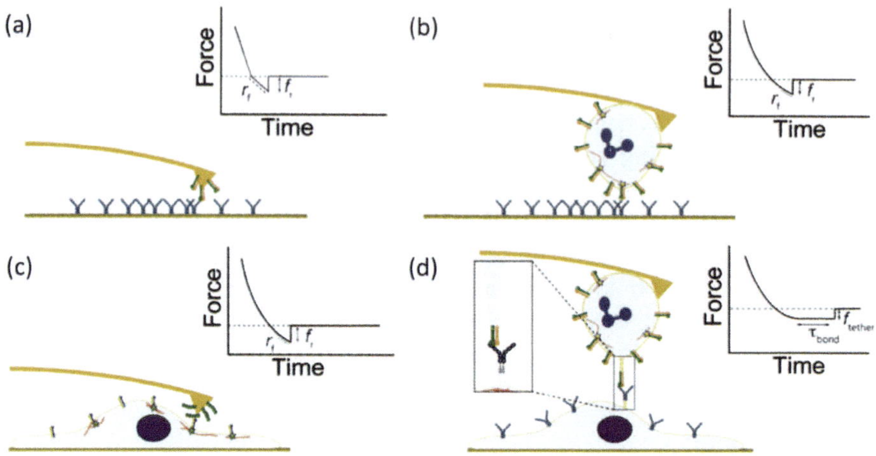

Figure 11.4. Tip and substrate configurations for single-molecule measurements of cell adhesion. Schematic examples of retraction force–time curves are shown for each case. (a) Protein on tip and substrate. (b) Cell on cantilever and protein on substrate. (c) Cell on substrate and protein on tip. (d) Cells on substrate and cantilever. The force–time curves in a–c represent rupture events typical of force measurements at constant retraction speed in which the receptor and ligand are firmly attached to the substrate (tip, dish, or cell cytoskeleton). The rupture forces (f_r) are measured as the force jump relative to noncontact (grey dashed line) plus a viscous drag correction,[42,43] while the loading rate (r_f) is estimated from the slope prior to rupture. The dynamic force spectra are obtained by measuring the most probable rupture force at different loading rates (Fig. 11.5). The force–time curve in (d) shows an example of an adhesion event in which membrane tethers are formed through detachment of the receptor and ligand from the underlying cells' cytoskeletons. In that case, the force jump (f_{tether}) represents the force required to extract the tethers and it is applied to the receptor–ligand bond. The measured lifetime (τ_{bond}) is thus the lifetime of the bond at the applied tether force. Tether forces mainly depend on the friction between the membrane and the cytoskeleton and on the retraction speed. At different retraction speeds, the tether forces vary allowing us to estimate the bond lifetime at various force levels.[44]

mentioned before, it is important to use coatings that will not activate cells, such as certain antibodies, as this would modify the adhesive properties of the cells and, perhaps, the affinity of the adhesion molecules.[3] The main advantage of using living cells is that membrane proteins are in their native environment, being thus fully functional. The main practical drawback is that cells are complex systems and a wide variety of other proteins are normally expressed in the cell surface. This can lead to undesired or multiple binding to the ligand of interest, which is difficult to discriminate and isolate. In addition, measurements in cells present normally more unspecific binding. Thus, blocking strategies and control measurements are particularly important. A common and widely used procedure to block uncoated regions of the substrate is to use bovine serum albumin (BSA). The substrate with the desired ligand is incubated with 1% w/v BSA for ~1 hour at room temperature just before measurements. To block hydrophilic surfaces, such as nontreated polystyrene, 1% Pluronic F108 (BASF), a nontoxic triblock copolymer, has also been used, providing negligible unspecific adhesion. In addition, cells are compliant bodies. Thus, a small compression force will lead to a relatively big area of contact between the cell and the substrate, increasing proportionally the probability of bond formation.[39] For example, a cell of ~5 μm radius and 1 kPa Young's modulus compressed against a flat surface with a force of 50 pN would provide an area of contact of ~1 μm^2, assuming Hertzian elastic contact.[39] On such a relatively big contact area, it is difficult to form individual bonds. For this reason, it is sometimes necessary to reduce the ligand concentration to very low levels and always work at minimal compression forces. Another strategy to reduce the area of contact between the cell and the substrate is to increase the approaching speed. Given the viscoelastic nature of cells, the faster the approach velocity, the stiffer the cell will appear. Thus, at a same compression force, increasing the approach velocity can help to reduce the area of contact. Moreover, faster approach will provide shorter contact times, reducing also the probability of bond formation.

A common property of living cells, particularly blood cells, is their ability to form membrane tethers. These membrane nanotubes are formed when a molecule on the cell surface being pulled is released from the underlying cytoskeleton and thus membrane is allowed to flow following the pulled molecule.[40,41] When pulled at constant speed, the tether exerts a constant, friction force against the cantilever tip (Fig. 11.4d). As we will see in the next section, in DFS measurements, in which complexes are stretch at constant loading rates (force–time), the formation of tethers is not the optimal condition. However, we can take advantage of this property of cells to carry out alternative measurements.

When probing adhesion with AFM on adherent cells, such as fibroblasts or endothelial cells, we can use cell culture dishes or coverslips on which adherent cells are commonly grown (Fig. 11.4c). Suspended cells can also be immobilized on the substrate using similar approaches as the ones described before for immobilization on AFM cantilevers. The advantage of using this configuration on adherent cells is that there is no need to detach them from the substrate of culture, a procedure that may stress the cell. With the cell immobilized on the substrate, the AFM tip has to be coated with the biomolecule of interest. In that case, the tip geometry will define the area of contact. The commonly used pyramidal tips have the advantage of providing smaller contact areas at similar indentation forces than when using cells immobilized on the cantilever pressed against a flat substrate. For example, a pyramidal tip of semiopen angle of 35° indenting a cell of 1 kPa with a compression force of 50 pN will provide an area of contact ~0.3 μm². It is thus easier to achieve single-molecule conditions than in the previous configuration. To further control contact conditions, modified cantilever tips with flat-ended cylindrical shape provide a constant and known area of contact, independent of applied force and indentation.[32,45] In this configuration, the formation of tethers may also be a problem. However, tethers are normally less common in adherent cells than in blood or suspended cells.

Using immobilized cells both on the substrate and on the cantilever is probably the optimal configuration for cell–cell adhesion studies, such as in cadherin–cadherin adhesion measurements (Fig. 11.4d). Receptors and ligands are in their native environment, fully functional and properly oriented. However, the actual application of such a configuration is complicated, especially to obtain specific single-molecule events. As one cell indents the other, similar cell size and elasticity provides an even bigger area of contact, in which multiple bonds are easily formed. More importantly, cells express a wide variety of CAMs that can bind one or multiple ligands. Thus, it is more difficult to discriminate between possible nondesired adhesion events between molecules. A very elegant approach to prove specificity is by using the same cell line with knocked-out protein of interest, and then showing that no adhesion is found on it. This set-up has been applied before to study cell–cell adhesion at single[7,38] and multiple molecule levels.[15,46] It is also possible to block undesired molecules using specific antibodies.[46] Under this configuration, the probability of tether formation is even higher than in the previous ones, as tethers can be extracted from both cells (Fig. 11.4d).

In any of the four configurations, control measurements are crucial to prove the specificity of the interaction. Different type of control approaches can be used depending on the configuration used and the biomolecules involved. The most straightforward approach consists in blocking the specific receptor–ligand interaction with soluble ligand. Another approach is to use AFM tips or substrates in which one of the molecules of interest has not been immobilized. A more reliable approach is to block the receptor or the ligand with specific blocking antibodies. As mentioned before, some interactions require the presence of multivalent ions, such as Ca^{2+} or Mg^{2+}.[5,6,8] Chelating the required ions can also be used to prove specificity. In the case of living cells, an elegant approach is to knock out one of the proteins of interest. In practice, a combination of these control experiments is recommended.

11.3.2 Force Measurements

Any of the configurations described in the previous section can be used to carry our force measurements on single receptor–ligand complexes and various approaches have been considered. The first approach consisted in measuring the forces required to break the bond when pulled apart.[28,30,47] However, because of thermal agitation and the statistical nature of bond dissociation, this force is not unique, but spread, and depends on the rate of force application.[48,49] The current approaches have the final goal of characterizing the free energy landscape of the interaction in terms of the intrinsic dissociation lifetime at zero force (τ_0), and the potential width (γ) and height ($\Delta G^{\ddagger}$). More importantly, the characterization of the interaction will provide a description on how force affects the lifetime of the bond. The theoretical framework is based on the idea that a mechanical force will distort the energy landscape of the interaction, lowering the energy barrier and facilitating dissociation. This concept was originally applied to biological bonds by Bell in 1978.[50] In this section, we will describe the experimental approaches used to characterize biological interactions.

11.3.2.1 Dynamic force spectroscopy

DFS is possibly the most widely used approach to characterize the adhesion strength of biological bonds. The adhesion strength of several biological complexes has been measured using DFS, including eukaryotic receptors, such as cadherins, integrins and selectins, bacterial receptors, such as FimH and mucin, and even virus proteins.[5-8,26,37,39,51-54] In addition, it has been also used to determine the unfolding kinetics of various proteins.[55,56] The approach consists in measuring the rupture forces (f_r) of receptor–ligand interactions by applying a constant loading rate, i.e. rate at which force is applied (r_f) (Fig. 11.4; for practical issues, it is sometimes useful to use force–time, instead of force–distance, curves since the slope before rupture is a direct estimate of the applied loading rate). At a given loading rate and because of thermal agitation, rupture forces are not unique but follow a certain probability distribution[48,57] (Fig. 11.5b). The situation is even more complicated as the rupture force will also depend on the dynamics of loading, given the statistical nature of the dissociation kinetics. As mentioned before, biological bonds can be characterized by an intrinsic lifetime at zero force (τ_0), which is the inverse of the characteristic rate at which the complex spontaneously dissociates, i.e. the dissociation rate ($k_{off} = 1/\tau_0$). If pulled faster than k_{off}, the bond resists detachment, giving rise to a measurable force.[10,49,53] As a result, the most probable rupture force depends on the loading rate. Thus, DFS experiments require the acquisition of many rupture events at various loading rates (Fig. 11.5). In practice this involves the acquisition of sets of force–distance curves at various retraction speeds, usually ranging three or more orders of magnitude (Fig. 11.5a). The rupture force distributions will provide a measure of the most probable rupture force at each loading rate (Fig. 11.5b). A plot of the most probable rupture force versus loading rate is named the *dynamic force spectrum* of the interaction (Fig. 11.5c). In the next section we will describe how to obtain the intrinsic parameters of the interaction from such dynamic force spectra and the force-dependent lifetime (Fig. 11.5d).

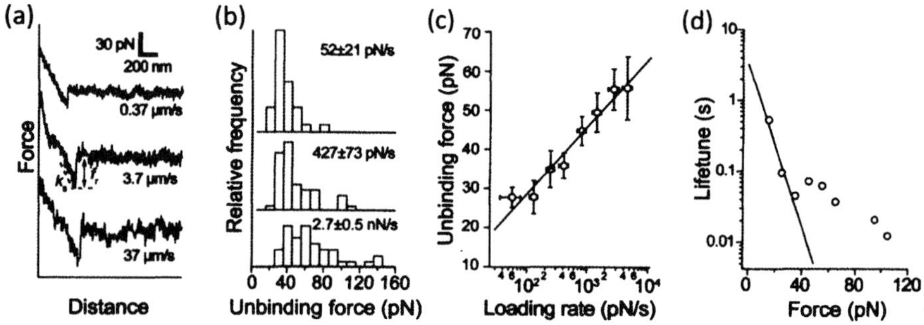

Figure 11.5. Dynamic force spectroscopy of integrin $\alpha_4\beta_1$ binding to vascular cell adhesion molecule-1 measured on living monocytic cells. (a) Representative examples of force curves showing single-molecule rupture events at three different retraction speeds. The determination of the rupture force (f_r) and the effective stiffness (k_s) is shown. The loading rate can be extracted by multiplying k_s by the retraction speed. (b) Rupture force distributions at three loading rates. (c) Dynamic force spectrum of the interaction, i.e. most probable rupture (±SEM) forces as a function of loading rate (open circles). The solid line represents the best fit to the Bell–Evans model Eq. (11.6), that leads to γ = 5.5 ± 0.5 Å and τ_0 = 3.3 ± 1.3 seconds. (d) Force-dependent lifetime (solid line) calculated using the fitted parameters in Eq. (11.2). The open symbols show the results obtained from directly computing the lifetime-force response from the rupture force distribution at moderate loading rate (middle histogram in a) using the new approach introduced by Dudko and coworkers, Eq. (11.7).[58]

It is important to mention that pulling from a receptor–ligand complex at constant retraction speed not always leads to constant loading rate. For example, when using long linkers to tether the biomolecules to the tip, constant retraction speeds give rise to nonlinear responses that can be described by different models, such as the freely jointed chain model.[25,26] In the case of living cells, the situation is even more complex. As mentioned before, some receptors are linked to the cytoskeleton via different molecules. If this link is stronger than the receptor–ligand bond, the force response at a constant pulling speed will be fairly linear given the mechanical properties of the cytoskeletal cortex. This will lead to constant loading rates. In contrast, if the link with the cytoskeleton ruptures or if there is not link at all, the force response will be nonlinear and membrane tethers may eventually form (Fig. 11.4d).[40,41] A force plateau preceding bond rupture has been interpreted as a signature of tether formation and elegantly confirmed recently by lateral inspection of the cell detachment process.[59] Thus, for some receptors, the application of different loading rates might be difficult on living cells, given the mechanical properties of the membrane and the underlying cytoskeleton. The formation of membrane tethers is not ideal to carry out DFS measurements. However, it is possible to take advantage of this *a priori* inconvenience.

11.3.2.2 Force clamp measurements

A more direct approach to determine the effect of force in the kinetics of receptor–ligand interactions is the application of what has been called *force clamp* technique, in analogy with patch clamp electrophysiology measurements. In force clamp measurements, a constant force is applied to the bond and the time until it disrupts is measured. This measured lifetime at a certain clamping force will not be unique but will follow an exponential distribution with an average value. Applying different levels of force and determining the corresponding average lifetime will directly lead to lifetime versus force plots. On the so-called *slip* bonds, in which an applied force will decrease the lifetime, fitting the Bell model (described in the next section) to the lifetime versus force data will allow us to determine the intrinsic parameters of the interaction. Force clamp measurements, however, are particularly useful to detect possible *catch bonds*, in which moderate forces counterintuitively increase the lifetime of the bond.[22,23]

As mentioned before, living cells have the capacity of forming membrane tethers. The physiological relevance of tethers is found, for example, in the rolling of leukocytes on the vascular wall, which reduces their speed at the early stages of the leukocyte adhesion cascade.[3] During AFM force measurements at constant pulling speed tethers may form, giving rise to characteristic force–distance profiles in which a force plateau precedes a jump in force (Fig. 11.4d). It has been suggested that tethers form when the receptor detaches from the cytoskeleton.[40] In that case, a membrane tube is pulled from the cell surface exerting a friction force. This force, being dissipative in nature, will depend on the pulling speed but also on the mechanical properties of the membrane and its interaction with the cytoskeleton. It has been suggested that the main responsible to this force is indeed the friction between the membrane, which flows toward the tether, with membrane proteins and the underlying cytoskeleton.[41,60] Thus, varying the pulling speed will lead to different levels of the force plateau. It is important to emphasize that this force (Fig. 11.4d) is not the force required to rupture the bond at the applied loading rate, since this is virtually zero given the zero slope of the plateaus. In contrast, this tether force is the force supported by the bond, which binds the receptor on the membrane to the ligand on the opposing substrate. Thus, the lifetime of the tethers, i.e. the lifetime of the force plateau, is a direct measure of the lifetime of the bond at the applied force. Therefore, measuring tether lifetimes at different pulling speeds (different force plateau values) is a physiological alternative to force clamp measurements. This approach has been recently applied using the micropipette aspiration technique[61] and the AFM[44] on different cell adhesion complexes.

11.3.3 Theoretical Framework

Biological interactions are specific interactions mediated by a complex and dynamic combination of hydrogen bonding, van der Waals, hydrophobic, steric and electrostatic interactions.[10] This combination of forces leads to an equilibrium state that is normally simplified by a one-dimensional free energy landscape with a deep minimum and a barrier along the reaction coordinate (x) (Fig. 11.6). The mechanical strength of the interaction is then determined by bringing the system out of equilibrium by applying a pulling force that distorts the energy landscape. The most established theoretical description of forced unbinding of biological bonds originated with the classical work by Bell.[50] Bell's approach was later reformulated by Evans and Ritchie and based on the reaction rate theory developed by Kramers.[48,57] The model assumes an intermolecular potential in which the dissociation dynamics is described as a diffusive process with an intrinsic dissociation lifetime τ_0, which is determined by a dissipative term (D), two length scales related to local curvatures of the energy landscape at the minimum and top of the barrier (l_c and l_{ts}, respectively) and the height $\Delta G^{\ddagger}$ of the dominant energy barrier.[10,49,57]

$$\tau_0 = \frac{l_c l_{ts}}{D} e^{\Delta G^{\ddagger}/k_B T} \tag{11.1}$$

The multiplicative term before the exponential is known as the diffusive relaxation time (t_D), the inverse of the attempt frequency, which is governed by molecular damping $\zeta = k_B T/D$. For biological molecules in liquid this relaxation time is very short, $t_D \sim 10^{-10}$ to 10^{-9} seconds. However, given the exponential dependence of the bond lifetime on the barrier height, it leads to relatively slow intrinsic dissociation rates, $\tau_0 \sim 1$ second, for a bond with a barrier of $\Delta G^{\ddagger} \sim 21\ k_B T$. The application of a force deforms the energy landscape along the pulling coordinate, by lowering the energy barrier (Fig. 11.6a). Assuming the barrier is sharp enough so the application of a force does not change the position and shape of the transition state, we obtain $G(F) = \Delta G^{\ddagger} - F\gamma$. Thus, when force is applied, the ligand can escape more easily from the well, which leads to a faster dissociation rate or, equivalently, a shorter lifetime. Therefore, the lifetime depends on the applied force and on the distance to the transition state (γ).

$$\tau(F) = \frac{1}{k(F)} = \tau_0 e^{-F\gamma/k_BT} \tag{11.2}$$

Eq. (11.2) is equivalent to Bell's equation, except for the definition and interpretation of τ_0 (or k_{off}) in Eq. (11.1).[48,50,57] The potential width provides a first description of how the bond resists the application of force. The wider the potential, the more the force will affect the dissociation kinetics.

As already mentioned, biological bonds are the result of a combination of different interactions. This leads to energy landscapes more complex than the one we just described, presenting more than one energy barrier (Fig. 11.6b). In this case, the outermost energy barrier will dominate the lifetime of the bond, until force drives the outer barrier below the inner one. This will happen when force exceeds a critical level (F_c), determined by the difference between the relative barrier heights ($\Delta\Delta G^{\ddagger}_{1,2}$). When force reaches $F_c = \Delta\Delta G^{\ddagger}_{1,2}/\gamma_1$ the inner barrier will dominate dissociation, leading to a different force dependence of lifetime. The lifetime will then present consecutive exponential regimes with characteristic lifetimes τ_{01}, τ_{02} and widths γ_1, γ_2... (Fig. 11.6b). This behaviour was first observed on the streptavidin–biotin complex by Merkel and coworkers using the biomembrane force probe.[47] It appeared later to be a common signature of biological bonds.[5,7,33,37,62,63]

The model described by Eq. (11.2) has been recently reformulated into a unified form by some authors,[58,64,65] assuming not a sharp barrier, but a certain potential shape in which not only the height of the barrier changes when force is applied but also the position along the reaction coordinate. Using Dudko, Hummer and Szabo approach,[66] the force dependence of the lifetime can be described by

$$\tau(F) = \tau_0 \left(1 - \frac{F\gamma}{b\Delta G^{\ddagger}}\right)^{1-b} e^{\frac{\Delta G^{\ddagger}}{k_BT}\left(1 - (1 - \frac{F\gamma}{b\Delta G^{\ddagger}})^b\right)} \tag{11.3},$$

where b is a parameter that selects the particular model of the potential well. Being $b = 3/2$ for a linear-cubic potential and $b = 2$ for a cusp potential. Notice that choosing $b = 1$ returns Eq. (11.2). The advantage of this approach is that it also provides an estimate of the barrier height and the approximate shape of the energy landscape, thus being less phenomenological. As we described in the previous section, force clamp measurements of receptor–ligand bonds provide direct data of lifetime versus force. Thus, Eqs. (11.2) and (11.3) can be directly fitted to determine the interaction parameters.

11.3.3.1 Dynamic force spectra

In practice, the application of the force is not instantaneous but force is ramped up at a constant rate. In this case of a bond pulled by a constantly increasing force, constant loading rate, the height of the energy barrier diminishes with time. Thus, the rupture force will depend on the loading rate. Three loading regimes can be differentiated in the plots of most probable rupture force versus loading rate.[10,48,57] At very low loading rates, i.e. near equilibrium conditions, the attempt rate is much faster that the loading rate and dissociation is governed by the activation energy and the thermal energy available. Thus, the most probable rupture force is not affected by the loading rate.[67] At very high loading rates, above the adiabatic limit, only accessible in molecular dynamics simulations, the applied increasing force reaches the maximum rupture force ($F_{max} = \Delta G^{\ddagger}/\gamma$) much faster than the intrinsic dissociation rate. In this regime, the rupture force will have this constant value with a linear contribution due to fluid damping and viscous friction.[48,57] At intermediate loading rates, at which most AFM measurements are carried out, the loading rate is comparable with the forced dissociation rate. Thus, the most probable rupture force will strongly depend on the loading rate. In addition, given the wide difference between the timescales for protein relaxation ($t_D \sim 10^{-10}$ to 10^{-9} seconds) and AFM measurements ($>10^{-5}$ seconds), the dissociation times during forced rupture of the bonds become continuous functions that lead to probability distributions of rupture forces, $p(F)$ (Fig. 11.5b). In this regime, the probability $S(t)$ that the bond is still intact at time t can be described by a first order rate equation $S^{\Upsilon} \equiv dS/dt = -S(t)/\tau(F(t))$, assuming negligible rebinding. $p(F)$ is related to the survival probability by $-S^{\Upsilon}dt = p(F)dF$ and is given by

$$p(F) = \frac{e^{-\int_0^F [F(f)\tau(f)]^{-1} df}}{r_f \tau(F)} \tag{11.4}$$

At constant loading rate, we can derive analytically the distribution of rupture forces using Eqs. (11.2) and (11.4)*

$$p(F) = \frac{e^{F\gamma/k_BT}}{\tau_0} \exp\left(\frac{k_B T}{\tau_0 \gamma r_f}[1 - e^{F\gamma/k_BT}]\right) \tag{11.5}$$

The most probable rupture force (F^*) is located at the maximum of the probability distribution [Eq. (11.5)]. Thus, imposing $dp(F)/dF = 0$ leads to

* Using Eq. 11.3 it is also possible to determine the distribution of rupture forces. We refer the reader to the original articles.[64, 66]

$$F^* = \frac{k_B T}{\gamma} \ln\left(\frac{\gamma \tau_0 r_f}{k_B T}\right) \tag{11.6}$$

which describes how rupture forces depend on the applied loading rate (Fig. 11.5c). Fitting Eq. (11.6) to the measured dynamic force spectrum allows us to determine the parameters of the interaction and to calculate the force-dependent lifetime (Fig. 11.5d). In the case of multiple barriers, the most probable rupture force will follow a series of linear regimes with the logarithm of the loading rate, to which Eq. (11.6) can be fitted separately.[5,48,68] The resulting force dependence of lifetime would lead to a series of exponential regimes described by $\tau(F) = \Sigma_i \tau_{0i}$ (Fig. 11.6b), τ_{0i} being the intrinsic lifetime of each barrier. Even if the observation of series of linear regimes in the dynamic force spectrum has been mainly interpreted as successive barriers along the dissociation coordinate, other interpretations such as intermediate states are also plausible.[69]

Recent developments have shown more direct approaches than DFS to determine force-dependent lifetime[58,70] (Fig. 11.5d open symbols). Dudko and coworkers recently showed that Eq. (11.4) can be inverted to directly compute the dissociation lifetime as a function of force, $\tau(F)$, from the distribution of rupture forces at a constant loading rate*

$$\tau(F) = \int_F^\infty \frac{p(f)}{r_f p(F)} df \tag{11.7}$$

By doing so, Eqs. (11.2) or (11.3) can be fitted to the resulting lifetime versus force plots to determine the interaction parameters.[58] We have applied this method in the example given in Fig. 11.5 to directly estimate the force-dependent lifetimes (open symbols in Fig. 11.5d). It can be seen that the first three points fit very well with the lifetime calculated from the fitted parameters from DFS. However, the following points deviate considerably from the expected trend. This can be due to the presence of an inner barrier not detected in the spectrum that would appear at higher loading rates or to the effect of possible multiple bonds rupture events in the probability distribution of rupture forces. A more detailed analysis and validation of the different methods should be carried out on receptor–ligand interactions. Even if this approach appears to be quite convenient, we have described the most established method of using the dynamic force spectrum and the Bell–Evans model [Eq. (11.6)] to derive the interaction parameters.

Oberbarnscheidt and coworkers recently developed an alternative and elegant approach, in which force curve data are directly used to extract force-

* See Eq. 10 in Ref. 56.

dependent dissociation lifetimes. The approach interprets each point in the force curve as an individual force clamp experiment. Thus, determining the total duration for each acting force allows us to calculate an almost continuous force-dependent lifetime.[70] A more sophisticated approach is the application of the Jarzynski equality to directly reconstruct the energy landscape of the interaction from force curves. This procedure has been applied in measurements of protein unfolding and receptor–ligand interactions.[71,72] The application of such model-free methods allows us to directly determine the effect of force on bond lifetimes and brings us the possibility of validating energy landscape models.

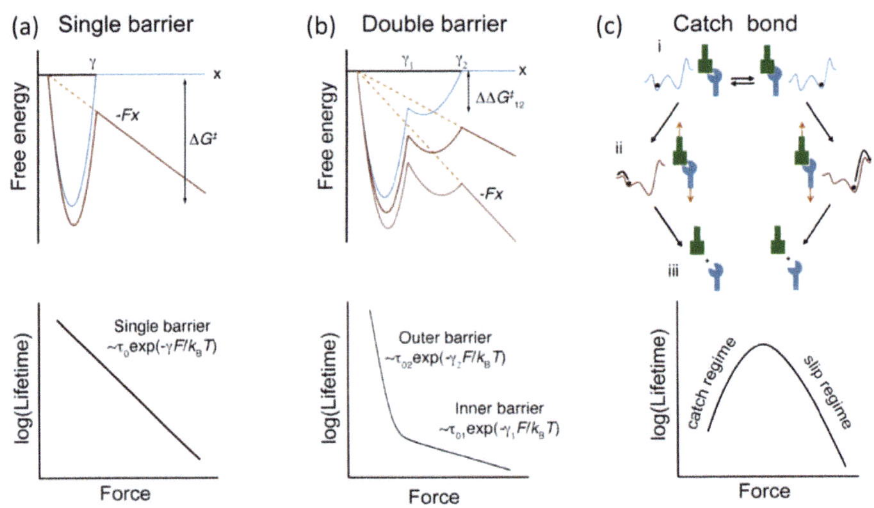

Figure 11.6. Effect of force on the energy landscape (top) and dissociation kinetics (bottom) of receptor–ligand interactions. (a) Single energy barrier at position (γ) and height ($\Delta G^{\ddagger}$) under no force (blue solid line) and under an applied constant force (red solid line). The orange dotted line represents the force applied to the bond. The force-dependent lifetime presents a single exponential decay [Bell model, Eq. (11.2)]. (b) Energy landscape with two energy barriers under no force (blue solid line) and under applied force (red solid line). The dotted line represents the force applied to the bond. The lifetime presents two exponential decays. The outer barrier governs the dissociation kinetics at low forces ($F < \Delta\Delta G^{\ddagger}_{1,2}/\gamma_1$), while the inner barrier governs the dissociation at high forces. (c) One of the possible mechanisms for catch bond behaviour. Two low energy states with two dissociation pathways (blue solid line). The cartoon reflects a hypothetical allosteric effect of force, which would change the conformational state of the receptor favouring the active, slow state. Low force would tilt the energy landscape (red solid line) increasing the population of the slower, active state, resulting in longer lifetimes (catch regime in the lifetime plot). Above a force threshold, the bond would behave as a slip bond, with force accelerating dissociation.

11.3.3.2 Energy landscape roughness

Given the combination of the various types of interactions (van der Waals, electrostatic, etc.), the energy landscape of biological bonds has been described as being rugged and composed of multiple hierarchical barriers and local minima. Zwanzig theoretically showed that the intrinsic dissociation lifetime increased importantly in a rough energy landscape.[73] The pioneer works by Frauenfelder showed that the energy landscape of haemoglobin presented a hierarchy of conformational states.[74] The theory developed by Zwanzig was recently adapted to describe the effect of energy landscape roughness on single-molecule force spectroscopy measurements.[75] The authors proposed an approach to calculate the roughness amplitude by measuring the dynamic force spectra of molecular interactions at various temperatures. Assuming a Gaussian distribution of roughness amplitudes independent of the position along the reaction coordinate, they showed that a constant term proportional to the squared roughness amplitude (ε) appeared in the expression of the most probable rupture force [Eq. (11.6)]. Equating the most probable rupture forces at two different temperatures they derived an expression for ε.[75] Nevo and coworkers adapted the approach to take into account possible variations with temperature of the potential width, obtaining an extended expression for ε.[76] In the same work, the authors experimentally tested the theory for the first time in the unbinding of the nuclear transport receptor importin β and GTPase Ran,[76] finding an unexpectedly high value of $\varepsilon \sim 5.7\ k_BT$. Since then, the energy landscape roughness of different systems has been determined. The unfolding of the actin cross-linking protein filamin[77] yielded a value of $\varepsilon \sim 4\ k_BT$, while it ranged from ~4 to ~7 k_BT from unfolding measurements of the transmembrane helices of bacteriorhodopsin.[78] Unbinding measurements on the well-studied streptavidin–biotin complex revealed two roughness values of ~5.5 and 7.5 k_BT along the dissociation pathway, corresponding to the inner and outer transition barriers, respectively.[34] The similar values of roughness observed in such different systems suggests a common origin, perhaps because of the oversimplification of a three-dimensional (3D) energy landscape to a single dimension.

11.3.3.3 Catch bonds

We have described biological bonds in a simplistic way by assuming an energy landscape with a single reaction coordinate being, thus, unidimensional. However, receptor–ligand complexes are complex structures in which proteins can undergo conformational changes that may vary their binding state. Not only energy landscapes present multiple barriers or rough potentials, but also they are probably better described as a multidimensional free energy surface presenting more than one dissociation pathway. In that case, the pulling direction might not be a good reaction coordinate leading to richer force-dependent lifetimes. An interesting case for cell adhesion bonds is the so-called *catch-slip bonds*.[79] In the catch regime, low forces increase the lifetime of the bond, while, above a certain force threshold, the bond behaves as a slip bond. This catch-slip bond behaviour has been already observed in P- and L-selectin binding to P-selectin glycoprotein ligand-1 (PSGL-1), on FimH bacterial receptor binding to mannose, on myosin binding to actin, on platelet glycoprotein Ibα-von Willebrand factor, and, more recently, on integrin $\alpha_5\beta_1$ binding to fibronectin.[22,35,52,80–84]

Various explanations have been suggested for this type of bonds. In the particular case of a two-dimensional (2D) free energy surface with one minimum and two dissociation pathways, it is possible that an applied force would lower the energy barrier along one of the pathways, while raising the other. Then, low forces could have the overall effect of slowing down the dissociation process instead of accelerating it.[58,85] Another explanation is that the number of dissociation pathways could be reduced when low forces are applied, eventually increasing the lifetime of the bond.[79,85,86] Figure 11.6c pictures another possibility of an energy landscape with two minima, i.e. two bound states, from either of which the bond can dissociate (Fig. 11.6c). If the energy barrier between the two states is high enough that the transition between the two is slower than the unbinding time of one of the bound states, then two different lifetimes would arise at a given level of force. This behaviour has been observed in FimH receptor binding to mannose. In such a landscape, if the longer lived bound state is also wider but with a deepest barrier, force could favour the population of this state, leading to an increase of the lifetime with force. Again, above a threshold, force would also lower the dissociation through this pathway, behaving then as a catch-slip bond, with a biphasic lifetime versus force response (Fig. 11.6c).[85]

It is important to notice that force clamp measurements are ideal to detect catch bond regimes as they provide a direct measure of force-dependent lifetimes. The application of alternative techniques, such as optical tweezers, to access ultralow force regimes may show catch bond behaviours for CAMs that have still not been observed with the AFM. Intriguingly, catch bonds are more common than expected initially. Moreover, they appear to be a common signature of force supporting complexes. It is tempting to hypothesize that perhaps all cell adhesion complexes behave as catch bonds, with a maximal lifetime at the range of forces more adequate for their function. Therefore, L-selectins would have a maximum lifetime at the force required to extract a membrane tether at physiological flow velocities, optimizing rolling. In turn, integrin $\alpha_5\beta_1$ binding to fibronectin would last longer at the forces exerted by the actomyosin cytoskeleton during cell migration, unbinding faster when force is released. Indeed, Forero and coworkers have recently shown that the uncoiling mechanism of *E. coli* type I fimbrae is optimized for catch bond behavior.[87]

11.4 LEUKOCYTE ADHESION TO THE VASCULAR ENDOTHELIUM

Leukocytes circulating in the blood vessels need to exit the bloodstream to enter specific tissues or areas of inflammation.[3] This finely tuned process is mediated by the interactions between leukocyte adhesion molecules, typically selectins, selectin ligands and integrins, and their adhesive partners expressed on the inner surface of the blood vessel wall. If this adhesive process is not under proper control, it could lead to severe diseases such as autoimmune diseases, asthma and atherosclerosis.[88] We will discuss later how the AFM approaches may provide better understanding of the molecular mechanism of leukocyte adhesion.

11.4.1 Selectins

The selectins are calcium-dependent, type I transmembrane glycoproteins that bind to sialylated carbohydrate moieties present on target proteins (Fig. 11.1).[89] Consisting of three members, P-selectin, E-selectin and L-selectin, selectins initiate leukocyte rolling on vascular endothelial cells during inflammatory responses. Structurally, the three selectins are very similar to each other and are composed of five types of domains (Fig. 11.1), starting from the N-terminal: a calcium-dependent lectin domain, an epidermal growth factor–like domain, a series of short-consensus-repeat domains, a transmembrane domain and a cytoplasmic tail (Fig. 11.1). The ligands for selectins are various glycoproteins, including PSGL-1, E-selectin ligand-1, glycosylation-dependent CAM-1 and CD34. Each of these ligands has a conjugated carbohydrate containing four sugar groups called sialyl Lewis X (sLeX), which forms the major binding site for all selectins. This interaction can be further strengthened by some amino acid residues within the ligands.[90]

Selectin–ligand interactions have been studied using AFM and other single-molecular approaches (Table 11.1). One of the earlier studies conducted by Fritz *et al.*, using a PSGL-1/IgG functionalized tip interacting with P-selectin-coated cover slide, reported unbinding forces of 110–170 pN, bond lifetime of 50 seconds and a Bell model barrier width of 2.5 Å.[91] Later work by Hanley *et al.*, using a more physiological ligand expressed on polymorphonuclear leukocytes, revealed a similar barrier width but much shorter lifetime (5 seconds).[92] Work by Evans' (using micropipette) and Moy's groups reported that selectin–ligand dissociation overcomes two activation barriers.[62,63] Compared with activated integrins, a majority of selectin–ligand bonds have shorter lifetime, which perhaps correlate with the rolling behaviour selectin mediates (Fig. 11.7). A series of studies by the group of Cheng Zhu indicated small pulling force could prolong the lifetime of P-selectin–ligand bonds, whereas force higher than 20 pN shortened the bond lifetime. Such a catch-slip bond behaviour could account for the "shear threshold" effect in selectin-mediated cell rolling.[23] To explain the catch bond phenomenon, the same group later proposed a sliding rebinding model, where force triggers two molecules sliding against each other, thereby increasing the affinity by initiating more intermolecular interactions.[93]

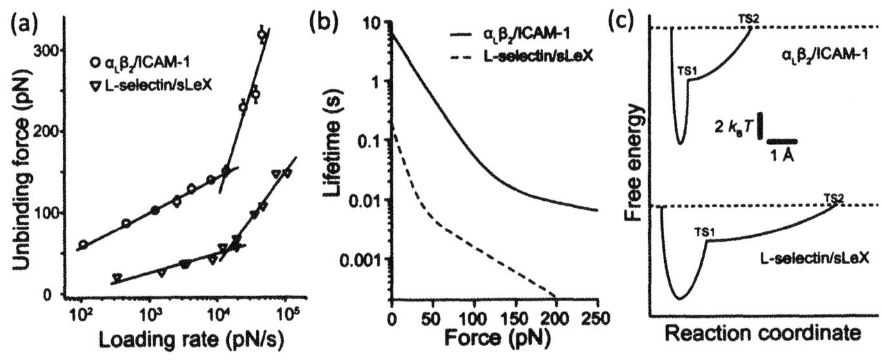

Figure 11.7. Adhesion strength of leukocyte adhesion complexes, L-selectin/sLeX and high affinity integrin $\alpha_L\beta_2$/ICAM-1. (a) Dynamic force spectra of the interactions (symbols, mean ± SEM) showing two loading rate regimes, a signature of two barriers. Solid lines represent the best fits of the Bell–Evans model [Eq. (11.6)] to each regime. The fitted parameters are shown in Table 11.1*. (b) Force-dependent lifetimes calculated from the fitted parameters. (c) Energy landscape of the interactions. The energy levels of the bound states were arbitrary chosen. The energy difference between the inner (1) and outer (2) barriers was obtained from $\Delta\Delta G^{\ddagger}_{1,2} = -k_BT\ln(\tau_{02}/\tau_{01})$. Compared with the integrin interaction, the selectin well was wider, i.e. more affected by force, and with shorter lifetime at zero force, being thus better adapted for its rolling function.[5,62]

* The fitted intrinsic lifetimes of the inner barriers were 0.001 seconds for L-selectin/sLeX and 0.025 seconds for $\alpha_L\beta_2$/ICAM-1.

11.4.2 Integrins

Integrins are heterodimeric transmembrane molecules, held together by non-covalent interactions and constitutively expressed in a wide variety of cells. Integrins mediate cell adhesion by binding to components of the extracellular matrix or to another cell by binding to members of the IgSF. The N-terminal region of all α subunits is made up of seven repeats that form a "β-propeller" structure. In half of the integrins, a 200-residue, Rossmann fold "I-domain" is inserted between the β-propeller repeats 2 and 3. A divalent cation coordination site, designated the metal ion-dependent adhesion site in the I-domain, binds negatively charged residues in ligands. A similar "β I-domain" structure is found in the N-terminal of the β subunit, which is directly involved in ligand binding in integrins that lack I-domains in the α subunits (both types of integrins shown in Fig. 11.1). Other domains of the α and β chain are important in regulating integrins' global conformation, affinity and the bidirectional signals crossing through the cell membrane.[94]

Affinity regulation is an important functional feature of all integrins. The strength of the integrin–ligand bond is drastically increased when the integrin molecule is activated through intracellular signals. Although the detailed molecular mechanism of affinity regulation is still obscure, it is shown that integrin activation is associated with a dramatic change of its overall global conformation. One of the most popular hypotheses is the separation of the two "legs" of the integrin; this separation results from a binding of activation adaptor molecules into the cytoplasmic tails during inside-out signaling.[94]

Because of its unique capability of characterizing in situ the strength of single molecular interactions, AFM became an ideal tool to probe the different activation states of integrins. Zhang *et al.* and Li *et al.* were able to observe the activation process of integrin $\alpha_L\beta_2$ and $\alpha_5\beta_1$, respectively. It has been found that resting integrins form short-lived receptor–ligand complexes of a fraction of seconds, whereas after activation, their lifetime increases about 100 folds.[5,33] Moreover, Kong *et al.* proposed recently that integrin $\alpha_5\beta_1$ also formed catch bonds when interacting with fibronectin,[22] indicating a more general mechanism for protein conformation of higher affinity induced by force.

It is noteworthy that a number of studies used AFM tips functionalized with integrin expressing cells.[39,95,96] This approach ensures that the heterodimeric integrin is under conditions close to the native environment, which has allowed researchers to monitor adhesion following cell activation in a whole cell level.

Table 11.1. A partial list of reported dynamic force spectroscopy studies of leukocyte selectins and integrins

Ligand–receptor pair	Approach	Rupture force (pN)	Intrinsic lifetime, τ_0 (s)	Potential width, γ (Å)	References
Selectin-ligand:					
P-selectin/PSGL-1	(1)	110–170	50	2.5	91
P-selectin/PSGL-1	(3)	30–220	5	1.4	92
P-selectin/sLeX	(1)	20–220	3	0.8, 4.5	62
E-selectin/sLeX	(1)	40–160	3	0.9, 5	62
L-selectin/sLeX	(1)	20–140	0.2	0.8, 4.5	62
Integrin–ligand:					
$\alpha_L\beta_2$/ICAM-1	(2)(4)	20–320	6	0.24, 2.1	5
$\alpha_L\beta_2$/ICAM-1	(2)	50–300	50	0.31, 4.5	6
$\alpha_M\beta_2$/ICAM-1	(3)	50–200	4	1.8	97
$\alpha_4\beta_1$/VCAM-1	(1)(2)	15–130	25	1, 5.5	37
$\alpha_5\beta_1$/fibronectin	(1)	40–170	83	0.9, 4	33
$\alpha_5\beta_1$/GRGDSP	(1)	32 ± 2		N/A	31
$\alpha_2\beta_1$/collagen	(2)	40–100	0.8	2.3	98

Notes: Experimental Approaches: (1) Protein on tip and substrate, (2) Cell attached on cantilever and protein on substrate, (3) Cell on substrate and proteins on tip, (4) Cells on substrate and tip; for integrins with different activation states, only the lifetime for the high affinity state was included.

11.5 CONCLUSIONS AND FUTURE DIRECTIONS

We have shown that the AFM is a well-established tool to probe the adhesion strength of biomolecular interactions. Unlike bulk studies, such as surface plasmon resonance, that provide averaged lifetimes and energies, AFM allows us to detect alternative dissociation pathways with possible intermediate states at the single-molecule level. Moreover, the AFM and other force techniques enable us to detect richer force-dependent behaviours such as those observed in catch bonds.

In practice, there are still some points in force measurements of adhesion strength that would require further study. For example, it is still under debate the possible effect of the probe stiffness in the measured forces.[36,67,99] This question is especially relevant in cell adhesion, since cells have the capacity of changing their mechanical properties in response to external signals.[39,100] Additional open questions include the effect of possible rebinding during pulling and the consequence of multiple bonds in the averaged rupture forces.[99,101] Although we have focused on the dissociation kinetics of cell adhesion bonds, the association kinetics might be even more relevant in some cases, such as in leukocyte binding to the endothelium. It has been reported that the association kinetics of biomolecules immobilized on a surface (2D), and even more on two opposing surfaces, vary significantly from those measured in 3D.[102] Some approaches have been already applied to determine the 2D association rates, using AFM and other force techniques, but an established method is still lacking.[102–105]

Regarding the biological relevance of force measurements using AFM, although it is useful to describe a biological interaction in terms of a well-defined energy landscape, it does not provide much information about the nature of the interaction itself, neither about its molecular determinants. Addressing this question requires complementary studies using molecular dynamic simulations, specific monoclonal antibodies or site-directed mutagenesis to be combined with force measurements. The molecular basis of the adhesion strength of some adhesion receptors have been studied using various force techniques combined with site-directed mutations or deletion constructs showing the contribution of specific amino acids to the shape of the energy landscape.[33,37,52,62,106] Molecular dynamics simulations provided an irreplaceable and valuable tool to determine the precise dissociation mechanism of molecular interactions in the early stages of force measurements and continue to shed light to the field.[57,107] An interesting question that would require further attention is the possible memory of adhesion receptors.[108]

An important and not very well-studied interaction relevant for cell adhesion is the binding between the cytoplasmic domains of CAMs and the proteins that mediate linking with the cytoskeleton. Although some approaches have been described using other techniques,[40] it is still a challenge to be able to measure the interaction forces in living cells in a controlled and specific manner.

The final goal of single-molecule studies is to understand how cell adhesion works. To describe whole cell adhesion, we will require appropriate and reliable models to jump from the single-molecule to the multiple binding of compliant, dynamic bodies that cells are.[109,110] Moreover, a standardized, objective and probe-independent method to experimentally quantify multiple molecule cell adhesion under static and dynamic loading is still to be developed. Thus, improved methodologies and theoretical developments are required to extrapolate single-molecule measurements to understand the mechanisms of whole cell adhesion. Being far from clearly understood, cell adhesion is a field for future research to which the AFM can still contribute importantly.

References

1. Hynes, R. O. (1999) Cell adhesion: old and new questions, *Trends Cell Biol.*, **9**, M33–M37.
2. Juliano, R. L. (2002) Signal transduction by cell adhesion receptors and the cytoskeleton: functions of integrins, cadherins, selectins, and immunoglobulin-superfamily members, *Annu. Rev. Pharmacol. Toxicol.*, **42**, 283–323.
3. Springer, T. A. (1994) Traffic signals for lymphocyte recirculation and leukocyte emigration: the multistep paradigm, *Cell*, **76**, 301–314.
4. Hukkanen, E. J., Wieland, J. A., Gewirth, A., Leckband, D. E., and Braatz, R. D. (2005) Multiple-bond kinetics from single-molecule pulling experiments: evidence for multiple NCAM bonds, *Biophys. J.*, **89**, 3434–3445.
5. Zhang, X. H., Wojcikiewicz, E., and Moy, V. T. (2002) Force spectroscopy of the leukocyte function-associated antigen-1/intercellular adhesion molecule-1 interaction, *Biophys. J.*, **83**, 2270–2279.
6. Wojcikiewicz, E. P., Abdulreda, M. H., Zhang, X., and Moy, V. T. (2006) Force spectroscopy of LFA-1 and its ligands, ICAM-1 and ICAM-2, *Biomacromolecules*, **7**, 3188–3195.
7. Panorchan, P., Thompson, M. S., Davis, K. J., Tseng, Y., Konstantopoulos, K., and Wirtz, D. (2006) Single-molecule analysis of cadherin-mediated cell-cell adhesion, *J. Cell Sci.*, **119**, 66–74.
8. Shi, Q., Chien, Y.-H., and Leckband, D. (2008) Biophysical properties of cadherin bonds do not predict cell sorting, *J. Biol. Chem.*, **283**, 28454–28463.
9. Baumgartner, W., Hinterdorfer, P., Ness, W., Raab, A., Vestweber, D., Schindler, H., and Drenckhahn, D. (2000) Cadherin interaction probed by atomic force microscopy, *Proc. Natl. Acad. Sci. USA*, **97**, 4005–4010.
10. Leckband, D., and Israelachvili, J. (2001) Intermolecular forces in biology, *Q. Rev. Biophys.*, **34**, 105–267.
11. Takagi, J., Petre, B. M., Walz, T., and Springer, T. A. (2002) Global conformational rearrangements in integrin extracellular domains in outside-in and inside-out signaling, *Cell*, **110**, 599–611.

12. Kim, M., Carman, C. V., and Springer, T. A. (2003) Bidirectional transmembrane signaling by cytoplasmic domain separation in integrins, *Science*, **301**, 1720–1725.

13. Carman, C. V., and Springer, T. A. (2003) Integrin avidity regulation: are changes in affinity and conformation underemphasized? *Curr. Opin. Cell Biol.*, **15**, 547–556.

14. Schmitz, J., and Gottschalk, K. E. (2008) Mechanical regulation of cell adhesion, *Soft Matter*, **4**, 1373–1387.

15. Krieg, M., Arboleda-Estudillo, Y., Puech, P. H., Kafer, J., Graner, F., Muller, D. J., and Heisenberg, C. P. (2008) Tensile forces govern germ-layer organization in zebrafish, *Nat. Cell Biol.*, **10**, 429–436.

16. Morfill, J., Neumann, J., Blank, K., Steinbach, U., Puchner, E. M., Gottschalk, K. E., and Gaub, H. E. (2008) Force-based analysis of multidimensional energy landscapes: application of dynamic force spectroscopy and steered molecular dynamics simulations to an antibody fragment-peptide complex, *J. Mol. Biol.*, **381**, 1253–1266.

17. Hutter, J. L., and Bechhoefer, J. (1993) Calibration of atomic-force microscope tips, *Rev. Sci. Instrum.*, **64**, 1868–1873.

18. Sader, J. E., Larson, I., Mulvaney, P., and White, L. (1995) Method for the calibration of atomic force microscope cantilevers, *Rev. Sci. Instrum.*, **66**, 3789–3798.

19. Burnham, N. A., Chen, X., Hodges, C. S., Matei, G. A., Thoreson, E. J., Roberts, C. J., Davis, M. C., and Tendler, S. J. B. (2003) Comparison of calibration methods for atomic-force microscopy cantilevers, *Nanotechnology*, **14**, 1–6.

20. Butt, H. J., Cappella, B., and Kappl, M. (2005) Force measurements with the atomic force microscope: technique, interpretation and applications, *Surf. Sci. Rep.*, **59**, 1–152.

21. Cappella, B., and Dietler, G. (1999) Force-distance curves by atomic force microscopy, *Surf. Sci. Rep.*, **34**, 1–104.

22. Kong, F., Garcia, A. J., Mould, A. P., Humphries, M. J., and Zhu, C. (2009) Demonstration of catch bonds between an integrin and its ligand, *J. Cell Biol.*, **185**, 1275–1284.

23. Marshall, B. T., Long, M., Piper, J. W., Yago, T., McEver, R. P., and Zhu, C. (2003) Direct observation of catch bonds involving cell-adhesion molecules, *Nature*, **423**, 190–193.

24. Verbelen, C., Gruber, H. J., and Dufrene, Y. F. (2007) The NTA-HiS(6) bond is strong enough for AFM single-molecular recognition studies, *J. Mol. Recognit.*, **20**, 490–494.

25. Sulchek, T. A., Friddle, R. W., Langry, K., Lau, E. Y., Albrecht, H., Ratto, T. V., DeNardo, S. J., Colvin, M. E., and Noy, A. (2005) Dynamic force spectroscopy of parallel individual Mucin1-antibody bonds, *Proc. Natl. Acad. Sci. USA*, **102**, 16638–16643.

26. Rankl, C., Kienberger, F., Wildling, L., Wruss, J., Gruber, H. J., Blaas, D., and Hinterdorfer, P. (2008) Multiple receptors involved in human rhinovirus attachment to live cells, *Proc. Natl. Acad. Sci. USA*, **105**, 17778–17783.

27. Abdulreda, M. H., Bhalla, A., Rico, F., Berggren, P.-O., Chapman, E. R., and Moy, V. T. (2009) Pulling force generated by interacting SNAREs facilitates membrane hemifusion, *Integr. Biol.*, **1**, 301–310.

28. Lee, G. U., Kidwell, D. A., and Colton, R. J. (1994) Sensing discrete streptavidin biotin interactions with atomic-force microscopy, *Langmuir*, **10**, 354–357.

29. Moy, V. T., Florin, E. L., and Gaub, H. E. (1994) Intermolecular forces and energies between ligands and receptors, *Science*, **266**, 257–259.

30. Florin, E. L., Moy, V. T., and Gaub, H. E. (1994) Adhesion forces between individual ligand-receptor pairs, *Science*, **264**, 415–417.

31. Lehenkari, P. P., and Horton, M. A. (1999) Single integrin molecule adhesion forces in intact cells measured by atomic force microscopy, *Biochem. Biophys. Res. Commun.*, **259**, 645–650.

32. Rico, F., Roca-Cusachs, P., Sunyer, R., Farre, R., and Navajas, D. (2007) Cell dynamic adhesion and elastic properties probed with cylindrical atomic force microscopy cantilever tips, *J. Mol. Recognit.*, **20**, 459–466.

33. Li, F. Y., Redick, S. D., Erickson, H. P., and Moy, V. T. (2003) Force measurements of the $\alpha_5\beta_1$ integrin-fibronectin interaction, *Biophys. J.*, **84**, 1252–1262.

34. Rico, F., and Moy, V. T. (2007) Energy landscape roughness of the streptavidin-biotin interaction, *J. Mol. Recognit.*, **20**, 495–501.

35. Marshall, B. T., Sarangapani, K. K., Lou, J. H., McEver, R. P., and Zhu, C. (2005) Force history dependence of receptor-ligand dissociation, *Biophys. J.*, **88**, 1458–1466.

36. Walton, E. B., Lee, S., and Van Vliet, K. J. (2008) Extending Bell's model: how force transducer stiffness alters measured unbinding forces and kinetics of molecular complexes, *Biophys. J.*, **94**, 2621–2630.

37. Zhang, X. H., Craig, S. E., Kirby, H., Humphries, M. J., and Moy, V. T. (2004) Molecular basis for the dynamic strength of the integrin $\alpha_4\beta_1$/VCAM-1 interaction, *Biophys. J.*, **87**, 3470–3478.

38. Benoit, M., Gabriel, D., Gerisch, G., and Gaub, H. E. (2000) Discrete interactions in cell adhesion measured by single-molecule force spectroscopy, *Nat. Cell Biol.*, **2**, 313–317.

39. Wojcikiewicz, E. P., Zhang, X., Chen, A., and Moy, V. T. (2003) Contributions of molecular binding events and cellular compliance to the modulation of leukocyte adhesion, *J. Cell Sci.*, **116**, 2531–2539.

40. Evans, E., Heinrich, V., Leung, A., and Kinoshita, K. (2005) Nano- to microscale dynamics of P-selectin detachment from leukocyte interfaces. I. Membrane separation from the cytoskeleton, *Biophys. J.*, **88**, 2288–2298.

41. Brochard-Wyart, F., Borghi, N., Cuvelier, D., and Nassoy, P. (2006) Hydrodynamic narrowing of tubes extruded from cells, *Proc. Natl. Acad. Sci. USA*, **103**, 7660–7663.

42. Alcaraz, J., Buscemi, L., Puig-de-Morales, M., Colchero, J., Baro, A., and Navajas, D. (2002) Correction of microrheological measurements of soft samples with atomic force microscopy for the hydrodynamic drag on the cantilever, *Langmuir*, **18**, 716–721.

43. Janovjak, H. J., Struckmeier, J., and Muller, D. J. (2005) Hydrodynamic effects in fast AFM single-molecule force measurements, *Eur. Biophys. J.*, **34**, 91–96.

44. Krieg, M., Helenius, J., Heisenberg, C. P., and Muller, D. J. (2008) A bond for a lifetime: employing membrane nanotubes from living cells to determine receptor-ligand kinetics, *Angew. Chem. Int. Ed. Engl.*, **47**, 9775–9777.

45. Obataya, I., Nakamura, C., Han, S., Nakamura, N., and Miyake, J. (2005) Nanoscale operation of a living cell using an atomic force microscope with a nanoneedle, *Nano Lett.*, **5**, 27–30.

46. Zhang, X. H., Chen, A., De Leon, D., Li, H., Noiri, E., Moy, V. T., and Goligorsky, M. S. (2004) Atomic force microscopy measurement of leukocyte-endothelial interaction, *Am. J. Physiol. Heart Circ. Physiol.*, **286**, H359–H367.

47. Evans, E., Ritchie, K., and Merkel, R. (1995) Sensitive force technique to probe molecular adhesion and structural linkages at biological interfaces, *Biophys. J.*, **68**, 2580–2587.

48. Evans, E., and Ritchie, K. (1997) Dynamic strength of molecular adhesion bonds, *Biophys. J.*, **72**, 1541–1555.

49. Evans, E. (2001) Probing the relation between force—Lifetime—and chemistry in single molecular bonds, *Annu. Rev. Biophys. Biomol. Struct.*, **30**, 105–128.

50. Bell, G. I. (1978) Models for specific adhesion of cells to cells, *Science*, **200**, 618–627.

51. Andre, G., Leenhouts, K., Hols, P., and Dufrene, Y. F. (2008) Detection and localization of single LysM-peptidoglycan interactions, *J. Bacteriol.*, **190**, 7079–7086.

52. Yakovenko, O., Sharma, S., Forero, M., Tchesnokova, V., Aprikian, P., Kidd, B., Mach, A., Vogel, V., Sokurenko, E., and Thomas, W. E. (2008) FimH forms catch bonds that are enhanced by mechanical force due to allosteric regulation, *J. Biol. Chem.*, **283**, 11596–11605.

53. Hinterdorfer, P., and Dufrene, Y. F. (2006) Detection and localization of single molecular recognition events using atomic force microscopy, *Nat. Methods*, **3**, 347–355.

54. Wojcikiewicz, E. P., Koenen, R. R., Fraemohs, L., Minkiewicz, J., Azad, H., Weber, C., and Moy, V. T. (2009) LFA-1 binding destabilizes the JAM-A homophilic interaction during leukocyte transmigration, *Biophys. J.*, **96**, 285–293.

55. Rief, M., Gautel, M., Oesterhelt, F., Fernandez, J. M., and Gaub, H. E. (1997) Reversible unfolding of individual titin immunoglobulin domains by AFM, *Science*, **276**, 1109–1112.

56. Muller, D. J., Sapra, K. T., Scheuring, S., Kedrov, A., Frederix, P. L., Fotiadis, D., and Engel, A. (2006) Single-molecule studies of membrane proteins, *Curr. Opin. Struct. Biol.*, **16**, 489–495.

57. Izrailev, S., Stepaniants, S., Balsera, M., Oono, Y., and Schulten, K. (1997) Molecular dynamics study of unbinding of the avidin-biotin complex, *Biophys. J.*, **72**, 1568–1581.
58. Dudko, O. K., Hummer, G., and Szabo, A. (2008) Theory, analysis, and interpretation of single-molecule force spectroscopy experiments, *Proc. Natl. Acad. Sci. USA*, **105**, 15755–15760.
59. Chaudhuri, O., Parekh, S. H., Lam, W. A., and Fletcher, D. A. (2009) Combined atomic force microscopy and side-view optical imaging for mechanical studies of cells, *Nat. Methods*, **6**, 383–387.
60. Sheetz, M. P., Sable, J. E., and Dobereiner, H. G. (2006) Continuous membrane-cytoskeleton adhesion requires continuous accommodation to lipid and cytoskeleton dynamics, *Annu. Rev. Biophys. Biomol. Struct.*, **35**, 417–434.
61. Girdhar, G., and Shao, J. Y. (2007) Simultaneous tether extraction from endothelial cells and leukocytes: observation, mechanics, and significance, *Biophys. J.*, **93**, 4041–4052.
62. Zhang, X. H., Bogorin, D. F., and Moy, V. T. (2004) Molecular basis of the dynamic strength of the sialyl Lewis X-selectin interaction, *Chemphyschem*, **5**, 175–182.
63. Evans, E., Leung, A., Hammer, D., and Simon, S. (2001) Chemically distinct transition states govern rapid dissociation of single L-selectin bonds under force, *Proc. Natl. Acad. Sci. USA*, **98**, 3784–3789.
64. Husson, J., and Pincet, F. (2008) Analyzing single-bond experiments: influence of the shape of the energy landscape and universal law between the width, depth, and force spectrum of the bond, *Phys. Rev. E*, **77**, 026108.
65. Friddle, R. W. (2008) Unified model of dynamic forced barrier crossing in single molecules, *Phys. Rev. Lett.*, **100**, 138302
66. Dudko, O. K., Hummer, G., and Szabo, A. (2006) Intrinsic rates and activation free energies from single-molecule pulling experiments, *Phys. Rev. Lett.*, **96**, 108101
67. Friddle, R. W., Podsiadlo, P., Artyukhin, A. B., and Noy, A. (2008) Near-equilibrium chemical force microscopy, *J. Phys. Chem. C Nanomater. Interfaces*, **112**, 4986–4990.
68. Merkel, R., Nassoy, P., Leung, A., Ritchie, K., and Evans, E. (1999) Energy landscapes of receptor-ligand bonds explored with dynamic force spectroscopy, *Nature*, **397**, 50–53.
69. Derenyi, I., Bartolo, D., and Ajdari, A. (2004) Effects of intermediate bound states in dynamic force spectroscopy, *Biophys. J.*, **86**, 1263–1269.
70. Oberbarnscheidt, L., Janissen, R., and Oesterhelt, F. (2009) Direct and model free calculation of force-dependent dissociation rates from force spectroscopic Data, *Biophys. J.*, **97**, L19–L21.
71. Preiner, J., Janovjak, H., Rankl, C., Knaus, H., Cisneros, D. A., Kedrov, A., Kienberger, F., Muller, D. J., and Hinterdorfer, P. (2007) Free energy of membrane protein unfolding derived from single-molecule force measurements, *Biophys. J.*, **93**, 930–937.

72. Liu, W., Montana, V., Parpura, V., and Mohideen, U. (2008) Comparative energy measurements in single molecule interactions, *Biophys. J.*, **95**, 419–425.

73. Zwanzig, R. (1988) Diffusion in a rough potential, *Proc. Natl. Acad. Sci. USA*, **85**, 2029–2030.

74. Ansari, A., Berendzen, J., Bowne, S. F., Frauenfelder, H., Iben, I. E. T., Sauke, T. B., Shyamsunder, E., and Young, R. D. (1985) Protein states and protein quakes, *Proc. Natl. Acad. Sci. USA*, **82**, 5000–5004.

75. Hyeon, C. B., and Thirumalai, D. (2003) Can energy landscape roughness of proteins and RNA be measured by using mechanical unfolding experiments? *Proc. Natl. Acad. Sci. USA*, **100**, 10249–10253.

76. Nevo, R., Brumfeld, V., Kapon, R., Hinterdorfer, P., and Reich, Z. (2005) Direct measurement of protein energy landscape roughness, *EMBO Rep.*, **6**, 482–486.

77. Schlierf, M., and Rief, M. (2005) Temperature softening of a protein in single-molecule experiments, *J. Mol. Biol.*, **354**, 497–503.

78. Janovjak, H., Knaus, H., and Muller, D. J. (2007) Transmembrane helices have rough energy surfaces, *J. Am. Chem. Soc.*, **129**, 246–247.

79. Dembo, M., Torney, D. C., Saxman, K., and Hammer, D. (1988) The reaction-limited kinetics of membrane-to-surface adhesion and detachment, *Proc. R. Soc. Lond. Biol. Sci.*, **234**, 55–83.

80. Thomas, W. E., Trintchina, E., Forero, M., Vogel, V., and Sokurenko, E. V. (2002) Bacterial adhesion to target cells enhanced by shear force, *Cell*, **109**, 913–923.

81. Friedland, J. C., Lee, M. H., and Boettiger, D. (2009) Mechanically activated integrin switch controls $\alpha_5\beta_1$ function, *Science*, **323**, 642–644.

82. Guo, B., and Guilford, W. H. (2006) Mechanics of actomyosin bonds in different nucleotide states are tuned to muscle contraction, *Proc. Natl. Acad. Sci. USA*, **103**, 9844–9849.

83. Alon, R., Chen, S., Fuhlbrigge, R., Puri, K. D., and Springer, T. A. (1998) The kinetics and shear threshold of transient and rolling interactions of L-selectin with its ligand on leukocytes, *Proc. Natl. Acad. Sci. USA*, **95**, 11631–11636.

84. Yago, T., Lou, J., Wu, T., Yang, J., Miner, J. J., Coburn, L., Lopez, J. A., Cruz, M. A., Dong, J. F., McIntire, L. V., McEver, R. P., and Zhu, C. (2008) Platelet glycoprotein Ibalpha forms catch bonds with human WT vWF but not with type 2B von Willebrand disease vWF, *J. Clin. Invest.*, **118**, 3195–3207.

85. Thomas, W. E., Vogel, V., and Sokurenko, E. (2008) Biophysics of catch bonds, *Annu. Rev. Biophys.*, **37**, 399–416.

86. Wei, Y. (2008) Entropic-elasticity-controlled dissociation and energetic-elasticity-controlled rupture induce catch-to-slip bonds in cell-adhesion molecules, *Phys. Rev. E*, **77**, 031910–031916.

87. Forero, M., Yakovenko, O., Sokurenko, E. V., Thomas, W. E., and Vogel, V. (2006) Uncoiling mechanics of type I fimbriae are optimized for catch bonds, *PLoS Biol.*, **4**, e298.

88. Kubes, P. (2002) The complexities of leukocyte recruitment, *Semin. Immunol.*, **14**, 65–72.

89. McEver, R. P. (2002) Selectins: lectins that initiate cell adhesion under flow, *Curr. Opin. Cell Biol.*, **14**, 581–586.

90. Rosen, S. D., and Bertozzi, C. R. (1994) The selectins and their ligands, *Curr. Opin. Cell Biol.*, **6**, 663–673.

91. Fritz, J., Katopodis, A. G., Kolbinger, F., and Anselmetti, D. (1998) Force-mediated kinetics of single P-selectin/ligand complexes observed by atomic force microscopy, *Proc. Natl. Acad. Sci. USA*, **95**, 12283–12288.

92. Hanley, W., McCarty, O., Jadhav, S., Tseng, Y., Wirtz, D., and Konstantopoulos, K. (2003) Single molecule characterization of P-selectin/ligand binding, *J. Biol. Chem.*, **278**, 10556–10561.

93. Lou, J., and Zhu, C. (2007) A structure-based sliding-rebinding mechanism for catch bonds, *Biophys. J.*, **92**, 1471–1485.

94. Luo, B. H., Carman, C. V., and Springer, T. A. (2007) Structural basis of integrin regulation and signaling, *Annu. Rev. Immunol.*, **25**, 619–647.

95. Franz, C. M., Taubenberger, A., Puech, P. H., and Muller, D. J. (2007) Studying integrin-mediated cell adhesion at the single-molecule level using AFM force spectroscopy, *Sci. STKE*, **2007**, pl5.

96. Hosseini, B. H., Louban, I., Djandji, D., Wabnitz, G. H., Deeg, J., Bulbuc, N., Samstag, Y., Gunzer, M., Spatz, J. P., and Hammerling, G. J. (2009) Immune synapse formation determines interaction forces between T cells and antigen-presenting cells measured by atomic force microscopy, *Proc. Natl. Acad. Sci. USA*, **106**, 17852–17857.

97. Yang, H., Yu, J., Fu, G., Shi, X., Xiao, L., Chen, Y., Fang, X., and He, C. (2007) Interaction between single molecules of Mac-1 and ICAM-1 in living cells: an atomic force microscopy study, *Exp. Cell Res.*, **313**, 3497–3504.

98. Taubenberger, A., Cisneros, D. A., Friedrichs, J., Puech, P. H., Muller, D. J., and Franz, C. M. (2007) Revealing early steps of $\alpha_2\beta_1$ integrin-mediated adhesion to collagen type I by using single-cell force spectroscopy, *Mol. Biol. Cell*, **18**, 1634–1644.

99. Tshiprut, Z., Klafter, J., and Urbakh, M. (2008) Single-molecule pulling experiments: when the stiffness of the pulling device matters, *Biophys. J.*, **95**, L42–L44.

100. Schmitz, J., Benoit, M., and Gottschalk, K. E. (2008) The viscoelasticity of membrane tethers and its importance for cell adhesion, *Biophys. J.*, **95**, 1448–1459.

101. Guo, S., Ray, C., Kirkpatrick, A., Lad, N., and Akhremitchev, B. B. (2008) Effects of multiple-bond ruptures on kinetic parameters extracted from force spectroscopy measurements: revisiting biotin-streptavidin interactions, *Biophys. J.*, **95**, 3964–3976.

102. Zhu, C., Bao, G., and Wang, N. (2000) Cell mechanics: mechanical response, cell adhesion, and molecular deformation, *Annu. Rev. Biomed. Eng.*, **2**, 189–226.

103. Guo, S., Lad, N., Ray, C., and Akhremitchev, B. B. (2009) Association kinetics from single molecule force spectroscopy measurements, *Biophys. J.*, **96**, 3412–3422.

104. Hinterdorfer, P., Baumgartner, W., Gruber, H. J., Schilcher, K., and Schindler, H. (1996) Detection and localization of individual antibody-antigen recognition events by atomic force microscopy, *Proc. Natl. Acad. Sci. USA*, **93**, 3477–3481.

105. Chesla, S. E., Selvaraj, P., and Zhu, C. (1998) Measuring two-dimensional receptor-ligand binding kinetics by micropipette, *Biophys. J.*, **75**, 1553–1572.

106. Zhang, F., Marcus, W. D., Goyal, N. H., Selvaraj, P., Springer, T. A., and Zhu, C. (2005) Two-dimensional kinetics regulation of $\alpha_L\beta_2$-ICAM-1 interaction by conformational changes of the α_L-inserted domain, *J. Biol. Chem.*, **280**, 42207–42218.

107. Sotomayor, M., and Schulten, K. (2007) Single-molecule experiments in vitro and in silico, *Science*, **316**, 1144–1148.

108. Zarnitsyna, V. I., Huang, J., Zhang, F., Chien, Y.-H., Leckband, D., and Zhu, C. (2007) Memory in receptor-ligand-mediated cell adhesion, *Proc. Natl. Acad. Sci. USA*, **104**, 18037–18042.

109. Seifert, U. (2002) Dynamic strength of adhesion molecules: role of rebinding and self-consistent rates, *Europhys. Lett.*, **58**, 792–798.

110. Erdmann, T., Pierrat, S., Nassoy, P., and Schwarz, U. S. (2008) Dynamic force spectroscopy on multiple bonds: experiments and model, *Europhys. Lett.*, **81**, 48001.

Chapter 12

MAPPING MEMBRANE PROTEINS ON LIVING CELLS USING THE ATOMIC FORCE MICROSCOPE

Atsushi Ikai[a] **and Rehana Afrin**[a,b]
[a] *Innovation Laboratory, Tokyo Institute of Technology, 4259 Nagatsuta, Midori-ku, Yokohama, 226-8501, Japan*
[b] *Biofrontier Center, Tokyo Institute of Technology, 4259 Nagatsuta, Midori-ku, Yokohama, 226-8501, Japan*
ikai.a.aa@m.titech.ac.jp

12.1 INTRODUCTION

Intrinsic membrane proteins are firmly anchored to the lipid bilayer membrane of the cell surface by a clever combination of hydrophobic and hydrophilic nature of peptide side chains. The transmembrane segment of the protein is highly hydrophobic and its extra-membranous domains are just as hydrophilic as most of the soluble proteins in an aqueous medium.[1] Membrane proteins are difficult to probe in solution in free state, without the help of detergents, and only a few data on the binding constants of such proteins to the lipid bilayer membrane are available.[2] Because of their very stable anchoring, membrane proteins almost never come out of the lipid bilayer. Otherwise, cells with no means of synthesizing new proteins such as red blood cells would continuously lose their membrane proteins without any means of replenishing them, resulting in a much higher rate of their dysfunctionalization than their present life time. The firm anchoring of membrane proteins can be exploited to manipulate them on the live cell membrane, and such manipulations of live cells

Life at the Nanoscale: Atomic Force Microscopy of Live Cells
Edited by Yves Dufrêne

www.panstanford.com

include mapping of cell surface receptors, probing the mechanical linkage of membrane receptors to the intracellular structures or extraction of membrane proteins without using detergents. In this chapter, we describe methods for mechanically mapping the presence (or absence) of particular proteins (mainly specific receptors) on cell membranes using the force mode of the atomic force microscope (AFM). In such experiments, an AFM probe is linked to membrane proteins, either with biochemical specificity or without, and forces of various magnitudes are applied to the probe through a force transducer to observe the effect of applied forces to the cell and cell membrane proteins. Besides AFM,[3,4] force transducers capable of manipulating membrane proteins at the single-molecule level include laser tweezers (or optical traps),[5,6] the biomembrane force probe[7] and magnetic tweezers.[8]

The effect of externally applied forces to membrane-bound proteins has been a subject of intense research in the field of biomechanics from the emerging[9] to the present matured phase.[10] Among early attempts to guide the experimental effort with a sound theoretical base, Bell published his seminal paper in 1978[11] explaining the on and off kinetics of protein–protein interactions under the influence of an applied force. He also gave an estimate of the force required for "uprooting" the typical intrinsic membrane protein glycophorin A from lipid bilayers as 250 pN. The idea of uprooting a firmly anchored membrane protein by force from the biological membrane was an eye-opening example of the potential of nanotechnology to many biologists.

Thanks to the invention of laser tweezers and AFM, what was envisaged by Bell in 1978 has become experimentally verifiable, and many mechanical experiments on single molecules of DNA,[12-14] RNA,[15] polypeptides,[16,17] synthetic polymers,[18] polysaccharides[19] and proteins[20-27] have been reported, to name a few examples. In addition to such single-molecule experiments, measurement of the force required to separate intermolecular complexes has been done on, for example, complexes between biotin–avidin,[28] antibody–antigen,[29] lectin–carbohydrate,[30] transferrin–transferrin receptor,[31] GroEL-unfolded protein[32] and so on.

In force measurements, a key feature is that the tensile strength (the maximum force to break composite structures by the application of tensile force) is not a constant for a given sample, but it depends on the loading rate in a predictable way.[7,33] By taking advantage of the loading rate dependence of the tensile force, one can determine two parameters related to the energy diagram of the unbinding reaction, i.e., the dissociation rate constant under zero external force (natural dissociation rate constant, k^0_{diss}) and the distance for the complex to reach the activated state from its equilibrium state, often designated as Δx or just x.[33,34] A detailed treatment of the force spectroscopy of single and multiple bond dynamics was given by Evans and Williams.[35]

In the next sections, we present early work in which cell surface molecules were mapped using modified AFM probes, discuss the use of the colloidal probe method for mapping vitronectin (VN) and prostaglandin receptors, describe measurements of the binding strength of transferin receptors and discuss the issue of unbinding versus uprooting of membrane proteins. We then provide a brief literature survey of recent protein mapping studies and discuss the possibility of mapping intracellular mRNA.

12.2 PIONEERING EXAMPLE OF CELL SURFACE MAPPING

As an early example of cell surface mapping, using an AFM probe coated with a sugar-specific lectin, concanavalin A, Gad *et al.*[36] mapped the presence or absence of mannan molecules over the surface of live *Saccharomyces cerevisiae* cells (Fig. 12.1). They first immobilized round yeast cells on a glass surface covered with covalently immobilized concanavalin A which is reactive with mannan on the cell surface. Cells were adsorbed to the glass surface in a highly packed condition, which was convenient for later mapping of mannan. After imaging closely packed yeast cells with a bare probe, a new probe that was coated with covalently immobilized concanavalin A was brought on top of one of the cells, and the interaction between the probe and the cell surface was measured by the force volume method. This AFM mode simultaneously gave the topography and 16 × 16 force curves recorded over 3 μm × 3 μm regions. The force curves obtained in this way showed extensions of very flexible molecules up to 1 to a few micrometres, which was interpreted as the extension of almost randomly coiled mannan molecules covalently attached on the yeast cell wall. The downward deflection of the cantilever as shown in Fig. 12.1a was considered to be due to the tensile force exerted by the flexible mannan chains to the cantilever. The sharp upward jumps (observed three times here) were interpreted as the unbinding events of concanavalin A from the surface mannans. The magnitude of the jump E measured in nanometres was converted to the unbinding force F by multiplying it with the force constant of the cantilever k, assuming the Hookean linear behaviour of the cantilever. The spring constant k was determined to be approximately 0.025 nN/nm by pushing a small cantilever cut out from a thin gold foil. Since the nominal spring constant supplied by the manufacture was 0.02 nN/nm, the measured value was within an allowable error range.

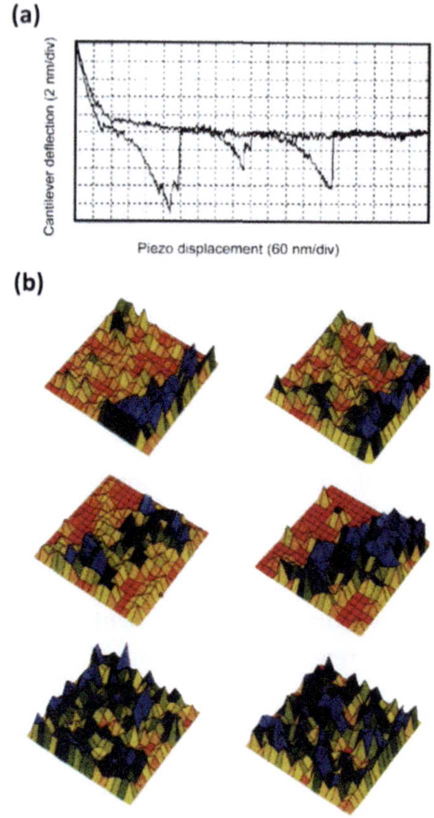

Figure 12.1. Pulling mannans on living yeast cells (*Saccharomyces cerevisiae*). (a) A typical force curve showing the pulling of mannan with a concanavalin A-coated AFM probe. The spring constant of the cantilever was determined as ~0.025 nN/nm; the largest deflection of about 9 nm in this figure corresponded to ~225 pN. (b) Mapping of the mannan distribution based on the strongest interactions with the concanavalin A-coated probe in each area of 3 μm × 3 μm. Maps on three different areas are given in pairs. The second map in each pair was obtained 10 minutes after the first map. Each horizontal pair represents successive mapping results on the same area. Colour code for interaction strength is as follows: red (0–50 pN), orange (50–100 pN), yellow (100–150 pN), light green (150–200 pN), dark green (200–250 pN), blue (>250 pN). Reproduced with permission from Gad *et al.*[36]

The mean rupture force of the concanavalin A-mannan bond was approximately in the range of 70–200 pN. The interaction force was reduced after the addition of free mannose verifying that the measured forces were specific for the concanavalin A versus mannan interaction. Using the force volume mode, they produced force maps of concanavalin A versus mannan interactions plotting the largest force obtained in each of 16 × 16 small areas over 3 μm × 3 μm regions. The mapping results, shown in Fig. 12.1b, reveal that the mannan distribution on the cell surface was not uniform at all but highly varied over the cell surface. The authors reproduced the local variations of the relative surface density of mannan based on the force curve measurements on six different positions on a yeast cell surface, repeating the measurements twice on all spots every 10 minutes to confirm the reproducibility of the mapping results. Mapping results after longer time intervals showed gradual changes in the local mannan density partly because of the drift on the scanning area under the AFM probe and/or damages that would have taken place to the immobilized concanavalin A on the AFM probe.

Two other convenient methods of yeast cell immobilization were developed. Gad *et al.*[37] confined yeast cells by half embedding them in a thin layer of agarose gel so that the cells would not be rolled about under the imaging force of the AFM probe. Under imaging with a bare probe, most cells were proved to be alive from the growth of their height in the time lapse AFM images. Another method of immobilizing yeast cells proposed by Kasas *et al.*[38] is based on the use of porous membranes, which is also suitable for manipulations of cell surfaces with AFM.

12.3 MAPPING OF VITRONECTIN AND PROSTAGLANDIN RECEPTORS

The radius of a typical AFM probe at the very tip is in the order of 10–50 nm. A sharp tip with small radius is good for high resolution mapping of membrane proteins over a wide area of the cell surface. We sometimes need a probe with a larger diameter to increase the contact area with the cell surface. Kim *et al.*[39] used a colloidal probe of 5 μm in diameter to increase the area to be scanned by the force volume mode of the AFM. With this probe, the mean indentation depth was about 165 nm and the contact area was calculated to be 2.6 μm². To obtain the contact area (S), they assumed the Hertz contact model and used the following equation.[40,41]

$$S = 2\pi R^2 \int_0^T \sin\theta \, d\theta, \tag{12.1}$$

where $T = \sin^{-1}(d/R)^{1/2}$, R and d are the probe radius and the depth of indentation, respectively.

Mapping of membrane proteins using a colloidal probe cannot measure the unbinding force of a single ligand–protein pair. Therefore, Kim *et al.* used the integrated area in the force-extension curve calling it the separation work in the unit of J. Figure 12.2 displays the result of mapping over 4 × 4 small areas of 4 μm × 4 μm regions on the surface of living fibroblasts (Balb 3T3) using a colloidal probe that was coated with fibronectin searching for its specific receptor, integrins. First, the data showed reproducible mapping patterns between consecutive maps (Fig. 12.2a) but with a gradual change of the mapping pattern with time. Then, the unbinding work was measured over a wide area of a living cell by repeating the measurement each time shifting the scanning area (Fig. 12.2b).

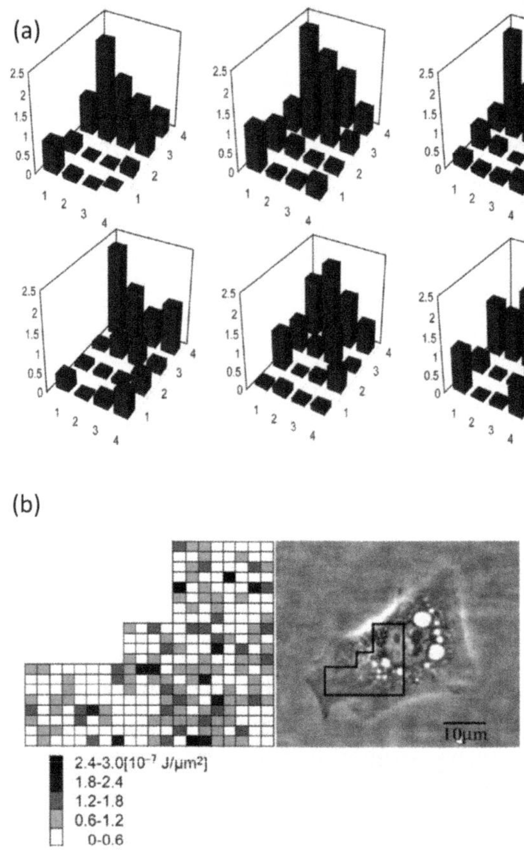

Figure 12.2. (a) Six time lapse mapping results of the same area obtained on a living Balb/3T3 fibroblast cell by scanning with an AFM probe coated with fibronectin every 1 minute from 1 to 4. Horizontal scales are in micrometre and the vertical axis is the separation work in 10^{-17} J/μm². (b) Composite mapping results on a living cell. The framed area on the right image (phase contrast) was divided into several sub-areas and each area was scanned with fibronectin-coated probe. The results from multiple areas were assembled to create the composite map on the left. Reproduced with permission from Kim *et al.*[39]

In their next work, Kim *et al.* used a colloidal probe coated with vitronectin (VN) molecules to map the location of VN receptors on the surface of live osteoblastic cells[42] and of prostaglandin receptors on living CHO cells[43] (details will be given later on). Since the method was based on the interactions between a fairly large number of ligand molecules on a colloidal probe and a corresponding number of receptors on the cell surface, a coarse grained map was available over a large area and the result was useful to make correspondence with that of the conventional labelling method using fluorescent antibodies and to map receptors on the live cell surface with high confidence. Previously available methods based on fluorescence labelling relied on fixing of the cell, which is fatal. Instead of counting the number of receptor molecules on the cell surface, Kim *et al.* used the separation work as defined earlier by Eq. (12.1).[42]

With this method, the distribution of VN receptors on a living murine osteoblastic cell was successfully measured. First, the distribution of the integrin $\alpha_V\beta_5$ subunit was confirmed by conventional immunohistochemistry after fixing the cell. To visualize the distribution of the receptor on a living cell by an independent and potentially a more quantitative method, the AFM was used with a micro-bead attached to the cantilever end to increase the area of contact and VN was immobilized on the micro-bead, and from the resulting force curve, separation work was calculated and displayed as described earlier.

Kim *et al.* further studied the distribution of EP3 receptors on a living CHO cell.[43] Green fluorescent protein (GFP) was fused to the extracellular region of the EP3 receptor on a CHO cell and a micro-bead was coated with anti-GFP antibody. The interactions between the antibodies and GFP molecules on the cell surface were recorded, and the result indicated that EP3 receptors were distributed on the CHO cell surface not uniformly but in small patches coincident with the result of immuno-histochemical observations. Repeated measurements on the same area of the cell surface gave confirmation that it was unlikely that the receptors were extracted from the cell membrane during the experiments. The separation work required to break a single molecular pair was estimated to be about 1.5×10^{-18} J. Using this value, the number of EP3 receptor on the CHO cell surface was estimated to be about 1×10^4 under the assumption that the area of the cell surface was about 5000 μm^2.

The colloidal probe method can cover a wide area of approximately 20 μm × 20 μm. Much wider areas may be covered by systematically shifting the area for scanning or by using colloidal probes of larger diameter, but as the number of ligand receptor interactions pairs increases, it will be harder to dissociate the probe from the cell surface, making it impossible to map receptor molecules.

To determine the sensitivity of the colloidal probe method of mapping cell surface proteins, Kim *et al.* first estimated the density of specific molecules over a glass surface.[44] Since direct counting of the protein molecules after immobilization on the glass surface was not feasible, they measured the number density of cross-linkers that were fixed on the substrate surface. First the amino-silanized glass was reacted with the covalent cross-linker, Sulfo-LC-SPDP (Sulfosuccinimidyl 6-(3'-[2-pyridyldithio]-propionamido) hexanoate), in a buffer solution. The amount of Sulfo-LC-SPDP that reacted with amino groups on the glass surface was determined by measuring the optical density of the buffer that contained the chromogenic by-product, pyridine-2-thione, at 343 nm. The results showed that the surface-immobilized SPDP showed a clear saturation behaviour. They chose a condition where the number density of the cross-linker on the surface was about $1 \times 10^4/\mu m^2$. They then reacted the activated glass surface with synthetic peptide named VN7 and coated the glass bead on an AFM cantilever with anti-VN7 antibody. By changing the concentration of antibody in the modifying solution, they prepared AFM probes with different relative densities of antibody and measured the adhesion property of each of the probe with the VN7-coated glass surface. They concluded that it is easy to differentiate the peptide density of 100 times difference and possibility of differentiating 10 times difference under favourable conditions. To estimate the number density of protein molecules subsequently immobilized on the substrate surface by reacting with SPDP, they assumed that the reaction efficiency of this step is 100%. This assumption may not be correct, and a more direct way of counting the immobilized peptides and protein on the surface should be tried.

Hinterdorfer *et al.* developed a new method of imaging by AFM, named "recognition imaging", based on a specific interactions between the ligand immobilized on an AFM probe and specific receptors on solid or biological surfaces (see Chapter 7).[45–47] The method is based on the detection of decreasing amplitude of cantilever oscillation where ligand–receptor interaction sets in. The AFM probe is coated with ligand molecules with a long tether of polyethylene glycol spacer and raster scans the sample surface in a similar manner as in the tapping mode imaging of AFM.[48,49] By using recognition imaging, Chtcheglova *et al.*[50,51] mapped the localization of vascular endothelial-cadherin on gently fixed microvascular endothelial cells and ergtoxin-1 receptors on hERG HEK 293 cells within 2 μm square regions. The method is capable of giving a simultaneous topographic image of the fixed cell surface. Their method has a finer resolution compared with Kim's method, but the area for mapping was confined to a narrower one of approximately 2 μm × 2 μm. It is interesting to observe that the high density interaction areas in their mapping, when overlapped on the topography map, coincide with the grooves between ridges probably because of the actin fibres inside the cell.

12.4 BINDING STRENGTH OF TRANSFERRIN RECEPTORS

Yersin *et al.* measured the strength of binding of transferrin to its receptor immobilized on a mica surface as well as on a cell membrane.[31] Transferrin is a serum protein with the maximum capacity to bind or unbind two ferric ions and deliver them from the liver to the peripheral tissue cells. Since ferric ions are not readily soluble in aqueous medium, the role of transferrin as their transporter is vital for the physiology of the whole body. The fully loaded transferrin with ferric ions, holo-transferrin, finds its specific receptors on the peripheral cell surface and binds to one of them. The affinity of transferrin for Fe^{3+} is extremely high at pH 7.4, but it decreases at a lower pH. The structure of transferrin–transferrin receptor complex was solved.[52]

The transferrin–receptor complex is then passively internalized into the cell following the general endocytosis pathway and, once it is internalized, it is exposed to a mildly acidic condition of the endosome. The conformation of transferrin is altered under the condition, and ferric ions are released although the transferrin–receptor complex remains intact. The complex is then returned to the neutral pH of the cell surface where apo-transferrin is dissociated from the receptor. In summary, both holo- and apo-transferrin bind to the same receptor with different affinity under acidic and neutral conditions. Yersin *et al.* measured the unbinding force of both holo- and apo-transferrin under different solution pHs. Force measurements were conducted both *in vitro* on a mica surface and *in vivo* on a live cell surface. The results obtained on a mica surface at pH 7.4 are given in Fig. 12.3a,b.

The force curves given in Fig. 12.3a have a long plateau-like extension of up to 30 nm because of the polyethylene glycol part of the covalent cross-linkers and show an increase of the tensile force towards the final rupture, bringing the cantilever to its zero force level. The final rupture event was taken as representing forced unbinding of holo-transferrin from its receptor immobilized on a mica surface. The distribution of measured force is given in Fig. 12.3b as histograms. The three histograms represent the results obtained (top) in the absence and (middle) in the presence of an excessive amount of free receptor molecules and (bottom) after washing out the free receptors. The mean unbinding forces were, respectively, 63 ± 8 pN, 61 ± 8 and 61 ± 8 pN and similar experiments were repeated on apo-transferrin at pH 5.3, and the resulting unbinding forces were 44 ± 5 pN in the absence of excess receptors. No binding was recorded at a neutral pH for apo-transferrin.

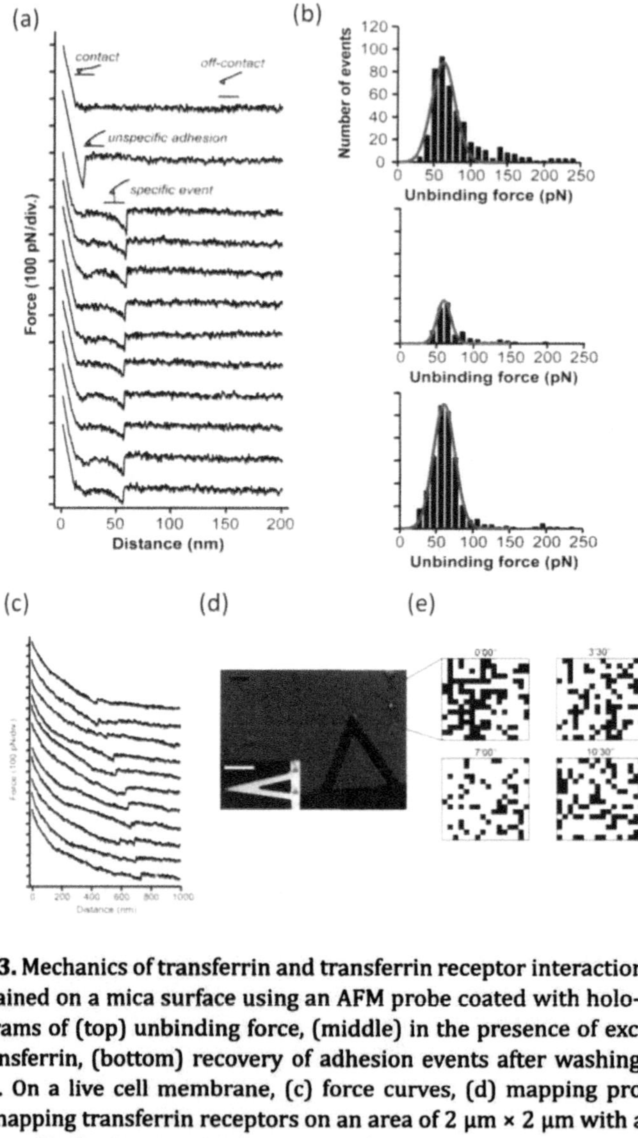

Figure 12.3. Mechanics of transferrin and transferrin receptor interactions. (a) Force curves obtained on a mica surface using an AFM probe coated with holo-transferrin. (b) Histograms of (top) unbinding force, (middle) in the presence of excess amount of holo-transferrin, (bottom) recovery of adhesion events after washing out excess transferrin. On a live cell membrane, (c) force curves, (d) mapping procedure, (e) results of mapping transferrin receptors on an area of 2 μm × 2 μm with a time lapse of 3.5 minutes. Dark squares represent areas where strong interaction was observed. Reproduced with permission from Yersin *et al.*[31]

The loading rate dependency of the mean of the force histograms was also reported for both types of transferrin. What was most striking was that the loading rate dependency plot of holo-transferrin had a break in the middle of the plot whereas the one for apo-transferrin was a straight line within the range of experimentally explored loading rate. From the slope and the intercept of the plot with the axes, Δx and k^0_{off} were determined and the energy diagram of the reaction was reconstructed. The results obtained on live HeLa cells were similar to those obtained on the mica surface including the pattern of loading rate dependency. They also mapped the locations where strong ligand–receptor interactions were observed with a time lapse of 3.5 minutes over a 2 μm × 2 μm area, and the result is given in Fig. 12.3c–e.

12.5 UNBINDING AGAINST UPROOTING OF MEMBRANE PROTEINS

A key question arises while mapping experiments are being performed, i.e., "which is a more likely event, unbinding of ligand from its receptor or uprooting of receptor from the cell membrane?" This is a difficult question to answer because the force to unbind a ligand–receptor pair and that of uprooting a membrane protein seems to be in a similar range of several tens of pNs to 400 pNs.

Experimentally, uprooting, i.e., extracting intrinsic proteins from the cell membrane, can be done by using bifunctional covalent cross-linkers. Afrin *et al.* modified an amino-silanized AFM probe with the bifunctional covalent cross-linkers, disuccinimidyl suberate, and brought the modified probe in contact with the surface of living Balb 3T3 fibroblast cells. After allowing the cross-linkers to react with the amino groups on the cell surface, the cantilever was pulled up and force curves were recorded for many such trials. The final rupture force of the force curves obtained from such trials were collected, and a mean value was obtained from the Gaussian fitting curve to the force histogram as approximately 450 pN,[53] which was much less than the force to sever a covalent bond. The mean force value was a little higher than the value of 250 pN as predicted by Bell for extraction of a dimer of glycophorin A from the lipid bilayer[11] but still within a fair agreement within a factor of two.

In their later work, Afrin *et al.* used a similar method as mentioned earlier to extract red blood cell membrane proteins after deglycosylation of the cell surface[54] and obtained a value of 150 pN from the major and ~70 pN from the minor peak as mean values of membrane protein uprooting. After a brief heating of the cells to denature spectrins, the mean force became 70–80 pN, indicating that the larger force of 150 pN corresponded to uprooting of membrane proteins that were initially linked to the spectrin-based cytoskeletal structure. Since covalent cross-linkers do not react with selected types of membrane proteins, correspondence of the measured force to uprooting events of specific kinds of membrane proteins could not be established. Because major proteins on the red blood cell surface are limited to glycophorin A, Band 3 (anion transporter) and stomatin, each comprising nearly one million copies per cell, the mean rupture force must correspond to the uprooting of these proteins. Figure 12.4 shows representative force curves and the histogram of the final rupture forces.[54]

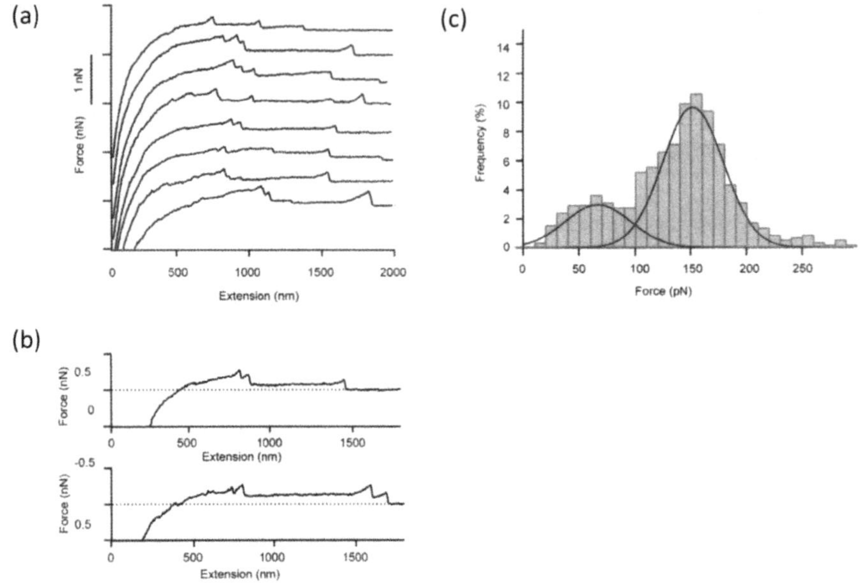

Figure 12.4. Force curves obtained on a red blood cell using an AFM probe coated with covalent cross-linkers. (a) Force-extension curves obtained on native red cell surface, (b) selected force-extension curves to emphasize the non-linear increase of the force towards the final rupture event, (c) histogram of rupture forces with two peaks with Gaussian fitting curves. Reproduced with permission from Afrin *et al.*[54]

The results of covalent cross-linking experiments indicated that if the unbinding force of a ligand–receptor pair is in the range of 60–100 pN, the probability of uprooting the receptor protein cannot be neglected. Although there is no clear picture of how the proteins are extracted from the membrane, two possible ways can be imagined. In the first case, proteins are freed from the lipid bilayer and extracted as lipid free entities, while, in the second case, the proteins are extracted together with membrane lipids. In the latter case, breakage of lipid tether trailing behind the ligand–receptor complex on the AFM probe would take place at any point between the protein and the cell surface. As a conclusion, it is safe to perform receptor mapping on a live, unfixed cell surface using a ligand that can be unbound with a relatively weak force of 10–50 pN from its receptor.

12.6 PROTEIN MAPPING: A BRIEF LITERATURE SURVEY

In the past years, AFM-based protein mapping has been applied in different laboratories to a wide variety of cells. Horton *et al.* reviewed mapping efforts on receptor distribution over a live cell surface together with the basic principles of AFM and its application to biological ligand receptor interaction analysis.[55] In this review, an example of mapping of vasoactive intestinal peptide, one of the neuropeptides found in bone, was presented. Recent progress in molecular recognition studies has also been reviewed by Hinterdorfer and Dufrêne.[47,56]

Dupres *et al.*[57] determined the adhesion forces between the heparin-binding haemagglutinin adhesin (HBHA) produced by *Mycobacterium tuberculosis* and heparin (see Chapter 15). They mapped the distribution of HBHA molecules on the surface of living mycobacteria and found that the adhesin is not randomly distributed over the mycobacterial surface, but concentrated into nanodomains. Robert *et al.*[58] asked the question of the biological relevance of the specific bond properties revealed by single-molecule studies and gave positive answers to the questions. These are the questions we always have to remember. (i) Which parameters do we need to know to predict the behaviour of an encounter between receptors and ligands, (ii) which information is actually yielded by single-molecule studies and (iii) is it possible to relate this information to molecular structure? Dazzi *et al.*[59] reported the construction of an AFM probe which finds the local transient deformation induced by an infrared pulsed laser tuned at a sample absorbing wavelength. The method is suitable for the identification of biological materials situated near or on the AFM probe.

Qiu *et al.*[60] mapped the presence of the human pluripotent stem cell marker, TRA-1-81 antigen, on a human embryonic stem cell. Popov *et al.*[61] mapped ceramides distribution in artificial lipid monolayers. Mapping of glycolipids is important in relation with the characterization of segregated lipid structures such as lipid rafts. Gunning *et al.*[62] mapped the surface of living Caco-2 human intestinal epithelial cells using a colloidal AFM probe modified with wheat germ agglutinin which was reactive to the glycosylated extracellular domain III of the epidermal growth factor receptor. They reported the value of 125 pN as the mean unbinding force from the cell surface and found non-homogeneous distribution of the receptors on the cell surface. The unbinding force of 125 pN was approximately twice as high as the value obtained by Afrin *et al.*[54] on the red blood cell surface when they used the same lectin to pull glycophorin A. Unbinding force of *Psathyrella velutina* lectin from glycophorin A was also in the range of 70 pN according to Yan *et al.*[30]

Ludwig *et al.*[63] summarized the use of AFM-based technology in combination with others in the study of tissue remodelling especially that of elasticity mapping and identification of proteolytic activity. Dazzi *et al.*[64] proposed the use of an infrared spectro-microscopy method based on a photo-thermal effect, which is able to localize single viruses, including when they are located inside the bacteria they have infected. Kim *et al.*[65] probed the distribution of olfactory marker protein (OMP) on a tissue section of vomeronasal organ using a glass bead that was coated with anti-OMP antibodies. Francius *et al.*[66] reported the result of pulling polysaccharides on live cells in a similar manner to Gad *et al.*[36] Verbelen *et al.*[67] mapped lipoarabinomannans on *Mycobacteria.* Kumar *et al.*[68] measured the change in the nanomechanical and topographical change in the form of mapping on live bacterial cells, *Brevibacterium*, under stress from environmental heavy metal ions. Xu *et al.*[69] used higher resonance frequencies of the cantilever to probe a wider range of mechanical and dynamical properties and identified the different components in biological materials. Plomp *et al.*[70] used immunolabelling technique on the cell surface and later used an AFM to identify protein compositions. Carberry *et al.*[71] used a high speed scanning AFM to map a wide area with a high lateral resolution suitable for imaging as well as component mapping.

12.7 MAPPING OF INTRACELLULAR mRNA

Besides mapping surface proteins, AFM can also probe intracellular molecules. An interesting example is the localization of intracellular mRNA, as recently performed by Osada *et al.*[72] and Uehara *et al.*[73,74] They developed a method to retrieve a small portion of intracellular mRNAs without compromising the viability of the targeted cell.[72–74] Their AFM-based method was to push a bare AFM probe into a live cell by applying a compressive force of 10–100 nN and, after keeping it inside of the cell for a short time, pull it out for subsequent amplification and quantitation of retrieved mRNAs into cDNA through quantitative RT-PCR and PCR procedures. The amplified DNAs were analyzed for their identification by gel electrophoresis. When analyzed for mRNA of the housekeeping protein of β-actin, the successful detection rate was about 97% ($n \approx 170$). This method can be used for the mapping of (temporary) mRNA localization inside of a living cell. As is given in Fig. 12.5, the amount of β-actin mRNA was high near the nucleus in the inactive state of the cell and, after activation, mRNA concentration increased in a more peripheral region in the direction of cell movement. A correlation between the AFM-based method and the conventional *in situ* hybridization method using fluorescently labelled complementary DNA was established,[74,75] as shown in Fig. 12.6.

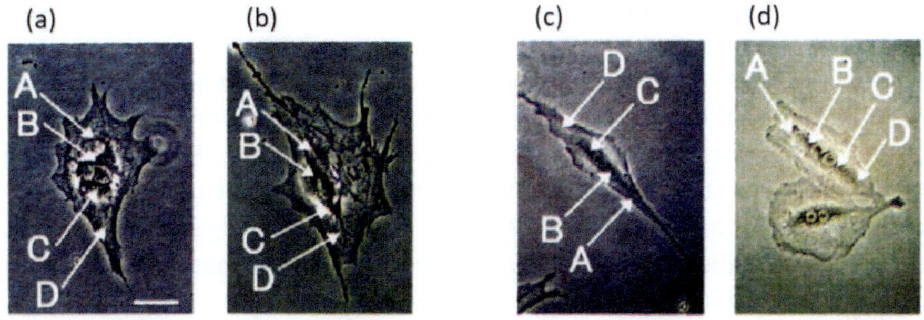

	(a)		(b)		(c)		(d)
locus	# of β-actin mRNA	locus	# of β-actin mRNA	locus	# of β-actin mRNA	locus	# of β-actin mRNA
A	0	A	0	A	20	A	100
B	9	B	58	B	14	B	126
C	50	C	19	C	8	C	80
D	0	D	0	D	0	D	0

Figure 12.5. Results of β-actin mRNA mapping in live cells. mRNA was retrieved from four different sites (inset labels A, B, C and D in each figure) of live cells and amplified through quantitative RT-PCR and PCR procedures. Cells in (a) and (b) are in inactive state with no nutrients in the culture medium, while those in (c) and (d) are active cells moving in the direction of inset (A). Reproduced with permission from Uehara *et al.*[73]

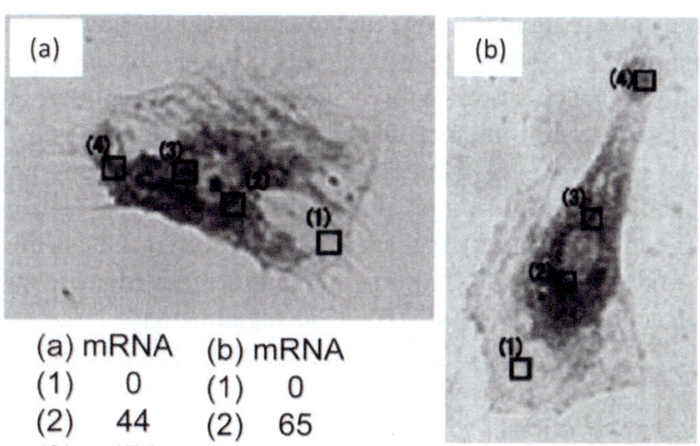

(a)	mRNA	(b)	mRNA
(1)	0	(1)	0
(2)	44	(2)	65
(3)	471	(3)	300
(4)	2	(4)	0

Figure 12.6. AFM-based quantitation of β-actin mRNA and fluorescence *in situ* hybridization. Results from two cells (a) and (b) are listed in the table at the bottom giving the number of mRNA from four sites. Dark colour in the figures represents areas where the results of *in situ* hybridization was highly positive. Reproduced with permission from Uehara *et al.*[75]

12.8 CONCLUSIONS

The use of AFM-based technology to map specific proteins on live cells together with other types of physical and chemical mapping will clearly be useful and important in future cell biology research. For this purpose, the initial barrier of the technology for a beginner to overcome must be lowered in terms of cost and training. AFM is still a relatively expensive instrument, and it takes a while to acquire the basic knowledge for its operation compared with most other instruments used in biological laboratories. In a sense, it is inevitable because the AFM is one of few instruments that enable us to touch single atoms and molecules with a virtual hand. Maybe it is true that nothing of this sort comes cheap and easy, but it is still vital to make the technology more accessible to many.

Future prospects of membrane protein mapping using AFM include combining the method with the mapping of other local properties on the cell surface, such as elasticity mapping, surface charge mapping, etc. Through such combinations, broader application fields will be opened. A good example is the use of AFM in elasticity mapping of the cross section of hair to detect the effect of hair caring materials such as taurin on damaged hair.[76] If such measurement is done in parallel with protein/polysaccharide mapping, the result would considerably improve the interpretation of hair care treatment.

Combination of protein mapping with charge distribution mapping determined by Kelvin probe force microscopy is another interesting possibility because non-uniform protein distribution on a cell membrane should affect the charge distribution as well.[77,78] The resulting non-uniform charge distribution would be an important factor in cell–cell interactions and 2D/3D construction of tissues and organs. Recent advancement of the resolution and reliability of scanning ion conductance microscopy seems to hold promise to map protein distribution together with topography of the cell surface.[79] Force measurement has been shown to be possible using this technology.[80] The least invasive nature of the method is an attractive feature for biological applications.

Acknowledgements

This work was supported by Grants-in-Aid for Exploratory Research for RA (19651058) and for Scientific Research (S) (No. 15101004) and Creative Scientific Research (19GS0418) to AI from the Japan Society for the Promotion of Science (JSPS).

References

1. Kyte, J. (2006) *Structure in Protein Chemistry*, 2nd edn, Garland Science, New York.
2. Reyes Mateo, C., Javier Gomez, J. V., and Ros, J. M. G. (2008) *Protein-Lipid Interactions: New Approaches and Emerging Concepts*, 2nd edn, Springer, New York.
3. Binnig, G., Quate, C. F., and Gerber, C. (1986) Atomic force microscope, *Phys. Rev. Lett.*, **56**, 930–933.
4. Butt, H. J., Cappella, B., and Kappl, M. (2005) Force measurements with the atomic force microscope: technique, interpretation and applications, *Surf. Sci. Rep.*, **59**, 1–152.
5. Ashkin, A. (1970) Acceleration and trapping of particles by radiation pressure, *Phys. Rev. Lett.*, **24**, 156–159.
6. Ashkin, A., Dziedzic, J. M., Bjorkholm, J. E., and Chu, S. (1986) Observation of a single beam gradient force optical trap for dielectric particles, *Opt. Lett.*, **11**, 288–290.
7. Merkel, R., Nassoy, P., Leung, A., Ritchie, K., and Evans, E. (1999) Energy landscapes of receptor-ligand bonds explored with dynamic force spectroscopy, *Nature*, **397**, 50–53.
8. Schmidt, F. G., Ziemann, F., and Sackmann, E. (1996) Shear field mapping in actin networks by using magnetic tweezers, *Eur. Biophys. J.*, **24**, 348–353.
9. Fung, Y. (1993) *Biomechanics: Motion, Flow, Stress, and Growth*, Springer-Verlag, New York
10. Boal, D. (2002) *Mechanics of the Cell*, Cambridge University Press, Cambridge, UK.
11. Bell, G. I. (1978) Models for the specific adhesion of cells to cells, *Science*, **200**, 618–627.
12. Smith, S. B., Finzi, L., and Bustamante, C. (1992) Direct mechanical measurements of the elasticity of single DNA molecules by using magnetic beads, *Science*, **258**, 1122–1126.
13. Lee, G. U., Chrisey, L. A., and Colton, R. J. (1994) Direct measurement of the forces between complementary strands of DNA, *Science*, **266**, 771–773.
14. Smith, S. B., Cui, Y., and Bustamante, C. (1996) Overstretching B-DNA: the elastic response of individual double-stranded and single-stranded DNA molecules, *Science*, **271**, 795–799.
15. Hansma, H. G., Oroudjev, E., Baudrey, S., and Jaeger, L. (2003) TectoRNA and "kissingloop" RNA: atomic force microscopy of self-assembling RNA structures, *J. Microsc.*, **212**, 273–279.
16. Lantz, M. A., Jarvis, S. P., Tokumoto, H., Martynski, T., Kusumi, T., Nakamura, C., and Miyake, J. (1999) Stretching the a-helix—a direct measure of the hydrogen bond energy of a single peptide molecule, *Chem. Phys. Lett.*, **315**, 61–68.

17. Idiris, A., Alam, M. T., and Ikai, A. (2000) Spring mechanics of alpha-helical polypeptide, *Protein Eng.*, **13**, 763–770.

18. Oesterhelt, F., Rief, M., and Gaub, H. (1999) Single molecule force spectroscopy by afm indicates helical structure of poly(ethylene-glycol) in water, *New J. Phys.*, **1**, 6.1–6.11.

19. Marszalek, P. E., Li, H., Oberhauser, A. F., and Fernandez, J. M. (2002) Chair-boat transitions in single polysaccharide molecules observed with force-ramp AFM, *Proc. Natl. Acad. Sci. USA*, **99**, 4278–4283.

20. Mitsui, K., Hara, M., and Ikai, A. (1996) Mechanical unfolding of alpha2-macroglobulin molecules with atomic force microscope, *FEBS Lett.*, **385**, 29–33.

21. Rief, M., Gautel, M., Oesterhelt, F., Fernandez, J., and Gaub, H. (1997) Reversible unfolding of individual titin immunoglobulin domains by afm, *Science*, **276**, 1109–1112.

22. Oberhauser, A. F., Marszalek, P. E., Carrion-Vazquez, M., and Fernandez, J. M. (1999) Single protein misfolding events captured by atomic force microscopy, *Nat. Struct. Biol.*, **6**, 1025–1028.

23. Oesterhelt, F., Oesterhelt, D., Pfeiffer, M., Engel, A., Gaub, H. E., and Müller, D. J. (2000) Unfolding pathways of individual bacteriorhodopsins, *Science*, **288**, 143–146.

24. Alam, M. T., Yamada, T., Carlsson, U., and Ikai, A. (2002) The importance of being knotted: effects of the C-terminal knot structure on enzymatic and mechanical properties of bovine carbonic anhydrase II, *FEBS Lett.*, **519**, 35–40.

25. Hertadi, R., and Ikai, A. (2002) Unfolding mechanics of holo- and apocalmodulin studied by the atomic force microscope, *Protein Sci.*, **11**, 1532–1538.

26. Carrion-Vazquez, M., Oberhauser, A. F., Fisher, T. E., Marszalek, P. E., Li, H., and Fernandez, J. M. (2000) Mechanical design of proteins studied by single-molecule force spectroscopy and protein engineering, *Prog. Biophys. Mol. Biol.*, **74**, 63–91.

27. Carrion-Vazquez, M., Li, H., Lu, H., Marszalek, P., Oberhauser, A., and Fernandez, J. (2003) The mechanical stability of ubiquitin is linkage dependent, *Nat. Struct. Biol.*, **10**, 738–743.

28. Florin, E., Moy, V., and Gaub, H. (1994) Adhesion forces between individual ligand receptor pairs, *Science*, **264**, 415–417.

29. Allen, S., Chen, X., Davies, J., Davies, M., Dawkes, A., Edwards, J., Roberts, C., Sefton, J., Tendler, S., and Williams, P. (1997) Detection of antigen-antibody binding events with the atomic force microscope, *Biochemistry*, **36**, 7457–7463.

30. Yan, C., Yersin, A., Afrin, R., Sekiguchi, H., and Ikai, A. (2009) Single molecular dynamic interactions between glycophorin A and lectin as probed by atomic force microscopy, *Biophys. Chem.*, **144**, 72–77.

31. Yersin, A., Osada, T., and Ikai, A. (2008) Exploring transferrin-receptor interactions at the single-molecule level, *Biophys. J.*, **94**, 230–240.

32. Sekiguchi, H., Arakawa, H., Taguchi, H., Ito, T., Kokawa, R., and Ikai, A. (2003) Specific interaction between GroEL and denatured protein measured by compression-free force spectroscopy, *Biophys. J.*, **85**, 484–490.

33. Evans, E., and Ritchie, K. (1999) Strength of a weak bond connecting flexible polymer chains, *Biophys. J.*, **76**, 2439–2447.

34. Evans, E., and Ritchie, K. (1997) Dynamic strength of molecular adhesion bonds, *Biophys. J.*, **72**, 1541–1555.

35. Evans, E. A., and Williams, P. (2002) Dynamic force spectroscopy, in *Physics of Biomolecules and Cells* (ed. Flyvbjerg, H., Jülicher, F., Ormos, P., and David, F.), Springer Press, Berlin, Germany, pp. 147–204.

36. Gad, M., Itoh, A., and Ikai, A. (1997) Mapping cell wall polysaccharides of living microbial cells using atomic force microscopy, *Cell Biol. Int.*, **21**, 697–706.

37. Gad, M., and Ikai, A. (1995) Method for immobilizing microbial cells on gel surface for dynamic AFM studies, *Biophys. J.*, **69**, 2226–2233.

38. Kasas, S., and Ikai, A. (1995) A method for anchoring round shaped cells for atomic force microscope imaging, *Biophys. J.*, **68**, 1678–1680.

39. Kim H., Arakawa, H., Osada, T., and Ikai, A. (2002) Quantification of fibronectin and cell surface interactions by AFM, *Colloids Surf. B Biointerfaces*, **25**, 33–43.

40. Sneddon, I. (1965) The relation between load and penetration in the axisymmetric boussinesq problem for a punch of arbitrary profile, *Int. J. Eng. Sci.*, **3**, 187–195.

41. Ikai, A. (1996) STM and AFM of bio/organic molecules and structures, *Surf. Sci. Rep.*, **26**, 261–332.

42. Kim, H., Arakawa, H., Osada, T., and Ikai, A. (2003) Quantification of cell adhesion force with AFM: distribution of vitronectin receptors on a living MC3T3-E1 cell, *Ultramicroscopy*, **97**, 359–363.

43. Kim, H., Arakawa, H., Hatae, N., Sugimoto, Y., Matsumoto, O., Osada, T., Ichikawa, A., and Ikai, A. (2006) Quantification of the number of EP3 receptors on a living CHO cell surface by the AFM, *Ultramicroscopy*, **106**, 652–662.

44. Kim, H., Tsuruta, S., Arakawa, H., Osada, T., and Ikai, A. (2004) Quantitative analysis of the number of antigens immobilized on a glass surface by AFM, *Ultramicroscopy*, **100**, 203–210.

45. Hinterdorfer, P., Baumgartner, W., Gruber, H. J., Schilcher, K., and Schindler, H. (1996) Detection and localization of individual antibody-antigen recognition events by atomic force microscopy, *Proc. Natl. Acad. Sci. USA*, **93**, 3477–3481.

46. Hinterdorfer, P. (2002) Molecular recognition studies using the atomic force microscope, *Methods Cell Biol.*, **68**, 115–139.

47. Hinterdorfer, P., and Dufrêne, Y. F. (2006) Detection and localization of single molecular recognition events using atomic force microscopy, *Nat. Methods.*, **3**, 347–355.

48. Zhong, Q., Inniss, D., Kjoller, K., and Elings, V. B. (1993) Fractured polymer/silica fiber surface imaged studied by tappingmode atomic force microscopy, *Surf. Sci.*, **290**, L688–L692.

49. Hansma, P. K., Cleveland, J. P., Radmacher, M., Walters, D. A., Hillner, V., Bezanilla, M., Fritz, M., Vie, D., Hansma, H. G., Prater, C. B., Massie, J., Fukunaga, L., Gurley, G., and Elings, V. (1994) Tapping mode atomic force microscopy in liquids, *Appl. Phys. Lett.*, **64**, 1738–1740.

50. Chtcheglova, L. A., Waschke, J., Wildling, L., Drenckhahn, D., and Hinterdorfer, P. (2007) Nano-scale dynamic recognition imaging on vascular endothelial cells, *Biophys. J.*, **93**, L11–L13.

51. Chtcheglova, L. A., Atalar, F., Ozbek, U., Wildling, L., Ebner, A., and Hinterdorfer, P. (2008) Localization of the ergtoxin-1 receptors on the voltage sensing domain of hERG K+ channel by AFM recognition imaging, *Pflugers Arch.*, **456**, 247–254.

52. Yang, F., Lum, J. B., McGill, J. R., Moore, C. M., Naylor, S. L., Bragt, P. H., van Baldwin, W. D., and Bowman, B. H. (1984) Human transferrin: cDNA characterization and chromosomal localization, *Proc. Natl. Acad. Sci. USA*, **81**, 2752–2756.

53. Afrin, R., Arakawa, H., Osada, T., and Ikai, A. (2003) Extraction of membrane proteins from a living cell surface using the atomic force microscope and covalent crosslinkers, *Cell Biochem. Biophys.*, **39**, 101–117.

54. Afrin, R., and Ikai, A. (2006) Force profiles of protein pulling with or without cytoskeletal links studied by AFM, *Biochem. Biophys. Res. Commun.*, **348**, 238–244.

55. Horton, M., Charras, G., and Lehenkari, P. (2002) Analysis of ligand receptor interactions in cells by atomic force microscopy, *J. Recept. Signal Transduct.*, **22**, 169–190.

56. Dufrêne, Y. F., and Hinterdorfer, P. (2008) Recent progress in AFM molecular recognition studies, *Pflugers Arch.*, **456**, 237–245.

57. Dupres, V., Menozzi, F. D., Locht, C., Clare, B. H., Abbott, N. L., Cuenot, S., Bompard, C., Raze, D., and Dufrêne, Y. F. (2005) Nanoscale mapping and functional analysis of individual adhesins on living bacteria, *Nat. Methods*, **2**, 515–520.

58. Robert, P., Benoliel, A.-M., Pierres, A., and Bongrand, P. (2007) What is the biological relevance of the specific bond properties revealed by single-molecule studies? *J. Mol. Recognit.*, **20**, 432–447.

59. Dazzi, A., Prazeres, R., Glotin, F., and Ortega, J. M. (2007) Analysis of nano-chemical mapping performed by an AFM-based ("AFMIR") acousto-optic technique, *Ultramicroscopy*, **107**, 1194–1200.

60. Qiu, D., Xiang, J., Li, Z., Krishnamoorthy, A., Chen, L., and Wang, R. (2008) Profiling TRA-1-81 antigen distribution on a human embryonic stem cell, *Biochem. Biophys. Res. Commun.*, **369**, 735–740.

61. Popov, J., Vobornik, D., Coban, O., Keating, E., Miller, D., Francis, J., Petersen, N. O., and Johnston, L. J. (2008) Chemical mapping of ceramide distribution in sphingomyelin rich domains in monolayers, *Langmuir*, **24**, 13502–13508.

62. Gunning, A. P., Chambers, S., Pin, C., Man, A. L., Morris, V. J., and Nicoletti, C. (2008) Mapping specific adhesive interactions on living human intestinal epithelial cells with atomic force microscopy, *FASEB J.*, **22**, 2331–2339.

63. Ludwig, T., Kirmse, R., Poole, K., and Schwarz, U. S. (2008) Probing cellular microenvironments and tissue remodeling by atomic force microscopy, *Pflugers Arch.*, **456**, 29–49.

64. Dazzi, A., Prazeres, R., Glotin, F., Ortega, J. M., Al-Sawaftah, M., and Frutos, M. de (2008) Chemical mapping of the distribution of viruses into infected bacteria with a photothermal method, *Ultramicroscopy*, **108**, 635–641.

65. Kim, H., Asgari, F., Kato-Negishi, M., Ohkura, S., Okamura, H., Arakawa, H., Osada, T., and Ikai, A. (2008) Distribution of olfactory marker protein on a tissue section of vomeronasal organ measured by AFM, *Colloids Surf. B Biointerfaces*, **61**, 311–314.

66. Francius, G., Alsteens, D., Dupres, V., Lebeer, S., De Keersmaecker, S., Vanderleyden, J., Gruber, H. J., and Dufrêne, Y. F. (2009) Stretching polysaccharides on live cells using single molecule force spectroscopy, *Nat. Protoc.*, **4**, 939–946.

67. Verbelen, C., Christiaens, N., Alsteens, D., Dupres, V., Baulard, A. R., and Dufrêne, Y. F. (2009) Molecular mapping of lipoarabinomannans on mycobacteria, *Langmuir*, **25**, 4324–4327.

68. Kumar, G. V. P., and Irudayaraj, J. (2009) SERS in salt wells *Chemphyschem*, **10**, 2670–2673.

69. Xu, X., Melcher, J., Basak, S., Reifenberger, R., and Raman, A. (2009) Compositional contrast of biological materials in liquids using the momentary excitation of higher eigenmodes in dynamic atomic force microscopy, *Phys. Rev. Lett.*, **102**, 060801-1–060801-4.

70. Plomp, M., and Malkin, A. J. (2009) Mapping of proteomic composition on the surfaces of bacillus spores by atomic force microscopy-based immunolabeling, *Langmuir*, **25**, 403–409.

71. Carberry, D. M., Picco, L., Dunton, P. G., and Miles, M. J. (2009) Mapping real-time images of high-speed AFM using multitouch control, *Nanotechnology*, **20**, 434018-1–434018-5.

72. Osada, T., Uehara, H., Kim, H., and Ikai, A. (2003) mRNA analysis of single living cells, *J. Nanobiotechnology*, **1**, 2-1–2-8.

73. Uehara, H., Osada, T., and Ikai, A. (2004) Quantitative measurement of mRNA at different loci within an individual living cell, *Ultramicroscopy*, **100**, 197–201.

74. Uehara, H., Ikai, A., and Osada, T. (2009) Detection of mRNA in single living cells using AFM nanoprobes, *Methods Mol. Biol.*, **544**, 599–608.

75. Uehara, H., Kunitomi, Y., Ikai, A., and Osada, T. (2007) mRNA detection of individual cells with the single cell nanoprobe method compared with in situ hybridization, *J. Nanobiotechnology*, **5**, 7–14.

76. Kitano, H., Yamamoto, A., Niwa, M., Fujinami, S., Nakajima, K., Nishi, T., and Naito, S. (2009) Young's modulus mapping on hair cross-section by atomic force microscopy, *Compos. Interfaces.*, **16**, 1–12.

77. Nonnenmacher, M., O'Boyle, M. P., and Wickramasinghe, H. K. (1991) Kelvin probe force microscopy, *Appl. Phys. Lett.*, **58**, 2921–2923.

78. Sinensky, A. K., and Belcher, A. M. (2007) Label-free and high-resolution protein/DNA nanoarray analysis using Kelvin probe force microscopy, *Nat. Nanotechnol.*, **2**, 653–659.

79. Korchev, Y. E., Bashford, C. L., Milovanovic, M., Vodyanoy, I., and Lab, M. J. (1997) Scanning ion conductance microscopy of living cells, *Biophys. J.*, **73**, 653–658.

80. Sánchez, D., Johnson, N., Li, C., Novak, P., Rheinlaender, J., Zhang, Y., Anand, U., Anand, P., Gorelik, J., Frolenkov, G. I., Benham, C., Lab, M., Ostanin, V. P., Schäffer, T. E., Klenerman, D., and Korchev, Y. E. (2008) Noncontact measurement of the local mechanical properties of living cells using pressure applied via a pipette, *Biophys. J.*, **95**, 3017–3027.

Chapter 13

PROBING BACTERIAL ADHESION USING FORCE SPECTROSCOPY

Terri A. Camesano

Department of Chemical Engineering, Life Science and Bioengineering Center at Gateway Park, Worcester Polytechnic Institute, 100 Institute Rd., Worcester, MA 01609 USA

terric@wpi.edu

13.1 INTRODUCTION

Atomic force microscopy (AFM) has become a significant tool for studying the complex interface between bacteria and surfaces. Bacterial adhesion is important to applications ranging from prevention of infection on biomaterials, vaccine development, groundwater protection from mobile pathogens, biomineralization, food safety, biosensors and bioenergy. The ability to use AFM now allows researchers a method to quantify the forces of adhesion and to fully characterize the properties of bacterial molecules that mediate the adhesion process.

AFM is a unique instrument in that it can be used not only to capture high-resolution images of biological samples, but to measure nanoscale interaction forces on biological samples and to probe biomolecules with excellent resolution. By making force measurements, the AFM can be used to study the chemical and mechanical properties of the sample surface such as elasticity, adhesion and even forces between single molecules. A review of using AFM for force measurements has been given by Butt *et al.*[1] During force mode, the piezoelectric crystal of the scanner stops moving in the x and y directions and only the movement in the z direction is recorded. As the probe approaches and retracts from the sample, a plot is generated that reflects the deflection of the cantilever as a function of probe–sample position.

Life at the Nanoscale: Atomic Force Microscopy of Live Cells
Edited by Yves Dufrêne

www.panstanford.com

When the AFM probe is very far away from the sample, there is no deflection of the cantilever since the probe–sample separation distance is too large to experience any attractive or repulsive van der Waals or electrostatic forces. As the probe continues to approach the sample, the cantilever bends towards the surface, sometimes jumping into contact if there is an attractive force from the sample. After contact is made, the cantilever bends until it reaches the specified force limit that is to be applied, and this region is known as the constant compliance region. After applying the desired force to the cantilever while the probe is in contact with the sample, the process is reversed and the second half of the force cycle takes place, which is known as the retraction portion. As the probe continues to be retracted from the surface, the force exerted by the bending of the cantilever overcomes the adhesion force between the probe and the sample, and the probe "snaps off" the substrate. Once the probe has retracted more than a few hundred nanometres, the cantilever returns to its original position where there is no deflection and the force cycle is complete and ready to start the next measurement.

The large peak observed in the retraction portion of the cycle can provide valuable information since it is possible to determine the rupture force required to break the bond of adhesion between two substrates. If we were studying surfaces such as bacteria or other cells, it is possible to obtain more than one peak in the retraction curve since there could be multiple polymers attaching to the AFM probe during the force cycle. As each one of these polymers detaches from the probe, a new peak will appear on the force profile.

13.2 ELASTICITY OF BACTERIAL POLYMERS

A bacterial surface is composed of many molecules, including lipopolysaccharides (LPS, only for Gram-negative bacteria), proteins and the less well-defined extracellular polymeric substances (EPS), which can include capsules or vesicles. The LPS extend away from the outer membrane and are often the first contact between a bacterium and a substrate (Fig. 13.1). In addition, some bacteria can produce specialized structures such as fimbriae, which recognize receptors and bind to mammalian cells, or flagella, which allow the bacteria to be motile. AFM provides an efficient tool to measure the adhesive interactions of bacterial polymers at the nano- and pico-Newton levels, since interactions between bacterial cells and either modified or unmodified AFM probes can be measured in a variety of solutions. Surface-modified probes can be helpful for studying hydrophobic or hydrophilic interactions, for probing the nature of the electrostatic interaction between the microbe and the probe, or for measuring a specific type of interaction, such as ligand–receptor bonds.

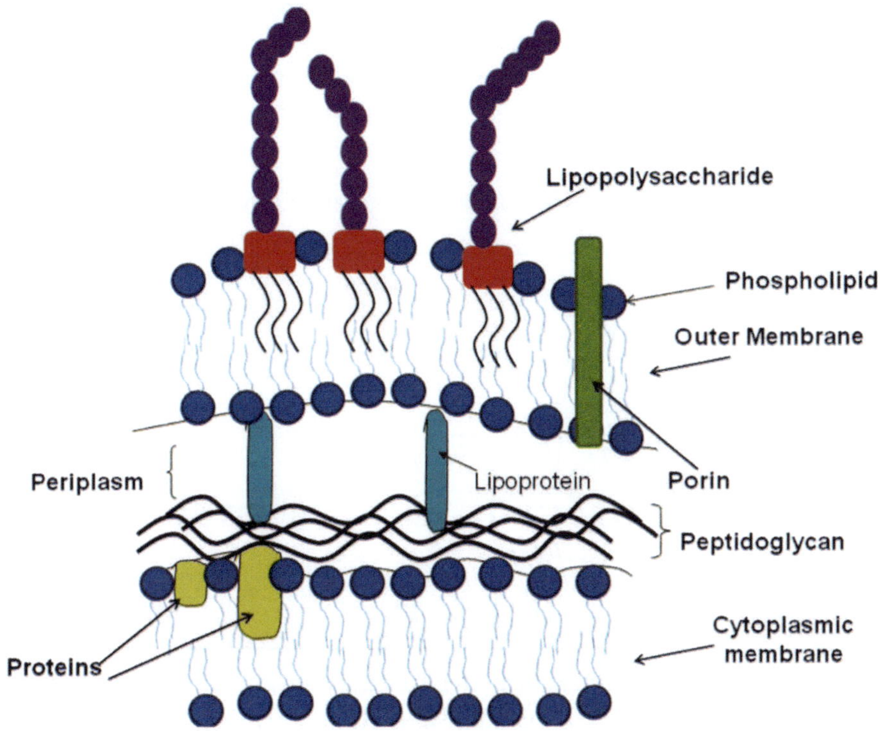

Figure 13.1. Schematic of the outer surface layers of the Gram-negative bacterial membrane, which include the cytoplasmic membrane and outer membrane.

A powerful approach for characterizing the elasticity of bacterial biopolymers is to use AFM to stretch these molecules, and to model their mechanical properties. This technique was first applied to isolated and purified biomolecules, such as DNA and proteins. In an early experiment, complementary strands of DNA were brought into proximity (one molecule on the AFM probe, and one on the substrate surface), and after forming a double helix, the molecules could be unzipped with AFM, giving rise to a characteristic "peak" in the AFM retraction profiles.[2,3] The work was extended towards the study of proteins, in which the AFM stretching causes a mechanical denaturation of the macromolecule,[4,5] as well as for characterizing mechanical properties of polysaccharides in isolation or in well-defined mixtures.[6–8]

Applying polymer models to the macromolecules on bacterial cells can be much more challenging, because one needs to be able to separate the polymer properties from other effects due to the curvature of the bacterium, and also there are multiple types of molecules present at once. In addition, the density of molecules is a fixed parameter on a living cell, while this could be controlled and optimized when working with isolated molecules. Despite the difficulties, researchers have stretched polymers on bacteria and were able to determine elastic properties of these molecules on bacteria,[9–15] yeast cells[16,17] and fungal spores.[10,18–22] A summary of the common models is provided in Table 13.1.

Table 13.1. Summary of statistical mechanical models of polymer elasticity

Model	Expression	Fitting parameters	Ref.
FJC	$h(F) = L_c\left[\coth\left(\frac{Fl_k}{kT}\right) - \frac{kT}{Fl_k}\right]$	L_c and l_k	23 and 24
FJC+	$h(F) = L_c \coth\left(\frac{Fl_k}{kT}\right) - \left(\frac{kT}{Fl_k}\right)\left[1 + \frac{F}{L_c \kappa_s}\right]$	L_c, l_k and κ_s	23 and 24
WLC	$F(h) = \frac{kT}{l_p}\left[\frac{h}{L_c} + \frac{1}{4\left(1 - \frac{h}{L_c}\right)^2} - \frac{1}{4}\right]$	L_c and l_p	23–25
WLC+	$F(h) = \frac{kT}{l_p}\left[\frac{1}{4}\left(1 - \frac{h}{L_c} + \frac{F}{\phi}\right)^2 + \frac{h}{L_c} - \frac{F}{\phi} - \frac{1}{4}\right]$	L_c, l_p and φ	23 and 24

FJC is the freely jointed chain model, FJC+ is extensible freely jointed chain, WLC is wormlike chain, WLC+ is extensible wormlike chain, *F* is pulling force, *h* the separation distance between the tip and the polymer, L_c the contour length, k_s the Boltzmann constant, *T* temperature, l_k the segment length, l_p the persistence length, κ the segment spring elasticity and φ the specific stiffness of the polymer.

The two most commonly used models to interpret polymer elasticity measurements are the freely jointed chain (FJC) and wormlike chain (WLC) models and variations on these forms. The FJC model considers that the polymer is composed of *n* rigid segments, each of a Kuhn length, l_k, connected by freely rotating pivots with equal probabilities for rotation in all directions. Essentially, this model treats the polymer as an aggregate of many independent segments.[26] The FJC cannot describe the chain if it is extended to its full contour length, because this would mean the chain is infinitely rigid. Also, this model considers only entropic effects and so the polymer cannot become stretched beyond its contour length. To account for these limitations, the FJC+ model was developed, known as the extensive freely jointed chain model.[24] The FJC+ model accounts for elastic deformations of bonds and bond angles that are neglected in the non-elastic FJC model. The polymer is modelled as consisting of n elastic springs, and a third fitting parameter (κ_s) is added to account for segment elasticity. Since an enthalpic term is included, forces at large extensions can be described, and the polymer can extend beyond its contour length. This may become significant during AFM experiments, in which bonds are overstretched because of the design of the experiment.

Another extremely common model for ideal stiff polymers is the WLC model.[26] According to this model, the polymer chain is continuously curved with a random direction for the curvature, according to the principle of self-avoidance. This model accounts for chain stiffness in terms of the microscopic persistence length, l_p. The bending energy of the curved chain gives rise to energetic and enthalpic factors, and the chain cannot extend beyond its contour length.[26] Finally, the extensible wormlike chain model (WLC+) is considered, which adds the stiffness of the chain as a third fitting parameter to the WLC model.[27] Enthalphic stretching, in which the segment length can continue to increase under stretching before the bond will break, has been observed experimentally in the high force regime.[26]

As an example, we tested the validity of different polymer models in describing the characteristics of biopolymers present on an environmentally isolated Gram-negative bacterium, *Pseudomonas putida* KT2442. Changing the salt concentration of the buffer solution allowed us to observe changes in polymer conformation, which could be captured with the various models.[9] The WLC model was unable to describe the biopolymers on *P. putida*, because we predicted persistence lengths that were too small to be realistic. Both the FJC and FJC+ models were generally applicable and gave good fits with the data (Fig. 13.2). However, at some very high forces, we found that the FJC+ model deviated more from the experimental values. Therefore, our comparisons are based on the application of the FJC model. We observed a transition in the flexibility of the biopolymers (as estimated by l_k values) as the salt concentration increased from that of ultrapure water to 0.01 M KCl.[28] The Kuhn length increased from 0.15 nm to 1.0 nm as salt concentration increased from that of ultrapure water to 0.01 M KCl.[9] These results showed that the biopolymers on a bacterial surface have mechanical properties similar to those of isolated polysaccharides. The Kuhn lengths have been calculated for a number of bacteria and can vary by more than an order of magnitude, from 0.15 nm (as we observed for *P. putida*) to 1.7 nm when the FJC+ model was applied to a *Lactobacillus rhamnosus* GG (LGG) mutant.[29]

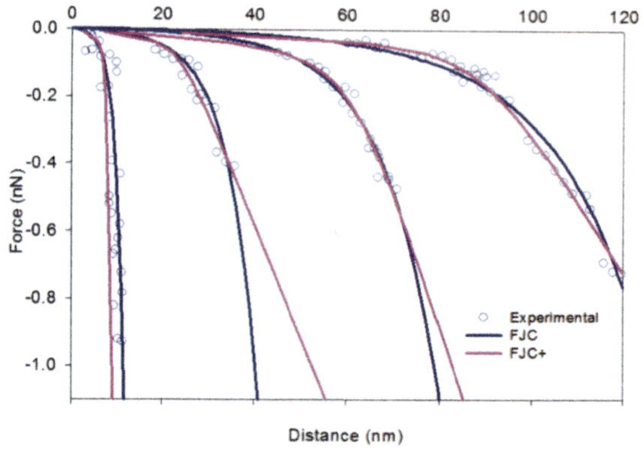

Figure 13.2. AFM was used to stretch molecules on *P. putida* KT2442, and these data were fit with either the freely jointed chain (FJC) or extensible freely jointed chain (FJC+) models. Experimental conditions and model equations are described elsewhere (Refs. 25 and 56).

In this same study, Francius *et al.* used the FJC+ model to measure Kuhn lengths and segment elasticities on two types of LGG bacteria.[29] The modelling was helpful in allowing the authors to conclude that the wild-type LGG bacteria have two types of surface polysaccharides, one group rich in mannose that is characterized by moderate extensions, and a group rich in galactose, the latter which can have much longer extensions. These results were used to discuss how LGG bacteria may bind to intestinal tissue and interact with immune receptors in the host.

Bacterial proteins and proteoglycans can usually be more appropriately fitted with the WLC model. For example, the WLC model could describe the mechanical properties of mucilage material isolated from marine diatoms,[30] surface proteins of *Staphylococcus aureus*[31] and the unfolding of an *Escherichia coli* transmembrane protein.[32] However, none of the models may work for very rigid biomolecules, such as pili, as was observed by Touhami *et al.* when studying *Pseudomonas aeruginosa* expressing type IV pili.[11] As none of the available polymer models could fit the extension profiles from the AFM data, the authors speculated that this was due to the very stiff nature of pili, in comparison with other bacterial macromolecules.

13.3 BACTERIAL INTERACTIONS WITH BIOMATERIALS

On implantation, a medical device is immediately coated with physiological molecules (fluids, peptides, etc.), forming a conditioning film.[33,34] Regardless of the material's surface chemistry at implantation, a gradual build-up of these molecules changes the surface to one easily colonized by microbes. AFM has been used to measure the interaction of bacteria with biomaterials for over 20 years, with an early study using a probe modified with a lawn of *E. coli* to probe the interactions of that bacterium with a polymer-coated substrate.[35] Methods have improved and become more facile, so that now it is relatively easy to measure the interactions between a bacterium or a biofilm and virtually any type of biomaterial.

The technique in which bacteria are attached to an AFM probe or to a tipless cantilever is referred to as biological force microscopy and is an extension of colloidal probe microscopy (CPM). Ducker *et al.* developed CPM in 1991, in which a colloidal sphere was glued to an AFM cantilever.[36,37] The advantage of CPM is that the contact area between the sample and the probe is much better defined than when a sharp probe is used, and therefore it is easier to apply models that require the contact area.

The first microbes used as biological probes were yeast, which are generally larger and more round than bacteria, properties that probably facilitated their attachment to the cantilever.[16] An example of a yeast cell (*Candida parapsilosis*) attached to an AFM cantilever is shown in Fig. 13.3. However, the methodologies developed for yeast were quickly extended towards bacterial studies.[38-40] The advantage of using a biological force probe is that any kind of substrate can be probed. Therefore, this technique has potential to be used for the study of bacterial interactions with biomaterials.

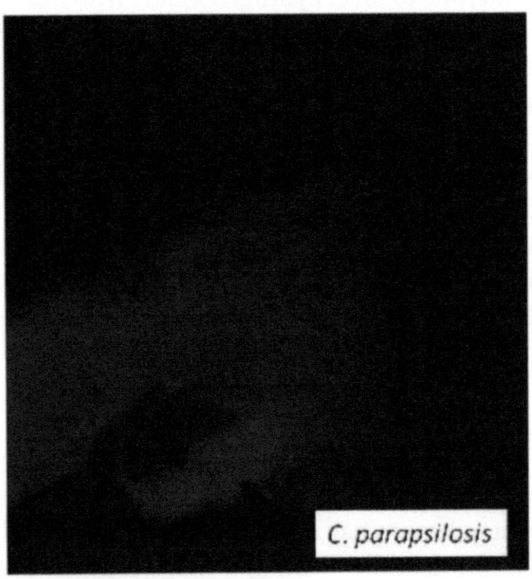

Figure 13.3. A cell probe was created by attaching *C. parapsilosis* to a silicon nitride AFM cantilever.

One disadvantage of this method is that a chemical linker is required to attach the bacteria to the AFM cantilever or probe. Some common agents applied are poly-L-lysine and hexadecanethiol. Control experiments must be performed to help ensure that the linking chemical does not alter the biological interaction forces observed.

Bacterial adhesion between model biomaterials can be easily quantified. For example, Boks *et al.* immobilized four different strains of *Staphylococcus epidermidis* on AFM probes and then made measurements of interaction forces on hydrophilic glass and a glass surface coated with a hydrophobic chemical, dimethyldichlorosilane (DDS).[41] On the hydrophobic (DDS coated) surface, the adhesion of all four bacterial strains was very rapid and was not time-dependent. This was attributed to hydrophobic interactions. However, when the bacterial strains were interacted with the hydrophilic glass, the bond strength increased over time, and this growth was attributed to the process of hydrogen bond formation. This finding has implications for the development of biomaterials or coatings that resist bacterial colonization. Since a number of types of interactions can occur within a single system, the timescales must be considered when designing a material to inhibit a particular type of interaction.

Hydrogen bonds have been considered an important part of the interaction between bacteria and surfaces, and several studies have suggested that a means to inhibit bacterial adhesion should focus on a method that disrupts the ability of the microbe to form hydrogen bonds with the substrate.[42,43]

A well-established method used for understanding bacterial interactions with biomaterials is to study the forces between a bacterium and a surface coated with biological molecules that form conditioning films on biomaterials. For example, we studied the interaction of two strains of *P. aeruginosa* with glass and bovine serum albumin (BSA)-coated glass.[43] Biomaterials that are implanted into the body rapidly accumulate a layer of proteins.[44] If this initial attachment occurs, bacteria can grow into a biofilm that is very difficult to eradicate, causing failures of biomedical devices that usually require implant removal.[45] Although biomaterial coatings represent an active area of research,[46,47] an antimicrobial coating will only be successful if it also resists colonization by serum proteins. The model system we used included the wild-type strain, *P. aeruginosa* PAO1, which contains two types of saccharides in the LPS, the A and B bands. A band represents neutral sugars, while B band is the serotype-specific O-antigen. A mutant strain, AK1401, was also examined. The mutant strain has a shorter A-band polymer, and lacks the B band entirely. For both strains, we found that hydrogen bonds control the association between *P. aeruginosa* and protein-coated surfaces. The mean adhesion force (F_{adh}) between BSA and AK1401 was 1.12 nN, compared with

0.40 nN for F_{adh} between BSA and PAO1. The higher adhesion of the mutant strain was believed to be due to absence of B-band saccharides and the shorter A-band unit on strain AK1401, which allowed for the lipid A and core region to be more exposed than in the parent strain. The lipid A and core region have strong affinity for BSA because of hydrogen bonding. We did not find that electrostatic or steric interactions were dominant in controlling *P. aeruginosa* interactions with BSA. This work also demonstrated that affinity of a bacterium for a protein coating depends on molecular properties of the bacterial surface molecules. In other words, it may be very difficult to design surfaces that are resistant to all types of bacterial colonization, because subtle differences in bacterial surface molecules control whether or not they will find a particular surface attractive. This finding suggests that a biomaterial should have more than one type of functionality, perhaps incorporating multiple mechanisms of preventing bacterial adhesion on a given biomaterial.

13.4 MICROBE–MICROBE INTERACTIONS

In the natural environment, bacteria are much more commonly found associated with a solid surface, rather than in free-floating form.[48] AFM has been very useful in measuring the forces the bacteria experience during the initial adhesion process, which is an established strategy for preventing biofilm formation. However, bacterial interactions with an inert surface are different than their interaction with another bacterium, or another type of biological cell.

As an example, we studied the adhesive interactions between a single cell of *C. parapsilosis* that was immobilized to an AFM cantilever, and either silicone rubber or silicone coated with a *P. aeruginosa* biofilm.[49] Using a *C. parapsilosis* cell probe, the interaction with a silicone substrate was adhesive, with forces of 2.3 ± 0.25 nN in the approach portion of the force cycle. However, when the *C. parapsilosis* probe was contacted with a *P. aeruginosa* biofilm, the attractive force from the approach curve decreased to 2.0 ± 0.40 nN. We also saw an unusual repulsive force (2.0 nN) in the AFM approach curve at longer distances of ~75 nm (Fig. 13.4). This repulsion may be attributed to steric and electrostatic interactions between the two microbial polymer brushes. The attractive forces are too large (and occur at distances too long) to be van der Waals or electrostatic interactions. We think that EPS molecules from *C. parapsilosis* were forming specific adhesive interactions with functional groups on the silicone rubber. Although we did not model such forces, they likely are hydrogen bonding interactions. The forces were slightly decreased when the *C. parapsilosis* contacted the *P. aeruginosa* (bacterial) biofilm, perhaps because steric and electrostatic repulsive forces were counteracting the hydrogen bonding.

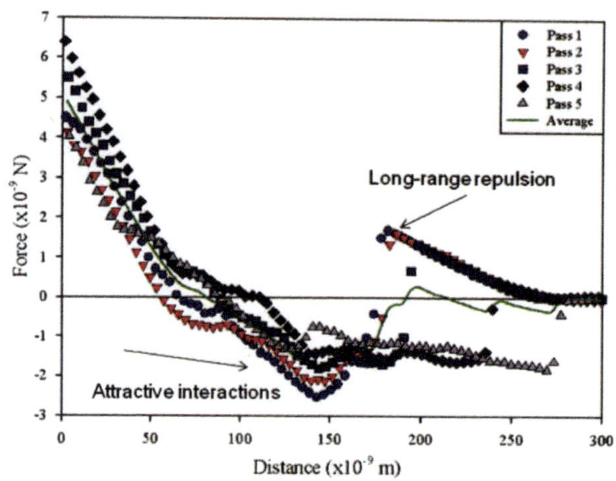

Figure 13.4. The cell probe shown in Fig. 13.3 was contacted with a well-developed biofilm of *P. aeruginosa*. The figure shows representative data from five approach cycles (labelled as pass 1 through pass 5), and the average of those five measurements. More information about this system is available elsewhere (Ref. 49).

Another study of cell–cell interactions for marine bacteria showed long-range repulsion in the approach curves but did not show the presence of any attractive forces in the approach curves.[50] In this case, the authors used a cell probe to study the interaction of *Desulfovibrio desulfuricans* with a biofilm of the same bacteria, and a similar experiment was performed with a *Pseudomonas* sp. It was interesting to note that the retraction profiles did not show any attraction between the two biological samples, because this seems to be unusual in AFM experiments with bacteria. The authors speculated that electrostatic repulsion was responsible for mitigating any effects of polymer adhesion, which are usually present when bacteria are probed with AFM.

13.5 FUTURE APPLICATIONS AND RESEARCH NEEDS

The use of AFM to study bacterial adhesion is now a widely accepted and commonly used technique. Biological applications in general are now much easier to study since it has become easier to work with AFM in liquids, the spring constant of the probe can be measured more accurately using simple methods, quantitative models are now available to help interpret biological data and many studies have been published describing how to functionalize AFM cantilevers with cells or biologically relevant molecules. Scientists interested in biology who choose AFM as part of their research can also more easily combine AFM with other forms of microscopy, such as side-view fluorescence microscopy[51] or total internal reflection fluorescence microscopy.[52] In addition to direct integration, AFM can be used side by side with complementary techniques, such as confocal laser scanning microscopy[53] and quartz crystal microbalance with dissipation monitoring.[54]

One of the key challenges that remain is in the ability to separate biological interaction forces from effects due to the compliance of the bacterium. There are still some cases for which the "origin" of the force curve made on a bacterial cell can be difficult to determine. To briefly summarize, the first and most widely accepted approach to set the origin of a displacement curve was specified as the point when the cantilever deflection is linear with respect to sample displacement at high force.[36] Although this method is simple, it was tested with silica spheres and is generally valid on hard surfaces. Force measurements on bacteria often show that cantilever deflection is a non-linear function of displacement. The Hertz theory can be used to estimate the point of zero,[55] provided that there are no adhesion forces observed. This is often not valid because AFM experiments with bacteria are very likely to show adhesion between the bacterium and substrate. Specifically when attraction is observed in the approach curve, it was suggested that the point at which the probe contacts the surface can be considered the position of zero separation.[56] Again, this method may not work since attraction may not be observed. An alternate method proposed was to consider the origin by taking the derivative of the force with respect to the distance, and finding the point at which the derivative becomes non-zero.[57] An issue with this method is that the typical amount of data scatter makes it hard to unambiguously determine the zero point. More studies on model systems can help clarify this issue for future AFM users.

Another challenge lies in the ability to apply quantitative models to AFM force profiles on bacteria. It has been very attractive to use DLVO-type models to describe bacterial interaction forces, but thus far, such models have failed to capture the behaviour of AFM experiments.[58] A recent study has reported better agreement between extended DLVO predictions and AFM force profiles for the interactions of hydrophobic bacteria *Acinetobacter venetianus* RAG1 and *Rhodococcus erythropolis* 20S-E1-c with alkanethiol-functionalized (hydrophobic) AFM probes (gold-coated silicon).[59] The authors simply added a term for steric interactions to the typical DLVO interactions. For this specific case of two hydrophobic surfaces interacting with one another, the modelling was consistent with experimental measurements from the AFM. However, retraction force data were not shown so we are unable to evaluate the extent of polymer bridging and adhesion forces that were present. Further, this particular case seems to be the exception among the many AFM studies on bacteria in which forces were observed that were not consistent with DLVO theory, because of heterogeneity, roughness, surface polymers, specific interactions or elastic responses of the cells. Also, many surfaces of interest are not going to be hydrophobic, and there will be other types of interactions present that simply cannot be explained with currently available models. Therefore, research is needed to develop extensions or modifications of existing models, or perhaps entirely new models, so that bacterial adhesion forces measured with AFM can be described more accurately and interpreted in terms of the governing physical phenomena.

References

1. Butt, H. J., Cappella, B., and Kappl, M. (2005) Force measurements with the atomic force microscope; technique, interpretation and applications, *Surf. Sci. Rep.*, **59**, 1–152.
2. Lee, G. U., Kidwell, D. A., and Colton, R. J. (1994) Sensing discrete streptavidin-biotin interactions with atomic force microscopy, *Langmuir*, **10**, 354–357.
3. MacKerell, A. D. J., and Lee, G. U. (1999) Structure, force, and energy of a double-stranded DNA oligonucleotide under tensile loads, *Eur. Biophys. J.*, **28**, 415–426.
4. Mitsui, K., Hara, M., and Ikai, A. (1996) Mechanical unfolding of alpha2-macroglobulin molecules with atomic force microscope, *FEBS Lett.*, **385**, 29–33.
5. Rief, M., Gautel, M., Oesterhelt, F., Fernandez, J. M., and Gaub, H. E. (1997) Reversible unfolding of individual titin immunoglobin domains by AFM, *Science*, **276**, 1109–1111.
6. Marszalek, P. E., Oberhauser, A. F., Pang, Y.-P., and Fernandez, J. M. (1998) Polysaccharide elasticity governed by chair-boat transitions of the glucopyranose ring, *Nature*, **396**, 661–666.
7. Marszalek, P. E., Li, H., and Fernandez, J. M. (2001) Fingerprinting polysaccharides with single-molecule force spectroscopy, *Nat. Biotechnol.*, **19**, 258–262.
8. Marszalek, P. E., Li, H. B., Oberhauser, A. F., and Fernandez, J. M. (2002) Chair-boat transitions in single polysaccharide molecules observed with force-ramp AFM, *Proc. Natl. Acad. Sci. USA*, **99**, 4278–4283.
9. Abu-Lail, N. I., and Camesano, T. A. (2002) Elasticity of *Pseudomonas putida* KT2442 biopolymers probed with single-molecule force microscopy, *Langmuir*, **18**, 4071–4081.
10. Dufrêne, Y. F., Boonaert, C. J. P., van der Mei, H. C., Busscher, H. J., and Rouxhet, P. G. (2001) Probing molecular interactions and mechanical properties of

microbial cell surfaces by atomic force microscopy, *Ultramicroscopy*, **86**, 113–120.

11. Touhami, A., Jericho, M. H., Boyd, J. M., and Beveridge, T. J. (2006) Nanoscale characterization and determination of adhesion forces of *Pseudomonas aeruginosa* pili by using atomic force microscopy, *J. Bacteriol.*, **188**, 370–377.
12. Yao, X. W., Burke, S., Stewart, S., Jericho, M. H., Pink, D., Hunter, R., and Beveridge, T. J. (2002) Atomic force microscopy and theoretical considerations of surface properties and turgor pressures of bacteria, *Colloids Surf. B Biointerfaces*, **23**, 213–230.
13. Fang, H. H. P., Chan, K. Y., and Xu, L.-C. (2000) Quantification of bacterial adhesion forces using AFM, *J. Microbiol. Methods*, **40**, 89–97.
14. Van der Aa, B. C., and Dûfrene, Y. F. (2002) In situ characterization of bacterial extracellular polymeric substances by AFM, *Colloids Surf. B Biointerfaces*, **23**, 173–182.
15. Kim, H. A., Osada, T., Ikai, A. (2002) Quantification of cell adhesion interactions by AFM: effects of LPS/PMA on the adhesion of C6 glioma cell to collagen type I, *Appl. Surf. Sci.*, **188**, 493–498.
16. Bowen, W. R., Hilal, N., Lovitt, R. W., and Wright, C. J. (1998) Direct measurement of the force of adhesion of a single biological cell using an atomic force microscope, *Colloids Surf. A Physicochem. Eng. Asp.*, **136**, 231–234.
17. Gad, M., Itoh, A., and Ikai, A. (1997) Mapping cell wall polysaccharides of living microbial cells using atomic force microscopy, *Cell Biol. Int.*, **21**, 697–706.
18. Dufrêne, Y. F. (2000) Direct characterization of the physicochemical properties of fungal spores using functionalized AFM probes, *Biophys. J.*, **78**, 3286–3291.
19. Van der Aa, B. C., Michel, R. M., Asther, M., Zamora, M. T., Rouxhet, P. G., and Dufrêne, Y. F. (2001) Stretching cell surface macromolecules by atomic force microscopy, *Langmuir*, **17**, 3116–3119.
20. Dufrêne, Y. F., Boonaert, C. J. P., Gerin, P. A., Asther, M., and Rouxhet, P. G. (1999) Direct probing of the surface ultrastructure and molecular interactions of dormant and germinating spores of *Phanerochaete chrysosporium*, *J. Bacteriol.*, **181**, 5350–5354.
21. Van der Aa, B. C., Asther, M., and Dufrêne, Y. F. (2002) Surface properties of *Aspergillus oryzae* spores investigated by atomic force microscopy, *Colloids Surf. B Biointerfaces*, **24**, 277–284.
22. Boonaert, C. J. P., Rouxhet, P. G., and Dufrêne, Y. F. (2000) Surface properties of microbial cells probed at the nanometer scale with atomic force microscopy, *Surf. Interface Anal.*, **30**, 32–35.
23. Janshoff, A., Neitzert, M., Oberdorfer, Y., and Fuchs, H. (2000) Force spectroscopy of molecular systems—single molecule spectroscopy of polymers and biomolecules, *Angew. Chem. Int. Ed. Engl.*, **39**, 3213–3237.
24. Ortiz, C., and Hadziioannou, G. (1999) Entropic elasticity of single polymer chains of poly(methylacrylic acid) measured by atomic force microscopy, *Macromolecules*, **32**, 780–787.

25. Rief, M. F., Fernandez, J. M., Gaub, H. E. (1998) Elasticity coupled two-level systems as a model for biopolymer extensibility, *Phys. Rev. Lett.*, **81**, 4764–4767.
26. Wei, H., and van de Ven, T. G. M. (2008) AFM-based single molecule force spectroscopy of polymer chains: theoretical models and applications, *Appl. Spectrosc. Rev.*, **43**, 111–133.
27. Odijk, T. (1995) Stiff chains and filaments under tension, *Macromolecules*, **28**, 7016–7018.
28. Abu-Lail, N. I., and Camesano, T. A. (2003) Role of ionic strength on the relationship of biopolymer conformation, DLVO contributions, and steric interactions to bioadhesion of *Pseudomonas putida* KT2442, *Biomacromolecules*, **4**, 1000–1012.
29. Francius, G., Lebeer, S., Alsteens, D., Wildling, L., Gruber, H. J., Hols, P., de Keersmaecker, S., Vanderleyden, J., and Dufrene, Y. F. (2008) Detection, localization, and conformational analysis of single polysaccharide molecules on live bacteria, *ACS Nano*, **2**, 1921–1929.
30. Higgins, M. J., Sader, J. E., Mulvaney, P., and Wetherbee, R. (2003) Probing the surface of living diatoms with atomic force microscopy: the nanostructure and nanomechanical properties of the mucilage layer, *J. Phycol.*, **39**, 722–734.
31. Yongsunthon, R. L., Lower, S. K. (2006) Force spectroscopy of bonds that form between a Staphylococcus bacterium and silica or polystyrene substrates, *J. Electron Spectros. Relat. Phenomena*, **150**, 228–234.
32. Lower, B. H., Yongsunthon, R., Vellano, F. P., and Lower, S. K. (2005) Simultaneous force and fluorescence measurements of a protein that forms a bond between a living bacterium and a solid surface, *J. Bacteriol.*, **187**, 2127–2137.
33. Bos, R., van der Mei, H. C., and Busscher, H. J. (1999) Physico-chemistry of initial microbial adhesive interactions—its mechanisms and methods for study, *FEMS Microbiol. Rev.*, **23**, 179–230.
34. Habash, M. B., Van der Mei, H. C., Busscher, H. J., and Reid, G. (1999) The effect of water, ascorbic acid, and cranberry derived supplementation on human urine and uropathogen adhesion to silicone rubber, *Can. J. Microbiol.*, **45**, 691–694.
35. Razatos, A., Ong, Y. L., Sharma, M. M., and Georgiou, G. (1998) Molecular determinants of bacterial adhesion monitored by atomic force microscopy, *Proc. Natl. Acad. Sci. USA*, **95**, 11059–11064.
36. Ducker, W. A., Senden, T. J., and Pashley, R. M. (1991) Direct measurement of colloidal forces using an atomic force microscope, *Nature*, **353**, 239–241.
37. Ducker, W. A., Senden, T. J., and Pashley, R. M. (1992) Measurement of forces in liquids using a force microscope, *Langmuir*, **8**, 1831–1836.
38. Lower, S. K., Tadanier, C. J., and Hochella, M. F., Jr. (2000) Measuring interfacial and adhesion forces between bacteria and mineral surfaces with biological force microscopy, *Geochim. Cosmochim. Acta*, **64**, 3133–3139.
39. Emerson, R. J., Liu, Y., Bergstrom, T. S., Soto, E. R., Brown, C. A., McGimpsey, W. G., and Camesano, T. A. (2006) A microscale correlation between surface

chemistry, texture and the adhesive strength of *Staphylococcus epidermidis*, *Langmuir*, **22**, 11311–11322.

40. Cail, T. L., and Hochella, M. F. (2005) The effects of solution chemistry on the sticking efficiencies of viable *Enterococcus faecalis*: an atomic force microscopy and modeling study, *Geochim. Cosmochim. Acta*, **69**, 2959–2969.

41. Boks, N. P., Busscher, H. J., van der Mei, H. C., and Norde, W. (2008) Bond-strengthening in Staphylococcal adhesion to hydrophilic and hydrophobic surfaces using atomic force microscopy, *Langmuir*, **24**, 12990–12994.

42. Atabek, A., and Camesano, T. A. (2007) An atomic force microscopy study of the effect of lipopolysaccharides and extrapolymeric substances on the adhesion of *Pseudomonas aeruginosa*, *J. Bacteriol.*, **189**, 8503–8509.

43. Atabek, A., Liu, Y., Pinzón-Arango, P. A., and Camesano, T. A. (2008) Importance of LPS structure on protein interactions with *Pseudomonas aeruginosa*, *Colloids Surf. B Biointerfaces*, **67**, 115–121.

44. Gristina, A. G. (1987) Biomaterial-centered infection: microbial adhesion versus tissue integration, *Science*, **237**, 1588–1595.

45. Heard, S. O. (2001) Catheter-related infection: diagnosis, prevention, and treatment, *Ann. Acad. Med. Singap.*, **30**, 419–429.

46. Gottenbos, B., Van der Mei, H. C., Klatter, F., Nieuwenhuis, P., and Busscher, H. J. (2002) In vitro and in vivo antimicrobial activity of covalently coupled quaternary ammonium silane coatings on silicone rubber, *Biomaterials*, **23**, 1417–1423.

47. Boulmedais, F., Frisch, B., Etienne, O., Lavalle, P., Picart, C., Ogier, J., Voegel, J.-C., Schaaf, P., and Egles, C. (2004) Polyelectrolyte multilayer films with pegylated polypeptides as a new type of anti-microbial protection for biomaterials, *Biomaterials*, **25**, 2003–2011.

48. Costerton, J. W. (1995) Overview of microbial biofilms, *J. Ind. Microbiol.*, **15**, 137–140.

49. Emerson, R. J., and Camesano, T. A. (2004) Nanoscale investigation of pathogenic microbial adhesion to a biomaterial, *Appl. Environ. Microbiol.*, **70**, 6012–6022.

50. Sheng, X., Ting, Y. P., and Pehkoven, S. O. (2007) Force measurements of bacterial adhesion on metals using a cell probe atomic force microscope, *J. Colloid Interface Sci.*, **310**, 661–669.

51. Chaudhuri, O., Parekh, S. H., Lam, W. A., and Fletcher, D. A. (2009) Combined atomic force microscopy and side-view optical imaging for mechanical studies of cells, *Nat. Methods*, **6**, 383–387.

52. Trache, A., and Lim, S. M. (2009) Integrated microscopy for real-time imaging of mechanostransduction studies in live cells, *J. Biomed. Opt.*, **14**, 034024, 1–13.

53. Lau, P. C. Y., Lindhout, T., Beveridge, T. J., Dutcher, J. R., and Lam, J. S. (2009) Differential lipopolysaccharide core capping leads to qualitative and correlated modifications of mechanical and structural properties in *Pseudomonas aeruginosa* biofilms, *J. Bacteriol.*, **191**, 6618–6631.

54. Strauss, J., Kadilak, A., Cronin, C., Mello, C. M., and Camesano, T. A. (2010) Binding, inactivation, and adhesion forces between antimicrobial peptide cecropin P1 and pathogenic *E. coli*, *Colloids Surf. B Biointerfaces*, **75**, 156–164.

55. Radmacher, M., Fritz, M., Kacher, C. M., Cleveland, J. P., and Hansma, P. K. (1996) Measuring the viscoelastic properties of human platelets with the atomic force microscope., *Biophys. J.*, **70**, 556–567.

56. Ong, Y.-L., Razatos, A., Georgiou, G., and Sharma, M. M. (1999) Adhesion forces between *E. coli* bacteria and biomaterial surfaces, *Langmuir*, **15**, 2719–2725.

57. Li, X., and Logan, B. E. (2004) Analysis of bacterial adhesion using a gradient force analysis method and colloid probe atomic force microscopy, *Langmuir*, **20**, 8817–8822.

58. Camesano, T. A., and Logan, B. E. (2000) Probing bacterial electrosteric interactions using atomic force microscopy, *Environ. Sci. Technol.*, **34**, 3354–3362.

59. Dorobantu, L. S., Bhattacharjee, S., Foght, J. M., and Gray, M. R. (2009) Analysis of force interactions between AFM tips and hydrophobic bacteria using DLVO theory, *Langmuir*, **25**, 6968–6976.

Chapter 14

FORCE SPECTROSCOPY OF MINERAL–MICROBE BONDS

Brian H. Lower and Steven K. Lower
Ohio State University, Columbus, Ohio, USA
Lower.9@osu.edu

14.1 BONDS BETWEEN MICROBES AND MINERALS

There are estimated to be 10^{30} prokaryotic cells on Earth, with as many as 97% living on, or in close proximity to, minerals in soil and subsurface environments.[1] Bonds between microorganisms and minerals are, therefore, ubiquitous in nature. Whether a cell forms a bond with a mineral depends solely on the interplay of attractive or repulsive forces that exist within the space between a cell and mineral surface. Until recently, it was largely impossible to probe the forces and structures within this molecular to nanometre scale space. However, this changed with the invention of the atomic force microscope (AFM)[2] and its use to measure forces between two surfaces.[3–6]

Within the past decade, there has been an explosion of research focused on the forces and bonds associated with microorganisms. This chapter will provide a brief overview of studies that have explored forces and bonds at the interface between microorganisms and minerals. This chapter will also draw upon our own expertise by presenting an example of how AFM can be used to gain an appreciation of one particular mineral–microbe bond, that being the protein-mediated bond between an iron-reducing bacterium and an Fe(III)-containing mineral. Before presenting this material, this chapter will provide a brief consideration of the fundamental forces that delineate and define the nature of any bond that forms between a living cell and a mineral surface.

Life at the Nanoscale: Atomic Force Microscopy of Live Cells
Edited by Yves Dufrêne

www.panstanford.com

14.2 FORCES BETWEEN A MICROORGANISM AND A MINERAL SURFACE

It is well established that there are four distinct forces at work in our present universe. These are the strong nuclear force, the weak nuclear force, the force of gravity and the electromagnetic force.[7] The two nuclear forces act over very short distances and therefore dominate interactions that occur within atoms. Electromagnetic and gravitation forces act over much larger distances from the atomic to near infinity. As such, these latter two forces dominate the domain of anything that is larger than an atom's nucleus, for example, a microorganism and a mineral.

Electromagnetic forces, as opposed to gravitation, will obviously control the interactions that occur when a very small bacterium makes contact with a mineral surface. Nonetheless, it is an interesting, albeit somewhat academic, exercise to determine which of these two forces (gravitational vs. electromagnetic) dominates as a function of a particle's size.

The force (F) of gravity between the two particles is described by Newton's equation: $F(D) = (-Gm_1m_2)/D^2$, where G is the universal gravitational constant (6.67×10^{-11} m^3 kg^{-1} s^{-2}), m_x is the mass (in kg) of particle x and D is the distance (in m) between the particles. This force is always attractive. The negative sign, which is customarily excluded from Newton's universal law of gravitation, is used here to indicate attraction as is often the convention in the biophysical literature.

The electromagnetic force, on the other hand, can be divided into several force types including van der Waals, electrostatic, steric and solvation forces.[8] The van der Waals force occurs between all particles and will therefore be considered as the reference force type for electromagnetic forces. Like the force of gravity, the van der Waals force is attractive (in most instances). The van der Waals force between two particles is described as follows: $F(D) = (-H_a r_x)/6D^2$), where H_a is the Hamaker constant (in J), D is the distance (in m) between the particles and r_x (in m) is $(r_1 \times r_2)/(r_1 + r_2)$, in which r is the radius of particle 1 or 2. For the interaction between a sphere and flat plane (e.g., a bacterium and a flat mineral surface), r_x reduces to the radius of the sphere (i.e., the bacterium).

These two equations are similar in several ways: both are attractive, a constant is present in the numerator of both equations and both predict an inverse square relationship between force and distance. The radius of a particle can be substituted for mass in the gravitational equation by assuming a given density (e.g., the density of water) and spherical shape for a particle. This allows one to determine the gravitational force between two particles in terms of size, like the van der Waals force, rather than mass. In so doing, one can determine the theoretical interplay between the gravitational and van der Waals forces for two particles that are in contact with one another.

Figure 14.1 compares these forces for two spherical particles of a given size (e.g., two particles each having a radius of 1 mm) and a density equivalent to that of water, which is approximately the density of biological cells. For particles here on Earth, obviously the greatest gravitation force will be the one between a particle and the Earth itself. Therefore, Fig. 14.1 also shows the predicted force of gravitation between the Earth and a particle of a given size. As the size of the two interacting particles decreases to less than 10^{-3} to 10^{-5} m, electromagnetic forces, like the van der Waals force, become the preeminent force. Not surprisingly, a bacterium with a density of ~1 g cm^{-3} and length scale of 10^{-6} m exists solely within the realm of electromagnetic forces.

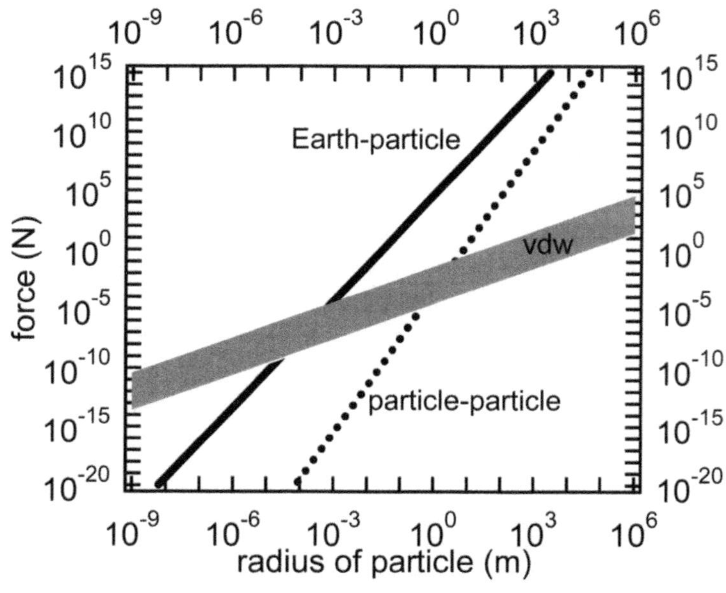

Figure 14.1. Log–log plot of the theoretical gravitational and van der Waals forces between two similarly sized particles that are in contact with one another. The dashed line describes the gravitational force between two particles of the same size. To determine the gravitational force, a particle's mass was converted to radius by assuming the particle was a solid homogeneous sphere with a density of 1 g cm^{-3}. The interacting distance was the sum of the radii of the two interacting particles. The shaded region outlines the boundaries of the expected van der Waals force between the two particles of the same size. Values for the Hamaker constant ranged from 10^{-20} to 10^{-21} J, which is appropriate for biological and inorganic solids,[7] and contact was defined as an effective separation between particles of 0.2 to 2 nm.[9] The solid line is the gravitational attraction between a particle of a given radius and the Earth (mass = 5.97×10^{24} kg). For the Earth-particle gravitational force, the interaction distance was set as 6.4×10^6 m (i.e., radius of the Earth).

Incidentally, this comparison of gravitation and electromagnetic forces also helps to place a perspective on the *magnitude* of forces that exist between a bacterium and mineral surface. As attested to in chapters throughout this book, the world of bacteria is dominated by forces less than one nanoNewton. What does it feel like to experience a force of 1 nN? If you, the reader, place this book 1.5 m away from your body then there is, according to Newton's Law of gravitation, a force of precisely 1 nN between you (70 kg) and the book (0.5 kg). This is the magnitude of the force that dictates whether a microorganism will either form a bond or break a bond with another surface.

14.3 SOME EXAMPLES FROM THE LITERATURE

The first papers dealing with force measurements on bacteria began to appear in the literature about a decade ago.[10-15] All of these works utilized AFM, which is still the instrument of choice for force measurement on cells. Since these first manuscripts, countless numbers of papers have been published on forces between bacteria and minerals. Table 14.1 provides a list of some of these publications. This table is not meant to be an exhaustive review of every publication on mineral–microbe forces. Rather, this table is meant to show a range of studies related to bonds and forces between microorganisms and minerals.

It is important to note that many of the papers listed in Table 14.1 are not "microbe–mineral" papers in the strictest sense. Minerals are, by definition, naturally occurring crystalline solids. Many papers make use of man-made substances such as silicon or silicon nitride (i.e., the composition of the typical AFM tip) rather than true minerals, and deal with the intermolecular forces detected upon the approach of a cell towards a mineral, not the bonds that form after a cell makes contact with a mineral. Nonetheless, Table 14.1 is a good starting point for those who are interested in these topics.

14.4 AN EXAMPLE OF MINERAL–MICROBE FORCE SPECTROSCOPY

14.4.1 Interactions Between an Iron-Reducing Bacterium and Fe(III) Minerals

In the late 1990s, we began to use AFM to measure intermolecular and adhesion forces between *Escherichia coli* and the minerals muscovite, goethite and graphite. This resulted in a publication by Lower *et al.* (2000).[14]

In many ways, this paper was a proof of concept. *E. coli* was a model Gram-negative bacterium, and the surface properties of the minerals varied in terms of their surface charge and hydrophobicity. However, our goal was to study a mineral–microbe system that was, arguably, far more interesting than the adhesion of *E. coli* to various minerals. We began to use AFM to study the interactions between *Shewanella*, an iron-reducing bacterium, and iron oxyhydroxide minerals.

Shewanella is a Gram-negative bacterium that can gain energy by shuttling electrons to not only oxygen but also ferric iron. The real interest in this creature comes from its novel ability to transfer electrons to Fe(III) that is within the crystalline structure of minerals like hematite (Fe_2O_3) or goethite (FeOOH). In other words, this microorganism is able to breathe on solid state iron minerals when oxygen is absent from a system.

In the late 1980s and early 1990s, a number of other groups proposed that cytochrome proteins catalyze the terminal transfer of electrons from the bacterium to Fe(III) in a mineral.[16-20] However, this hypothetical electron transfer event was challenging to prove because it was essentially hidden within the small interfacial space between a living cell and mineral.

Table 14.1. Studies of forces and bonds between microorganisms and minerals

Microorganism	Mineral or material	Ref.
Gram-negative bacteria		
Acidithiobacillus ferrooxidans	Silicon nitride	21
Burkholderia sp.	Muscovite	15
Burkholderia cepecia	Silicon nitride	13
Escherichia coli	Muscovite, graphite, goethite	14
E. coli	Silica or quartz	22, 23
E. coli	Silicon nitride or silicon	10, 24, 25
Haemophilus influenza	Silicon nitride or silicon	26
Klebsiella terrigena	Silicon nitride	27
Pseudomonas aeruginosa	Mica	28
P. aeruginosa	Silicon nitride or silicon	28, 29
Pseudomonas putida	Silicon nitride	13
Shewanella oneidensis	Diaspore, goethite, hematite	30-33
Shewanella putrefaciens	Silicon nitride	34
Gram-positive bacteria		
Bacillus subtilis	Quartz	23
Enterococcus faecalis	Glass	35
Staphylococcus aureus	Silica	36
S. aureus	Silicon nitride	37
Staphylococcus epidermidis	Gold	38

Streptococcus mitis	Silicon nitride	39
Archaea		
Methanospirillum hungatei	Silicon nitride	40
Eukarya fungi		
Aspergillus niger	Mica	12
Pharerochaete chrysosporium	Silicon nitride	11
Saccharomyces cerevisiae	Mica	41
Eukarya diatom or protista		
Corynebacterium glutamicum	Silicon nitride	42
Craspedostauros australis	Silicon nitride	43
Cryptosporidium parvum	Silica	44
Phaeodactylum tricornutum	Silicon nitride	45
Tovanium undulatum	Silicon nitride	46

Therefore, we used AFM to essentially create, and then separate, an interface between a living cell of *Shewanella oneidensis* and the mineral goethite in aqueous solution.[30] This work, as well as the complementary AFM studies with purified proteins,[31-33] allowed us to directly probe this particular microbe–mineral interface and shed new light onto an important question in environmental and geological microbiology. The force spectra that helped us unravel this question will be discussed in more detail later.

14.4.2 Probing Bonds Between a Mineral Surface and a Living Bacterium

As shown in Table 14.1, the vast majority of force studies on microorganism make use of essentially one material, silicon nitride (or silicon), because this is the composition of commercial AFM probes. It would have been relatively easy to measure forces between *S. oneidensis* and an AFM tip, but silicon nitride is not an appropriate electron acceptor. Therefore, it was unlikely to elicit a biological response from *S. oneidensis*. For this work, we needed to attach the cell, in a living state, to the AFM cantilever. This biologically active probe was then used to probe different minerals and determine whether Fe(III)-containing minerals stimulated a response from the bacterium.

Figure 14.2 shows living bacteria on the end of an AFM cantilever. Cantilevers like this one were used on the minerals FeOOH and AlOOH. These two minerals were selected because they have virtually identical surface properties (e.g., surface charge and hydrophobicity), but only the iron-containing mineral can serve as a terminal electron acceptor for *S. oneidensis*. Figure 14.2 shows the resulting force spectra. Some spectra exhibited distinct sawtooth-shaped force-signatures in the retraction curves. These force-signatures were observed for *S. oneidensis* only when it was in contact with FeOOH (not AlOOH) under anaerobic (not oxygenated) conditions.[30,31] We hypothesized that these force-signatures originated from the unfolding of cytochrome proteins, presumably to shuttled electrons to Fe(III), which formed bonds between the bacterium and mineral.

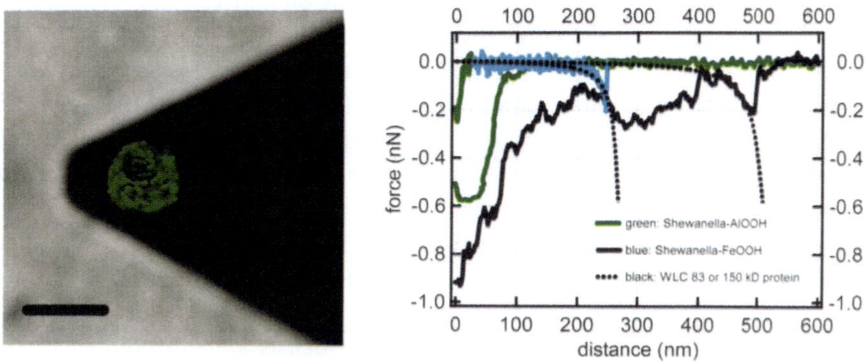

Figure 14.2. (Left) Biologically active force probe showing living bacteria on the end of an AFM cantilever.[14] Cells are fluorescent green because of expression of an intracellular green fluorescent protein. Scale bar is ~10 μm. (Right) Force spectra for a living *Shewanella oneidensis* bacterium on each of two minerals: goethite (FeOOH, light and dark blue) and diaspore (AlOOH, light and dark green) immersed in an anaerobic solution.[30,31] Black curves correspond to the modelled force-extension relationship for two outer membrane proteins (83 and 150 kD) as determined by the worm-like chain model.

14.4.3 Measuring Interactions Between Minerals and Pure Proteins

We turned to the well-established worm-like chain (WLC) model to try to determine whether the non-linear force-signatures observed in Fig. 14.2 might be due to the unravelling of proteins that form a bond between the bacterium and mineral surface. The WLC equation is as follows: $F(x) = (k_B T/b)\,[0.25(1 - x/L)^{-2} + x/L - 0.25]$, where F is force (N), x is separation or the extension distance of the protein (m), k_B is Boltzmann's constant (1.381×10^{-23} J K^{-1}), T is the temperature (298 K), L is the contour length of the polypeptide of interest (in m) and b is its persistence length. For proteins, the persistence length is often taken as the length scale of a single amino acid, ~0.4 nm.[47-49] By applying this equation to retraction profiles, one is able to back out the contour length of the protein that forms the bond between two surfaces. The size (in kD) of this protein can then be estimated by dividing the contour length by 0.4 nm (the length scale of an individual amino acid) and multiplying by 110 Da per amino acid.

By applying the WLC model to experimentally measured force spectra, we were able to identify two putative proteins as potential candidates involved in the formation of a bond with Fe(III) minerals (see Fig. 14.2). These proteins had contour lengths of approximately 290 and 540 nm, which corresponded to polypeptides with masses of approximately 80 and 150 kD, respectively.

Fortunately, at this same time, the genome of *S. oneidensis* had been determined,[50] and we had begun to use two-dimensional gel electrophoresis to characterize the outer membrane proteins produced by *S. oneidensis*.[31,32] Partly on the basis of the whole-cell force spectra noted earlier, we decided to focus our efforts on two outer membrane proteins from *S. oneidensis:* MtrC and OmcA. These two proteins contained heme groups, which meant they could catalyze electron transfer reactions. Furthermore, each protein had a molecular mass of ~80 kD, which was consistent with the putative protein force-signatures in the AFM force spectra.

Therefore, MtrC and OmcA from *S. oneidensis* were produced in large enough quantities to use in the AFM. Each protein was linked to a gold substrate. The AFM tip was coated with a thin film of an Fe(III) mineral, and this mineral-coated tip was used to probe MtrC and OmcA (Fig. 14.3). The resulting force spectra were the first of their kind for a bacterial cytochrome protein that formed a bond with a crystalline Fe(III) mineral.[33] Further, the spectra for the pure protein could be compared directly with the spectra collected on living cells of *S. oneidensis*.

Recall that the whole-cell force spectra lead us to hypothesize that proteins of ~80 and 150 kD were responsible for the bond between *S. oneidensis* and Fe(III) minerals. However, *S. oneidensis* decorates its outer surface with many different proteins, not just the two proteins of interest. The ultimate test of our hypothesis came when we compared the force spectra for a living cell of *S. oneidensis* with those collected on individual proteins that were purified from the outer membrane of *S. oneidensis*.

Figure 14.3 compares the whole-cell and protein force spectra. It is important to note that the cell and protein data were collected on two completely different AFMs. There are significant similarities between the cell and protein spectra. Specifically, there are two distinct, nonlinear, sawtooth-shaped force-signatures at approximately 300 and 550 nm. The shorter sawtooth observed for a living cell of *S. oneidensis* is consistent with the forced unfolding of either MtrC or OmcA. The longer sawtooth is still a bit of a challenge to explain. It is too long to correspond to the mechanical unfolding of a monomer of MtrC or OmcA. However, a dimer of either protein, linked end to end, would have a contour length of approximately the same length as the longer sawtooth. Indeed, this same sawtooth was observed when force spectra were collected on purified MtrC (see Fig. 14.3).

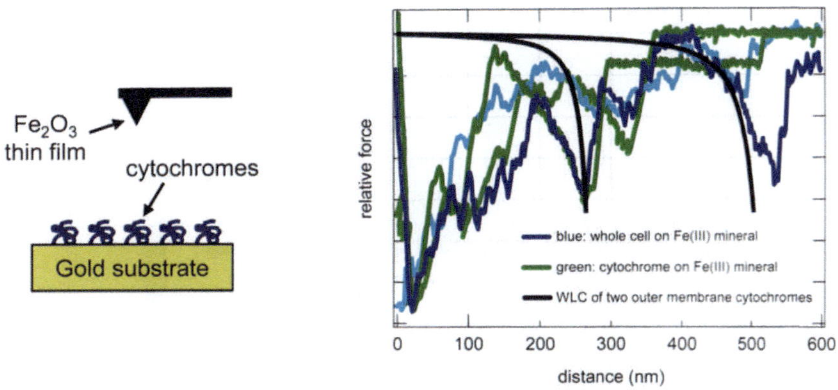

Figure 14.3. (Left) Schematic of mineral thin-film probe used on protein molecules that were purified from *S. oneidensis*. (Right) Force spectra for Fe_2O_3-probe on OmcA or MtrC (light and dark green).[33] Blue (both light and dark) curves correspond to force spectra for the *S. oneidensis*–FeOOH pair. Black curves correspond to the modelled force-extension relationship for a monomer (shorter curve whose extended length is ~300 nm) and dimer of MtrC or OmcA as determined by the worm-like chain model.

14.4.4 Tuning into a Force-Signature to Identify Specific Proteins

The aforementioned example shows how force spectra can provide the critical piece of information that allows us to understand phenomena that occur within the space between a microorganism and mineral surface. But, force spectra such as these embody only one-dimensional events that occur at a very specific location on a microorganism (or mineral). If a one-dimensional force spectrum yields a distinct force-signature, then it can be mapped in three-dimensional space to actually show the location (and number) of particular macromolecules across the surface of a living bacterium. Such information is virtually impossible to gain with any other instrument or technique. Optical microscopy lacks the resolution to detect single-molecule events, and high-resolution electron microscopy cannot be performed on living cells in solution.

As noted earlier, a distinct force-signature was observed when MtrC or OmcA formed a bond between *S. oneidensis* and a Fe(III)-containing mineral. Therefore, we attempted to use AFM to collect an image that shows the positions of MtrC or OmcA molecules on the surface of a living cell of *S. oneidensis* as it was resting on the surface of an Fe(III) mineral immersed in a deoxygenated solution.[51] Figure 14.4 shows one of the recognition or affinity maps for OmcA.

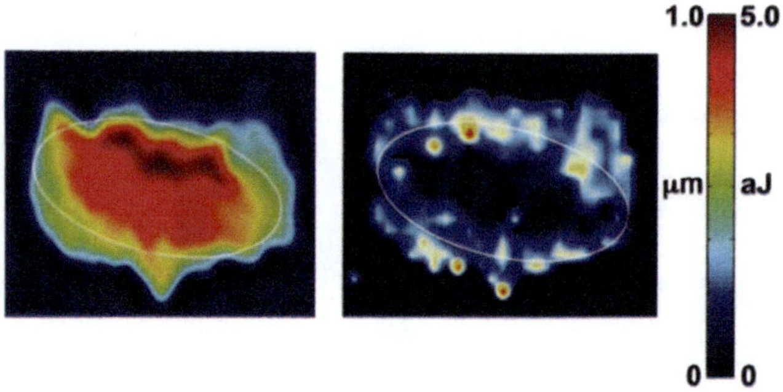

Figure 14.4. (Left) Topographic image of live *S. oneidensis* cell sitting on a Fe_2O_3 substrate. (Right) Complementary affinity map, also known as a recognition force microscopy image, collected with an anti-OmcA-functionalized tip.[51] Warm colours (e.g., red) in the right image show the position of putative OmcA molecules produced by the bacterium to form a bond with the mineral. The thin white oval outlines the approximate location of the bacterium on the Fe(III) mineral.

Briefly, this image was captured with so-called force–volume imaging using an antibody-functionalized AFM tip (anti-OmcA). The tip was used to collect a 32 × 32 grid of force curves across the cell and underlying mineral. The energy (in attoJoules) of each individual force curve was determined by integrating force with respect to distance. The energy of binding is shown in the affinity map of Fig. 14.4. Binding activity increases from cool to warm colours (e.g., blue indicates no bonds).

OmcA was not observed across the entire cell surface. Rather, it was observed only at the cell's perimeter. Presumably, this cytochrome was also located under the cell but hidden from the AFM tip as it scanned across the top of the cell. Indeed, whole-cell spectra (see earlier) demonstrate that OmcA can be located *between* the cell and mineral. This evidence strongly suggests that OmcA is localized to the interface between *S. oneidensis* and the Fe(III) mineral. It can therefore be inferred that OmcA (and MtrC) function in the transfer of electrons from *S. oneidensis* to Fe(III) in the crystal structures of minerals like FeOOH and Fe_2O_3.

14.5 SUMMARY

Most microorganisms on Earth live *on* solid surfaces such as pebbles in a stream, quartz grains in a subsurface aquifer, clay aerosols floating in the atmosphere or even apatite crystals in the human body. Intermolecular and intramolecular forces play a central role in each instance regardless of whether we, as humans, classify the interaction as environmental, geological or medical. AFM is a powerful tool, arguably the only tool that can be used to probe forces and bonds between living cells and mineral surfaces. This chapter provides an overview of papers that explore mineral–microbe forces and bonds. This chapter also highlights the use of AFM to study interactions between iron-reducing bacteria and Fe(III)-containing minerals. With 10^{30} prokaryotes on Earth,[1] there is plenty left to be explored. Future directions might include such things as the use of molecular dynamic simulations to corroborate experimentally measured force spectra,[52] or the development of force-based models of ligand–receptor pairs to compliment structural-based models determined with x-ray crystallography and nuclear magnetic resonance.[53,54]

Acknowledgements

This work was supported by grants from the U.S. National Science Foundation and the U.S. Department of Energy.

References

1. Whitman, W. B., Coleman, D. C., and Wiebe, W. J. (1998) Prokaryotes: the unseen majority, *Proc. Natl. Acad. Sci. USA*, **95**, 6578–6583.
2. Binnig, G., Quate, C. F., and Gerber, C. (1986) Atomic force microscope, *Phys. Rev. Lett.*, **56**, 930–933.
3. Weisenhorn, A. L., Hansma, P. K., Albrecht, T. R., and Quate, C. F. (1989) Forces in atomic force microscopy in air and water, *Appl. Phys. Lett.*, **54**, 2651–2653.
4. Ducker, W. A., Senden, T. J., and Pashley, R. M. (1991) Direct measurement of colloidal forces using an atomic force microscope, *Nature*, **353**, 239–241.
5. Ducker, W. A., Senden, T. J., and Pashley, R. M. (1992) Measurements of forces in liquids using a force microscope, *Langmuir*, **8**, 1831–1836.
6. Weisenhorn, A. L., Maivald, P., Butt, H.-J., and Hansma, P. K. (1992) Measuring adhesion, attraction, and repulsion between surfaces in liquids with an atomic-force microscope, *Phys. Rev. B*, **45**, 11226–11232.
7. Israelachvili, J. (1992) *Intermolecular and Surface Forces*, 2nd edn, Academic Press, London.
8. Israelachvili, J. N., and McGuiggan, P. M. (1988) Forces between surfaces in liquids, *Science*, **241**, 795–800.
9. Leckband, D., Israelachvili, J. (2001) Intermolecular forces in biology, *Q. Rev. Biophys.*, **34**, 105–267.
10. Razatos, A., Ong, Y.-L., Sharma, M. M., and Georgiou, G. (1998) Molecular determinants of bacterial adhesion monitored by atomic force microscopy, *Proc. Natl. Acad. Sci. USA*, **95**, 11059–11064.

11. Dufrene, Y. F., Boonaert, C. J. P., Gerin, P. A., Asther, M., and Rouxhet, P. G. (1999) Direct probing of the surface ultrastructure and molecular interactions of dormant and germinating spores of *Phanerochaete chrysosporium*, *J. Bacteriol.*, **181**, 5350–5354.

12. Bowen, W. R., Lovitt, R. W., and Wright, C. J. (2000) Direct quantification of *Aspergillus niger* spore adhesion in liquid using an atomic force microscope, *J. Colloid Interface Sci.*, **228**, 428–433.

13. Camesano, T. A., and Logan, B. E. (2000) Probing bacterial electrosteric interactions using atomic force microscopy, *Environ. Sci. Technol.*, **34**, 3354–3362.

14. Lower, S. K., Tadanier, C. J., and Hochella, M. F. (2000) Measuring interfacial and adhesion forces between bacteria and mineral surfaces with biological force microscopy, *Geochim. Cosmochim. Acta*, **64**, 3133–3139.

15. Lower, S. K., Tadanier, C. J., and Hochella, M. F. (2001) Dynamics of the mineral-microbe interface: use of biological force microscopy in biogeochemistry and geomicrobiology, *Geomicrobiol. J.*, **18**, 63–76.

16. Myers, C. R., and Nealson, K. H. (1988) Bacterial manganese reduction and growth with manganese oxide as the sole electron acceptor, *Science*, **240**, 1319–1321.

17. Arnold, R. G., Hoffmann, M. R., DiChristina, T. J., and Picardal, F. W. (1990) Regulation of dissimilatory Fe(III) reductase activity in *Shewanella putrefaciens*, *Appl. Environ. Microbiol.*, **56**, 2811–2817.

18. Myers, C. R., and Nealson, K. H. (1990) Respiration-linked proton translocation coupled to anaerobic reduction of manganese(IV) and iron(III) in *Shewanella putrefaciens* MR-1, *J. Bacteriol.*, **172**, 6232–6238.

19. Lovley, D. R. (1991) Dissimilatory Fe(III) and Mn(IV) reduction, *Microbiol. Rev.*, **55**, 259–287.

20. Myers, C. R., and Myers, J. M. (1992) Localization of cytochromes to the outer membrane of anaerobically grown *Shewanella Putrefaciens* MR-1, *J. Bacteriol.*, **174**, 3429–3438.

21. Taylor, E. S., and Lower, S. K. (2008) Thickness and surface density of extracellular polymers on *Acidithiobacillus ferrooxidans*, *Appl. Environ. Microbiol.*, **74**, 309–311.

22. Hanna, A., Berg, M., Stout, V., and Razatos, A. (2003) Role of capsular colanic acid in adhesion of uropathogenic *Escherichia coli*, *Appl. Environ. Microbiol.*, **69**, 4474–4481.

23. Kang, S., and Elimelech, M. (2009) Bioinspired single bacterial cell force spectroscopy, *Langmuir*, **25**, 9656–9659.

24. Lower, B. H., Yongsunthon, R., Vellano, F. P., and Lower, S. K. (2005) Simultaneous force and fluorescence measurements of a protein that forms a bond between a living bacterium and a solid surface, *J. Bacteriol.*, **187**, 2127–2137.

25. Abu-Lail, N. I., and Camesano, T. A. (2006) Specific and nonspecific interaction forces between *Escherichia coli* and silicon nitride, determined by Poisson statistical analysis, *Langmuir*, **22**, 7296–7301.

26. Arce, F. T., Carlson, R., Monds, J., Veeh, R., Hu, F. Z., Stewart, P. S., Lal, R., Ehrlich, G. D., and Avci, R. (2009) Nanoscale structural and mechanical properties of nontypeable *Haemophilus influenzae* biofilms, *J. Bacteriol.*, **191**, 2512–2520.

27. Vadillo-Rodrigues, V., Busscher, H. J., Norde, W., De Vries, J., Dijkstra, R. J. B., Stokroos, I., and van der Mei, H. C. (2004) Comparison of atomic force microscopy interaction forces between bacteria and silicon nitride substrata for three commonly used immobilization methods, *Appl. Environ. Microbiol.*, **70**, 5441–5446.

28. Touhami, A., Jericho, M. H., Boyd, J. M., and Beveridge, T. J. (2006) Nanoscale characterization and determination of adhesion forces of *Pseudomonas aeruginosa* pili by using atomic force microscopy, *J. Bacteriol.*, **188**, 370–377.

29. Atabek, A., and Camesano, T. A. (2007) Atomic force microscopy study of the effect of lipopolysaccharides and extracellular polymers on adhesion of *Pseudomonas aeruginosa*, *J. Bacteriol.*, **189**, 8503–8509.

30. Lower, S. K., Hochella, M. F., and Beveridge, T. J. (2001) Bacterial recognition of mineral surfaces: nanoscale interactions between *Shewanella* and α-FeOOH, *Science*, **292**, 1360–1363.

31. Lower, B. H., Hochella, M. F., and Lower, S. K. (2005) Putative mineral-specific proteins synthesized by a metal reducing bacterium, *Am. J. Sci.*, **305**, 687–710.

32. Lower, S. K. (2005) Directed natural forces of affinity between a bacterium and mineral, *Am. J. Sci.*, **305**, 752–765.

33. Lower, B. H., Shi, L., Yongsunthon, R., Droubay, T. C., McCready, D. E., and Lower, S. K. (2007) Specific bonds between an iron oxide surface and outer membrane cytochromes MtrC and OmcA from *Shewanella oneidensis* MR-1, *J. Bacteriol.*, **189**, 4944–4952.

34. Gaboriaud, F., Gee, M. L., Strugnell, R., and Duval, J. F. L. (2008) Coupled electrostatic, hydrodynamic, and mechanical properties of bacterial interfaces in aqueous media, *Langmuir*, **24**, 10988–10995.

35. Cail, T. L., and Hochella, M. F. (2005) The effects of solution chemistry on the sticking efficiencies of viable *Enterococcus faecalis*: an atomic force microscopy and modeling study, *Geochim. Cosmochim. Acta*, **69**, 2959–2969.

36. Yongsunthon, R., and Lower, S. K. (2006) Force spectroscopy of bonds that form between a *Staphylococcus* bacterium and silica or polystyrene substrates, *J. Electron Spectros. Relat. Phenomena*, **150**, 228–234.

37. Touhami, A., Jericho, M. H., and Beveridge, T. J. (2004) Atomic force microscopy of cell growth and division in *Staphylococcus aureus*, *J. Bacteriol.*, **186**, 3286–3295.

38. Emerson, R. J., Bergstrom, T. S., Liu, Y. T., Soto, E. R., Brown, C. A., McGimpsey, W. G., and Camesano, T. A. (2006) Microscale correlation between surface chemistry, texture, and the adhesive strength of *Staphylococcus epidermidis*, *Langmuir*, **22**, 11311–11321.

39. Vadillo-Rodriguez, V., Busscher, H. J., Norde, W., de Vries, J., and van der Mei, H. C. (2003) On relations between microscopic and macroscopic physicochemical

properties of bacterial cell surfaces: an AFM study on *Streptococcus mitis* strains, *Langmuir*, **19**, 2372–2377.

40. Xu, W., Mulhern, P. J., Blackford, B. L., Jericho, M. H., Firtel, M., and Beveridge, T. J. (1996) Modeling and measuring the elastic properties of an archaeal surface, the sheath of Methanospirillum hungatei, and the implication for methane production, *J. Bacteriol.*, **178**, 3106–3112.
41. Bowen, W. R., Lovitt, R. W., and Wright, C. J. (2001) Atomic force microscopy study of the adhesion of *Saccharomyces cerevisiae*, *J. Colloid Interface Sci.*, **237**, 54–61.
42. Scheuring, S., Stahlberg, H., Chami, M., Houssin, C., Rigaud, J. L., and Engel, A. (2002) Charting and unzipping the surface layer of *Corynebacterium glutamicum* with the atomic force microscope, *Mol. Microbiol.*, **44**, 675–684.
43. Higgins, M. J., Crawford, S. A., Mulvaney, P., and Wetherbee, R. (2002) Characterization of the adhesive mucilages secreted by live diatom cells using atomic force microscopy, *Protist*, **153**, 25–38.
44. Considine, R. F., Dixon, D. R., and Drummond, C. J. (2002) Oocysts of *Cryptosporidium parvum* and model sand surfaces in aqueous solutions: an atomic force microscope (AFM) study, *Water Res.*, **36**, 3421–3428.
45. Dugdale, T. M., Willis, A., and Wetherbee, R. (2006) Adhesive modular proteins occur in the extracellular mucilage of the motile, pennate diatom *Phaeodactylum tricornutum*, *Biophys. J.*, **90**, L58–L60.
46. Dugdale, T. M., Dagastine, R., Chiovitti, A., and Wetherbee, R. (2006) Diatom adhesive mucilage contains distinct supramolecular assemblies of a single modular protein, *Biophys. J.*, **90**, 2987–2993.
47. Mueller, H., Butt, H.-J., and Bamberg, E. (1999) Force measurements on myelin basic protein adsorbed to mica and lipid bilayer surfaces done with the atomic force microscope, *Biophys. J.*, **76**, 1072–1079.
48. Muller, D. J., Baumeister, W., and Engel, A. (1999) Controlled unzipping of a bacterial surface layer with atomic force microscopy, *Proc. Natl. Acad. Sci. USA*, **96**, 13170–13174.
49. Oberhauser, A. F., Marszalek, P. E., Carrion-Vazquez, M., and Fernandez, J. M. (1999) Single protein misfolding events captured by atomic force microscopy, *Nat. Struct. Biol.*, **6**, 1025–1028.
50. Heidelberg, J. F., Paulsen, I. T., Nelson, K. E., Gaidos, E. J., Nelson, W. C., Read, T. D., Eisen, J. A., Seshadri, R., Ward, N., Methe, B., Clayton, R. A., Meyer, T., Tsapin, A., Scott, J., Beanan, M., Brinkac, L., Daugherty, S., DeBoy, R. T., Dodson, R. J., Durkin, A. S., Haft, D. H., Kolonay, J. F., Madupu, R., Peterson, J. D., Umayam, L. A., White, O., Wolf, A. M., Vamathevan, J., Weidman, J., Impraim, M., Lee, K., Berry, K., Lee, C., Mueller, J., Khouri, H., Gill, J., Utterback, T. R., McDonald, L. A., Feldblyum, T. V., Smith, H. O., Venter, J. C., Nealson, K. H., and Fraser, C. M. (2002) Genome sequence of the dissimilatory metal ion-reducing bacterium *Shewanella oneidensis*, *Nat. Biotechnol.*, **20**, 1118–1123.
51. Lower, B. H., Yongsunthon, R., Shi, L., Wildling, L., Gruber, H. J., Wigginton, N. S., Reardon, C. L., Pinchuk, G. E., Droubay, T. C., Boily, J. F., and Lower, S. K.

(2009) Antibody recognition force microscopy shows that outer membrane cytochromes OmcA and MtrC are expressed on the exterior surface of *Shewanella oneidensis* MR-1, *Appl. Environ. Microbiol.*, **75**, 2931–2935.

52. Lower, B. H., Lins, R. D., Oestreicher, Z., Straatsma, T. P., Hochella, M. F., Shi, L., and Lower, S. K. (2008) In vitro evolution of a peptide with a hematite binding motif that may constitute a natural metal-oxide binding archetype, *Environ. Sci. Technol.*, **42**, 3821–3827.

53. Yongsunthon, R., Fowler, V. G., Lower, B. H., Vellano, F. P., Alexander, E., Reller, L. B., Corey, G. R., and Lower, S. K. (2007) Correlation between fundamental binding forces and clinical prognosis of *Staphylococcus aureus* infections of medical implants. *Langmuir*, **23**, 2289–2292.

54. Buck, A.W., Fowler, V. G., Yongsunthon, R., Liu, J., DiBartola, A.C., Que, Y.-A., Moreillon, P., and Lower, S. K. (2010) Bonds between fibronectin and fibronectin-binding proteins on *Staphylococcus aureus* and *Lactococcus lactis*. *Langmuir*, **26**, 10764-10770.

Chapter 15

SINGLE-MOLECULE FORCE SPECTROSCOPY OF MICROBIAL CELL ENVELOPE PROTEINS

Claire Verbelen, Vincent Dupres, David Alsteens, Guillaume Andre and Yves F. Dufrêne

Institute of Condensed Matter and Nanosciences, Université catholique de Louvain, Croix du Sud 2/18, B-1348 Louvain-la-Neuve, Belgium

Yves.Dufrene@uclouvain.be

15.1 PROBING THE MICROBIAL CELL ENVELOPE

Most microbes possess a well-defined cell envelope, consisting of a plasma membrane and of a cell wall, that presumably evolved in the course of evolution by selection in response to environmental and ecological pressures.[1] Because the envelope represents the boundary between the external environment and the cell, it plays several important roles, including determining cellular shape, growth and division, enabling the organisms to resist turgor pressure, acting as molecular sieves, interacting with drugs and mediating molecular recognition and cellular interactions.

The functions of the cell envelope are directly related to its composition. The wall mechanical strength in eubacteria is provided by peptidoglycan, consisting of glycan chains cross-linked by short peptide chains.[1,2] Archaebacteria possess stress-bearing wall components which may have different forms: peptidoglycan-like polymers, proteinaceous sheats, crystalline glycoprotein arrays (S-layers). In Gram-positive bacteria, anionic polymers (e.g., teichoic acids) are bound to the cytoplasmic membrane (lipoteichoic acids) and to the peptidoglycan layers (wall teichoic acids), while in Gram-negative bacteria, the thin peptidoglycan layer is overlayed by an outer membrane, i.e., an asymmetrical bilayer of phospholipids and lipopolysaccharides

Life at the Nanoscale: Atomic Force Microscopy of Live Cells
Edited by Yves Dufrêne

www.panstanford.com

containing membrane proteins (e.g., porins). Among Gram-positive bacteria, mycobacteria contain an unusual lipid monolayer (mycolic acids, glycolipids, complex lipids) with inserted porins which mimics the inner leaflet of the outer membrane of Gram-negative bacteria. The mycolic acids are bound to an underlying arabinogalactan polysaccharide layer that is, in turn, linked to peptidoglycan. For many bacterial strains, cell wall constituents are covered by additional surface layers in the form of polysaccharide capsules, surface appendages (fimbriae, pili, fibrils, flagella) or crystalline S-layers. Strong cell walls are formed in yeasts and filamentous fungi by the aggregation of polysaccharide polymers. In yeasts, these are made of a microfibrillar array of β1-3 glucan, overlaid by β1-6 glucan and mannoproteins. The walls of fungal hyphae consist of microfibrillar polysaccharides, chitin or cellulose, covered by layers of proteins and glucans. Fungal spores are often covered by an outer layer of regularly arranged proteins, referred to as rodlets. Although much progress has been made in elucidating the structure and biosynthesis of cell envelope constituents, their three-dimensional organization, assembly, and interactions remain poorly understood at the molecular level.

Since van Leeuwenhoek, microscopy and microbiology have been intimately connected. Light microscopy is a fundamental tool of microbiologists, enabling counting and identification of the cells as well as determination of their general morphological details. Valuable information on the cell wall organization, assembly and dynamics can be obtained using fluorescence microscopy,[3] but the resolution is generally limited to the wavelength of the light source. Our current view of the cell wall ultrastructure essentially relies on the tremendous development of electron microscopy techniques. Elegant techniques have been developed for transmission electron microscopy (TEM) such as the use of freeze-fracture and surface replica to visualize for example cell surface layers, and negative staining for studying purified structures such as flagella and fimbriae.[4,5] These approaches, however, are limited by the requirement of vacuum conditions during the analysis, i.e., native hydrated samples cannot be directly investigated unless sophisticated cryoTEM methods are employed.

Besides microscopy techniques, molecular biology and proteomic approaches have allowed to identify the main components of the cell envelope.[6] Here, difficulties often arise in solubilizing and separating the different constituents. Also, large ensembles of molecules and cells are probe, meaning information at the single-molecule level is not available. Hence, there is a clear need to complement traditional ensemble measurements with non-invasive single-cell and single-molecule techniques.[7,8]

Various force-measuring techniques are available to probe single molecules,[9-12] including flow chamber experiments, microneedles, the biomembrane force probe, the optical and magnetic tweezers and atomic force microscopy (AFM). These assays cover a wide range of forces and length scales that are relevant to biology, going from small intermolecular interactions to strong covalent bonds. As opposed to ensemble techniques, single-molecule experiments can detect, localize and analyze individual biomolecules in heterogeneous populations, thereby revealing rare events that would otherwise be hidden. Notably, owing to its tiny scanning force probe, AFM is the only force technique which can simultaneously localize, manipulate and force probe single-molecules on microbial cells, thereby enabling an important paradigm shift in microbiology. In this chapter, we discuss recent progress made in using AFM to measure the adhesive and mechanical properties of microbial cell envelope proteins.

15.2 BINDING STRENGTH OF CELL ADHESION PROTEINS

Cell adhesion proteins play essential roles in mediating cellular events such as pathogen–host interactions and represent privileged targets for anti-adhesion therapy. Advances in AFM-based single-molecule force spectroscopy (SMFS)[11,12] have allowed researchers to measure the specific binding strength of various cell adhesion molecules (Chapter 11), including selectins,[13,14] cadherins,[15] integrins,[16] proteoglycans[17] and bacterial adhesins.[18] Specific molecular recognition forces are measured by recording force-distance curves between the sample (cells or purified receptors) and an AFM tip modified with appropriate ligands, and then assessing the binding force between complementary molecules.[19] Notably, the spatial distribution of the individual adhesion molecules can be mapped (Chapter 12). To this end, force curves are recorded at multiple locations of the x, y plane to generate a spatially resolved force map in which the adhesion force values are displayed as gray pixels. In single-cell force spectroscopy (Chapter 10),[20,21] cells are attached to the cantilever to measure cell–cell or cell–substrate adhesion forces.

15.2.1 Functionalized Tips

An important prerequisite for successful molecular recognition experiments is to functionalize the AFM tip with ligands or receptors.[19] The forces which immobilize the molecules have to be stronger than the intermolecular force being studied and the attached biomolecules should have enough mobility so that they can freely interact with complementary molecules. It is also recommended to minimize the contribution of non-specific adhesion to the measured forces, and to attach the biomolecules at a low surface density to ensure single-molecule detection.

A good strategy to covalently anchor proteins on tips is to use a polyethylene glycol (PEG) crosslinker which provides motional freedom and prevents denaturation. Tips are first modified with amino groups, further reacted with PEG linkers carrying benzaldehyde functions that are then directly attach to proteins through their lysine residues.[22,23] Another approach is to use self-assembled monolayers of alkanethiols on gold tips.[18,19] Both methods make it possible to orientate the attached biomolecules via their C-terminal or N-terminal domains by linking recombinant histidine-tagged proteins onto tips coated with nitrilotriacetate groups.[18,24]

15.2.2 The NTA–His$_6$ System: A Powerful Platform for SMFS

The site-directed nitrilotriacetic acid (NTA)–polyhistidine (His$_n$) system has recently emerged as a powerful platform for SMFS studies. The NTA–His$_6$ binding chemistry, well known for affinity purification of recombinant proteins, involves the formation of a hexagonal complex between the tetradental ligand NTA and divalent metal ions like Ni^{2+}. Since NTA occupies four of the six coordination sites of Ni^{2+}, the two remaining sites are accessible to other Lewis bases, e.g., the histidines of tagged proteins. In the SMFS context, the NTA–His$_6$ approach offers the important advantage that the attached proteins remain fully functional and can be oriented via their C-terminal or N-terminal His tag. A pertinent question though is to ensure that the NTA–His bond is sufficiently strong for stable protein immobilization during force measurements. This issue was recently clarified by recording force–distance curves between AFM tips modified with Ni^{2+}-NTA-terminated alkanethiols and solid supports functionalized with His$_6$-Gly-Cys peptides (Fig. 15.1).[25] Consistent with the earlier work of Kienberger *et al.*,[24] the adhesion force histogram showed three maxima at rupture forces of 153 ± 57 pN, 316 ± 50 pN and 468 ± 44 pN, attributed to monovalent and multivalent interactions between a single His$_6$ moiety and one, two and three NTA groups, respectively (Fig. 15.1). The plot of adhesion force versus log of the loading rate revealed a linear regime, from which a kinetic off-rate constant of dissociation was deduced. The obtained value was in the range of that estimated for the multivalent interaction involving two NTA, using fluorescence measurements, and may account for an increased binding stability of the NTA–His$_6$ bond. Since the measured forces of ~150 pN are well above the 50–100 pN unbinding forces typically observed for receptor–ligand pairs, it was concluded that the NTA–His$_6$ system is a well-suited platform for the stable, oriented immobilization of proteins in SMFS studies.

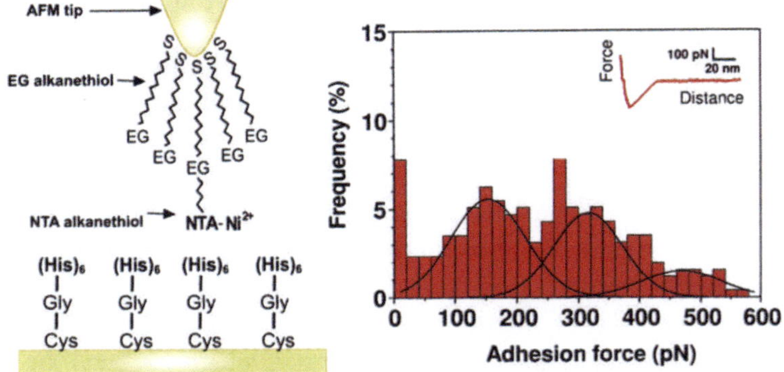

Figure 15.1. AFM force spectroscopy of the NTA–His_6 bond. Force histogram and typical force curve obtained between a NTA tip and a His_6 support in the presence of Ni^{2+}. Left: surface chemistry used to modify tip and support.

15.2.3 Cell Immobilization

For cell experiments, another crucial issue is to attach the cells on a solid support using non-destructive methods.[26] Unlike animal cells, microbes have a well-defined shape and have no tendency to spread on surfaces. As a result, the contact area between a cell and a support is very small, often leading to cell detachment by the scanning tip. Therefore, several approaches have been developed to promote cell attachment. A convenient method makes use of porous polymer membranes.[26,27] This approach allows us to image single bacterial, yeast and fungal cells under aqueous conditions while minimizing denaturation of the surface molecules (Fig. 15.2).

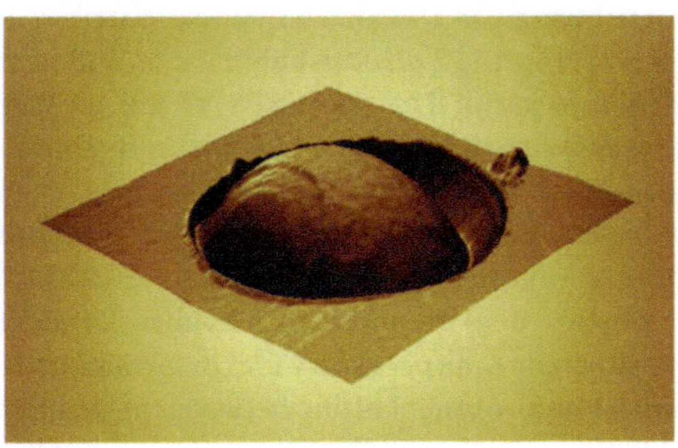

Figure 15.2. Microbial cell imaging by AFM requires attaching the cells firmly onto an appropriate support, which can be achieved by trapping the cells into a porous polymer membrane. Shown here is a single living yeast cell of ~5 μm diameter, decorated with a bud scar.

15.2.4 SMFS of Mycobacterial Cell Adhesion Proteins

Microbial infection is often initiated by the specific adhesion of pathogens to host tissues, via cell adhesion proteins referred to as adhesins. *Mycobacterium tuberculosis*, for instance, adheres to epithelial cells via the heparin-binding haemagglutinin adhesin (HBHA).[28-32] Although the three-dimensional structure of HBHA has not yet been determined, structural predictions suggest that the N-terminal domain of the protein is rich in α helices, whereas the C-terminal lysine-rich region is relatively unstructured.

An interesting treat of HBHA is its ability to work as a multifunctional adhesin. On the one hand, the C-terminal domain, which contains the entire heparin-binding domain, binds to heparan sulphate proteoglycan (HSPG) receptors on target cells. The direct role of HBHA in bacterial adherence to epithelial cells was confirmed using isogenic *M. tuberculosis* mutant strains.[31] *M. tuberculosis* or BCG strains in which the gene encoding HBHA (hbhA) is disrupted expressed reduced adherence to the human type II pneumocyte cell line A549 compared with the respective isogenic parent strain. On the other hand, the N-terminal moiety of HBHA is involved in the formation of multimeric structures and in homophilic HBHA–HBHA interactions.[32] Until recently, the molecular details underlying the multi-adhesive interactions of HBHA remained mysterious.

15.2.4.1 HBHA–heparin interactions

To shed new light into the HBHA binding forces, force curves were recorded between AFM tips modified with HBHA and model surfaces modified with heparin, used as a model sulphated glycoconjugate receptor (Fig. 15.3).[18] The adhesion force histogram revealed a bimodal distribution with average rupture forces of 50 pN and 117 pN, attributed to one and two binding events between HBHA and heparin. The specificity of the measured interaction was confirmed by showing a dramatic reduction in the number of adhesion events when working in the presence of free heparin. Both the adhesion frequency and adhesion force increased with contact time, consistent with the formation of multiple intermolecular bridges between HBHA and its receptor. The prolonged contact time required to establish strong HBHA–heparin interaction may reflect the time necessary for conformational changes within both molecules to allow an optimal fitting between the positive charges of the HBHA heparin-binding domain and the sulphate groups of heparin.

Next, the distribution of single HBHA was mapped on the surface of living mycobacteria, using heparin-modified tips.[18] High-resolution images of mycobacteria revealed a smooth and homogeneous surface, consistent with earlier scanning electron microscopy observations. Adhesion force maps recorded on cells with heparin tips revealed adhesion events in about half of the locations. The adhesion force magnitude was very close to the value expected for single HBHA–heparin interactions, indicating single HBHA were detected. This was confirmed by showing that a mutant strain lacking HBHA did not bind the heparin tip. Interestingly, the HBHA distribution was not homogeneous, but apparently concentrated into nanodomains which may promote adhesion to target cells by inducing the recruitment of receptors within membrane rafts. Besides providing novel molecular insights into cell adhesion mechanisms, we anticipate that, in the near future, these single-molecule recognition studies may help in the development of new drugs capable of blocking bacterial adhesion.

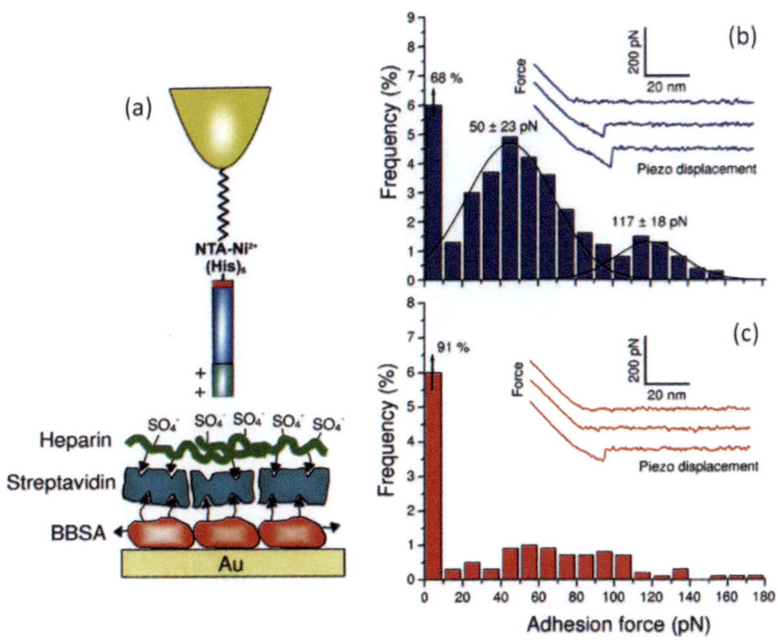

Figure 15.3. SMFS of the HBHA–heparin interaction. (a) Schematics of the surface chemistry used to functionalize the AFM tip and substrate with HBHA and heparin. Recombinant histidine-tagged HBHA were attached onto an Ni^{2+}-NTA tip, while biotinylated heparin was bound to a gold surface via streptavidin and biotinylated bovine serum albumin layers. (b) Representative force curves and adhesion force histogram obtained in PBS between a HBHA tip and a heparin surface. The adhesion force histogram revealed a bimodal distribution reflecting the binding strength of one and two adhesins. (c) Same experiment in the presence of free heparin (50 μg/ml) demonstrating a dramatic reduction of adhesion frequency due to the blocking of the HBHA adhesion sites. Adapted with permission from Dupres *et al.*[18]

15.2.4.2 HBHA–HSPG interactions

The next question we addressed is whether similar interactions occur between HBHA and HSPG receptors on host cells? To this end, we measured the forces between HBHA tips and living A549 pneumocyte cells (Fig. 15.4).[33] AFM imaging revealed that A549 cells were typically 50 μm in diameter and showed two prominent features, i.e., the central round nucleus surrounded by the flattened cytoplasm and membrane, and the underlying cytoskeleton structures (actin filaments). The specific binding force measured at moderate pulling velocity between single HBHA–HSPG pairs was about 50 pN, which

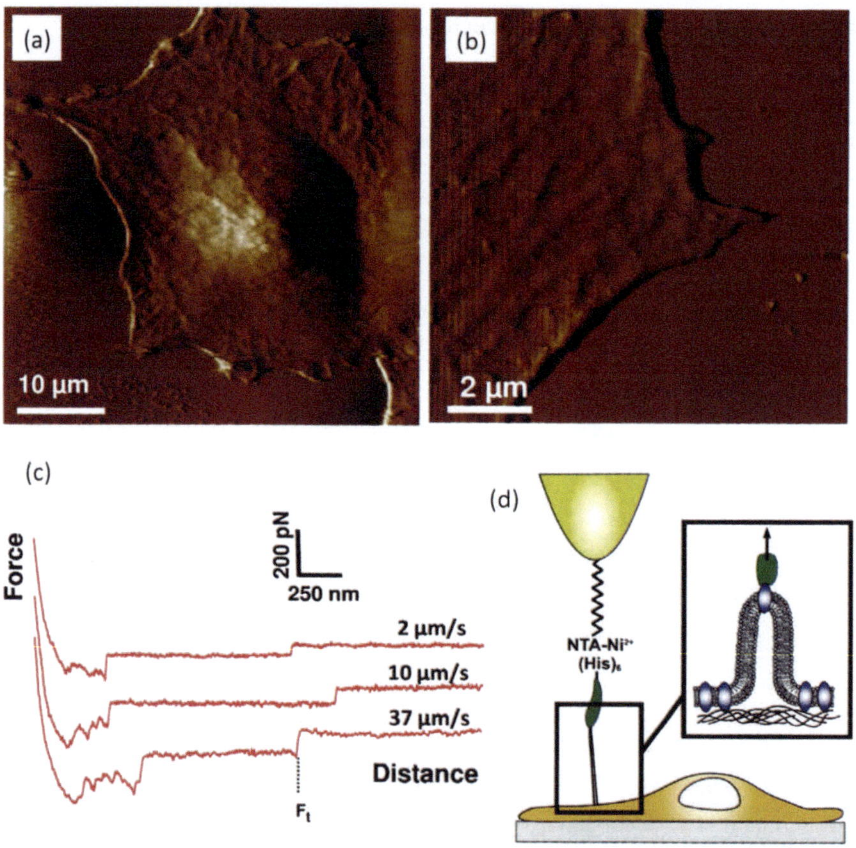

Figure 15.4. SMFS of the HBHA–HSPG interaction. (a, b) Deflection images of live A549 pneumocytes. (c) Representative force curves recorded between a HBHA tip and a A549 cell showing constant force plateaus. The tether extraction force F_t increases with the pulling velocity. (d) Suggested mechanism: the stressed receptors detached from the cytoskeleton, leading to the extraction of membrane tethers. Reprinted with permission from Dupres *et al.*[33]

is similar to the forces measured between HBHA and heparin molecules. Adhesion maps recorded across A549 cells showed fairly homogeneous contrast, indicating that the receptors were widely and homogeneously exposed. The specificity of the measured interaction was confirmed by showing a dramatic reduction of both the adhesion frequency and adhesion force values when the cell surface was treated with heparinase. Strikingly, at large pulling velocities, constant force plateaus were seen in most curves (Fig. 15.4c). This indicated that stressed HSPG receptors detached from the cytoskeleton, therefore leading to the extraction of membrane tethers or nanotubes (Fig. 15.4d). These membrane structures have been observed in liposomes and different cell types, including red blood cells, neutrophils, neurons, fibroblasts as well as mesendoderm, epithelial and endothelial cells.[34,35] Tether formation may play a role in pathogen–host interactions since the invasion mechanisms of pathogens such as *Salmonella* and *Shigella* are known to involve the production of large membrane projections and the formation of membrane-bound vacuoles.

15.2.4.3 Homophilic HBHA–HBHA interactions

The N-terminal domains of HBHA contain a predicted coiled-coil region involved in homophilic interactions that may potentially contribute to bacterial aggregation and to the formation of polymeric HBHA structures.[32] Until now, detailed information on the forces of such homophilic interactions was lacking. In this context, SMFS could reveal the molecular forces driving HBHA–HBHA interactions, both on model surfaces and on live mycobacteria.[36] Histidine-tagged proteins were attached, via their C-terminal or N-terminal end, to gold-coated AFM tips and supports (Fig. 15.5a). Force–distance curves recorded between HBHA exposing their N-terminal regions showed a bimodal distribution cantered at 68 ± 2 pN and 130 ± 14 pN (Fig. 15.5b). These maxima were attributed to the formation of one or two HBHA dimers, resulting from specific coiled-coils interactions. Most adhesion peaks displayed non-linear elongation forces that were best described by the worm-like chain (WLC) model, classically used to model the unfolding of polypeptide chains. Hence, elongation forces were attributed to the unfolding of α-helices of the coiled-coil domain.

When the forces between the lysine-rich C-terminal domains were measured, binding events showed a much broader distribution. The lack of a well-defined maximum, together with the larger average binding force values, suggests these forces do not primarily originate from specific coiled-coil interactions. Rather, the main contribution is probably due to multiple intermolecular electrostatic bridges between the cationic groups of the

C-terminal, lysine-rich region and anionic aspartates/glutamates of the protein.

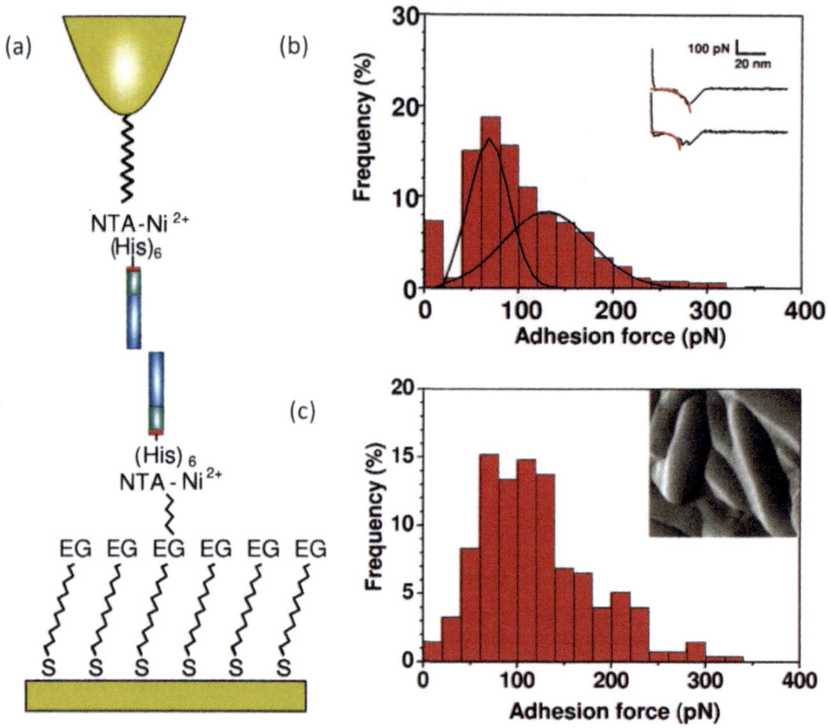

Figure 15.5. SMFS of the HBHA–HBHA interaction. (a) Surface chemistry used to measure the forces between HBHA exposing their N-terminal regions. (b) Adhesion force histogram and representative force curves measured between the N-terminal domains. Elongation forces were generally well described by the WLC model (red lines). (c) Similar data were obtained between a tip exposing the HBHA N-terminal region and living mycobacteria. The inset shows an AFM image of a few mycobacteria.

Homophilic HBHA interactions were also directly measured on the surface of living mycobacteria (Fig 15.5c). Force curves recorded over the bacterial surface with a tip exposing N-terminal regions exhibited a distribution reminiscent of that observed for model surfaces exposing N-terminal tails. Adhesion forces were therefore attributed to multimer formation due to specific coiled-coil interactions. The tip exposing C-terminal regions also showed binding forces in most areas, but a broader distribution was observed, suggesting these forces originate from electrostatic interactions. Adhesion forces were never detected on a bacterial mutant impaired in HBHA production.

These nanoscale measurements were shown to correlate with microscale mycobacterial aggregation assays. Cultures of native bacteria and of a mutant lacking HBHA were incubated with growing concentrations of HBHA and observed with an optical microscope. The addition of HBHA to the cell suspensions induced cellular aggregation in a dose-dependent manner. By contrast, the mutant strain did not significantly aggregate following addition of HBHA. These observations support the notion that mycobacterial aggregation involves homophilic HBHA–HBHA interactions, measured here for the first time at the single-molecule level.

In summary, the data surveyed here indicate that SMFS offers new avenues for elucidating the molecular mechanisms of bacterial adhesion, e.g., in the context of infection diseases, and for developing new anti-adhesion strategies for therapy.

15.3 MECHANICAL PROPERTIES OF CELL SURFACE PROTEINS

Another exciting area where SMFS has great promise is the elucidation of the folding and molecular elasticity of cell surface proteins. Pioneering studies showed the ability of SMFS to stretch and manipulate bacterial membrane proteins, thereby providing details of their unfolding pathways and of the forces that anchor them into the membrane.[37–39] These experiments were performed on membranes that were removed from the cellular environment which controls the protein assembly and functional state. Consequently, studying the molecular elasticity of proteins in living cells remains very challenging.

15.3.1 Unfolding Adhesion Proteins on Living Yeast Cells

SMFS was recently used to unfold single agglutinin-like sequence (Als) cell adhesion proteins from *Candida albicans*.[40] Als proteins possess four functional regions (Fig. 15.6a), i.e., an N-terminal immunoglobulin (Ig)-like region, which initiates cell adhesion, followed by a threonine-rich region (T), a tandem repeat (TR) region that participates in cell–cell aggregation and a stalk region projecting the molecule away from the cell surface. Soluble Als fragments containing six TR domains were attached on gold surfaces and picked up by their terminal Ig domain using an AFM tip (Fig. 15.6b). Force-extension curves showed sawtooth patterns with well-defined force peaks, each peak corresponding to the force-induced unfolding of an individual TR domain (Fig. 15.6c). Force peaks were well described by the WLC model, supporting further the interpretation of the sawtooth pattern. For the first six peaks, the change in contour length between consecutive peaks was 8.4 nm, which corresponds to the lengthening of the 36 amino acids of a single repeat. Urea strongly altered the shape of the unfolding peaks, confirming that disruption of the protein hydrogen bonds leads to a loss of mechanical stability. This observation correlates with the cellular behaviour since Als5p-mediated adhesion has been shown to be reversibly inhibited by urea and formamide.

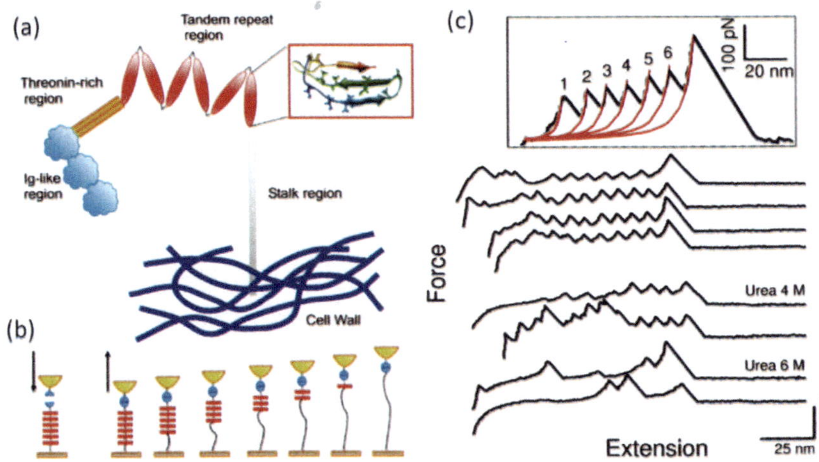

Figure 15.6. Unfolding single cell adhesion proteins. (a) Als5p contains a tandem repeat (TR) region comprising multiple glycosylated 36-amino acid repeats that are arranged in anti-parallel β-sheets. (b) Ig-T-TR_6 fragments were attached on a gold surface and stretched via their Ig domains using an Ig-T tip. (c) Force-extension curves obtained by stretching single Ig-T-TR_6 showed periodic features reflecting the sequential unfolding of the TR domains (upper traces). Force peaks were well described by the WLC model (inset; red line). Addition of urea dramatically altered the unfolding peaks (lower traces). Reprinted with permission from Alsteens *et al.*[40]

Remarkably, single Als proteins could also be unfolded on live cells. Force curves obtained on yeast cells expressing six repeats displayed sawtooth patterns similar to those found on isolated proteins, while cells expressing no repeat were unable to bind the AFM tip. The unfolding probability increased with the number of repeats and was correlated with the level of cell–cell adhesion, indicating these modular domains may play a role in fungal adhesion. The modular and flexible nature of Als conveys both strength and toughness to the protein, making it ideally suited for cell adhesion. These single-molecule measurements provide novel insights into the mechanical properties of adhesion molecules and may help us to elucidate their potential implication in diseases.

15.3.2 Measuring the Spring Behaviour of Yeast Membrane Sensors

Mechanosensors in living cells convert mechanical forces into biochemical signals.[41] In yeast, surface stresses acting on the cell wall and plasma membrane are detected by a group of five membrane sensors, i.e., Wsc1, Wsc2, Wsc3, Mid2 and Mtl2. Although much is known about the genetics and molecular biology of the sensors, how they probe extracellular signals remains mysterious. It is believed that these membrane proteins act as mechanosensors, activating stress pathways in response to physical changes in the cell wall. Yet, direct evidence for such a mechanism had never been provided.

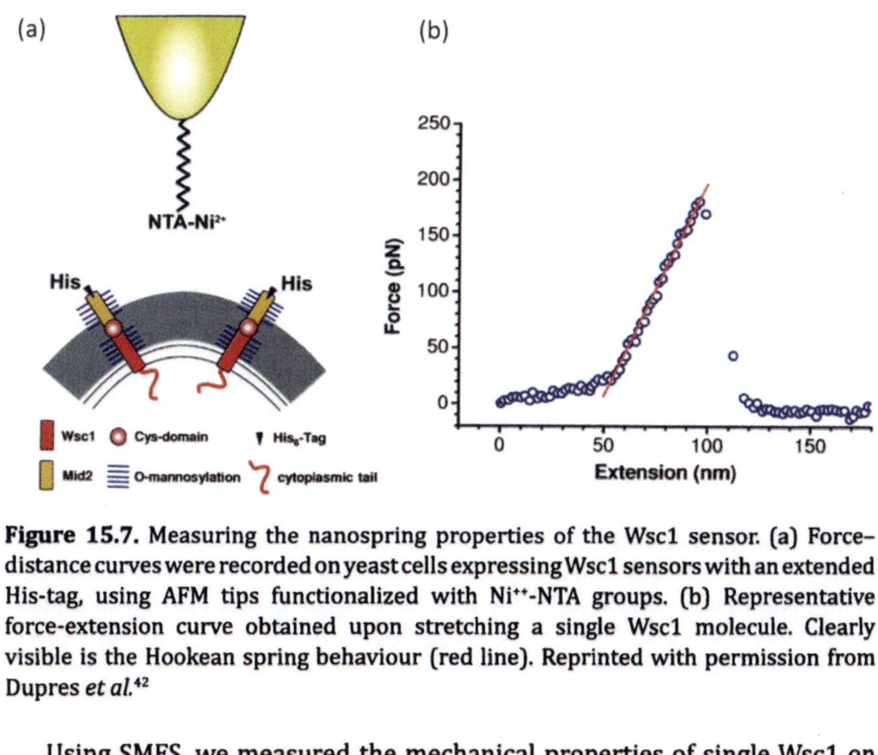

Figure 15.7. Measuring the nanospring properties of the Wsc1 sensor. (a) Force–distance curves were recorded on yeast cells expressing Wsc1 sensors with an extended His-tag, using AFM tips functionalized with Ni^{++}-NTA groups. (b) Representative force-extension curve obtained upon stretching a single Wsc1 molecule. Clearly visible is the Hookean spring behaviour (red line). Reprinted with permission from Dupres *et al.*[42]

Using SMFS, we measured the mechanical properties of single Wsc1 on live cells (Fig. 15.7).[42] Genetic manipulations could solve a major experimental constraint: in essence AFM is a surface technique, so how can it probe sensors that are embedded within the cell wall? Simple calculations indicate that native Wsc1 sensors extend ~80 nm above the plasma membrane. As the cell wall is ~110 nm thick, this means native sensors cannot reach the outermost cell surface. To elongate the molecule, extended Wsc1 proteins were designed by adding the extracellular part of the Mid2 protein (Fig. 15.7a). In addition, a His-tag was inserted to allow for specific detection with AFM tips terminated with NTA groups. With this strategy, single His-tagged elongated sensors were detected on *Saccharomyces cerevisiae*. Adhesion force maps revealed the localization of individual proteins, therefore confirming they were long enough to reach the cell surface. By contrast, His-tagged Wsc1 that were not elongated could not be detected, except in bud scars.

Notably, stretching single sensors revealed they behave like nanosprings, capable to resist high mechanical force without undergoing secondary structure unfolding (Fig. 15.7b). The sensor spring constant was estimated to be ~5 pN nm^{-1}, which is very close to the behaviour of ankyrin repeats. Lowering the salt concentration or increasing temperature resulted in a substantial reduction of the sensor spring constant, indicating that Wsc1 is sensitive to cell surface stress. Both a genomic pmt4 deletion and the insertion of a stretch of glycines in Wsc1 resulted in severe alterations in protein spring properties, supporting the important role of glycosylation at the extracellular serine/threonine-rich region. These findings have pharmaceutical implications since drugs used in the treatment of fungal infections are often directed against the protective fungal cell wall.

In the future, SMFS may help understanding how cell wall integrity is maintained or altered upon interaction of the microbes with drugs. More broadly, the combined method of genetic design and single-molecule measurements used in this study has great potential for investigating how proteins respond to forces in living cells and how mechanosensing events proceed *in vivo*.

15.4 CONCLUSION

Our current view of microbial cell envelopes owes much to the development of electron microscopy techniques. Yet, these methods cannot probe living cells in buffer solution. The data surveyed here demonstrate the power of AFM for imaging and force probing live cells down to molecular resolution. These single-molecule assays complement traditional proteomic and molecular biology approaches for the functional analysis of membrane and cell wall proteins and may help in the search for novel anti-microbial drugs. In particular, we anticipate that SMFS will find valuable applications in clinical microbiology and pathogenesis to investigate the interactions between microbial pathogens and host cells and to localize cell surface receptors, which may eventually help developing new therapeutic approaches.

Today, most commercial AFMs are user friendly, and reliable protocols are available for functionalizing tips and for immobilizing microbial cells. Also, procedures to measure the localization, binding strength and nanomechanics of cell surface molecules are well established. Nevertheless, newcomers should realize that accurate data collection and interpretation require strong expertise and a lot of patience, especially when analyzing complex cellular samples. A central issue is to be sure that functionalized tips are of good quality, permitting reliable and reproducible single-molecule measurements, and that their integrity is preserved during the course of the experiment.

A crucial challenge for future microbiological research will be to combine SMFS with other advanced scanning probe modalities, such as high-speed imaging (Chapter 8), single-cell force spectroscopy (Chapter 10) and near-field scanning optical microscopy (Chapter 9). Combining these methods with light microscopy techniques should provide novel insight into the organization, assembly and dynamics of cell walls. In particular, optical nanoscopy, in which the resolution is no longer limited by the wavelength of light, should allow us to resolve cellular surface structures with a few nanometre dimensions.[43,44]

Acknowledgements

Work in our team was supported by the National Foundation for Scientific Research (FNRS), the Foundation for Training in Industrial and Agricultural Research (FRIA), the Université catholique de Louvain (Fonds Spéciaux de Recherche), the Région wallonne, the Federal Office for Scientific, Technical and Cultural Affairs (Interuniversity Poles of Attraction Programme) and the Research Department of the Communauté française de Belgique (Concerted Research Action).

References

1. Beveridge, T. J., and Graham, L. L. (1991) Surface layers of bacteria, *Microbiol. Rev.*, **55**, 684–705.
2. Cabeen, M. T., and Jacobs-Wagner, C. (2005) Bacterial cell shape, *Nat. Rev. Microbiol.*, **3**, 601–610.
3. Daniel, R. A., and Errington, J. (2003) Control of cell morphogenesis in bacteria: two distinct ways to make a rod-shaped cell, *Cell*, **113**, 767–776.
4. Beveridge, T. J. (1999) Structures of gram-negative cell walls and their derived membrane vesicles, *J. Bacteriol.*, **181**, 4725–4733.

5. Matias, V. R. F., and Beveridge, T. J. (2005) Cryo-electron microscopy reveals native polymeric cell wall structure in *Bacillus subtilis* 168 and the existence of a periplasmic space, *Mol. Microbiol.*, **56**, 240–251.
6. Poetsch, A., and Wolters, D. (2008) Bacterial membrane proteomics, *Proteomics*, **8**, 4100–4122.
7. Brehm-Stecher, B. F., and Johnson, E. A. (2004) Single-cell microbiology: tools, technologies, and applications, *Microbiol. Mol. Biol. Rev.*, **68**, 538–559.
8. Dufrêne, Y. F. (2008) Towards nanomicrobiology using atomic force microscopy, *Nat. Rev. Microbiol.*, **6**, 674–680.
9. Bustamante, C., Macosko, J. C., and Wuite, G. J. L. (2000) Grabbing the cat by the tail: manipulating molecules one by one, *Nat. Rev. Mol. Cell Biol.*, **1**, 130–136.
10. Sotomayor, M., and Schulten, K. (2007) Single-molecule experiments in vitro and in silico, *Science*, **316**, 1144–1148.
11. Müller, D. J., and Dufrêne, Y. F. (2008) Atomic force microscopy as a multifunctional molecular toolbox in nanobiotechnology, *Nat. Nanotechnol.*, **3**, 261–269.
12. Müller, D. J., Helenius, J., Alsteens, D., and Dufrêne, Y. F. (2009) Force probing surfaces of living cells to molecular resolution, *Nat. Chem. Biol.*, **5**, 383–390.
13. Fritz, J., Katopodis, A. G., Kolbinger, F., and Anselmetti, D. (1998) Force-mediated kinetics of single P-selectin/ligand complexes observed by atomic force microscopy, *Proc. Natl. Acad. Sci. USA*, **95**, 12283–12288.
14. Zhang, X. H, Bogorin, D. F., and Moy, V. T. (2004) Molecular basis of the dynamic strength of the sialyl Lewis X-selectin interaction, *Chem. Phys. Chem.*, **5**, 175–182.
15. Baumgartner, W., Hinterdorfer, P., Ness, W., Raab, A., Vestweber, D., Schindler, H., and Drenckhahn, D. (2000) Cadherin interaction probed by atomic force microscopy, *Proc. Natl. Acad. Sci. USA*, **97**, 4005–4010.
16. Zhang, X. H., Wojcikiewicz, E., and Moy, V. T. (2002) Force spectroscopy of the leukocyte function-associated antigen-1/intercellular adhesion molecule-1 interaction, *Biophys. J.*, **83**, 2270–2279.
17. Dammer, U., Popescu, O., Wagner, P., Anselmetti, D., Guntherodt, H. J., and Misevic G. N. (1995) Binding strength between cell adhesion proteoglycans measured by atomic force microscopy, *Science*, **267**, 1173–1175.
18. Dupres, V., Menozzi, F. D., Locht, C., Clare, B. H., Abbott, N. L., Cuenot, S., Bompard, C., Raze, D., and Dufrêne, Y. F. (2005) Nanoscale mapping and functional analysis of individual adhesins on living bacteria, *Nat. Methods*, **2**, 515–520.
19. Hinterdorfer, P., and Dufrêne, Y. F. (2006) Detection and localization of single molecular recognition events using atomic force microscopy, *Nat. Methods*, **3**, 347–355.
20. Benoit, M., Gabriel, D., Gerisch, G., and Gaub, H. E. (2000) Discrete interactions in cell adhesion measured by single-molecule force spectroscopy, *Nat. Cell Biol.*, **2**, 313–317.

21. Helenius, J., Heisenberg, C. P., Gaub, H. E., and Muller, D. J. (2008) Single-cell force spectroscopy, *J. Cell Sci.*, **121**, 1785–1791.

22. Hinterdorfer, P., Baumgartner, W., Gruber, H. J., Schilcher, K., and Schindler, H. (1996) Detection and localization of individual antibody-antigen recognition events by atomic force microscopy, *Proc. Natl. Acad. Sci. USA*, **93**, 3477–3481.

23. Riener, C. K., Stroh, C. M., Ebner, A., Klampfl, C., Gall, A. A., Romanin, C., Lyubchenko, Y. L., Hinterdorfer, P., and Gruber, H. J. (2003) Simple test system for single molecule recognition force microscopy, *Anal. Chim. Acta*, **479**, 59–75.

24. Kienberger, F., Kada, G., Gruber, H. J., Pastushenko, V. Ph., Riener, C., Trieb, M., Knaus, H. G., Schindler, H., and Hinterdorfer, P. (2000) Recognition force spectroscopy studies of the NTA-His6 bond, *Single Mol.*, **1**, 59–65.

25. Verbelen, C., Gruber, H. J., and Dufrêne, Y. F. (2007) The NTA-His6 bond is strong enough for AFM single-molecular recognition studies, *J. Mol. Recognit.*, **20**, 490–494.

26. Dufrêne, Y. F. (2008) Atomic force microscopy and chemical force microscopy of microbial cells, *Nat. Protoc.*, **3**, 1132–1138.

27. Kasas, S., and Ikai, A. (1995) A method for anchoring round shaped cells for atomic force microscope imaging, *Biophys. J.*, **68**, 1678–1680.

28. Menozzi, F. D., Rouse, J. H., Alavi, M., LaudeSharp, M., Muller, J., Bischoff, R., Brennan, M. J., Locht, C. (1996) Identification of a heparin-binding hemagglutinin present in mycobacteria, *J. Exp. Med.*, **184**, 993–1001.

29. Menozzi, F. D., Bischoff, R., Fort, E., Brennan, M. J., and Locht, C. (1998) Molecular characterization of the mycobacterial heparin-binding hemagglutinin, a mycobacterial adhesion, *Proc. Natl. Acad. Sci. USA*, **95**, 12625–12630.

30. Pethe, K., Aumercier, M., Fort, E., Gatot, C., Locht, C., and Menozzi, F. D. (2000) Characterization of the heparin-binding site of the mycobacterial heparin-binding hemagglutinin adhesion, *J. Biol. Chem.*, **275**, 14273–14280.

31. Pethe, K., Alonso, S., Biet, F., Delogu, G., Brennan, M. J., Locht, C., and Menozzi, F. D. (2001) The heparin-binding haemagglutinin of *M. tuberculosis* is required for extrapulmonary dissemination, *Nature*, **412**, 190–194.

32. Delogu, G., and Brennan, M. J. (1999) Functional domains present in the mycobacterial hemagglutinin HBHA, *J. Bacteriol.*, **181**, 7464–7469.

33. Dupres, V., Verbelen, C., Raze, D., Lafont, F., and Dufrêne, Y. F. (2009) Force spectroscopy of the interaction between mycobacterial adhesins and heparan sulphate proteoglycan receptors, *ChemPhysChem*, **10**, 1672–1675.

34. Sun, M. Z., Graham, J. S., Hegedus, B., Marga, F., Zhang, Y., Forgacs, G., and Grandbois, M. (2005) Multiple membrane tethers probed by atomic force microscopy, *Biophys. J.*, **89**, 4320–4329.

35. Krieg, M., Helenius, J., Heisenberg, C. P., and Muller D. J. (2008) A bond for a lifetime: employing membrane nanotubes from living cells to determine receptor-ligand kinetics, *Angew. Chem. Int. Ed. Engl.*, **47**, 9775–9777.

36. Verbelen, C., Raze, D., Dewitte, F., Locht, C., and Dufrêne, Y. F. (2007) Single-molecule force spectroscopy of mycobacterial adhesin-adhesin interactions, *J. Bacteriol.*, **189**, 8801–8806.
37. Müller, D. J., Baumeister, W., and Engel, A. (1999) Controlled unzipping of a bacterial surface layer with atomic force microscopy, *Proc. Natl. Acad. Sci. USA*, **96**, 13170–13174.
38. Oesterhelt, F., Oesterhelt, D., Pfeiffer, M., Engel, A., Gaub, H. E., and Müller, D. J. (2000) Unfolding pathways of individual bacteriorhodopsins, *Science*, **288**, 143–146.
39. Scheuring, S., Stahlberg, H., Chami, M., Houssin, C., Rigaud, J. L., and Engel, A. (2002) Charting and unzipping the surface layer of *Corynebacterium glutamicum* with the atomic force microscope. *Mol. Microbiol.*, **44**, 675–684.
40. Alsteens, D., Dupres, V., Klotz, S. A., Gaur, N. K., Lipke, P. N., and Dufrêne, Y. F. (2009) Unfolding individual Als5p adhesion proteins on live cells, *ACS Nano*, **3**, 1677–1682.
41. Vogel, V., and Sheetz, M. (2006) Local force and geometry sensing regulate cell functions, *Nat. Rev. Mol. Cell Biol.*, **7**, 265–275.
42. Dupres, V., Alsteens, D., Wilk, S., Hansen, B., Heinisch, J. J., and Dufrêne, Y. F. (2009) The yeast Wsc1 cell surface sensor behaves like a nanospring *in vivo*, *Nat. Chem. Biol.*, **5**, 857–862.
43. Hell, S. W. (2007) Far-field optical nanoscopy, *Science*, **316**, 1153–1158.
44. Hell, S. W. (2009) Microscopy and its focal switch, *Nat. Methods*, **6**, 24–32.

Chapter 16

PROBING THE NANOMECHANICAL PROPERTIES OF VIRUSES, CELLS AND CELLULAR STRUCTURES

Sandor Kasas[a,b] and Giovanni Dietler[a]

[a] *Laboratoire de Physique de la Matière Vivante, Ecole Polytechnique Fédérale de Lausanne, CH-1015 Lausanne, Switzerland*

[b] *Département de Biologie Cellulaire et de Morphologie, Université de Lausanne, CH-1005 Lausanne, Switzerland*

Sandor.Kasas@EPFL.CH

16.1 ELASTICITY

16.1.1 Ideal Solids

Elasticity describes the ability of a material to recover its original shape after the withdrawal of an external deformation force. If the force (stress) is sufficiently small, then it is proportional to the deformation (strain). The elastic modulus of a body is expressed as the ratio of stress to strain:

$$\text{Elastic modulus} = \frac{\text{Stress}}{\text{Strain}} \tag{16.1}$$

16.1.2 Length Deformation

If a force is applied longitudinally to a body, the body will deform according to Eq. (16.1). Stress and strain are then qualified by the term tensile, and the elastic modulus, named Young's modulus (E), is given by

$$E = \frac{\text{tensile stress}}{\text{tensile strain}} \tag{16.2}$$

Life at the Nanoscale: Atomic Force Microscopy of Live Cells
Edited by Yves Dufrêne

www.panstanford.com

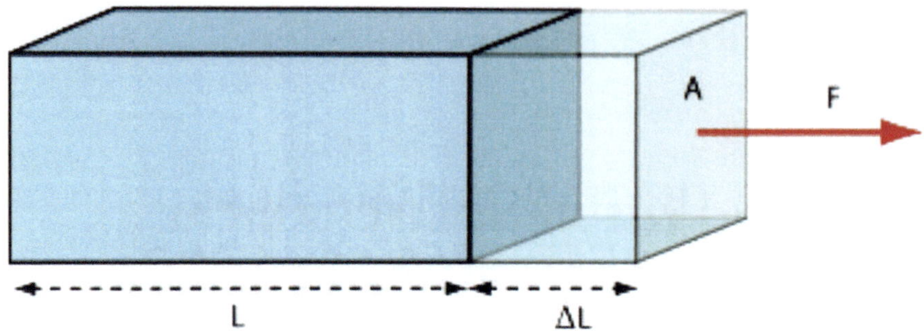

Figure 16.1. The relationship between force and deformation in a homogeneous, isotropic solid.

The behaviour of a metallic spring is a classical example of this phenomenon, which was first described in 1675 by Robert Hooke as "*Ut tensio, sic vis*" ("As the extension, so the force"). The relationship between force and deformation is most readily appreciated by considering a homogeneous, isotropic solid (viz., a solid with uniform mechanical properties in all directions), with the cross-sectional area of a square (see Fig. 16.1). If a small tensile force (F) is applied to the cross-sectional area (A), then the strain (viz., the relative length change [ΔL]) is proportional to the stress (viz., the force per unit area or pressure):

$$\frac{F}{A} = E\frac{\Delta L}{L} \qquad (16.3)$$

where F is applied force; A, cross-sectional area; ΔL, extension; L, length of the sample; and E, Young's modulus.

The proportionality constant in Eq. (16.3) is the Young's or the elastic modulus (E). It is expressed in Newtons per square metre (N/m^2) or Pascals (Pa). It should be borne in mind that Young's modulus is a material property which does not depend upon the size or the shape. Young's moduli for various materials are given in Table 16.1.

Table 16.1. Young's moduli for various materials

Material	Young's modulus (GPa)
Diamond	1200
Stainless steel	211
Glass	73
Wood	16
Plexiglas	3
Rubber	0.02

Actin	2.3
Collagen	2
Tubulin	1.9
Living cells	0.001–0.1

The Poisson's ratio (ν) expresses the shrinkage perpendicular to the elongation direction. More explicitly, it is the ratio between the transverse (ε_{trans}) and the axial strain (ε_{axial}):

$$\nu = \frac{\varepsilon_{trans}}{\varepsilon_{axial}} \tag{16.4}$$

Since strain is a dimensionless number ($\Delta L/L$ in Fig. 16.1), it follows that the Poisson's ratio is likewise dimensionless. Its value usually lies between 0.5 (incompressible rubber) and 0.3 (most metals). Cork has a Poisson's ratio close to 0, which accounts for its use as a plug for wine bottles.

16.1.3 Shear Deformation

Another important concept in elasticity is shear stress, which measures the tendency of one part of a solid to slide past a neighbouring portion, as depicted in Fig. 16.2.

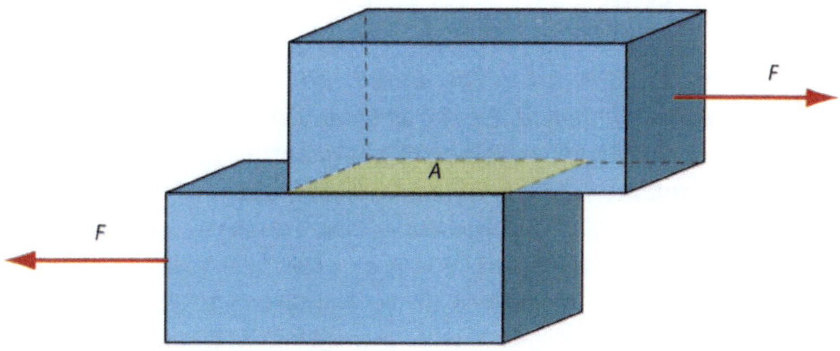

Figure 16.2. Shear stress in a solid.

Shear stress (N) is defined as the ratio between a force and a surface.

$$N = \frac{F}{A} \tag{16.5}$$

where N is shear stress (N/m^2); F, force (N); and A, area (m^2).

The shear strain [$\tan(\theta)$], as depicted in Fig. 16.3, is described as the ratio of x/h. For small deformations, $\tan(\theta) \approx \theta$ and is usually measured in radians.

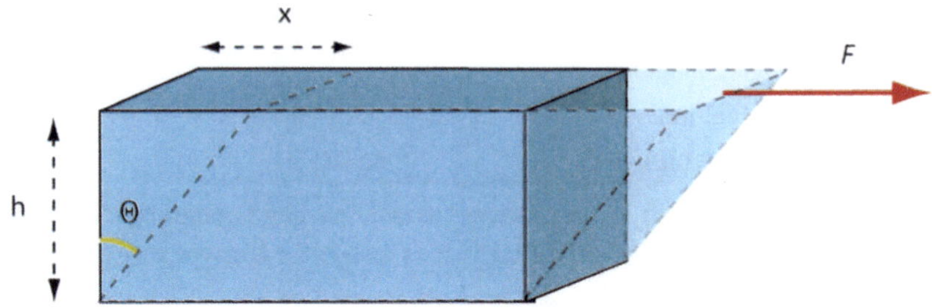

Figure 16.3. Shear deformation of a solid.

Shear modulus (G) is described as the ratio between shear stress (N) and shear strain (θ):

$$G = \frac{N}{\theta} \tag{16.6}$$

Shear modulus (G) is thus an analogue of Young's modulus and is expressed as N/m^2 or in Pascals. The shear moduli of steel (79 GPa) and rubber (0.02 GPa) represent two examples at opposite ends of the spectrum.

16.2 MEASURING THE YOUNG'S MODULUS WITH THE ATOMIC FORCE MICROSCOPE

Very soon after its invention, the atomic force microscope (AFM) was shown to be capable of measuring the mechanical properties of microscopic samples. The measurement is achieved by indenting (pushing) the AFM tip into the sample and monitoring online the deformation of the cantilever. The graph depicting the vertical deformation of the cantilever as a function of the tip-sample distance is usually referred to as a force–distance (FD) curve. The "force" is assumed to be equivalent to the cantilever deformation, since the behaviour of the cantilever is generally considered to follow linear elasticity rules, viz., the vertical deflection is deemed to be directly proportional to the force that is applied to the sample or to the force that the cantilever exerts on the sample. Figure 16.4 depicts the shapes of two typical FD curves: one for a hard, and the other for a soft sample.

The horizontal green line corresponds to the off-contact region of the curve, viz., to the distance between the sample and the AFM tip. In this off-contact region, the cantilever senses no force and is thus not deformed; it maintains its resting position. After the tip has touched the surface of the sample, the cantilever will deform according to the sample's stiffness. If the sample is hard, no indentation will occur and the path of the FD curve will be a straight line at an inclination of 45° (blue line in Fig. 16.4). If the sample is soft, the tip will indent the sample and the FD curve will be flatter and non-linear (orange curve in Fig. 16.4).

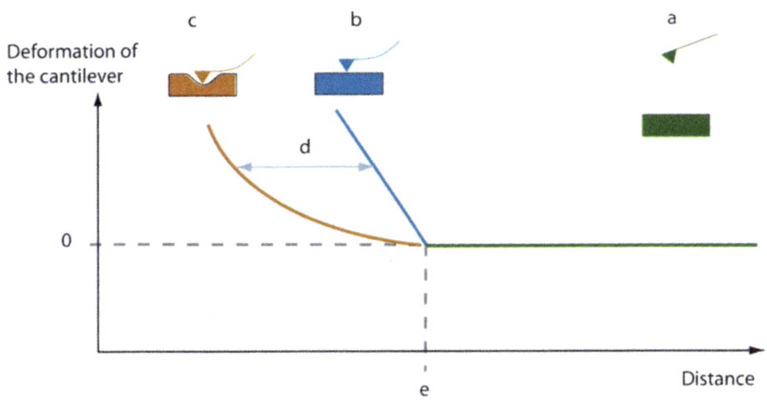

Figure 16.4. The shapes of two typical indentation curves: one for a hard (b: blue), and one for a soft (c: orange) sample. The portion of the trace than runs parallel to the *x*-axis (green) is referred to as the off-contact (a) region, and is common to both samples. The point of coincidence of the two curves (e) corresponds to the point of contact between the cantilever tip and the sample. "d" denotes the indentation distance, which is obtained by subtracting curve c from curve b.

As is apparent from Fig. 16.4, the in-contact segment of an FD curve provides information on the stiffness of the sample. Different mathematical models exist to extract this information from the FD curves. The oldest one, referred to as the Hertz model, assumes the tip to be spherical, and the sample to be of infinite dimensions, perfectly flat, isotropic and homogeneous. This model does not account for the existence of any adhesive, capillary, electrostatic or magnetic force between the tip and the sample. To obtain a numerical value of the sample's Young's modulus, the FD curve must be converted into an indentation curve, which describes the relationship between the indentation depth of the tip and the cantilever deflection. In other words, the indentation curve describes the force that must be applied to the tip to push it a given depth into the sample. The indentation curve is obtained by subtracting the FD curve for a soft sample from the FD curve for a hard sample (see Fig. 16.4). According to the Hertz model,

$$E = \frac{3(1-\nu^2)F}{4r^{1/2}\,\delta^{3/2}} \qquad (16.7)$$

where E is Young's modulus of the sample; ν, Poisson's ratio; F, Force applied by the cantilever; R, radius of cantilever tip; and δ, indentation depth.

An alternative model was developed by Sneddon in 1965. It considers indentation by rigid indenters with arbitrary axis-symmetry profiles. This model is commonly used in AFM indentation experiments. However, neither the Hertz nor the Sneddon model takes into account the effects of surface energy on the contact deformation. These surface forces cannot be neglected, therefore more sophisticated models, such as those propounded by Bradley, Derjaguin–Müller–Toporov and Johnson–Kendall–Roberts, have to be implemented. For an in-depth analysis of these models, the interested reader is referred to the review by Cappella and Dietler.[1]

It has to be emphasized that all these models consider idealized tips and samples which approximate the experimental conditions. Numerical approaches, such as finite-element or molecular dynamics modelling, permit a more precise analysis of the experimental situation, but at the cost of increased complexity and of sophisticated computational requirements.

16.3 FINITE ELEMENTS AND MOLECULAR DYNAMICS

Finite-element modelling is an analytical technique which was first developed for the field of structural engineering. It is based on the assumption that any structure can be subdivided into smaller regions (elements) for which the differential equations describing the deformation under a load can be solved numerically. By assembling the sets of equations for each region, the behaviour of the entire problem domain can be approximated. This method is particularly well adapted for samples that have a complex geometrical shape and in situations where a substantial deformation occurs during the indentation process. More and more investigators are now using finite-element modelling to interpret their AFM data. For a comprehensive overview of the applications of finite-element modelling in conjunction with the AFM, the interested reader is referred to the review by Ikai *et al.*[2]

Molecular dynamics modelling is another analytical tool, which involves calculating the interactions between single atoms as a function of time. Equations in which the Newtonian motion of each individual atom is integrated describe the dynamics of the particles. The technique is widely used to study the conformational changes of proteins. The drawback of this method is its high computational requirements. A flavour of these requirements may be obtained from the following example: a 50 ns simulation of the million atoms that comprise a single tobacco mosaic virus would require several dozen years using a single home computer.

16.4 PROBING THE MECHANICAL PROPERTIES OF VIRUSES

Viruses are amongst the smallest infectious agents, and they are the most abundant biological entity. They exist in almost all ecosystems and infect all types of organism, from animals and plants to bacteria and even archaea. They display a wide diversity in shape and although most viruses are much smaller than bacteria (sizes range: 10–300 nm), some (filoviruses) reach a length of more than 1 μm (Fig. 16.5). A complete virus particle (a virion) typically consists of a protective coat (capsid), which contains its genetic material (RNA or DNA). However, in some cases, such as the human immunodeficiency virus, a protective coat is lacking, and the genetic material is embedded within a lipid envelope. Most of the viruses that infect plants and animals have the form of an icosahedron, but not all. The tobacco mosaic virus, for example, has a helical structure. Yet other viruses have a highly complex geometry. The enterobacteria phage T4, for example, has an icosahedral capsid and a helical tail (Fig. 16.6).

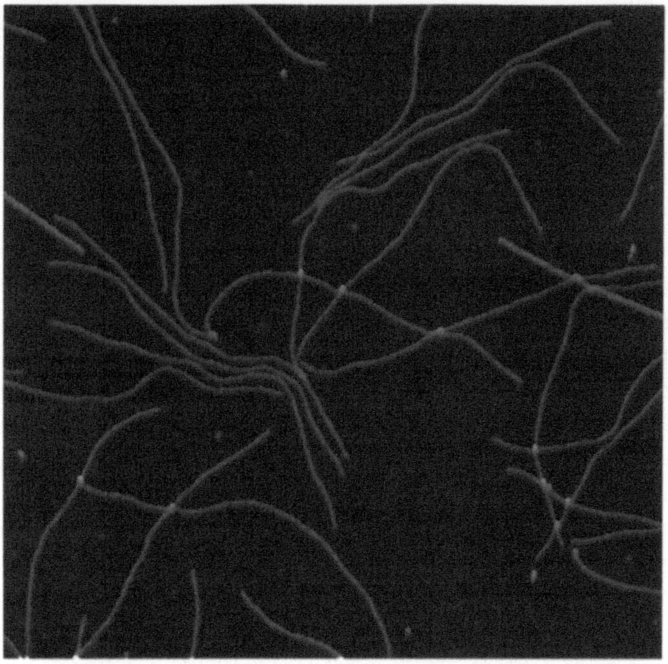

Figure 16.5. AFM image of filoviruses deposited on mica. Scanning distance: 1.5 μm. The image was recorded in air using the tapping mode (kindly provided by Dr. J. Adamcik).

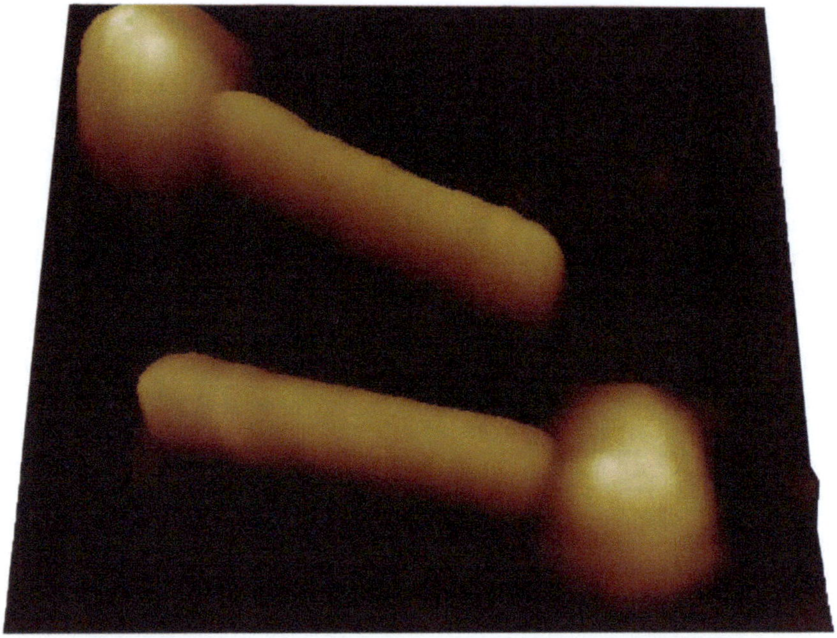

Figure 16.6. AFM image of two T4-phages deposited on mica. Scanning size: 280 nm. The image was recorded in air using the tapping mode (kindly provided by J. Adamcik).

The mechanical properties of the viral capsid are important for the survival and the infectivity of the organism. Moreover, this protective coating can serve as a "nanocontainer" for gene or drug delivery.[3] An understanding of the elastic properties of the capsid is a prerequisite for its use as a pharmacological vehicle.

Numerous studies have been conducted to explore the mechanical properties of different types of virus. For a comprehensive overview of the pertinent literature, the reader is referred to the review by Roos *et al.*[4] Recording FD curves for virus particles is nowadays a relatively straightforward task. However, it is not easy to extract Young's modulus from the recorded data, owing to the often complex geometry of the capsid and the small size of the particles. When a load is applied to the capsid, it deforms and new contacts are established with the underlying substrate. These new contacts modify, and render non-linear, the boundary conditions in the mathematical model, thereby complicating the calculation of Young's modulus. Moreover, when an AFM tip compresses a virus, it very rapidly "senses" also the underlying incompressible substrate, thereby introducing an additional complication to the FD curve.

As mentioned earlier, some viruses have a fairly simple geometrical shape, which renders relatively easy a theoretical calculation of their mechanical properties. For example, a Young's modulus of about 6 GPa has been calculated for the Tobacco mosaic virus.[5] If the virus to be studied has a complex shape, then numerical tools such as finite-element modelling must be implemented to interpret the FD curves. A detailed handling of the finite-element method in relation to viral capsids has been published by Gibbons and Klug[6] and Klug *et al.*[7] Michel *et al.*[8] have applied finite-element modelling to determine the elastic properties of both wild-type and single-point-mutant strains of the cowpea chlorotic mottle virus, in an empty as well as in an RNA-filled state, from AFM measurements. As anticipated, the RNA-filled capsids were less deformable than the empty ones. Interestingly, a single-point mutation sufficed to modify the stiffness of the capsid: Young's moduli for the wild-type and mutant forms were 140 MPa and 190 MPa, respectively. Using a similar set-up, Ivanovska *et al.*[9–11] have measured the elastic properties of empty and DNA-filled lambda-phages. Also in this study, the DNA-filled viruses were found to be stiffer than their empty counterparts, and they could withstand forces twice as high before irreversible damage occurred. Kol *et al.*[12,13] have explored the evolution of mechanical properties as a function of maturation stage using the human immunodeficiency and murine leukaemia viruses. The immature particles (930 MPa) were found to be 14-fold stiffer than the mature ones (115 MPa). These data suggest that the maturation phase of viruses may play an important role in their capacity to penetrate host tissues and thus in determining their infectivity.

Finite-element modelling was used in these studies to analyze the FD curves that were recorded for the investigated viruses. An analysis of this kind is based upon the assumption that the paradigm for continuum mechanics still applies on a nanometric scale. Molecular dynamics modelling yields a more accurate simulation of the experimental situation, but requires huge computational powers. Zink and Grubmüller[14] have conducted one such study. For this purpose, the icosahedral southern bean mosaic virus was used, and the simulation of its interaction with an AFM tip was conducted in an also simulated liquid medium. The model consisted of 4.5 millions atoms, 1 million of which stemmed from water molecules. The simulated time was approximately 100 ns. Indentation of the AFM tip was simulated at numerous grid points on the capsid and revealed the viral shell to exhibit a highly elastic behaviour. However, the curves varied greatly according to the point of indentation. This heterogeneity may reflect differences either in geometry or in the interaction forces operating between the proteins within the capsid. This type of simulation is currently the most accurate means of predicting the deformation of molecular structures under loading conditions.

However, the tool is still confined to investigators who have access to powerful computational resources.

16.5 CELLS

Difficulties in interpreting FD curves are also encountered with living cells, which have highly anisotropic and inhomogeneous structures. Moreover, the elastic modulus appears to vary on a region-specific basis. This phenomenon has been demonstrated by Hoffmann *et al.*[15] for living chicken cardiocytes. Mapping of these cells revealed stiff areas (Young's modulus: 100–200 kPa) to be embedded within softer ones (Young's modulus: 5–30 kPa). Additional difficulties are encountered during the AFM imaging of cells. These include viability and mechanical fixation to the imaging substratum. Some cells, such as erythrocytes, can be immobilized by attachment to a glass surface that has been coated with poly-L-lysine. However, this treatment is not appropriate for all cell types, since it can induce a reorganization of the membrane in the contact area.[16] Spherical cells can sometimes be mechanically lodged within the holes of millipore filters[17] or within microwells.[18] However, the latter option requires the microfabrication of custom-built chambers and an adaptation of the mathematical model of the FD curves. Another factor that influences the measurement of elastic modulus is cell thickness, which can vary topographically. However, this factor can be neglected if the tip does not indent more than 10% of the cell thickness.[19] The processing of data that are recorded at the cell periphery, or on lamellipodia or axons, can be particularly taxing.

Despite these difficulties, the first attempt to measure Young's modulus in cells was made in 1994 by Hoh and Schoenenberger.[20] By recording successive FD curves for canine kidney cells, these investigators monitored what occurs after exposure to glutaraldehyde. This agent, which cross-links proteins, is used as a chemical fixative in the preservation of biological samples. Two years later, Radmacher *et al.*[21] mapped the mechanical properties of human platelets using the force–volume mode. This mode of AFM imaging involves the recording of successive FD curves on a predefined area of the sample. Each pixel represents an FD curve. The authors identified areas of different stiffness on the cell surface: the centrally located pseudonucleus was the softest part (Young's modulus: 1.5–4 kPa), and the outer filamentous zone, which consists of actin filaments and microtubules, was the stiffest (Young's modulus: 10–40 kPa).

The cytoskeleton is an essential part of all eukaryotic cells. It is involved in several basic cellular functions, such as division, vesicular transport and displacement. It also plays a particularly important role in determining the shape and the mechanical properties of the cell. The cytoskeleton is essentially composed of actin, tubulin and intermediate filaments. In the year 2000, Rotsch and Radmacher[22] monitored the evolution of cellular stiffness after a drug-induced disruption of the cytoskeleton. Disruption of the actin cytoskeleton, which is located peripherally, resulted in a marked decrease in the average elastic modulus of the cell. On the other hand, disruption of the microtubules, which are located less peripherally, elicited no measurable change in cellular elasticity. More recent studies have revealed that even the more deeply located tubulin cytoskeleton can be monitored by AFM. Information concerning its stiffness is contained within the last portions of the FD curves.[23]

The effect of hormones on cellular stiffness has also been addressed in several studies. Oberleithner *et al.* have measured the effects of the blood pressure-regulating hormone aldosterone on vascular endothelial cells.[24,25] This drug was found to increase the stiffness of the cells by sodium uptake-induced swelling.[26]

Numerous AFM studies have provided evidence that changes in the mechanical properties of cells are correlated with their age, the stage of the cell cycle and the degree of differentiation. For example, the elasticity of human umbilical vein endothelial cells increase with age,[27] whilst the cardiomyocytes of young rats are softer (Young's modulus: 35 kPa) than those of older (Young's modulus: 42 kPa).[28] Collinsworth *et al.*[29] have shown the Young's modulus to increase during the differentiation of myoblasts (11 kPa) into myocytes (45 kPa). In most instances, changes in the mechanical properties of the cells reflect a reorganization of the cytoskeleton, especially of the actin and myosin components.

Another interesting field of research that has been pursued by AFM imaging is the influence of pathology on the mechanical properties of cells. Since any biochemical or structural modification is likely to induce a change in the mechanical properties of a cell, this field of study could potentially lead to the development of novel diagnostic tools. Several cell types and diseases have been explored, and significant changes in the mechanical properties have been observed. Dulinska *et al.*[16] have measured the Young's modulus for erythrocytes that were derived from patients suffering from spherocytosis, thalassaemia or G6PD deficiency. In all of these pathologies, the erythrocytes were found to be stiffer than their normal counterparts. The erythrocytes that had been derived from patients with a G6DP deficiency were the stiffest (Young's modulus: 90 kPa vs. 26 kPa [control]). Similarly, the erythrocytes of diabetic patients have been found to be stiffer than their normal counterparts.[30]

Lekka *et al.*[31] have studied the mechanical properties of normal and cancerous epithelial bladder cells. Young's modulus for the normal cells was found to be about one order of magnitude higher than that for the cancerous ones. The changes in the elastic properties of the cancerous cells were attributed to a reorganization of the cytoskeleton.

16.6 CELLULAR STRUCTURES

The mechanical properties of several different isolated cell structures have been studied using the AFM. The relevance of such studies lies in the fact that subcellular structures, such as the nucleus or components of the cytoskeleton, influence the mechanical properties of the cell as a whole. The cytoskeleton, in particular, has an important influence on the properties of a cell. A study of this structural component is of particular interest in the pharmacological industry, since several anti-mitotic drugs that are used in cancer therapy interfere with components of the cytoskeleton.

Among the cytoskeletal components, microtubules are the largest, with a diameter of about 20 nm. During cell division, they guide in the segregation of the chromosomes within the mitotic and meiotic spindles. They also serve as tracks for the intracellular movement of several components, such as vesicles and mitochondria. Their mechanical properties were first measured by AFM imaging in 1995.[32] For these experiments, the microtubules were deposited upon a substratum of mica, and the FD curves were recorded in the absence or presence of increasing concentrations of glutaraldehyde. Young's modulus was calculated to lie between 1 and 12 MPa. However, it later appeared that shear rather than Young's modulus had been calculated. In 2002, Young's modulus for suspended microtubules was measured for the first time in the AFM.[33] In these experiments, the shear modulus was found to be about 1.4 MPa. Subsequently, finite-element modelling of AFM measurements revealed the Young's modulus of microtubules to be about 0.8 GPa.[34] Measurements made by the same authors using the plastic regime (irreversible deformation) yielded a crude estimate of the rupture force, which was in the order of several hundred piconewtons.

Intermediate filaments are another important component of the cytoskeleton. They play an essential role in maintaining the shape and the mechanical stability of the cells. Using AFM, their bending and sliding properties have been measured by the deformation of single vimentin filaments that were suspended over a porous membrane;[35] Young's modulus was determined to be at least 0.9 GPa.

The mechanical properties of chromosomes have also been measured by AFM.[36,37] In the first study,[36] Young's modulus was found to vary between 0.05 and 0.1 MPa, depending on the nature of the imaging buffer: chromosomes that were immersed in acetate buffer were 10 times stiffer than those that were bathed in neutral or alkaline buffers. In the second study,[37] a Young's modulus was determined to be 0.4 MPa.

The mechanical properties of secretory vesicles have likewise been studied in the AFM. The molecular mechanism that underlies the release of secretory products from cellular vesicles is still incompletely understood, although the stiffness of these intracellular organelles is believed to play an important role in the process. In 1997, Laney *et al.*[38] measured the Young's modulus of cholinergic synaptic vesicles that had been isolated from the marine ray. The value lay between 200 kPa and 1.3 MPa, depending on the composition of the bathing medium.

Lipid rafts are small cholesterol-enriched membrane domains, which have been implicated in several physiological and pathological processes. Until recently, their properties were unknown. To ascertain whether the stiffness

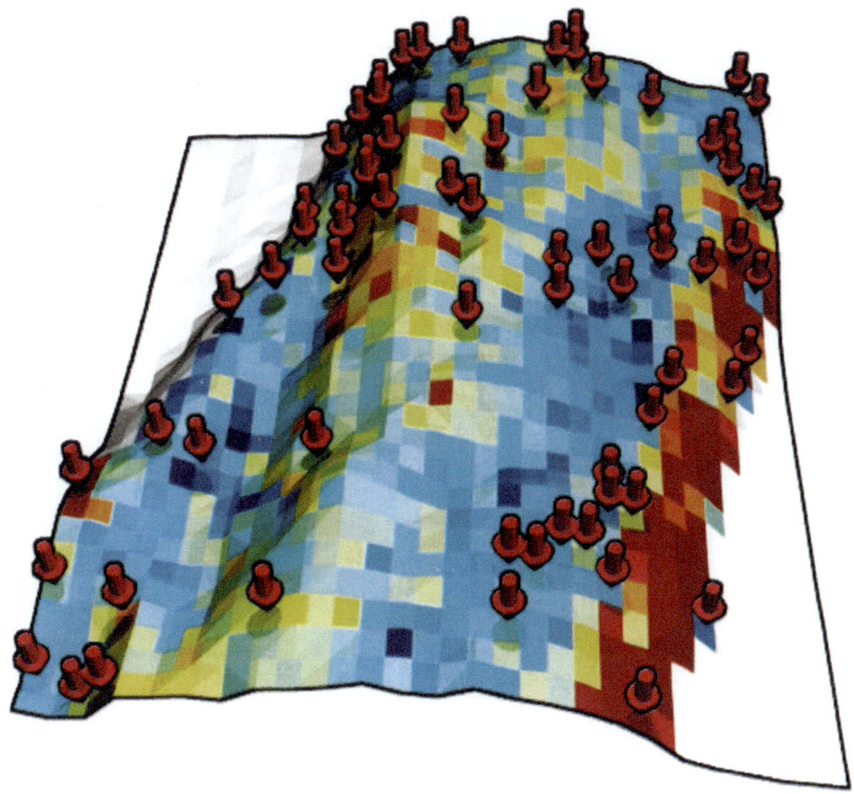

Figure 16.7. AFM image of an axon showing topography, elasticity and positions of raft domains. The stiffness of the membrane is colour-coded (blue: soft; red: hard). The arrows point to lipid rafts. Scan size: 2 μm (kindly provided by Dr. C. Roduit).

of lipid rafts differed from that of the surrounding membrane, Roduit *et al.*[39] recorded force–volume images of living neurones using an aerolysin-coated AFM tip. Aerolysin is known to interact with molecular domains that are specifically expressed within the lipid rafts. Consequently, during scanning with an aerolysin -coated AFM tip, it should be possible to detect specific interactions with the lipid rafts. By comparing the stiffness of the regions in which a specific interaction occurred with the stiffness of the surrounding domains, the authors were able to demonstrate that the lipid rafts were about 30% stiffer than the rest of the membrane (Fig. 16.7).

Finally, it has recently become possible to glean information concerning the mechanical properties of subcellular components within living cells from AFM imaging.[40] For this purpose, a more sophisticated analysis of the FD curves is called for. Instead of fitting the entire FD curve to the Hertz model, small segments are individually analyzed. The stiffness of each segment is then displayed as a colour-coded volume within a three-dimensional matrix. The location of the volume corresponds to that of the recording (Fig. 16.8). Using this technique, it is possible to distinguish cytoskeletal components beneath the cell membrane (coloured red in Fig. 16.8).

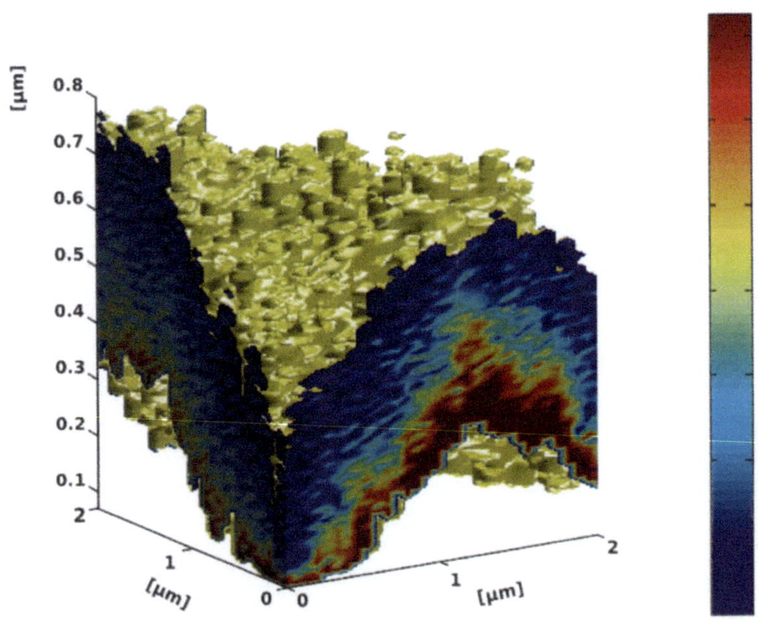

Figure 16.8. Stiffness tomography of an axon. This imaging mode reveals structures hidden in the bulk of the sample. The stiffness of the intracellular structures is colour-coded (blue: soft; red: hard). Scanning distance: 2 µm (kindly provided by Dr. C. Roduit).

16.7 CONCLUSIONS

In biological systems, structure is closely coupled with function, and this relationship holds true no less for mechanical than for other properties. An exploration of the nanomechanical properties of biological entities can facilitate our understanding of their functional roles per se as well as in a more global context. And since mechanical compliance can be influenced by pathological agents, an evaluation of this property might be put to use in a clinical context, either as a diagnostic tool or in monitoring the evolution of a disease. Recent developments in the AFM technology now render possible a fairly straightforward measurement of the mechanical properties of biological structures in the micrometer-to-nanometre size range. However, a number of stumbling blocks remain. Experimentally, these problems include an immobilization of the specimen on a rigid, flat substratum without compromising its responsiveness to deformational forces. Analytically, the automatic processing of FD curves and the interpretation of data are difficulties that have been only partially overcome. However, progress in this field has been steady, and there is no reason to suppose that the remaining hurdles will not be overcome in the not-too-distant future.

References

1. Cappella, B., and Dietler, G. (1999) Force-distance curves by atomic force microscopy, *Surf. Sci. Rep.*, **34**, 1–104.
2. Kim, H. C., Asgari, F., Kato-Negishi, M., Ohkura, S., Okamura, H., Arakawa, H., Osada, T., and Ikai, A. (2008) Distribution of olfactory marker protein on a tissue section of vomeronasal organ measured by AFM, *Colloids Surf. B Biointerfaces*, **61**, 311–314.
3. Douglas, T., and Young, M. (1999) Virus particles as templates for materials synthesis, *Adv. Mater.*, **11**, 679–681.
4. Roos, W. H., Ivanovska, I. L., Evilevitch, A., and Wuite, G. J. L. (2007) Viral capsids: mechanical characteristics, genome packaging and delivery mechanisms, *Cell. Mol. Life Sci.*, **64**, 1484–1497.
5. Schmatulla, A., Maghelli, N., and Marti, O. (2007) Micromechanical properties of tobacco mosaic viruses, *J. Microsc.*, **225**, 264–268.
6. Gibbons, M. M., and Klug, W. S. (2007) Nonlinear finite-element analysis of nanoindentation of viral capsids, *Phys. Rev., E* **75**, 031901
7. Klug, W. S., Bruinsma, R. F., Michel, J. P., Knobler, C. M., Ivanovska, I. L., Schmidt, C. F., and Wuite, G. J. L. (2006) Failure of viral shells, *Phys. Rev. Lett.*, **97**, 228101.
8. Michel, J. P., Ivanovska, I. L., Gibbons, M. M., Klug, W. S., Knobler, C. M., Wuite, G. J. L., and Schmidt, C. F. (2006) Nanoindentation studies of full and empty viral

capsids and the effects of capsid protein mutations on elasticity and strength, *Proc. Natl. Acad. Sci. USA*, **103**, 6184–6189.

9. Ivanovska, I., Wuite, G., Jonsson, B., and Evilevitch, A. (2007) Internal DNA pressure modifies stability of WT phage, *Proc. Natl. Acad. Sci. USA*, **104**, 9603–9608.
10. Ivanovska, I., Wuite, G., Jonsson, B., and Evilevitch, A. (2007) Internal DNA pressure effect on the phage capsid stability. Evolutionary optimization of phage, *Biophys. J.*, 49A-49A.
11. Ivanovska, I. L., Agirrezabala, X., Carrascosa, J. L., Schmidt, C. F., and Wuite, G. J. L. (2007) Deconstructing viral capsids reveals their strengths and weaknesses, *Biophys. J.*, 49A-49A.
12. Kol, N., Gladnikoff, M., Barlam, D., Shneck, R. Z., Rein, A., and Rousso, I. (2006) Mechanical properties of murine leukemia virus particles: Effect of maturation, *Biophys. J.*, **91**, 767–774.
13. Kol, N., Shi, Y., Tsvitov, M., Barlam, D., Shneck, R. Z., Kay, M. S., and Rousso, I. (2007) A stiffness switch in human immunodeficiency virus, *Biophys. J.*, **92**, 1777–1783.
14. Zink, M., and Grubmüller, H. (2009) Mechanical properties of the icosahedral shell of southern bean mosaic virus: a molecular dynamics study, *Biophys. J.*, **96**, 1350–1363.
15. Hofmann, U. G., Rotsch, C., Parak, W. J., and Radmacher, M. (1997) Investigating the cytoskeleton of chicken cardiocytes with the atomic force microscope, *J. Struct. Biol.*, **119**, 84–91.
16. Dulinska, I., Targosz, M., Strojny, W., Lekka, M., Czuba, P., Balwierz, W., and Szymonski, M. (2006) Stiffness of normal and pathological erythrocytes studied by means of atomic force microscopy, *J. Biochem. Biophys. Methods*, **66**, 1–11.
17. Kasas, S., and Ikai, A. (1995) A method for anchoring round shaped cells for atomic-force microscope imaging, *Biophys. J.*, **68**, 1678–1680.
18. Rosenbluth, M. J., Lam, W. A., and Fletcher, D. A. (2006) Force microscopy of nonadherent cells: a comparison of leukemia cell deformability, *Biophys. J.*, **90**, 2994–3003.
19. Mathur, A. B., Collinsworth, A. M., Reichert, W. M., Kraus, W. E., and Truskey, G. A. (2001) Endothelial, cardiac muscle and skeletal muscle exhibit different viscous and elastic properties as determined by atomic force microscopy, *J. Biomech.*, **34**, 1545–1553.
20. Hoh, J. H., and Schoenenberger, C. A. (1994) Surface-morphology and mechanical-properties of Mdck monolayers by atomic-force microscopy, *J. Cell Sci.*, **107**, 1105–1114.
21. Radmacher, M., Fritz, M., Kacher, C. M., Cleveland J. P., and Hansma, P. K. (1996) Measuring the viscoelastic properties of human platelets with the atomic force microscope, *Biophys. J.*, **70**, 556–567.
22. Rotsch, C., and Radmacher, M. (2000) Drug-induced changes of cytoskeletal structure and mechanics in fibroblasts: an atomic force microscopy study, *Biophys. J.*, **78**, 520–535.

23. Kasas, S., Wang, X., Hirling, H., Marsault, R., Huni, B., Yersin, A., Regazzi, R., Grenningloh, G., Riederer, B., Forro, L., Dietler, G., and Catsicas, S. (2005) Superficial and deep changes of cellular mechanical properties following cytoskeleton disassembly, *Cell Motil. Cytoskeleton*, **62**, 124–132.

24. Oberleithner, H. (2005) Aldosterone makes human endothelium stiff and vulnerable, *Kidney Int.*, **67**, 1680–1682.

25. Oberleithner, H., Riethmuller, C., Ludwig, T., Shahin, V., Stock, C., Schwab, A., Hausberg, M., Kusche, K., and Schillers, H. (2006) Differential action of steroid hormones on human endothelium, *J. Cell Sci.*, **119**, 1926–1932.

26. Oberleithner, H., Riethmuller, C., Schillers, H., MacGregor, G. A., de Wardener, H. E., and Hausberg, M. (2007) Plasma sodium stiffens vascular endothelium and reduces nitric oxide release, *Proc. Natl. Acad. Sci. USA*, **104**, 16281–16286.

27. Sato, H., Kataoka, N., Kajiya, F., Katano, M., Takigawa, T., and Masuda, T. (2004) Kinetic study on the elastic change of vascular endothelial cells on collagen matrices by atomic force microscopy, *Colloids Surf. B Biointerfaces*, **34**, 141–146.

28. Lieber, S. C., Aubry, N., Pain, J., Diaz, G., Kim, S. J., and Vatner, S. F. (2004) Aging increases stiffness of cardiac myocytes measured by atomic force microscopy nanoindentation, *Am. J. Physiol. Heart Circ. Physiol.*, **287**, H645–H651.

29. Collinsworth, A. M., Zhang, S., Kraus, W. E., and Truskey, G. A. (2002) Apparent elastic modulus and hysteresis of skeletal muscle cells throughout differentiation, *Am. J. Physiol. Heart Circ. Physiol.*, **283**, C1219–C1227.

30. Fornal, M., Lekka, M., Pyka-Fosciak, G., Lebed, K., Grodzicki, T., Wizner, B., and Styczen, J. (2006) Erythrocyte stiffness in diabetes mellitus studied with atomic force microscope, *Clin. Hemorheol. Microcirc.*, **35**, 273–276.

31. Lekka, M., Laidler, P., Gil, D., Lekki, J., Stachura, Z., and Hrynkiewicz, A. Z. (1999) Elasticity of normal and cancerous human bladder cells studied by scanning force microscopy, *Eur. Biophys. J.*, **28**, 312–316.

32. Vinckier, A., and Semenza, G. (1998) Measuring elasticity of biological materials by atomic force microscopy, *FEBS Lett.*, **430**, 12–16.

33. Kis, A., Kasas, S., Babic, B., Kulik, A. J., Benoit, W., Briggs, G. A. D., Schonenberger, C., Catsicas, S., andForro, L. (2002) Nanomechanics of microtubules, *Phys. Rev. Lett.*, **89**, 248101.

34. de Pablo, P. J., Schaap, I. A. T., MacKintosh, F. C., and Schmidt, C. F. (2003) Deformation and collapse of microtubules on the nanometer scale, *Phys. Rev. Lett.*, **91**, 098101.

35. Guzman, C., Jeney, S., Kreplak, L., Kasas, S., Kulik, A. J., Aebi, U., and Forro, L. (2006) Exploring the mechanical properties of single vimentin intermediate filaments by atomic force microscopy, *J. Mol. Biol.*, **360**, 623–630.

36. Ikai, A., Mitsui, K., Tokuoka, H., and Xu, X. M. (1997) Mechanical measurements of a single protein molecule and human chromosomes by atomic force microscopy, *Mater. Sci. Eng. C Biomim. Mater. Sens. Syst.*, **4**, 233–240.

37. Jiao, Y., and Schaffer, T. E. (2004) Accurate height and volume measurements on soft samples with the atomic force microscope, *Langmuir*, **20**, 10038–10045.

38. Laney, D. E., Garcia, R. A., Parsons, S. M., and Hansma, H. G. (1997) Changes in the elastic properties of cholinergic synaptic vesicles as measured by atomic force microscopy, *Biophys. J.*, **72**, 806–813.

39. Roduit, C., Van der Goot, G., De Los Rios, P., Yersin, A., Steiner, P., Dietler, G., Catsicas, S., Lafont, F., and Kasas, S. (2007) Elastic membrane heterogeneity of living cells revealed by stiff nanoscale membrane domains, *Biophys. J.*, **94**, 1521–1532.

40. Roduit, C., Sekatski, S., Dietler, G., Catsicas, S., Lafont, F., and Kasas, S. (2009) Stiffness tomography by atomic force microscopy, *Biophys. J.*, **97**, 674–677.

Chapter 17

LABEL-FREE MONITORING OF CELL SIGNALLING PROCESSES THROUGH AFM-BASED FORCE MEASUREMENTS

Charles M. Cuerrier, Elie Simard, Charles-Antoine Lamontagne, Julie Boucher, Yannick Miron and Michel Grandbois
Département de Pharmacologie, Université de Sherbrooke,
Sherbrooke, Québec, Canada, J1H5N4
Michel.Grandbois@usherbrooke.ca

17.1 LABEL-FREE MONITORING OF CELL SIGNALLING PROCESSES

Continued success in drug discovery largely relies on the development and evolution of assay technologies capable of accurate representation of cellular behaviour in response to specific stimulations. The development of such technologies is vital to the drug discovery process as it guides the selection of promising compounds as well as the early abandonment of potential failing drug candidates. Compared with the data obtained using cell-free assays, direct monitoring of drug-modulated signalling in live cell systems offers high content information, in conditions designed to closely resemble a physiological environment. These benefits have driven the use of whole cell systems in drugs screening and fundamental biomedical research. In cell-based assays, compounds are either screened on genetically modified cells expressing an exogenous receptor or on selected primary cells expressing the target of interest. In this context, a compound can simultaneously behave as an agonist or antagonist depending on which intracellular signalling pathway is monitored. Therefore, a lack of efficacy

Life at the Nanoscale: Atomic Force Microscopy of Live Cells
Edited by Yves Dufrêne

www.panstanford.com

for a preselected biochemical signalling event does not necessarily mean a lack of receptor activation. Thus, because of the ability of ligands to stabilize or stimulate subsets of receptor activities, a receptor activation screen should not rely on a single specific assay, but rather on an integrated approach to measure multiple signalling events simultaneously. Unlike label-dependent cell assays that measure specific cellular events, cell-based biosensing technologies such as those based on impedance mesurement[1–3] or evanescent field at surface[4,5] have the potential to provide an integrated cellular response. The recognition of this need for more in-depth analysis of biological activity beyond simple specificity, selectivity and affinity is clear. Hence, what is required is a high content information on the action of molecules on drug targets and, especially, a more precise profile of the molecular pathways induced by a particular candidate drug molecule. In this chapter, we present an experimental strategy based on atomic force microscope (AFM) force sensing on individual cells to report on the effects of external pharmacological stimuli on cellular functions. This kind of cell-based biosensing allows for label-free, multimodal and real-time monitoring of cellular responses and signalling pathways in recombinant cell models as well as primary cell cultures related to physiologically and pathophysiological models. Exposition of receptors present at the external surface of cells to external chemical or biochemical messenger entities (hormones, neurotransmitters, ions, light, scent, taste, etc) leads to the activation of a variety of intracellular molecular signalling pathways which are often associated with change in cell morphology, cell motility or specific proteins expression. Here we show that AFM-based force measurements in conjunction with fluorescence imaging of intracellular component can fingerprint the contribution of signalling pathways subsequent to the activation of specific receptor at the cell membrane.

17.2 AFM-BASED MONITORING OF CELLULAR RESPONSES

Because of its great sensitivity, AFM-based force measurement can detect small magnitude morphological changes, which are not readily detectable with conventional optical imaging techniques. Nevertheless, microscopic observations in conjunction with mechanical assays based on flexible substrates have provided a wealth of information pertaining to cell contraction and locomotion.[6–8] Nanoscaled approaches such as optical and magnetic traps,[6,7,9–11] micropipette aspiration[8] and AFM[12–18,23] were established as valuable tools for studying the mechanical response in individual cells in various physiologically relevant contexts. Force measurement with the

AFM was shown to be particularly suitable to cell mechanical studies as demonstrated in the monitoring of single cardiomyocytes pulsatile activity[15] as well as monitoring mechanical properties of various cell types.[19-21] The technique was also applied to the monitoring of metabolically driven cell membrane fluctuation in *Saccharomyces cerevisiae*[16,17] and to the monitoring of cell pulsation in different human cell phenotypes.[18,22] Cell membrane receptor activation involves spatiotemporal events, starting with ligand–receptor recognition, receptor activation, proteins recruitment and associated production of second messengers, internalization, cytoskeletal remodelling, adhesion contact remodelling as well as genes expression.

Each of these events is susceptible to produce a specific signature in the AFM force signal, mainly in terms of amplitude, kinetics and duration. In Fig. 17.1, we combined the AFM with an inverted phase contrast and

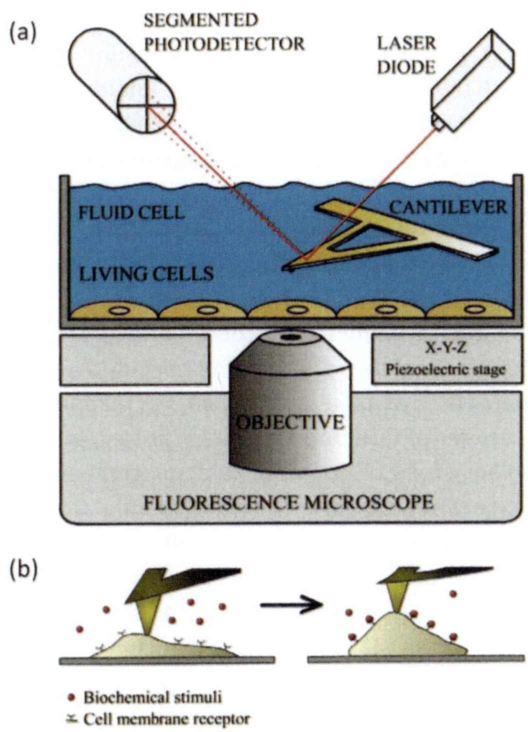

Figure 17.1. Monitoring of cell activity using an AFM-based force measurement instrument mounted over an inverted fluorescence microscope. (a) In a typical experiment, an AFM cantilever is allowed to contact the apical region of an individual cell. (b) The stimulation of receptors present at the cell surface or in the cytosol by selective agonist induces morphological or mechanical activity, which can be monitored in real time with the AFM.

fluorescence microscope to elucidate the cellular responses associated with the activation of cell receptors by selective biochemical stimuli. In a typical experiment, an AFM cantilever mounted with a tip of 50 nm in radius is allowed to contact the cell surface with a force inferior to 250 pN. Using an X–Y piezo stage, the tip can be precisely positioned over specific regions of an individual cell. In this procedure, the positioning precision is normally limited by the resolution of the optical microscope under which the AFM-based force measurement instrument is mounted. The highest part of the apical region (i.e. the nucleus) is often selected to minimize variation in the cellular response that could be associated with the varying geometry throughout the cell surface.

Activation of pathways leading to minute morphological changes in the cell body is detectable through AFM-based force experiments as demonstrated using HEK-293 cell line stably transfected with the angiotensin receptor (AT_1-R),[23] which is a model for cell responding to angiotensin II (AngII).[24] This receptor is well known to induce cellular response in endothelial cells and smooth muscle cells controlling blood flow and pressure. Figure 17.2 represents typical AngII-induced cellular response detected with the AFM cantilever, plotted as cell membrane displacement (nm) as a function of time (minutes). The response could be alternatively plotted in force unit (pN), using the spring constant of the cantilevers, which are usually chosen in the 0.005–0.015 N/m range for optimum sensitivity. Prior to stimulation, the baseline is recorded for several minutes to ensure thermal and mechanical stability. During this period, the signal exhibits an average height fluctuation of 0.70 ± 0.07 nm, corresponding to 7.0 ± 0.7 pN, which mainly consist of inherent cantilever noise. An important feature of these curves is the large cell membrane displacement (262 ± 52 nm, n = 6 independent experiments) observed immediately after the stimulation by AngII (100 nM), which occurs as a result of the cell developing a positive force against the tip. Simultaneously recording of phase contrast micrographs confirms that the change in the AFM signal originates from minute morphological changes occurring throughout the cell body (Fig. 17.2b). The comparison of the cell morphology before (–2 minutes) and at the maximum of the height signal (~2 minutes) shows a contraction of the cells bodies that is related to the AT_1 receptor stimulation. The extent of the contraction is assessed (Fig. 17.2c) by delineating the cell body before contraction as a standard for comparison at different stages of the cell response. For this particular cell, a contraction is observed at ~2 minutes, which leads to an elevation of the apical regions of the cells as measured with the AFM. This confirms that the signal detected with the AFM reflects the structural and mechanical changes in the cells. The variability in the

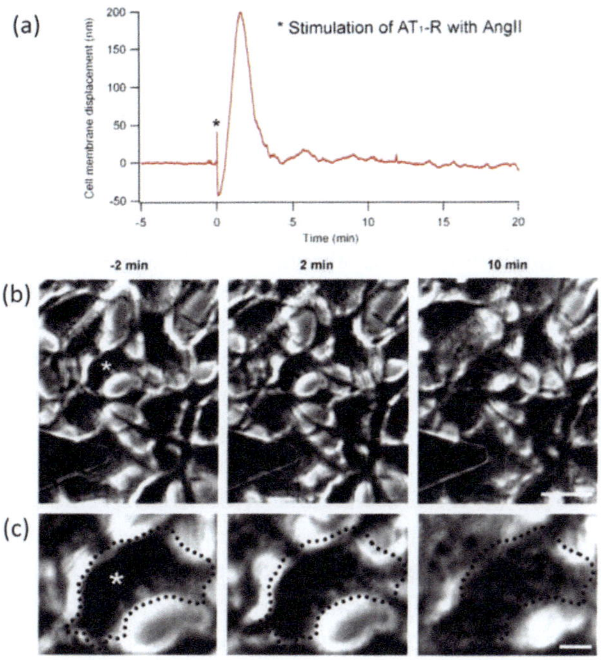

Figure 17.2. Simultaneous monitoring of AFM signal and phase contrast micrograph on AT_1-transfected HEK-293 cells stimulated by AngII. (a) In a typical experiment, an AFM cantilever is allowed to contact the apical region of an individual cell with a force inferior to 250 pN. A baseline is recorded for several minutes before the injection of 100 nM AngII (0 minute). The cell mechanical response is plotted as cell membrane displacement (nm) as a function of time. (b) Phase contrast micrographs recorded before (−2 minutes) and after AngII stimulation (2 and 10 minutes). The AFM cantilever can be seen in left lower corner of the micrograph. The micrograph at 2 minutes shows contraction whereas spreading is observed at 10 minutes. (c) Magnified view of the cell marked with a white asterisk in (b) to demonstrate cell body contraction using the doted contour of the cell prior to stimulation. The contour at −2 minutes is projected on the images at 2 and 10 minutes. Scale bars are 20 μm (b) and 5 μm (c). Reprinted with permission from Ref. 23.

height change measured with the AFM is most certainly due to intrinsic differences between individual cells composing the global cell population. Another important feature of the AT_1 receptor-dependent cellular response is the considerable change in morphology observed after the initial contractile response. Indeed, the phase contrast micrograph taken 10 minutes after AngII stimulation (Fig. 17.2b,c) clearly shows a significant spreading of the cell body as reported previously for this cell model.[25] Interestingly, these structural changes can be detected as an increased fluctuation in the AFM signal when comparing the signal before (−5 to 0 minutes) and after (5 to 20 minutes) AngII stimulation. Indeed, in these time intervals, the averaged fluctuation increases from 0.70 ± 0.07 nm to 6.30 ± 0.46 nm in amplitude. A tenfold increase in the cell surface fluctuations is most likely indicative of the cell cytoskeleton remodelling. Monitoring of cell response with the AFM requires the injection of significant volume (100–500 μl) of buffer containing a cell receptor agonist (i.e. AngII) into the AFM fluid cell. This injection of fluids could potentially influence the cells and thus the AFM signal monitoring through liquid flow disturbance. Indeed, a sharp spike is normally observed immediately after the injection of 500 μl of the buffer. However, this sharp discontinuity in the force signal is not to be mistaken with a cellular response such as the one presented in Fig. 17.2a. Figure 17.3 presents two control experiments conducted to confirm the contribution of the AT_1 receptor stimulation in the height response recorded with the AFM. As a control, HEK-293 cells transfected with the AT_1 receptor were stimulated with HBSS (vehicle for all AngII injection). As for the AngII stimulation experiment, the buffer HBSS is gently introduced with a micropipette into the fluid cell while the AFM signal is recorded. As expected, no significant height increase of the cell body was detected, nor any morphological changes observed. In an additional control, the stimulation of MOCK HEK-293 cell (transfected with an empty vector) with 100 nM AngII generated no contractile response, and no significant difference is observed in the averaged signal fluctuation before and after

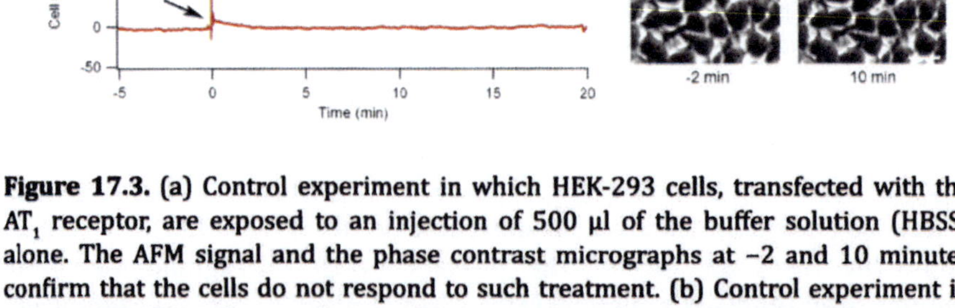

Figure 17.3. (a) Control experiment in which HEK-293 cells, transfected with the AT_1 receptor, are exposed to an injection of 500 μl of the buffer solution (HBSS) alone. The AFM signal and the phase contrast micrographs at −2 and 10 minutes confirm that the cells do not respond to such treatment. (b) Control experiment in which HEK-293 cells transfected with an empty vector (without the AT_1 receptor coding sequence) are exposed to AngII (in 500 μl HBSS). Injection spikes occur in both experiments but no cell responses are seen. Reprinted with permission from Ref. 23.

stimulation. Such control experiments confirm that the AFM makes it possible to detect the activation of AT_1 receptor (Fig. 17.2) and its effect on the cells mechanical homeostasis.

17.3 AFM-BASED MEASUREMENT AND FLUORESCENCE IMAGING OF CELL CONSTITUENTS

Because of its great sensitivity, the AFM can detect morphological changes of small magnitude, which are not readily detectable with conventional optical imaging techniques. The visualization of molecular, ionic, structural and morphological changes, occurring within the cell simultaneously to force signal monitoring, offers a significant advantage in the study of receptor-dependent cell responses.

Using appropriate fluorescent markers, it is possible to visualize the contractile response of the cell body, actin rearrangement, organelle movement, receptor internalization and chemical (pH, Ca^{2+}) changes occurring within the cell. The observation of these events and their quantification make possible a rational interpretation of the features seen in the AFM signal and provide independent cues confirming the contribution of specific signalling pathways in the morphological and mechanical changes in the cell occurring as a result of receptor activation. In this section, we illustrate how actin cytoskeleton visualization and the monitoring of intracellular calcium can be used in conjunction with AFM-based force experiments on receptor stimulated cells.

17.3.1 Contribution of the Actin Cytoskeleton in the AFM Force Signal

Contribution of actin cytoskeleton in the recorded AFM signal can be assessed using HEK-293 cells expressing the AT_1 receptor and co-transfected with GFP–actin to generate fluorescence micrographs at different times after stimulation of living cells.[23,25,26] Correspondence between the AFM signal and confocal micrographs, showing the actin cytoskeleton at different times before and after AngII stimulation, is presented in Fig. 17.4. The confocal micrograph of the basal section (Fig. 17.4b at −2 minutes), obtained prior to AT_1 stimulation, shows a morphologically stable cell with numerous actin structures.[25] A transversal confocal micrograph of a cell is presented in Fig. 17.4c and shows a relatively uniform distribution of the GFP–actin throughout the cell body prior to the stimulation (−2 minutes). This transversal presentation does not easily allow for the observation of the actin structures but allow for qualitative evaluation of GFP–actin distribution throughout the cell body, which is composed of both the polymeric (f-actin) and monomeric (g-actin) components. Consistent with the contraction observed in the phase contrast image (Fig. 17.2b,c), a notable reorganization of the actin structure towards the centre of the cell body is observed 2 minutes after the stimulation of the cell by AngII (Fig. 17.4b). In addition, the transverse view at 2 minutes shows an increase in actin content at the apical region of the cell (Fig. 17.4c, see arrowhead at 2 minutes), which is consistent with the height increase observed by AFM.

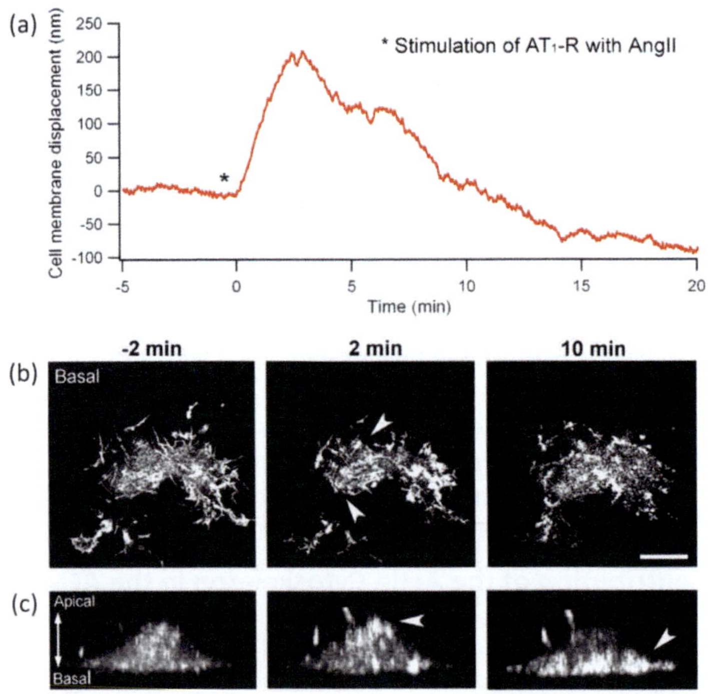

Figure 17.4. Confocal imaging of HEK-293 cells, co-transfected with AT_1-R and GFP–actin, in relation to the mechanical response measured with the AFM. (a) Mechanical response observed after stimulation with 100 nM AngII (same condition as Fig. 17.2a). (b) Confocal micrographs of 1 μm thick section recorded in the basal region of the cell showing actin structures. Micrographs are presented at selected time before and after AngII stimulation (−2, 2 and 10 minutes). The fluorescent background is attributed to the monomeric component of GFP–actin. The arrows indicate the apparent contraction of the cell body. (c) Transversal views of an individual cell, constructed from 55 sections of 250 nm, before and after AngII stimulation. The arrows show the redistribution of the GFP–actin to apical and basal regions of the cell at 2 and 10 minutes after AngII stimulation. Scale bar corresponds to 10 μm. Reprinted with permission from Ref. 23.

GFP–actin distribution in the late phase of the stimulation demonstrates an extensive rearrangement of the actin structures, most notably the loss of f-actin component (Fig. 17.4b at 10 minutes) in the apical region and the increase in actin density at the basal region of the cell (Fig. 17.4c, see arrowhead at 10 minutes). This reorganization of the actin cytoskeleton is concomitant with the spreading of the cell body and is most certainly responsible for the fluctuation in the AFM signal in the late phase of the stimulation. Essentially, these observations point towards the implication of the actin cytoskeleton in the fluctuating AFM signal observed in the late phase of the stimulation as well as the cell spreading observed in the phase contrast micrographs. In this kind of experiment, the contribution of the actin cytoskeleton is often discriminated by pretreating the cells with the actin depolymerizing drug latrunculin A before their stimulation by AngII. In this case, the mechanical response is largely abolished,[23] which confirms the essential contribution of the actin cytoskeleton in the development of the mechanical response.

17.3.2 Fluorescence Quantification of the Intracellular Calcium Level in Relation to the AFM Force Signal

Stimulation of AT_1 receptor is well known to be linked to the activation of several distinct signalling pathways, most notably the Gq pathway (Fig. 17.5a), which is a key regulator of the cytosolic calcium concentration.[27,28] Temporal relationship between the receptor activation and the mechanical response can also be established by measuring intracellular Ca^{2+} level as an independent biochemical indicator. Changes in intracellular Ca^{2+} concentration are easily detected using the fluorescent Ca^{2+} probe FURA-2/AM.[29,30] In this procedure, the cells are loaded with FURA-2/AM and the calcium quantified in individual cells by calculating the ratio of the fluorescence emission at 510 nm between the calcium-bound probe (excitation at 340 nm) and the calcium-free probe (excitation at 380 nm). Using an inverted fluorescence microscope,[24,31] cytosolic calcium release can be recorded simultaneously to force monitoring with a time resolution on the order of 1 second (Fig. 17.5b,c). Following the stimulation with AngII, an increase in cytosolic Ca^{2+} is observed and reaches a maximum within few minutes and typically returns to baseline level. A strong correlation exists between the mechanical and the Ca^{2+} signal, thus confirming that the mechanical response occurs as a direct consequence of AT_1-R activation. It can also be noted that the Ca^{2+} signal occurs prior to the onset in the force signal by approximately 30 seconds, which is consistent with the possibility that the mechanical signal occurs as a consequence of an increased intracellular Ca^{2+} level.

Intracellular Ca^{2+} level is well known to regulate several intracellular kinases and calcium-binding proteins. It is also directly involved in the polymerization of the g-actin monomer into polymeric f-actin filaments, making the actin content a dynamic component of the cytoskeleton. Indeed, several actin regulatory cytoskeletal proteins are regulated by calcium, such as calmodulin,[32] and CaMKII,[33,34] which are involved in motility and

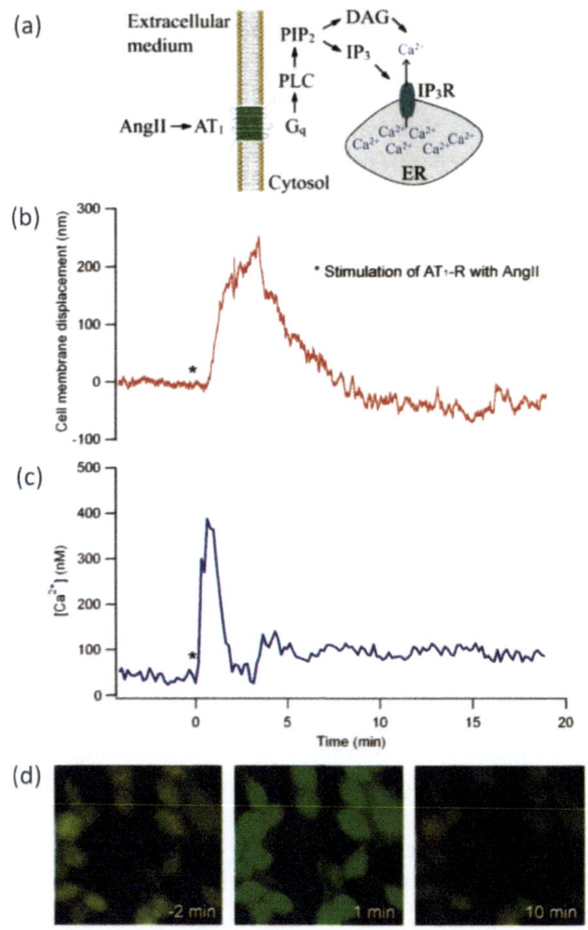

Figure 17.5. Simultaneous monitoring of intracellular calcium and cell mechanical fluctuation in AT_1-transfected HEK-293 cells stimulated by AngII. (a) Stimulation of the AT_1-receptor leads to the elevation of the cytoplasmic calcium pool. (b) Mechanical response recorded together with (c) intracellular Ca^{2+} concentration evaluated using the ratiometric fluorescent calcium indicator FURA 2/AM. (d) Fluorescence micrographs of the FURA-2/AM loaded cells at different time of AT_1-receptor stimulation. The fluorescence intensity in (c) was evaluated using the ratio of the emission at 510 nm for sequential excitation at 340 and 380 nm.

cell contraction, annexin,[35] involved in the interaction of cytoskeleton with the membrane and gelsolin,[36] which severs and caps actin filaments. Most importantly, calcium is involved in the molecular process of actomyosin contraction.[37] To delineate the contribution of the signalling pathways activated by a given receptor in the development of mechanical response in individual cells, one should decouple or isolate their contributions. This can be achieved using pharmacological inhibitors or activators of molecular components involved downstream of receptor activation. Alternatively, silencing RNA can also be used to evaluate the contribution of key structural and functional proteins.[38] As presented in introduction, force sensing over individual cells provides a direct readout for the contribution of signalling pathways in response to the activation of specific receptors at the cell surface.

17.4 AFM FORCE MEASUREMENT ON INDIVIDUAL CELLS AS A TOOL TO DELINEATE CELL SIGNALLING EVENTS

Stimulation of cell receptors leads to the activation of several distinct signalling pathways involving a variety of structural and functional intracellular proteins and biochemical second messengers. Receptor activation involves a large variety of biomolecular events, including ligand–receptor recognition, receptor activation, protein recruitment at specific site, production of second messengers, activation or inhibition of kinases and phosphatases, internalization of receptors, cytoskeleton remodelling, external adhesion contact remodelling and gene expression. The modulation of these events is susceptible to produce a specific signature in the force signal, mainly in terms of amplitude, kinetics and duration.

AT_1 receptor activation in various cell type, such as endothelial cells, vascular smooth muscle cells and cardiac myocytes, is well known to activate several distinct signalling pathways (Fig. 17.6a), including two pathways signalling through the G-proteins Gq and $G_{12/13}$.[27,28] Activation of the G-protein Gq leads to phospholipase C (PLC) activation, production of inositol triphosphate (IP_3) from the phosphatidylinositol phosphate lipidic membrane pool, activation of the IP_3 receptor causing a release of Ca^{2+} from the endoplasmic reticulum into the cytosol and concomitant activation of numerous kinases. In contrast, $G_{12/13}$ activation by the AT_1 receptor leads to the activation of the small GTP-binding protein RhoA,[39] a process facilitated by a guanine exchange factor. Once activated, RhoA then activates its effector RhoA kinase (RhoK). In muscle and nonmuscle cells, both Gq and $G_{12/13}$ are largely involved in the regulation of contraction, motility and actin-based cytoskeleton dynamics.[40–43] To test their implication in force generation observed following the stimulation of the AT_1 receptor, cells are pretreated with the inhibitor blebbistatin. This agent is known to inhibit myosin-II ATPase activity and lower actin–myosin affinity,[44] thus interfering directly with the actomyosin contraction mechanism. Figure 17.6b shows a representative mechanical response recorded by AFM where the initial contractile response is totally abolished by blebbistatin, which indicates that the actomyosin contractile machinery plays a prominent role in the development of the mechanical response following the stimulation of the AT_1 receptor.

Actomyosin-dependent contractile response requires the phosphorylation of the regulatory myosin light chain (MLC). As illustrated in Fig. 17.6a, phosphorylation-dependent activation of MLC was previously shown to involve two alternative signalling pathways. The first one involves MLCK, a kinase regulated through the G-protein Gq and the second is due to the direct phosphorylation of MLC by RhoK, which is regulated by the $G_{12/13}$ signalling pathway. The Rho-kinase also has the capability to inactivate the MLC phosphatase. To test whether MLCK contributes to the mechanical response detected by AFM, cells are pretreated with an MLCK inhibitor (ML-9). The AngII-induced mechanical response in Fig. 17.6c is similar to the one shown in Fig. 17.2 and therefore indicates that the inhibition of MLCK has virtually no effect on the mechanical response of the cell to AngII. This is confirmed by phase contrast images, which shows that the spreading behaviour of the cell is not affected by ML-9. In contrast, pretreatment of the cells with an inhibitor of the kinase RhoK (Y-27632) largely diminished the initial mechanical response (Fig. 17.6d). This result points towards a main contribution of RhoK in the regulation of the mechanical response induced by AngII. Additionally, the apparent decrease in the signal fluctuation normally observed after the stimulation (6.30 ± 0.46 nm for AngII vs. 2.23 ± 0.28 nm for Y-27632/AngII) as well as the decrease in the spreading of the cell observed in the phase contrast micrograph is consistent with the role of RhoK in cytoskeleton remodelling. Naturally, the contribution of several other signalling elements in the mechanical response remains to be evaluated since AT_1 receptor signalling is not limited to classical G-proteins but also to G-protein independent pathways involving a variety of protein effectors such as Cdc42, Jak, β-arrestin and Src.[45] Given the fact that AFM-based force measurement requires no labelling, the contribution of these components could be tested using appropriate modulators (inhibitor, activator, silencing RNA) for their activities.

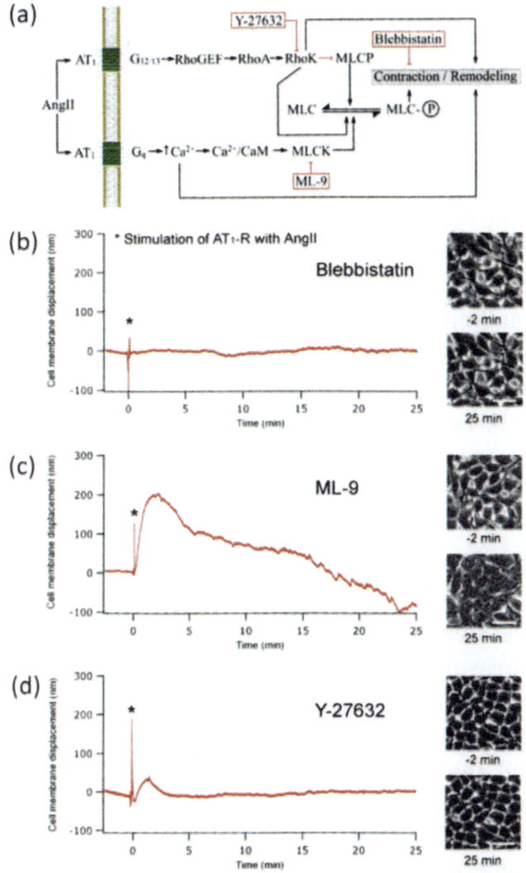

Figure 17.6. Identification of the signalling pathways involved in the mechanical response induced by the activation of the AT_1 receptor. (a) Schematization of the two independent signalling pathways known to be involved in contraction mediated by AT_1 receptor activation by AngII. The first involves Gq and ultimately leads to MLCK activation. The second proceeds through $G_{12/13}$, the activation of RhoK and the phosphorylation of MLC. In (b), (c) and (d) are presented the temporal mechanical responses to AngII after pretreatment with inhibitor targeting directly the actomyosin contraction (100 μM blebbistatin), MLCK (20 μM ML-9) and RhoK (10 μM Y-27632), respectively. Scale bars are 50 μm. Reprinted with permission from Ref. 23.

17.5 APPLICABILITY OF AFM-BASED MEASUREMENTS TO OTHER RECEPTOR/AGONIST SYSTEMS

AFM-based force measurement of cell receptor stimulation in living individual cells produces a robust signal that should allow for label-free detection of a large variety of cellular events. Membrane receptor such as those represented in the large family of G-protein-coupled receptors can be activated by a variety of extracellular signals, including neuropeptides, chemokines, biogenic amines, hormones, lipid-derived mediators, proteases, light, flavours and odours. Upon ligand binding, the membrane receptors transduce these signals into a quantity of intracellular responses that regulate cell functions via the heterotrimeric G-proteins.[46] Actin filaments provide the basic infrastructure for maintaining cell morphology and various functions. Cell cytoskeleton remodelling resulting in morphological changes is often observed after activation of signalling pathways involving calcium mobilization,[47,48] cAMP production[49] as well as small G-protein activation.[40,50,51] Activation of both the P2 purinergic receptors (P2Ys) by its agonist ATP in CHO cells[52] and the bradykinin receptor (B_2) by its agonist bradykinin in endothelial cells[53,54] is known to signal via Gq like the AT_1 receptor presented previously. Their activation is well documented to involve a large variety of biological processes related to vascular, immunological and intestinal functioning by the regulation of contraction, chemotaxis, proliferation, gap junction event *in vitro* and *in vivo*. These processes implicitly involve changes in morphological and mechanical parameters of the cells and are thus good candidate for AFM-based real-time measurements. Figure 17.7 presents AFM-based monitoring of cell activation showing a variation in the AFM signal, consistant with an initial contraction of the cell body followed by its remodelling. Although, the AT_1, P2Ys and the B_2 receptors are known to signal through the Gq pathway, comparison between the AFM data shows clear differences that could be quantified in term of amplitude, kinetic and duration of the response. These differences most likely have their origin in the efficacy of a receptor to generate a cell response at a given agonist concentration or in the contribution of alternative signalling pathways such as Gi/s or the small G-protein family selectively involved with these receptors. Naturally, the cell type is also very likely to play a determinant role in the development of a morphological and/or mechanical response following the activation of a receptor. Indeed, mechanically competent cells such as cardiomyocytes or smooth muscle cells are more likely to generate high amplitude mechanical responses owing to the presence of a functional actomyosin contractile machinery, when compared with endothelial or epithelial cell lines. However, as demonstrated for the AT_1 receptor, the contribution of these factors could be delineated using appropriate tools from pharmacological and molecular biology toolbox in conjunction with appropriate cell models.

Several other types of receptors either found at the cell membrane or in the cytosol are involved in morphological, motile or mechanical response in cells. These receptors include the receptors tyrosine kinase family

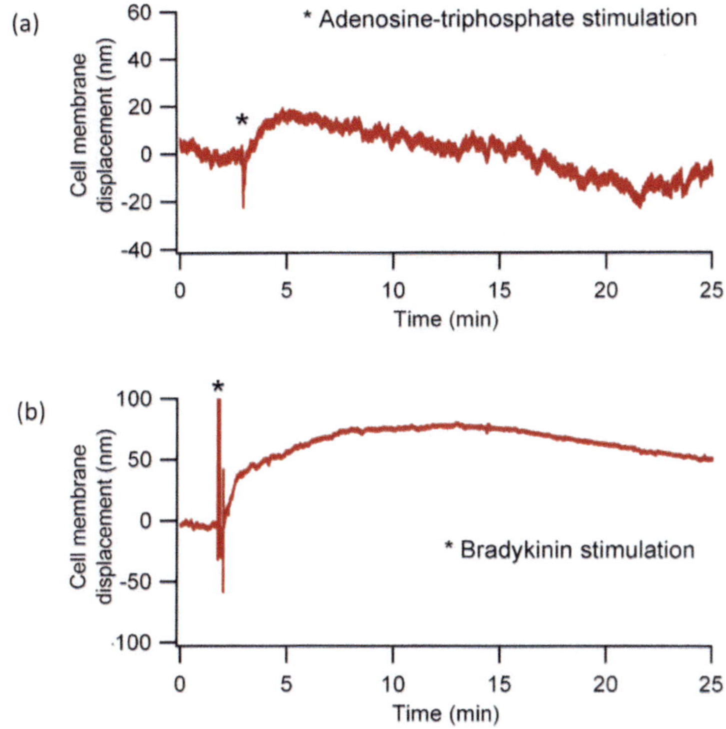

Figure 17.7. AFM measurement of the stimulation of two G-protein coupled receptors: (a) the purinergic receptors P2Ys by the agonist ATP (10 μM) in CHO cell and (b) the bradykinin receptor (B_2) by its agonist bradykinin (1 μM) in the immortalized endothelial cell line EA.hy 926.

mostly associated with the recognition of growth factor like VEGF (vascular endothelial growth factor),[55,56] ligand-gated ion channels like the $GABA_A$ chloride channel, the major inhibitory neurotransmitter in the mammalian central nervous system[57,58] or cytosolic/nuclear receptors which act mostly as a transcription factor like the receptor for testosterone.[59,60] Apart from these, there is also a multitude of receptors deprived of endogenous enzymatic activity nor coupled to G-protein such as the integrin receptor family which binds elements of the extracellular matrix and the Toll-like receptors family that are able to bind specific patterns found on pathogen proteins to initiate immune responses. Those receptors are able to recruit different intracellular protein to trigger specific signalling pathways though scaffolding and adaptor proteins such as the cytosolic small G-protein and various kinases and phosphatases. Alternatively any exogenous agents affecting intracellular calcium or cyclic AMP level or any factor affecting the activity of intracellular lipases or kinases will most likely influence the mechanical and morphological state of a given cell and should thus be detectable by AFM. This is demonstrated by AFM experiments in various receptor and cellular contexts (Fig. 17.8). First the stimulation of the toll-like receptor in endothelial cells using the lipopolysaccharide (LPS, an endotoxin found at the surface of Gram-negative bacteria involved in inflammatory response of the endothelium) produces a very strong contractile-like response followed by an important reorganization of the cell body as previously observed in various physiological contexts.[61] Stimulation of ligand-gated ion channels such as the ryanodine receptor found in a variety of muscle cells is well known to trigger large amplitude contractile responses because of cytosolic calcium mobilization. Ryanodine receptors are also expressed in other marginally mechanically competent cells such as the HEK-293.[23,62] In Fig. 17. 8b, ryanodine receptors are stimulated with caffeine, which increases the sensitivity of the receptor for intracellular calcium. Here again the AFM response is consistent with the contraction of the cell body occurring as a result of the activation of this receptor. Considering the central role played by the second messenger calcium, one can assume that any factor influencing its cytosolic level will be susceptible to generate mechanical activity at the cellular level. Figure 17.8c shows AFM-based monitoring of the highly motile glioblastoma cell line U251. Because of their motile behaviour, the baseline recorded on these cells usually exhibit sustained mechanical activity, which can nevertheless be stimulated by affecting cytosolic calcium level with the membrane ionophore ionomycin.[63] Hence, the results presented earlier clearly illustrate that AFM-based force experiment performed on individual cells represents a powerful tool to probe cellular processes involving morphological or mechanical activity, as controlled by a very large variety of ligands, receptors, enzymes or second messenger at the cellular level.

The examples presented in this chapter show that AFM-based force measurements allow for the detection of a large variety of cellular events originating from G-protein-coupled receptors, receptor tyrosine kinase and ligand-gated ion channels which are activated by a variety of extracellular signals. As mentioned earlier, the modulation of these events is susceptible to produce specific signatures in the force signal, mainly in terms of amplitude, kinetic and duration, that could be use to evaluate the mechanism of action and the potential of new drugs. In conclusion, future studies should aim at using a variety of cell models, receptor/ligands systems and pharmacological modulators to mechanically "fingerprint" the principal signalling pathways and deconvolute them from each other through AFM-based force experiments in conjunction with fluorescence imaging and monitoring of cell components.

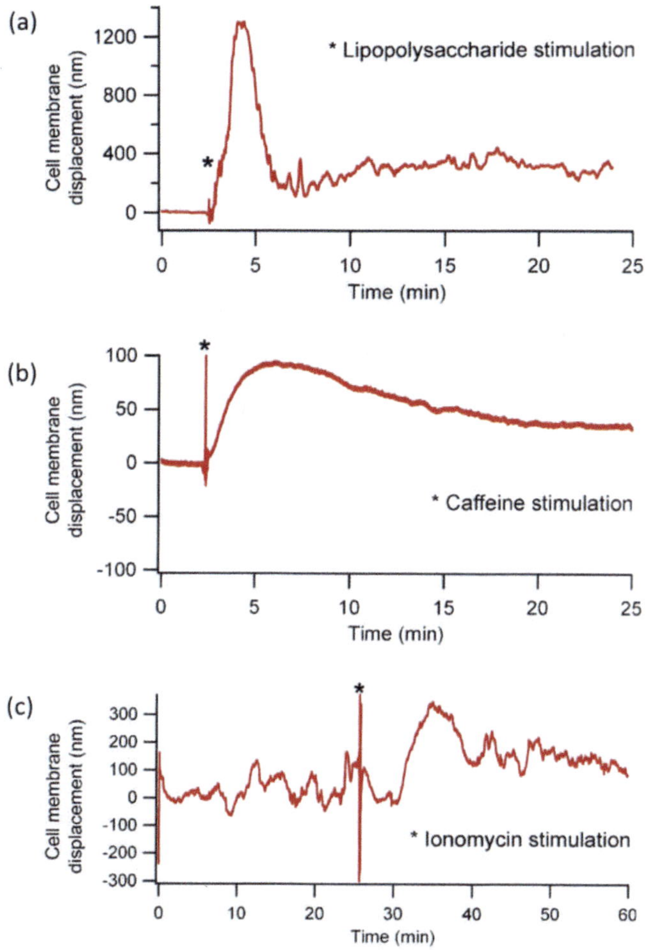

Figure 17.8. (a) AFM measurements of the stimulation of individual cells through (a) the toll-like receptor by the bacterial endotoxin LPS (100 μM), (b) the ligand-gated ion channel ryanodine by caffeine (1 μM) in HEK-293 cells and (c) the exposition of the U251 glioblastoma cell line to the ionophore ionomycin (1 μM).

References

1. Asphahani, F., and Zhang, M. (2007) Cellular impedance biosensors for drug screening and toxin detection, *Analyst*, **132**, 835–841.

2. Ciambrone, G. J., Liu, V. F., Lin, D. C., McGuinness, R. P., Leung, G. K., and Pitchford, S. (2004) Cellular dielectric spectroscopy: a powerful new approach to label-free cellular analysis, *J. Biomol. Screen.*, **9**, 467–480.

3. Verdonk, E., Johnson, K., McGuinness, R., Leung, G., Chen, Y., Tang, H. R., Michelotti, J. M., and Liu, V. F. (2006) Cellular dielectric spectroscopy: a label-free comprehensive platform for functional evaluation of endogenous receptors, *Assay Drug Dev. Technol.*, **4**, 609–619.
4. Fang, Y. (2006) Label-free cell-based assays with optical biosensors in drug discovery, *Assay Drug Dev. Technol.*, **4**, 583–595.
5. Fang, Y., Ferrie, A. M., Fontaine, N. H., Mauro, J., and Balakrishnan, J. (2006) Resonant waveguide grating biosensor for living cell sensing, *Biophys. J.*, **91**, 1925–1940.
6. Tanase, M., Biais, N., and Sheetz, M. (2007) Magnetic tweezers in cell biology, *Methods Cell Biol.*, **83**, 473–493.
7. Sterba, R. E., and Sheetz, M. P. (1998) Basic laser tweezers, *Methods Cell Biol.*, **55**, 29–41.
8. Evans, E., and Yeung, A. (1989) Apparent viscosity and cortical tension of blood granulocytes determined by micropipet aspiration, *Biophys. J.*, **56**, 151–160.
9. Chen, J., Fabry, B., Schiffrin, E. L., and Wang, N. (2001) Twisting integrin receptors increases endothelin-1 gene expression in endothelial cells, *Am. J. Physiol., Cell Physiol.*, **280**, C1475-C1484.
10. Svoboda, K., and Block, S. M. (1994) Biological applications of optical forces, *Annu. Rev. Biophys. Biomol. Struct.*, **23**, 247–285.
11. Hosu, B. G., Sun, M., Marga, F., Grandbois, M., and Forgacs, G. (2007) Eukaryotic membrane tethers revisited using magnetic tweezers, *Phys. Biol.*, **4**, 67–78.
12. Charras, G. T., and Horton, M. A. (2002) Single cell mechanotransduction and its modulation analyzed by atomic force microscope indentation, *Biophys. J.*, **82**, 2970–2981.
13. Sen, S., Subramanian, S., and Discher, D. E. (2005) Indentation and adhesive probing of a cell membrane with AFM: theoretical model and experiments, *Biophys. J.*, **89**, 3203–3213.
14. Rotsch, C., Jacobson, K., and Radmacher, M. (1999) Dimensional and mechanical dynamics of active and stable edges in motile fibroblasts investigated by using atomic force microscopy, *Proc. Natl. Acad. Sci. USA*, **96**, 921–926.
15. Domke, J., Parak, W. J., George, M., Gaub, H. E., and Radmacher, M. (1999) Mapping the mechanical pulse of single cardiomyocytes with the atomic force microscope, *Eur. Biophys. J.*, **28**, 179–186.
16. Pelling, A. E., Sehati, S., Gralla, E. B., Valentine, J. S., and Gimzewski, J. K. (2004) Local nanomechanical motion of the cell wall of *Saccharomyces cerevisiae*, *Science*, **305**, 1147–1150.
17. Pelling, A. E., Sehati, S., Gralla, E. B., and Gimzewski, J. K. (2005) Time dependence of the frequency and amplitude of the local nanomechanical motion of yeast, *Nanomedicine*, **1**, 178–183.
18. Pelling, A. E., Veraitch, F. S., Pui-Kei Chu, Nicholls, B. M., Hemsley, A. L., Mason, C., and Horton, M. A. (2007) Mapping correlated membrane pulsations and fluctuations in human cells, *J. Mol. Recognit.*, **20**, 467–475.

19. Hillebrand, U., Hausberg, M., Lang, D., Stock, C., Riethmuller, C., Callies, C., and Bussemaker, E. (2008) How steroid hormones act on the endothelium-insights by atomic force microscopy, *Pflugers Arch.*, **456**, 51–60.

20. Schafer, A., and Radmacher, M. (2005) Influence of myosin II activity on stiffness of fibroblast cells, *Acta Biomater.*, **1**, 273–280.

21. Oberleithner, H., Riethmuller, C., Ludwig, T., Shahin, V., Stock, C., Schwab, A., Hausberg, M., Kusche, K., and Schillers, H. (2006) Differential action of steroid hormones on human endothelium, *J. Cell. Sci.*, **119**, 1926–1932.

22. Pamir, E., George, M., Fertig, N., and Benoit, M. (2008) Planar patch-clamp force microscopy on living cells, *Ultramicroscopy*, **108**, 552–557.

23. Cuerrier, C., Benoit, M., Guillemette, G., Jr. Gobeil, F., and Grandbois, M. (2009) Real-time monitoring of angiotensin II-induced contractile response and cytoskeleton remodeling in individual cells by atomic force microscopy, *Pflugers Arch.*, **457**, 1361–1372.

24. Auger-Messier, M., Arguin, G., Chaloux, B., Leduc, R., Escher, E., and Guillemette, G. (2004) Down-regulation of inositol 1,4,5-trisphosphate receptor in cells stably expressing the constitutively active angiotensin II N111G-AT(1) receptor, *Mol. Endocrinol.*, **18**, 2967–2980.

25. Auger-Messier, M., Turgeon, E. S., Leduc, R., Escher, E., and Guillemette, G. (2005) The constitutively active N111G-AT1 receptor for angiotensin II modifies the morphology and cytoskeletal organization of HEK-293 cells, *Exp. Cell Res.*, **308**, 188–195.

26. Westphal, M., Jungbluth, A., Heidecker, M., Mühlbauer, B., Heizer, C., Schwartz, J., Marriott, G., and Gerisch, G. (1997) Microfilament dynamics during cell movement and chemotaxis monitored using a GFP–actin fusion protein, *Curr. Biol.*, **7**, 176–183.

27. Inagami, T., Kambayashi, Y., Ichiki, T., Tsuzuki, S., Eguchi, S., and Yamakawa, T. (1999) Angiotensin receptors: molecular biology and signalling, *Clin. Exp. Pharmacol. Physiol.*, **26**, 544–549.

28. de Gasparo, M., Catt, K. J., Inagami, T., Wright, J. W., and Unger, T. (2000) International union of pharmacology. XXIII. The angiotensin II receptors, *Pharmacol. Rev.*, **52**, 415–472.

29. Paredes, R. M., Etzler, J. C., Watts, L. T., Zheng, W., and Lechleiter, J. D. (2008) Chemical calcium indicators, *Methods*, **46**, 143–151.

30. Roe, M. W., Lemasters, J. J., and Herman, B. (1990) Assessment of Fura-2 for measurements of cytosolic free calcium, *Cell Calcium*, **11**, 63–73.

31. O'Connor, N., and Silver, R. B. (2007) Ratio imaging: practical considerations for measuring intracellular Ca^{2+} and pH in living cells, *Methods Cell Biol.*, 415–433.

32. Chin, D., and Means, A. R. (2000) Calmodulin: a prototypical calcium sensor, *Trends Cell Biol.*, **10**, 322–328.

33. Bers, D. M., and Grandi, E. (2009) Calcium/calmodulin-dependent kinase II regulation of cardiac ion channels, *J. Cardiovasc. Pharmacol.*, **54**, 180–187.

34. Vest, R. S., Davies, K. D., O'Leary, H., Port, J. D., and Bayer, K. U. (2007) Dual mechanism of a natural CaMKII inhibitor, *Mol. Biol. Cell*, **18**, 5024–5033.

35. Lemmon, M. A. (2008) Membrane recognition by phospholipid-binding domains, *Nat. Rev. Mol. Cell Biol.*, **9**, 99–111.

36. McGough, A. M., Staiger, C. J., Min, J., and Simonetti, K. D. (2003) The gelsolin family of actin regulatory proteins: modular structures, versatile functions, *FEBS Lett.*, **552**, 75–81.

37. Gordon, A. M., Homsher, E., and Regnier, M. (2000) Regulation of Contraction in Striated Muscle, *Physiol. Rev.*, **80**, 853–924.

38. Kumar, L. D., and Clarke, A. R. (2007) Gene manipulation through the use of small interfering RNA (siRNA): from in vitro to in vivo applications, *Adv. Drug Deliv. Rev.*, **59**, 87–100.

39. Ohtsu, H., Suzuki, H., Nakashima, H., Dhobale, S., Frank, G. D., Motley, E. D., and Eguchi, S. (2006) Angiotensin II signal transduction through small GTP-binding proteins: mechanism and significance in vascular smooth muscle cells, *Hypertension*, **48**, 534–540.

40. Park, J. K., Lee, S. O., Kim, Y. G., Kim, S. H., Koh, G. Y., and Cho, K. W. (2002) Role of rho-kinase activity in angiotensin II-induced contraction of rabbit clitoral cavernosum smooth muscle, *Int. J. Impot. Res.*, **14**, 472–477.

41. Emmert, D. A., Fee, J. A., Goeckeler, Z. M., Grojean, J. M., Wakatsuki, T., Elson, E. L., Herring, B. P., Gallagher, P. J., and Wysolmerski, R. B. (2004) Rho-kinase-mediated Ca^{2+}-independent contraction in rat embryo fibroblasts, *Am. J. Physiol. Cell Physiol.*, **286**, C8–C21.

42. Neves, S. R., Ram, P. T., and Iyengar, R. (2002) G protein pathways, *Science*, **296**, 1636–1639.

43. Kamm, K. E., and Stull, J. T. (2001) Dedicated myosin light chain kinases with diverse cellular functions, *J. Biol. Chem.*, **276**, 4527–4530.

44. Kovacs, M., Toth, J., Hetenyi, C., Malnasi-Csizmadia, A., and Sellers, J. R. (2004) Mechanism of blebbistatin inhibition of myosin II, *J. Biol. Chem.*, **279**, 35557–35563.

45. Hunyady, L., and Catt, K. J. (2006) Pleiotropic AT1 receptor signalling pathways mediating physiological and pathogenic actions of angiotensin II, *Mol. Endocrinol.*, **20**, 953–970.

46. Harden, T. K., Boyer, J. L., and Dougherty, R. W. (2001) Drug analysis based on signalling responses to G-protein-coupled receptors, *J. Recept. Signal Transduct.*, **21**, 167–190.

47. Hansson, K. M., Johansen, K., Wetterö, J., Klenkar, G., Benesch, J., Lundström, I., Lindahl, T. L., and Tengvall, P. (2007) Surface plasmon resonance detection of blood coagulation and platelet adhesion under venous and arterial shear conditions, *Biosens. Bioelectron.*, **23**, 261–268.

48. Yanase, Y., Suzuki, H., Tsutsui, T., Uechi, I., Hiragun, T., Mihara, S., and Hide, M. (2007) Living cell positioning on the surface of gold film for SPR analysis, *Biosens. Bioelectron.*, **23**, 562–567.

49. Kostenko, S., Johannessen, M., and Moens, U. (2009) PKA-induced F-actin rearrangement requires phosphorylation of Hsp27 by the MAPKAP kinase MK5, *Cell. Signal.*, **21**, 712–718.

50. Fukata, Y., Amano, M., and Kaibuchi, K. (2001) Rho-Rho-kinase pathway in smooth muscle contraction and cytoskeletal reorganization of non-muscle cells, *Trends Pharmacol. Sci.*, **22**, 32–39.

51. Sahai, E., Olson, M. F., and Marshall, C. J. (2001) Cross-talk between Ras and Rho signalling pathways in transformation favours proliferation and increased motility, *EMBO J.*, **20**, 755–766.

52. Iredale, P. A., and Hill, S. J. (1993) Increases in intracellular calcium via activation of an endogenous P2-purinoceptor in cultured CHO-K1 cells, *Br. J. Pharmacol.*, **110**, 1305–1310.

53. Van Kolen, K., and Slegers, H. (2006) Integration of P2Y receptor-activated signal transduction pathways in G protein-dependent signalling networks, *Purinergic Signal.*, **2**, 451–469.

54. Chretien, L., Laporte, S. A., Escher, E., Leduc, R., and Guillemette, G. (1999) The bradykinin B2 receptor couples less efficiently than the angiotensin AT1 receptor to the G protein Gq in transiently transfected COS-7 cells, *Receptors Channels*, **6**, 425–433.

55. Geer, P. V., Hunter, T., and Lindberg, R. A. (1994) Receptor protein-tyrosine kinases and their signal transduction pathways, *Annu. Rev. Cell Biol.*, **10**, 251–337.

56. Robinson, C., and Stringer, S. (2001) The splice variants of vascular endothelial growth factor (VEGF) and their receptors, *J. Cell. Sci.*, **114**, 853–865.

57. Barnard, E. A., Skolnick, P., Olsen, R. W., Mohler, H., Sieghart, W., Biggio, G., Braestrup, C., Bateson, A. N., and Langer, S. Z. (1998) International union of pharmacology. XV. Subtypes of γ-aminobutyric acidA receptors: classification on the basis of subunit structure and receptor function, *Pharmacol. Rev.*, **50**, 291–314.

58. Collingridge, G. L., Olsen, R. W., Peters, J., and Spedding, M. (2009) A nomenclature for ligand-gated ion channels, *Neuropharmacology*, **56**, 2–5.

59. Lu, N. Z., Wardell, S. E., Burnstein, K. L., Defranco, D., Fuller, P. J., Giguere, V., Hochberg, R. B., McKay, L., Renoir, J., Weigel, N. L., Wilson, E. M., McDonnell, D. P., and Cidlowski, J. A. (2006) International Union of Pharmacology. LXV. The pharmacology and classification of the nuclear receptor superfamily: glucocorticoid, mineralocorticoid, progesterone, and androgen receptors, *Pharmacol. Rev.*, **58**, 782–797.

60. Hynes, R. O. (2002) Integrins: bidirectional, allosteric signalling machines, *Cell*, **110**, 673–687.

61. Kaneider, N. C., Leger, A. J., Agarwal, A., Nguyen, N., Perides, G., Derian, C., Covic, L., and Kuliopulos, A. (2007) "Role reversal" for the receptor PAR1 in sepsis-induced vascular damage, *Nat. Immunol.*, **8**, 1303–1312.

62. Luo, D., Sun, H., Xiao, R. P., and Han, Q. (2005) Caffeine induced Ca^{2+} release and capacitative Ca^{2+} entry in human embryonic kidney (HEK293) cells, *Eur. J. Pharmacol.*, **509**, 109–115.

63. Komuro, H., and Rakic, P. (1996) Intracellular Ca^{2+} fluctuations modulate the rate of neuronal migration, *Neuron*, **17**, 275–285.

Chapter 18

INVESTIGATING MAMMALIAN CELL NANOMECHANICS WITH SIMULTANEOUS OPTICAL AND ATOMIC FORCE MICROSCOPY

Yaron R. Silberberg,[a,b] **Louise Guolla**[c] **and Andrew E. Pelling**[b,c]

[a] *Laboratory of Plasma Membrane and Nuclear Signalling, Graduate School of Biostudies, Kyoto University 1-1, Yoshida-Konoecho, Sakyo-ku, Kyoto, 606-8501, Japan*

[b] *London Centre for Nanotechnology, University College London, 17-19 Gordon Street, London, WC1H 0AH, UK*

[c] *Department of Physics, University of Ottawa, MacDonald Hall, 150 Louis Pasteur, Ottawa, ON K1N 6N5, Canada*

a@pellinglab.net

18.1 CELLULAR STRUCTURE AND NANOMECHANICS

The living cell is embedded in a complex mechanical environment, in which its behaviour is constantly influenced by mechanical cues arriving from the extracellular matrix (ECM) and from neighbouring cells. These signals regulate various cellular processes including differentiation, gene expression, mitosis, development, gastrulations and apoptosis.[1–15] Hence, understanding the mechanisms that are involved in cellular transduction of forces is crucial for understanding how those forces affect the living cell. Advances in live cell staining and imaging techniques allow the observation of intracellular structures with high temporal and spatial resolution. In addition, tools such as atomic force microscopy (AFM)[16] allow for the high-precision measurement and application of forces in the nano- and pico-Newton scale.[17] The ability to visualize changes in the intracellular

Life at the Nanoscale: Atomic Force Microscopy of Live Cells
Edited by Yves Dufrêne

www.panstanford.com

architecture of the living cell in real time, in response to locally applied extracellular perturbations, together with quantified measurements of changes in cell elasticity, can provide insights into the immediate effect of stress on the behaviour of the cell and on the mechanism in which forces are transmitted through the cell.[11,18–20]

The cellular cytoskeleton and organelles are some of the major elements responsible for modulating and controlling the mechanical properties of the cell. Moreover, internal remodelling and deformation of this complex network is highly dependent of the mechanics, topography and biochemistry of the microenvironment.[1–13] The cytoskeleton is an elaborate network of filamentous protein fibres spread throughout the cytoplasm. The cytoskeleton provides mechanical stability and often regulates controlled and dynamic mechanical processes such as migration, chromosome separation during mitosis and muscle contractions. The cytoskeleton also forms a network of tracks on which cargos, both membrane-bound such as the Golgi and mitochondria and non-membrane-bound such as mRNA and protein, can be transported.[21,22] Three major types of filaments that make up the cytoskeleton include the actin filaments, intermediate filaments (IFs) and microtubules (MTs).[23]

Actin filaments (Fig. 18.1a) are typically located below the plasma membrane and are cross-linked by a variety of proteins, including motor proteins such as myosin, which can generate forces and perform mechanical work. They are assembled from subunits called G-actin and are roughly 8 nm thick in diameter. The filaments are also linked to the plasma membrane through the Ezrin–Radixin–Moesin (ERM) proteins and membrane-spanning integrins, allowing signals from the ECM to be transmitted to the cytoskeleton, and vice versa.[24–27] MTs (Fig. 18.1b) are hollow, cylindrical filaments of approximately 25 nm in diameter, which are formed by the assembly of tubulin monomers. Individual MTs originate from a centrosome near the nucleus and can span the entire cell. They play an important role in organelle transport and organization, in cell division and chromosome distribution, and in mechanical stabilisation of the cell.[28] IFs (Fig. 18.1c), unlike actin filaments and MTs, are not polarised and are made of elongated polypeptide rods that are arranged in a coiled-coil structure of about 8–10 nm in diameter. They are located in two separate systems, one in the nucleus and one in the cytoplasm. Their main role is believed to be that of a passive mechanical absorber to provide structural reinforcement, particularly in cells that need to withstand strong mechanical stress such as epithelial cells.[29,30] Apart from the structural contribution, IFs also have cell-type-specific physiological roles and contribute to some gene-expression programmes.[29]

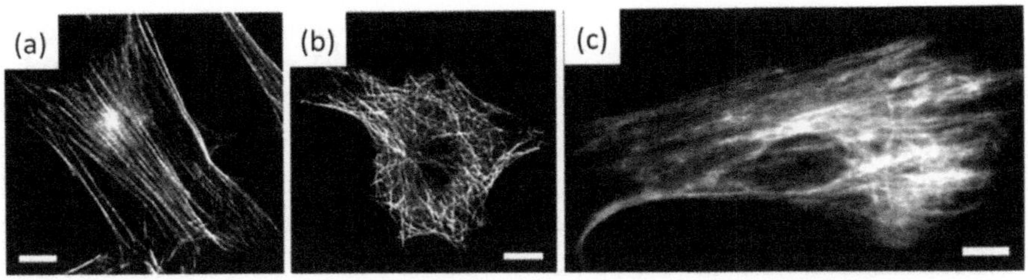

Figure 18.1. The cytoskeleton of mouse fibroblasts consists of actin (a), microtubules (b) and intermediate filaments (c). Scale bars = 10 μm.

18.2 APPROACHES TO STUDYING FORCE TRANSMISSION IN CELLS

Historically, interest in the mechanical properties of cells and tissues stems almost from the moment of their discovery. Using some of the first microscopes in the seventeenth century, motion of particles in and around cells was observed. From these microscopic movements, early scientists postulated that measurements could be taken that would allow for estimates of viscosity and other physical properties.[31] Technology at the time did not allow for quantitative measurements, and it was not until the early twentieth century that many physical properties began to be determined.[31] Many research groups around the world are investigating the phenomena of mechanotransduction and force transmission through cells, using a variety of techniques, and several different models now exist to explain the observed effects. Though the exact process of mechanotransduction and force transmission and their pathways have yet to be elucidated, there is consensus in which cellular structures appear to play an important part. Foremost among these are the cytoskeleton and its connections to the extracellular environment through the ERMs, focal adhesion complexes and mechanosensitive ion channels.

In the late 1980s, a variety of approaches were being employed to determine the mechanical properties of living cells and intracellular structures.[32-35] The most commonly used techniques at the time were micropipette aspiration,[34] a rudimentary cell poker[36,37] and application of a shear, twisting force using magnetic fields and ferromagnetic beads.[32,33,38,39] Micropipette aspiration involves suction of a portion of the cell into a tube with a diameter of a few micrometres (usually between 1 and 8 μm), using a known suction pressure (typically between 0.1 and 105 Pa). The geometry and known pressure are then used to determine the mechanical properties of the cell.[40] Early work investigated the viscoelasticity and cortical tension of red blood cells.[34]

Magnetic tweezers were later developed to utilize magnetic fields to generate forces on small paramagnetic beads with a typical size of 0.1–5 μm. Resulting displacements of the beads can then be used to deduce rheological properties of living cells. Beads were functionalized and bound to integrin receptors on the cell membrane to measure viscoelastic properties of fibroblast cells[41] and their response to deformation.[42] A series of experiments[38] using magnetic twisting cytometry clarified that applied force was transmitted through integrin receptors found at focal adhesions, which are directly connected to the cytoskeleton. Cells with RGD-coated ferromagnetic beads attached to integrin receptors experienced a force-dependent increase in stiffness, while beads attached to other receptors did not experience the same effect. It was also found that this effect was proportional to an increased number of connections to the ECM. Together, this indicates that integrins act as mechanoreceptors which transmit signals to the cytoskeleton from the ECM and directly modulates cell rigidity. Published evidence supports the transmission of force through focal adhesions using a combination of micromanipulation with glass needles and cells expressing green fluorescent protein (GFP) conjugated to actin.[43]

Advances in optical technology have also led to several interesting approaches to studying cell nanomechanics. Optical tweezers (laser traps) are a highly sensitive technique in which dielectric spherical beads are trapped at the focus of a laser beam.[44] The surface of the bead is functionalized and can be attached to a cell membrane or other molecules. The laser beam creates a field that "traps" the bead at the focal point, allowing measurement of forces acting on the bead. Using this method, forces such as those generated by single molecules such as kinesin motors[45] and cytoskeleton–integrin linkage[46] were successfully measured. The ability to apply a controlled and localized force to a cell demonstrated that increased force on focal adhesion complexes and stress fibres leads to an increased calcium ion influx near those focal adhesion complexes. This supports the theory that mechanosensitive ion channels can be activated by increased tension in the cytoskeleton.[47] It is also possible to use a focused laser with enough precision to sever a single cytoskeleton filament, known as laser ablation.[48] A series of laser ablation experiments[20] demonstrated that stress fibres will mechanically retract (as opposed to depolymerization) and force pseudo focal adhesions along the basal membrane as they slide along it. Evidence also supports the presence of a "tension sensor" protein, zyxin, which localizes to points of increased tension along the cytoskeleton and at adhesion sites, both new and old, and disappears immediately following a loss of tension. Finally, in a related technique, optical stretching[49] has been demonstrated to be an extremely powerful tool in the study of cell nanomechanics. Unlike an optical trap, the optical stretcher utilizes two unfocussed lasers to trap and stretch suspended cells in solution.

The rheological and mechanical properties of cells have been measured and directly linked to their function, metastatic potential and cytoskeletal architecture.[49-53]

With the development of the AFM it quickly became possible to apply known and controlled forces to very localized positions on living cells as well as measure their mechanical properties in physiological conditions. The AFM is an effective tool for investigating mechanical and material properties of biological samples in their native conditions. These include the investigation of cellular strain distribution and cytoskeleton disruption in response to stress,[54] and the extraction of Young's modulus.[55-58] During such experiments, living cells can be kept at physiological conditions by heating of the stage on which the culture dish is mounted or having the whole microscope apparatus inside a controlled incubator. pH levels can be monitored and adjusted during the experiment using buffered culture media. Recent technical developments have integrated traditional microscopy methods, such as fluorescence and laser scanning confocal microscopes with AFM systems. This has enabled the simultaneous measurement of material properties of living cells and their biological responses.[54,59-61] The combined AFM–fluorescence microscope apparatus can also be used to apply controlled mechanical perturbations on the living cell, while imaging the real-time deformations and/or displacements that occur intracellularily.

The AFM has found an extremely large number of very different applications in biology. Not only is the AFM capable of delivering and measuring small forces and mechanical dynamics, it is also an extremely powerful imaging tool. Now capable of sub-nanometre resolution imaging at high speeds (>30 fps) the AFM has found many uses in studying molecular structures in physiological environments with high temporal and spatial resolution. Moreover, the AFM is also highly sensitive to small forces and capable of delivering forces over several orders of magnitude (pN-nN). The AFM has been employed to detect local nanomechanical dynamics of living mammalian and bacterial cells undergoing important physiological processes, as well as detecting the onset and progression of disease states.[62,63] The shear number of imaging and force spectroscopy applications in artificial bilayers, mammalian cells, bacteria, multicellular complexes, tissues is beyond the scope of this particular chapter but have been reviewed previously.[17,64,65] Therefore, here, we limit our discussion to living mammalian cells and applications that utilize the AFM. Specifically, we will discuss the AFM as a tool to deliver temporally and spatially controlled localized nanomechanical forces to living mammalian cells while simultaneous optical measurements are performed to image biological responses at the single cell level.

The popularization of fluorescent tags, particularly through transfection or commercial dyes, became useful for direct visualization of the effect of applied force on the inner structure of the cell. Previous work combined fluorescence imaging techniques with force-application methods, to observe structural intracellular changes in response to extracellular perturbations. Among these studies are the observations of changes in the actin and MT cytoskeleton of live fibroblast cells in response to deformations produced by glass needles, which were visualized using GFP-tagged cytoskeletal proteins.[66] Deformations of the IF cytoskeleton were analysed by visualizing GFP–vimentin in live endothelial cells before and after the application of shear stress in a flow chamber.[67] In another study that combined the magnetic bead twisting technique with GFP–fluorescent imaging, application of forces to focal adhesions by the use of specifically coated beads resulted in displacements of actin filament bundles at distances of 20–30 μm from the beads.[68] A similar technique was used to visualize displacements of intracellular organelles such as mitochondria[69] and to analyse the propagation of forces to the nucleus by quantifying displacements of nucleolar structures in response to load.[70] Visualization of responses to extracellular perturbations is not limited to the tracking of natural organelles or cytoskeletal components: in a recent study, AFM was used to apply perturbations onto live, adherent cells, while quantifying stress propagation through the cell by tracking of integrin-bound fluorescent microspheres.[71]

Here, we will review some of our previous work[18,19,72–74] on the application of simultaneous AFM and fluorescence microscopy or laser scanning confocal microscopy (LSCM) in the context of living mammalian cells. Three examples will be presented which demonstrate the utility of simultaneous AFM and optical approaches to understand the origin and control of force transmission inside and through living mammalian cells to the underlying substrate.

18.3 CELLULAR NANOMECHANICS AND FORCE TRANSDUCTION THROUGH THE CELLULAR ARCHITECTURE

18.3.1 Mitochondrial Displacements in Response to Force

Mitochondria are semi-autonomous and highly dynamic organelles, which have the ability to change their shape and their location inside the living cell.[75] Localization and rearrangement of mitochondria in higher eukaryotes is known to be dependent on the MT. More recent research suggests that actin filaments have an important role as well, such as facilitating mitochondrial organization in yeast and vertebrate neurons,[76,77] and controlling mitochondrial movement and morphology.[78] Given the strong association of mitochondria with the cytoskeleton, it is predicted that forces locally applied by the AFM tip will affect their arrangement through mechanical transduction.[79–81]

Previously we have shown that nuclei and cytoskeleton deformations were observed following local AFM indentation.[72] Here, we review our work that demonstrates the effect of instantaneous displacement of fluorescently labelled mitochondria upon the static application of force with the AFM.[18,73] Mitochondria form dense three-dimensional (3D) networks around the nucleus and become flattened and more sparsely distributed at the edges of the cell. We examined how locally applied forces above the nucleus are physically transmitted over long distances to the cell edge. It was impossible to distinguish and separate two-dimensional (2D) versus 3D movement of mitochondria around the nucleus in response to applied force from the AFM tip because of the thickness of the cell. Therefore, we limited our analysis to the cell edge. In these regions, the cell is very flat, as little as 200 nm thick, and mitochondria are assumed to move perpendicular to the normal force delivered by the AFM tip over the nucleus, enabling accurate measurement of physical force transduction from the AFM tip. Furthermore, individual mitochondria can be resolved much more clearly in these regions, allowing for accurate image registration and tracking analysis.

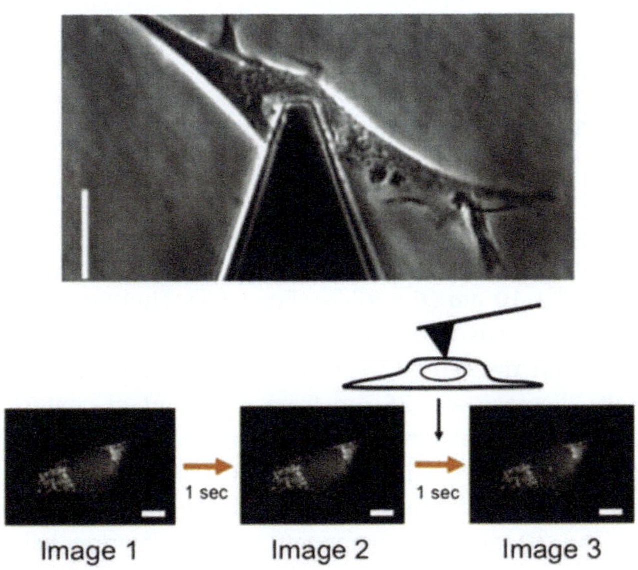

Figure 18.2. A typical phase-contrast image of the AFM tip and a living cell (scale bar = 10 μm).[18] A sequence of images is then acquired at 1 second intervals. Three images were picked for analysis: 2 images taken prior to AFM indentation (images 1 and 2) and the one image that followed the indentation (image 3). Changes between image 1 and 2 reflect basal mitochondrial movement, while changes between image 2 and image 3 reflect the force-induced movement resulting from AFM indentation.

Mitochondria are dynamic structures, which display basal movements driven by the cytoskeleton. Thus, to measure and distinguish baseline displacements from displacements caused by the AFM tip, we designed the following experiment that included a built-in control for each cell measured.[18] NIH3T3 cells were cultured in 60 mm culture dishes. Dishes were mounted on the temperature-controlled stage of a simultaneous AFM–fluorescence microscope that was used to deliver precise forces to living cells. Prior to image capture, the AFM tip was first optically positioned ~2 μm above the cell and the time to contact was approximately 250 ms (Fig. 18.2).

Then image capture was started at 1 frame/sec, and after collecting several images of basal movement of mitochondria, the tip was brought into contact with the cell at an applied force of 10 nN. The contact time of the tip was ~3 seconds, and the total imaging time was typically 10 seconds. By creating two fluorescence image overlays (images 1 + 2, prior to the perturbation and images 2 + 3, after perturbation) we are able to qualitatively observe that the AFM tip does indeed produce increased displacements of the mitochondria (Fig. 18.2). Besides the obvious displacement around the centre of the cell, displacements further away towards the cell edge are also visible. To produce a quantitative displacement analysis, we used the Particle Tracker plug-in for ImageJ. For each cell measured, displacements were calculated for the average basal displacements in addition to the average perturbed displacements of individual mitochondrial structures.

The results reveal that ~80% of cells displayed an increase in mitochondrial displacement over the basal movements within each cell. We found that the average basal displacement of mitochondria was 114 ± 6 nm. However, after pushing with the AFM tip, the average displacement increased to 160 ± 10 nm ($P < 8E\text{-}7$) (Fig. 18.3). Therefore, locally applied forces over the nucleus induced a statistically significant rearrangement of mitochondria at the cell edges, increasing ~40% following indentation at an average distance of ~26 μm from the point of contact. Moreover, mitochondria are often observed to move both towards and away from the point of contact (Fig. 18.4). In our analysis, it is clear that the mitochondria around the nucleus also moves in response to the tip; however, it is difficult to separate the 2D and 3D components of the motion using standard fluorescence microscopy, and we leave that analysis for a future study with confocal microscopy (see section 18.3.2).

To investigate the role of the cytoskeleton in transmitting force, we used the anti-cytoskeletal drugs cytochalasin D (CytD) and nocodazole to selectively disrupt both the actin and MT networks, respectively.[54] Cells were incubated for 30 minutes with each of the drugs (10 μM nocodazole,

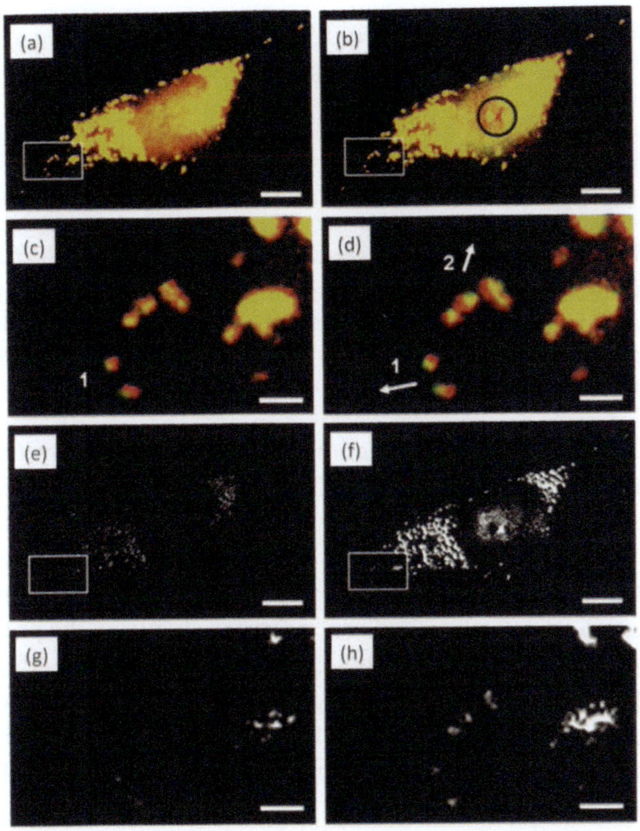

Figure 18.3. Comparison between basal and force-induced mitochondrial displacements.[18] The left column shows the basal displacement (control) and the right column shows the displacement following AFM indentation. (a) Overlay of consecutive fluorescent images 1 (red) and 2 (green), both acquired prior to AFM indentation. (b) Overlay of consecutive images 2 (red, before AFM indentation) and 3 (green, after perturbation). The yellow colour results from the red-green overlay, and is much denser around the nucleus where mitochondria are much sparser. The reflection image of the perturbing AFM tip can be seen in the centre of the nucleus (b, circle). (c–d) Magnified sections of the cell where motion of mitochondria in different directions can be visually observed. Arrows show direction of displacement of different mitochondrial structures (d1,2; the green colour shows the post-indentation image and thus the direction of displacement). Although some natural displacements are evident in the control image (c, 1), the displacement in the post-indentation image is higher and includes a larger number of organelles (d, 1 and 2). (e–h) Subtraction images of control (e) and post-indentation (f), and magnified images of the relevant sections (g–h). The magnitude of the post-perturbation displacement can be clearly seen, in comparison with the control. Scale bars are: a–b, e–f: 10 μm; c–d, g–h: 2 μm.

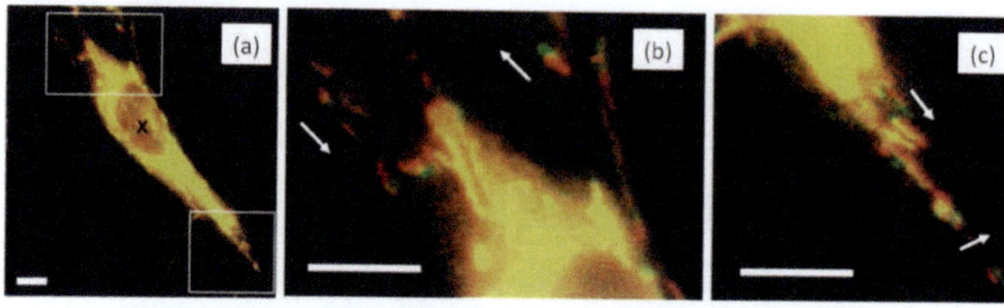

Figure 18.4. Mitochondrial displacements following AFM indentation. (a) An overlay of images taken before (red) and after (green) indentation. (b c) Magnified section of the cell, where mitochondrial structures clearly show displacements into different and, in some cases, opposite directions (b, arrows). Scale bars are 10 μm.

5 μM CytD), prior to experimentation. We found that the average natural displacement of mitochondria in cells treated with CytD was 56 ± 3 nm and 58 ± 3 nm ($P > 0.6$) after perturbation with the AFM tip (Fig. 18.5a). For nocodazole-treated cells, the average natural displacement was 57 ± 2 nm and 54 ± 2 nm ($P > 0.3$) after perturbation (Fig. 18.5a). Therefore, the results show no statistically significant difference between the pre- and post-perturbation displacements, in both cases. These results clearly show that mitochondrial displacements following a locally applied force are completely dependent on an intact actin and MT cytoskeletal network. However, the natural displacements of the mitochondria in cells pretreated with CytD and nocodazole are significantly different ($P <$ E-20) compared with untreated

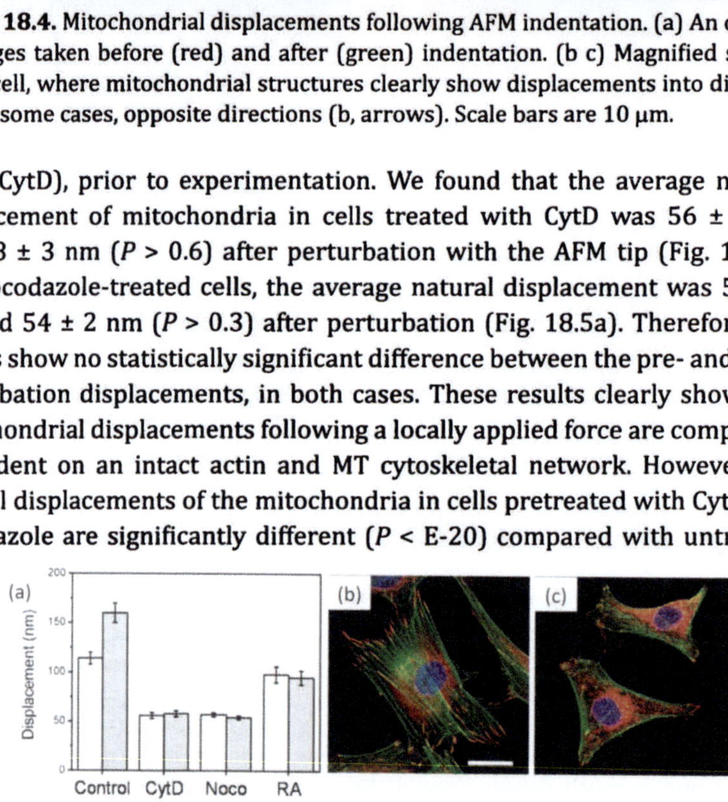

Figure 18.5. (a) Comparison of the difference in mean average displacement of mitochondria between the control (white bars) and the post-perturbation (grey bars) images for cells left untreated and treated with the CytD, nocodazole and RA. The average displacement of mitochondria in untreated cells increased ~40% in response to perturbation with the AFM tip. The natural displacement of mitochondria in cells treated with CytD and nocodazole was ~50% lower than control cells, and there was no significant increase in displacement in response to locally applied forces. (b) Focal adhesions (red) appear as point-like structures at the end of F-actin filaments (green) and act to anchor the cell to the substrate (scale bar = 10 μm). (c) After treatment with retinol, the number of focal adhesions per cell is greatly reduced throughout the cell contact area.

cells. The average natural displacement was ~50% lower in cells treated with either one of the two drugs, in comparison with the natural displacement in untreated cells. These data suggest that the cytoskeletal network has an important role to play in governing natural motions of mitochondria within living cells. It shows that natural mitochondrial motion is strongly dependent on both intact actin filaments and the MT network, confirming findings on the cytoskeleton's role in mitochondrial transport.[78,82,83]

To examine the role focal adhesions play in governing force transduction through the cytoplasm, we treated the cells with retinoic acid (RA). Retinoids are naturally occurring derivatives of retinol (vitamin A) and have an important role in gene regulation and control in a variety of cellular and tissue processes, including proliferation, cell differentiation and apoptosis.[84,85] These compounds also have wider functions reflected in their diverse effects on the regulation of specific genes,[86] including impacting on cell adhesion mediated by integrin cell adhesion receptors.[87] RA has been shown to stimulate keratinocyte growth in culture and also to inhibit the ECM molecules fibronectin (FN) and thrombospondin.[87] Similar results on FN inhibition were observed on 3T3 fibroblasts. Adhesion to the substrate was also reduced after treatments with RA, together with a decrease in attachment and spreading.[87,88] Treatment with 20 μM RA led to a distinct decrease in the number of focal adhesions by ~50% while leaving the cytoskeleton intact[74] (Fig. 18.5b,c). Concomitant with the decrease in FAs was a decrease in the basal movement of mitochondria and no effect of applied forces on mitochondrial displacements in a fashion similar to CytD and nocodazole-treated cells (Fig. 18.5a).

In each case of drug treatment, the cellular Young's moduli were also observed to decrease significantly[74] (Fig. 18.6). Moreover, force curves measured with pyramidal tips and cantilevers modified with 19 μm microspheres demonstrate that although the absolute value of the Young's modulus was dependent on tip geometry, the relative decrease in Young's modulus remains approximately constant (Fig. 18.6). These data demonstrate that the local and global mechanical properties of the cell are significantly impaired after treatment with the drugs. Importantly, it is clear that the cell requires an intact actin and MT cytoskeleton in addition to strong connections to the microenvironment via focal adhesions to maintain and regulate its stiffness. Moreover, all three of these elements of the cytoarchitecture are required for the transmission of force throughout the cell. The data presented thus far have revealed that the mechanical properties of the cell are regulated through the complex interplay of several architectural elements. By tracking the displacement of intracellular

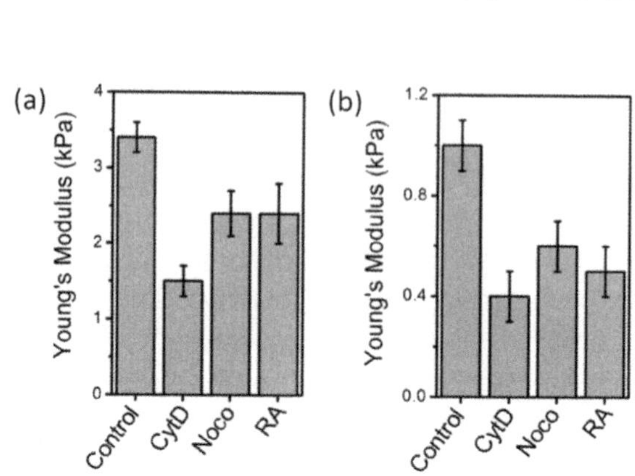

Figure 18.6. Average Young's modulus of NIH3T3 cells measured over the nucleus with (a) pyramidal tips and (b) 19 μm polystyrene sphere modified tips. Drug treatments clearly result in a mechanical softening of the cell. Although the absolute modulus of the cell is dependent on the tip geometry, the relative change after treatment with each drug is similar. The results demonstrate that the cells are becoming softer locally and globally, which has a clear impact on the transmission of force through the cytoarchitecture.

organelles, we can infer the transmission of force through the cytosol likely via the cytoskeleton. However, we clearly observe mitochondria moving both towards and away from the point of force on the nucleus. This implies that force transmission is a complex process and that the cell is not behaving as an isotropic and homogeneous material. In the next section, we will demonstrate the direct visualization of cytoskeletal deformation in response to applied loads from the AFM tip with simultaneous LSCM.

18.3.2 Force Transduction Through the Cytoskeleton

Utilizing simultaneous LSCM and AFM, we have demonstrated that it is possible to directly correlate cytoskeletal viscous deformation in response to applied mechanical loads.[19] Control of force dissipation was visualized by generating cells transiently expressing GFP tagged to the actin and MT cytoskeleton. In the previous section, we inferred that NIH3T3 cells transmit force via the cytoskeleton, resulting in the movement of mitochondria. NIH3T3 cells were transiently transfected with 1 μg of plasmid DNA encoding for actin–GFP. Utilizing a simultaneous AFM and fluorescence microscope (as described in section 18.3.1), we were able to identify a cell expressing actin–GFP and position the AFM tip above the nucleus. Images of the cell were then acquired before and after indentation with the AFM tip at a maximum force of 10 nN. Figure 18.7 shows the deformations in the actin cytoskeleton that resulted from AFM indentation. Images are coloured so that a red (before indentation) and green (after indentation) overlay can be created. As can be seen, changes in the actin fibres are visible at locations far from the indentation point. Comparing the natural and the indented states, some filaments at the cell edge assume a curved state following indentation (green), in comparison with their pre-indented stretched state (red) (Fig. 18.7). Significant deformation is taking place throughout the actin network in response to a point load over the nucleus. This is particularly important as we postulated that mitochondria move in response to this type of deformation. Moreover, the deformation is taking place over very short timescales (<5 seconds).

To investigate longer timescale (60 seconds) viscous deformation and relaxation processes through the cytoskeleton we performed stress-relaxation tests.[19,89] In these experiments, a cell was allowed to relax for 60 seconds under an initial contact force (2 nN) from the AFM tip (Fig. 18.8). We have shown previously[19] that the relaxation time and viscosity

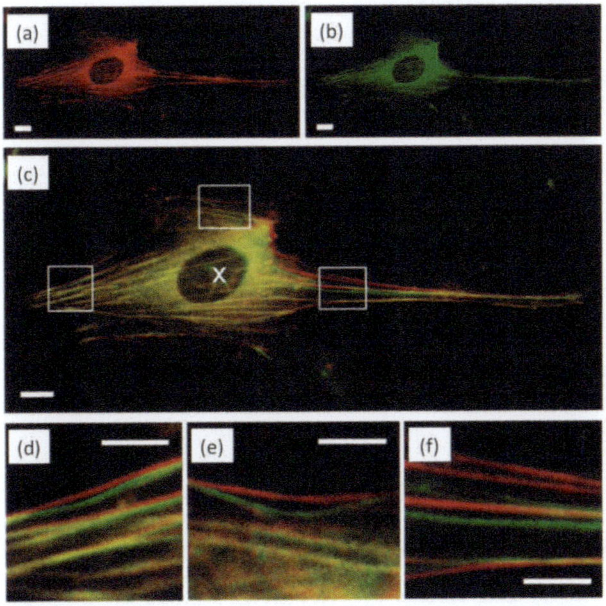

Figure 18.7. Deformation in the actin cytoskeleton following AFM indentation. Images of actin-GFP-transfected cells were taken prior (a, green) and after (b, red) AFM indentation, with 4 second interval between the two images. The overlayed images are shown in (c). Local deformations of the actin cytoskeleton can be clearly seen far from the indentation point (c, white cross). d, e and f are magnified areas, marked by the white squares in (c). Scale bars are: a–c, 10 μm; d–f, 5 μm.

of the cell can be determined by recording the cantilever deflection as it decreases during cell relaxation and internal remodelling of the cytoskeleton (Fig. 18.8). To qualitatively visualize the deformation and relaxation processes in the actin, MT and IF cytoarchitecture we employed cells (human foreskin fibroblasts cultured as described in section 18.3.1) transiently expressing GFP-actin, GFP-tubulin and GFP-vimentin, respectively (Fig. 18.9).

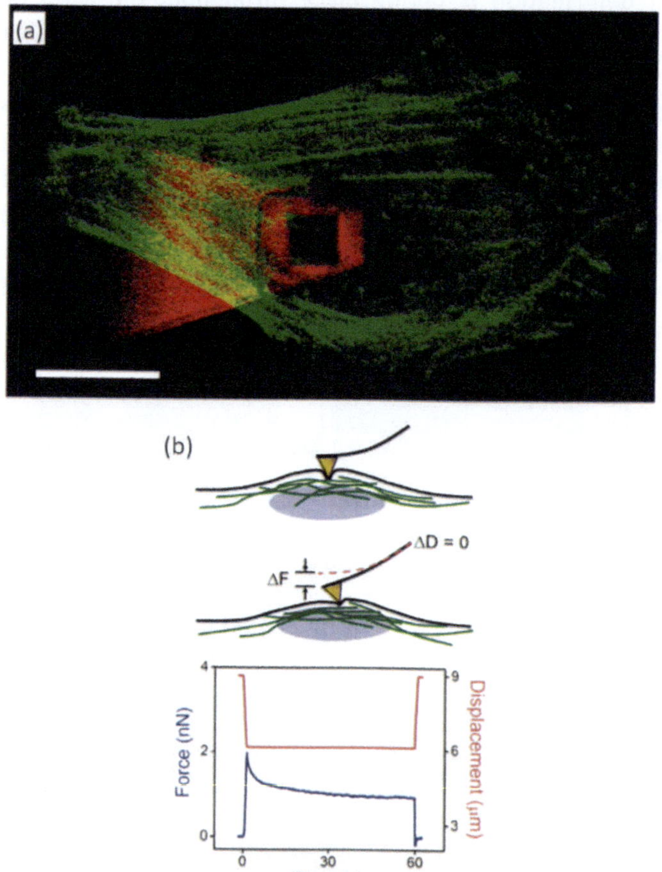

Figure 18.8. (a) LSCM image of a cell transiently expressing GFP-actin (green).[3] The AFM tip can be visualized by capturing the autofluorescence resulting from excitation with a 405 nm diode laser (scale bar = 10 μm). b) Stress-relaxation experiments can be performed in which the AFM tip is brought into contact with the cell at a specific setpoint force. The cells are then allowed to relax while the cantilever deflection is monitored as a function of time. This type of measurement yields the cellular viscosity. Confocal stacks acquired immediately before and after the experiment allow one to directly visualize cytoskeletal deformation in response to local forces.

A simultaneous AFM and LSCM was used to acquire confocal stacks before and after the stress-relaxation experiments allowing us to examine the 3D deformation and relaxation of the cytoarchitecture.[3] The two stacks were then subtracted to produce an image in which contrast is generated from the movement of specific structures relative to their initial positions. Several general phenomena were observed to occur during the viscous relaxation and deformation of the architecture in this cell type. F-actin filaments were not observed to significantly deform or remodel under 2 nN and up to 10 nN of force. This is in contrast to mouse NIH3T3 fibroblasts (Fig. 18.7) in which F-actin filaments were observed to deform readily.

The MT and IF networks clearly deform in response to force applied above the nucleus (Fig. 18.9), as evidenced by the formation of filamentous structures in three dimensions after subtraction. MT deformation notably occurs throughout the cell, including at the cell edge often >30 μm away from the point of force. Furthermore, filaments do not appear to move in a

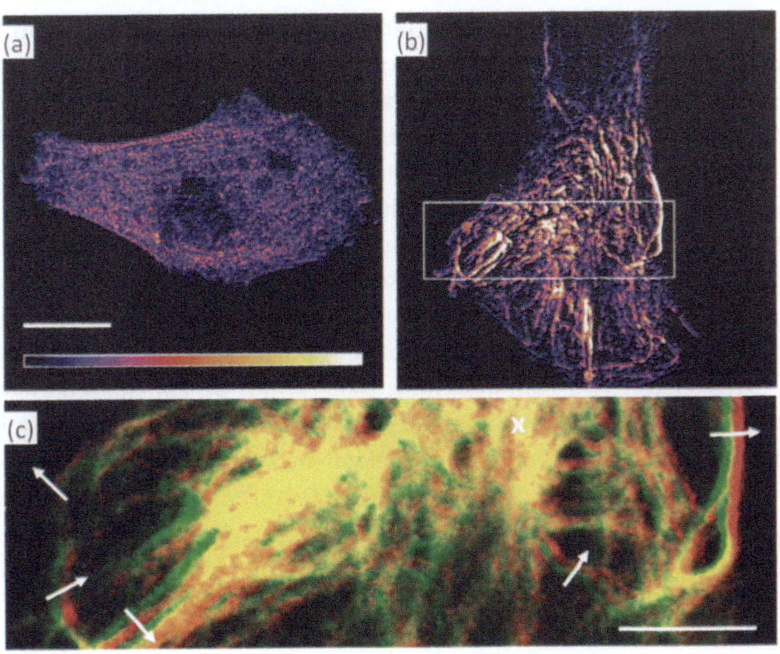

Figure 18.9. (a) A subtraction image of GFP–actin before and after the stress-relaxation experiment reveals no significant F-actin deformation in human fibroblast cells (scale bars = 10 μm).[3] However, the microtubule cytoskeleton (b) reveals significant deformation and as evidenced by filamentous contrast in the subtraction image. (c) A zoom of the area in (b) presented as a green-red overlay demonstrates how filaments move both towards and away from the contact point (white cross).

purely circular deformation profile away from the point of force (Fig. 18.9). Rather individual filaments were observed to move both towards and away from the point of force. This is in contrast to the IF network which tends to undergo a uniform outward deformation (data presented previously[19]) around the nucleus and filaments at the cell edge do not appear to be significantly deformed.

Several important characteristics are revealed through these relatively simple experimental approaches. First, forces are transduced rapidly through the cellular architecture. Cytoskeletal deformation occurs within seconds of a small point load and occurring many tens of microns away from the contact point. This has important implications to our interpretation of locally measured mechanical properties with AFM tips as the whole cell is responding to such point loads especially during force curve measurements. Secondly, there appears to be an important dependence of force transduction pathways on the species type of the cell. F-actin in human fibroblasts does not appear to deform significantly in response to point loads but the opposite is true for mouse fibroblast cells. This difference in force transduction pathway is likely due to the 3D arrangement in F-actin in these two cell types. F-actin tends to align along the bottom of the cell (under the nucleus) in human fibroblasts, but in mouse fibroblasts it is found around and above the nucleus. Therefore, force delivered via the AFM tip is more likely to be transmitted through the F-actin in mouse fibroblasts. This species type dependence should make it clear that "generalized" models of cell mechanics must somehow take into account cell type. Finally, in the case of MT deformation, it was observed that tubulin filaments deform both towards and away from the contact point. This is clear evidence that the cytoskeleton is a complex mesh that cannot be considered isotropic. Moreover, this type of behaviour was only observed in the MT cytoskeleton and not in the F-actin or IF cytoskeletal networks. These initial studies clearly indicate that much more work is required (such as simultaneous visualization of more than one filament system, quantitative filament tracking and finally modelling) to fully understand how force is transduced through the 3D cytoarchitecture.

18.3.3 Cellular Traction Forces in Response to Mechanical Loading

The development of traction force microscopy (TFM) approaches has allowed the investigation of cellular traction mechanics on substrates during migration and other physiological processes.[90–99] In early studies, cells were grown on silicone gels where gel wrinkling corresponds to the magnitude of cellular traction forces.[90–92] To quantify traction forces, cells are often grown on soft deformable substrates which are embedded with fiduciary fluorescent tracking particles.[94,99] In many TFM applications, bead displacements are measured during cell migration. As the material properties of the deformable substrate are known and controllable, these bead displacements can be converted into forces, allowing local maps of traction force to be created.[94,99] Several important early studies have demonstrated the usefulness and biological relevance of TFM in the study of cellular nanomechanics.[90–101] Typically, substrates of polyacrylamide, gelatin (GE) or polydimethylsiloxane pillars have been used successfully and have revealed striking examples of how living cells respond and affect their local mechanical environments.[94–96,99,102]

Here, we present a method in which a biocompatible glutaraldehyde cross-linked GE (GXG) substrate, with 200 nm fluorescent beads, can be poured directly into a standard tissue culture dish (or onto any other substrate) in a simple one-step approach (Fig. 18.10). The GXG substrate has a high melting point (>60°C) allowing for mammalian cell culture, it is completely biocompatible without further surface functionalization (but able to be functionalized if necessary), it is optically clear allowing for fluorescence microscopy and the substrate stiffness can be controlled by varying the percentage of GE. Finally, we demonstrate the application of simultaneous traction and atomic force microscopy (TAFM).

Biocompatible GXG gels for TAFM were produced from 5% solutions of GE. 200 nm red or green fluorescent microspheres were mixed thoroughly with the GE solution. Then the GE was cross-linked with glutaraldehyde and spread evenly over the surface of a 60 mm plastic culture dish. No functionalization of the surface was required for cell growth but typical surface molecules (poly-L-lysine, FN, gelatine) were found to be compatible with the GXG substrate (Fig. 18.10). GXG substrates were found to have a Young's modulus of ~28 kPa. C2C12 muscle myoblast cells were used as they are inherently sensitive to mechanical force. Mechano-stimulation of these cells is a critical step in the myogeneic pathway during muscle formation that involves the ability of these cells to apply and generate traction forces within their micro-environment. Therefore, we expect them to respond and alter their cellular traction force dynamics in response to mechanical stimulation with the AFM.

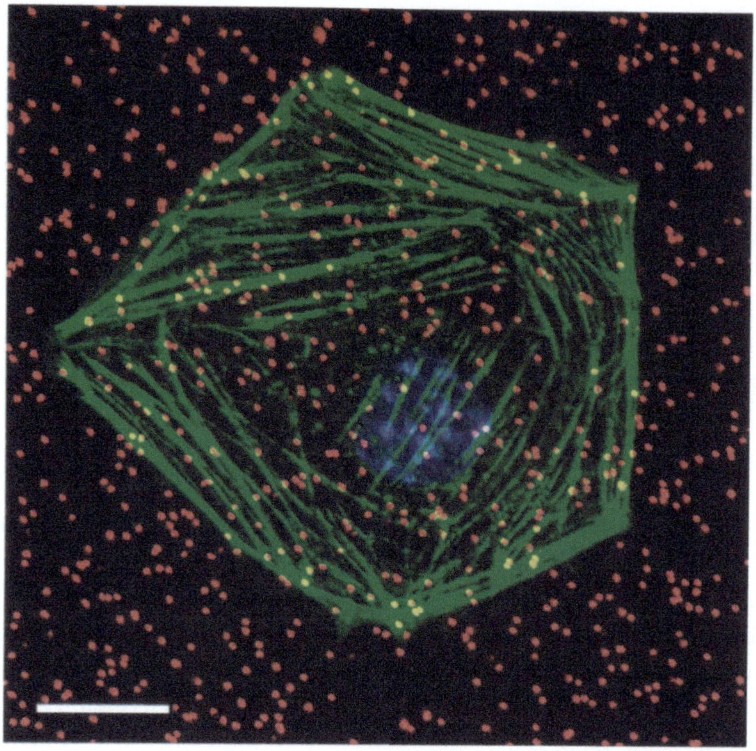

Figure 18.10. A fluorescence image of a C2C12 rat myoblast expressing Actin–GFP on a GXG substrate with embedded 200 nm red fluorescent beads (scale bar = 10 μm).

Dishes were mounted on the stage of a simultaneous AFM–fluorescence microscope and phase-contrast/fluorescence images of the cell and associated stressed and relaxed bead positions were captured automatically with a deep cooled CCD camera for TFM analysis. Experiments were designed to incorporate a built-in control for every cell measured. In the "control" experiment, a cell was chosen and a phase-contrast image was acquired followed by a series of fluorescent images of the surface beads every 30 seconds for 2 minutes. This was followed by the "stress" experiment by positioning of the AFM tip above the nucleus of the same cell and repeating the aforementioned procedure. The AFM tip was lowered onto the cell immediately after the t = 0 second image of the surface beads. In both "control" and "stress" experiments, the t = 0 second fluorescence image was treated as the "null" image and subsequent images were treated as "stressed" images. Therefore, each cell measured has a built-in control measurement which provides us with the natural cellular traction force dynamics and the perturbed dynamics in response to mechanical stimulation. We performed differential TFM analysis[103] in which we measured the change in traction forces as a function of time. This is in comparison with the absolute traction forces that are typically determined by removing the cells with trypsin after an experiment to measure the unstressed bead positions.[94,99] Foregoing the trypsin step allowed us to measure more cells per dish and quickly obtain a reliable statistical sample. Traction analysis was carried out using the LIBTRC-2.0 analysis libraries developed and kindly provided by Professor M. Dembo (Boston University).

Cells on the GXG gels described earlier did not display any significant traction force dynamics when left unperturbed. However, the cells demonstrated a significant increase in cellular traction force over time in response to applied loads. What is particularly important to notice is that applied forces to the cell nucleus are not merely transmitted through the cell and to the substrate in a circular deformation profile. In reality, the applied force is converted into biochemical signalling which results in localized "hot spots" randomly distributed over the cell contact area as seen in Fig. 18.11. These areas of large magnitude traction forces are discontinuous, heterogeneous and increase over time in response to a constant applied force to the nucleus. Consistent with our imaging of cytoskeletal deformation, force appears to be rapidly transduced throughout the cell (increase in cellular traction observed within 30 seconds) and applied forces are not simply transmitted through the cell as if it behaves as an isotropic and continuous medium.

To directly probe the origin of the cellular traction forces, we transiently transfected the cells with zyxin–RFP which is a protein found in stable focal adhesions and known to be mechanically regulated. Simultaneous imaging

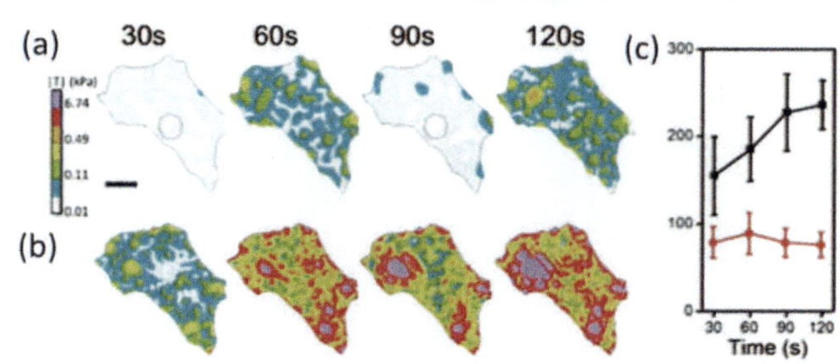

Figure 18.11. Traction force maps of a single cell over 2 minutes in the absence of any applied forces (a) and with a constant 10 nN force applied to the nucleus (b) (scale bar = 15 μm). From visual inspection, it is clear that the cell generates transient changes in traction forces in the absence of mechanical stimuli. However, a mechanical stimulus results in the generation of distinct "hot spots" in which traction forces increase rapidly. The average traction force per cell is plotted as a function of time in (c). Traction forces in control cells (red) do not vary significantly over time but rapidly increase in cells that are mechanically stimulated (black).

of zyxin–RFP, the green fluorescent beads and cell morphology allowed us to directly correlate changes in traction force magnitude and direction with focal adhesion remodelling (Fig. 18.12). In preliminary work, we have observed two major remodelling pathways of the focal adhesion structures at cell edges. The focal adhesions will disappear, appear to move outwards or grow larger towards the cell edge resulting in a traction force vector pointing outwards and away from the point of force. On the other hand, focal adhesions will appear to move inwards resulting in traction force vectors pointing towards the point of force. These remodelling pathways are in agreement with current models that describe focal adhesions centres for force transduction as described in the beginning of this chapter. This work clearly reveals that applied forces to living mammalian cells are rapidly transmitted through the cytoarchtiecture and results in fast remodelling of focal adhesion structures that generate cellular traction forces. Importantly, the applied force from the AFM tip is not simply transmitted in an isotropic manner through the cell and to the flexible substrate.

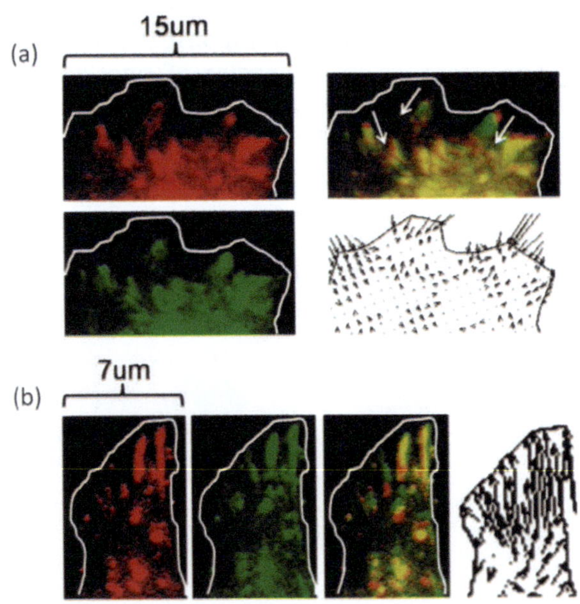

Figure 18.12. zyxin–RFP remodelling at cell edges (white lines) in response to applied loads. Images of zyxin–RFP are captured before (red) and after (green) 2 minutes of mechanical stimulation. Simultaneous imaging of bead movements allows us to correlate focal adhesion remodelling with the observed cellular traction forces. The results reveal that the inward (a) and the (b) outward movement of focal adhesions are two possible related mechanisms by which cellular traction forces can be generated.

18.4 CONCLUSIONS AND OUTLOOK

Three examples of recent work have been presented here in which the application of AFM and simultaneous optical imaging has yielded significant insights into our understanding of cellular nanomechanics. Moreover, using these approaches we are able to begin elucidating the architectural deformation and force transmission pathways through the cell in two and even three dimensions at relatively high speed. What is immediately clear is that localized nanomechanical forces are rapidly transmitted throughout the cellular architecture and the regulation of force transmission can be quite complex. Mitochondria found at cell edges (often greater than 30 μm away from the point of force on the nucleus) were observed to be displaced both towards and away from the contact point, indicating that they are somehow connected to a complex network within the cell. Treatment with drugs which result in the specific disassembly of actin, MTs and focal adhesions demonstrated that all three elements of the cytoarchitecture are required for the displacement of mitochondria in response to applied loads. The actin and MT cytoskeletons act as the tracks upon which mitochondria travel and respond directly to the application of forces to the cell. Moreover, both filament systems are required for the transmission of force to occur along with intact focal adhesions which enable the maintenance of cellular tension in the cytoskeleton. Loss of any one of these systems results in the impairment of force transduction and significant local and global decreases in cellular Young's modulus.

Creating cells which transiently express GFP-tagged cytoskeletal filaments (actin, tubulin and IFs) has allowed us to directly visualize the deformation of the cytoskeleton in two and three dimensions. Similar behaviours are observed here which agree with the results on mitochondrial displacements. All elements of the cytoskeleton appear to deform significantly and rapidly in response to applied loads. Furthermore, the deformation of the cytoskeleton occurs throughout the cell rather than at the local point where the cell has been mechanically stimulated. Moreover, tubulin filaments were observed to more both towards and away from the point of contact, indicating that force transmission through the cytoskeleton is highly complex. Finally, there appears to be a very important species type dependence to the force transmission pathways which govern cytoskeletal deformation which has not been taken into account in modern models of cell mechanics.

Finally, applied forces to cells are clearly not isotropically and homogenously transmitted through the cell and to the substrate. This was verified by measuring cellular traction forces in response to applied loads.

Again, there was no evidence of a circular transmission of force outwards and away from the AFM tip. Applied force was converted into a biochemical signal that resulted in focal adhesion remodelling. Traction force vectors were produced which were discontinuous and again demonstrated the transmission of force towards and away from the point of contact on the cell.

The forces and timescales examined in these studies are similar to those experienced by cells during typical force–distance curve measurements. This has important implications in our interpretations of such force curves as clearly the entire cell can respond rapidly and globally to localized contact forces. Moreover, the elements that control the cellular response are complex and appear to be species type dependent. This indicates that care must be taken in interpreting force curves, not only in which mechanical model is used to extract parameters of interest, but the molecular mechanism controlling the observed properties must be understood.

If anything, the work presented here has revealed that much remains unknown when it comes to understanding how the cell regulates and controls force transmission in two and three dimensions. With the developments of high-speed confocal imaging and new fluorophores it has become possible to image more than one element of the cytoarchitecture at a time and with very high temporal resolution. However, simply imaging structural responses is not enough. Close collaboration between disciplines is required to then develop predictive and time-dependent models that can account for the complexities observed experimentally. Understanding the biological mechanisms of force transduction and force sensitivity has a wide range of impacts in many field from a fundamental understanding of cellular mechanics to healthcare. It has become clear that stem cell differentiation, apoptosis, mitosis, myogenesis and many other critical physiological pathways are intimately linked to the cell's ability to sense and respond to the mechanics and mechanical forces found in their microenvironment.[1–15] The utility of simultaneous AFM and optical approaches is only now being realized in full detail, and with future technological advancements the applications may be limitless. The AFM literally provides us with a finger at the nanoscale which enables us to apply temporally and spatially controlled forces to live cells and tissues while imaging their structural and biochemical responses with the wealth of optical approaches now available. This approach to studying cell mechanics is still very much in its infancy, but as the simple examples presented here demonstrate, the wealth of new science in multiple disciplines (physics, biology, medicine, engineering) will be very exciting.

Acknowledgements

We gratefully acknowledge our co-workers who made essential contributions to the original work which was reviewed here: Professor Michael A. Horton, Dr. Gleb Yakubov, Dr. Farlan Veraitch, Dr. Chris Mason, David Yadin, Alexandra Hemsley and Carol Chu. This work was supported by the Biotechnology and Biological Sciences Research Council, the "Dr. Mortimer and Mrs. Theresa Sackler Trust" and the Nanotechnology IRC through an Exploratory Grant. YRS acknowledges the Japan Society for the Promotion of Science for a post-doctoral fellowship. LG thanks the Natural Sciences and Engineering Research Council for a graduate fellowship. AEP is a Canada Research Chair in Experimental Cell Mechanics.

References

1. Kunda, P., Pelling, A. E., Liu, T., and Baum, B. (2008) Moesin controls cortical rigidity, cell rounding, and spindle morphogenesis during mitosis, *Curr. Biol.*, **18**, 91–101.
2. Stolberg, S., and McCloskey, K. E. (2009) Can shear stress direct stem cell fate? *Biotechnol. Prog.*, **25**, 10–19.
3. Pelling, A. E., Veraitch, F. S., Chu, C. P., Mason, C., and Horton, M. A. (2009) Mechanical dynamics of single cells during early apoptosis, *Cell Motil. Cytoskeleton*, **66**, 409–422.
4. Paszek, M. J., Zahir, N., Johnson, K. R., Lakins, J. N., Rozenburg, G. I., Gefen, A., Reinhart-King, C. A., Margulies, S. S., Dembo, M., Boettiger, D., Hammer, D. A., and Weaver, V. M. (2005) Tensional homeostasis and the malignant phenotype, *Cancer Cell*, **8**, 241.
5. Engler, A. J., Sweeney, H. L., Discher, D. E., and Schwarzbauer, J. E. (2007) Extracellular matrix elasticity directs stem cell differentiation, *J. Musculoskelet. Neuronal. Interact.*, **7**, 335.
6. Engler, A. J., Sen, S., Sweeney, H. L., and Discher, D. E. (2006) Matrix elasticity directs stem cell lineage specification, *Cell*, **126**, 677–689.
7. Engler, A. J., Griffin, M. A., Sen, S., Bonnemann, C. G., Sweeney, H. L., and Discher, D. E. (2004) Myotubes differentiate optimally on substrates with tissue-like stiffness: pathological implications for soft or stiff microenvironments, *J. Cell Biol.*, **166**, 877–887.
8. Engler, A. J., Carag-Krieger, C., Johnson, C. P., Raab, M., Tang, H. Y., Speicher, D. W., Sanger, J. W., Sanger, J. M., and Discher, D. E. (2008) Embryonic cardiomyocytes beat best on a matrix with heart-like elasticity: scar-like rigidity inhibits beating, *J. Cell Sci.*, **121**, 3794–3802.
9. Chowdhury, F., Na, S., Li, D., Poh, Y. C., Tanaka, T. S., Wang, F., and Wang, N. (2010) Material properties of the cell dictate stress-induced spreading and differentiation in embryonic stem cells, *Nat. Mater.*, **9**, 82–88.

10. Adamo, L., Naveiras, O., Wenzel, P. L., McKinney-Freeman, S., Mack, P. J., Gracia-Sancho, J., Suchy-Dicey, A., Yoshimoto, M., Lensch, M. W., Yoder, M. C., Garcia-Cardena, G., and Daley, G. Q. (2009) Biomechanical forces promote embryonic haematopoiesis, *Nature*, **459**, 1131–1135.
11. Hogan, C., Dupre-Crochet, S., Norman, M., Kajita, M., Zimmermann, C., Pelling, A. E., Piddini, E., Baena-Lopez, L. A., Vincent, J. P., Itoh, Y., Hosoya, H., Pichaud, F., and Fujita, Y. (2009) Characterization of the interface between normal and transformed epithelial cells, *Nat. Cell Biol.*, **11**, 460–467.
12. Kim, S. J., Lee, J. K., Kim, J. W., Jung, J. W., Seo, K., Park, S. B., Roh, K. H., Lee, S. R., Hong, Y. H., Lee, Y. S., and Kang, K. S. (2008) Surface modification of polydimethylsiloxane (pdms) induced proliferation and neural-like cells differentiation of umbilical cord blood-derived mesenchymal stem cells, *J. Mater. Sci. Mater. Med.*, **19**, 2953–2962.
13. Veraitch, F. S., Scott, R., Wong, J. W., Lye, G. J., and Mason, C. (2008) The impact of manual processing on the expansion and directed differentiation of embryonic stem cells, *Biotechnol. Bioeng.*, **99**, 1216–1229.
14. Krieg, M., Arboleda-Estudillo, Y., Puech, P. H., Kafer, J., Graner, F., Muller, D. J., and Heisenberg, C. P. (2008) Tensile forces govern germ-layer organization in zebrafish, *Nat. Cell Biol.*, **10**, 429–436.
15. Puech, P. H., Taubenberger, A., Ulrich, F., Krieg, M., Muller, D. J., and Heisenberg, C. P. (2005) Measuring cell adhesion forces of primary gastrulating cells from zebrafish using atomic force microscopy, *J. Cell Sci.*, **118**, 4199–4206.
16. Binnig, G., Quate, C. F., and Gerber, C. (1986) Atomic force microscope, *Phys. Rev. Lett.*, **56**, 930–933.
17. Muller, D. J., Helenius, J., Alsteens, D., and Dufrene, Y. F. (2009) Force probing surfaces of living cells to molecular resolution, *Nat. Chem. Biol.*, **5**, 383–390.
18. Silberberg, Y. R., Pelling, A. E., Yakubov, G. E., Crum, W. R., Hawkes, D. J., and Horton, M.A. (2008) Mitochondrial displacements in response to nanomechanical forces, *J. Mol. Recognit.*, **21**, 30–36.
19. Pelling, A. E., Wilkinson, P. R., Stringer, R., and Gimzewski, J. K. (2009) Dynamic mechanical oscillations during metamorphosis of the monarch butterfly, *J. R. Soc. Interface*, **6**, 29–37.
20. Colombelli, J., Besser, A., Kress, H., Reynaud, E. G., Girard, P., Caussinus, E., Haselmann, U., Small, J. V., Schwarz, U. S., and Stelzer, E. H. (2009) Mechanosensing in actin stress fibers revealed by a close correlation between force and protein localization, *J. Cell Sci.*, **122**, 1665–1679.
21. Vale, R. D. (2003) The molecular motor toolbox for intracellular transport, *Cell*, **112**, 467–480.
22. Hirokawa, N. (1998) Kinesin and dynein superfamily proteins and the mechanism of organelle transport, *Science*, **279**, 519–526.
23. Alberts, B., Lewis, J., Raff, M., Roberts, K., and Watson, J. D. (2002) *Molecular Biology of the Cell*, Garland Science, New York.
24. Burridge, K., and Chrzanowska-Wodnicka, M. (1996) Focal adhesions, contractility, and signaling, *Annu. Rev. Cell Dev. Biol.*, **12**, 463–518.

25. Maniotis, A. J., Chen, C. S., and Ingber, D. E. (1997) Demonstration of mechanical connections between integrins, cytoskeletal filaments, and nucleoplasm that stabilize nuclear structure, *Proc. Natl. Acad. Sci. USA*, **94**, 849–854.
26. Bretscher, A., Edwards, K., and Fehon, R. G. (2002) Erm proteins and merlin: integrators at the cell cortex, *Nat. Rev. Mol. Cell Biol.*, **3**, 586–599.
27. McClatchey, A. I., and Fehon, R. G. (2009) Merlin and the erm proteins--regulators of receptor distribution and signaling at the cell cortex, *Trends Cell Biol.*, **19**, 198–206.
28. Ingber, D. E. (1997) Integrins, tensegrity, and mechanotransduction, *Gravit. Space Biol. Bull.*, **10**, 49–55.
29. Herrmann, H., Bar, H., Kreplak, L., Strelkov, S. V., and Aebi, U. (2007) Intermediate filaments: from cell architecture to nanomechanics, *Nat. Rev. Mol. Cell Biol.*, **8**, 562–573.
30. Fuchs, E., and Cleveland, D. W. (1998) A structural scaffolding of intermediate filaments in health and disease, *Science*, **279**, 514–519.
31. Pelling, A. E., and Horton, M. A. (2008) An historical perspective on cell mechanics, *Pflugers Arch.*, **456**, 3–12.
32. Valberg, P. A., and Feldman, H. A. (1987) Magnetic particle motions within living cells. Measurement of cytoplasmic viscosity and motile activity, *Biophys. J.*, **52**, 551–561.
33. Valberg, P. A., and Butler, J. P. (1987) Magnetic particle motions within living cells. Physical theory and techniques, *Biophys. J.*, **52**, 537–550.
34. Evans, E., and Yeung, A. (1989) Apparent viscosity and cortical tension of blood granulocytes determined by micropipet aspiration, *Biophys. J.*, **56**, 151–160.
35. Evans, E. (1989) Kinetics of granulocyte phagocytosis: rate limited by cytoplasmic viscosity and constrained by cell size, *Cell Motil. Cytoskeleton*, **14**, 544–551.
36. Petersen, N. O., McConnaughey, W. B., and Elson, E. L. (1982) Dependence of locally measured cellular deformability on position on the cell, temperature, and cytochalasin b, *Proc. Natl. Acad. Sci. USA*, **79**, 5327–5331.
37. Zahalak, G. I., McConnaughey, W. B., and Elson, E. L. (1990) Determination of cellular mechanical properties by cell poking, with an application to leukocytes, *J. Biomech. Eng.*, **112**, 283–294.
38. Wang, N., and Ingber, D. E. (1995) Probing transmembrane mechanical coupling and cytomechanics using magnetic twisting cytometry, *Biochem. Cell Biol.*, **73**, 327–335.
39. Bizal, C. L., Butler, J. P., and Valberg, P. A. (1991) Viscoelastic and motile properties of hamster lung and peritoneal macrophages, *J. Leukoc. Biol.*, **50**, 240–251.
40. Hochmuth, R. M. (2000) Micropipette aspiration of living cells, *J. Biomech.*, **33**, 15–22.
41. Bausch, A. R., Ziemann, F., Boulbitch, A. A., Jacobson, K., and Sackmann, E. (1998) Local measurements of viscoelastic parameters of adherent cell surfaces by magnetic bead microrheometry, *Biophys. J.*, **75**, 2038–2049.

42. Coughlin, M. F., and Stamenovic, D. (2003) A prestressed cable network model of the adherent cell cytoskeleton, *Biophys. J.*, **84**, 1328–1336.

43. Heidemann, S. R., Kaech, S., Buxbaum, R. E., and Matus, A. (1999) Direct observations of the mechanical behaviors of the cytoskeleton in living fibroblasts, *J. Cell Biol.*, **145**, 109–122.

44. Ashkin, A. (1997) Optical trapping and manipulation of neutral particles using lasers, *Proc. Natl. Acad. Sci. USA*, **94**, 4853–4860.

45. Kuo, S. C., and Sheetz, M. P. (1993) Force of single kinesin molecules measured with optical tweezers, *Science*, **260**, 232–234.

46. Choquet, D., Felsenfeld, D. P., and Sheetz, M. P. (1997) Extracellular matrix rigidity causes strengthening of integrin-cytoskeleton linkages, *Cell*, **88**, 39–48.

47. Hayakawa, K., Tatsumi, H., and Sokabe, M. (2008) Actin stress fibers transmit and focus force to activate mechanosensitive channels, *J. Cell Sci.*, **121**, 496–503.

48. Botvinick, E. L., Venugopalan, V., Shah, J. V., Liaw, L. H., and Berns, M. W. (2004) Controlled ablation of microtubules using a picosecond laser, *Biophys. J.*, **87**, 4203–4212.

49. Guck, J., Ananthakrishnan, R., Mahmood, H., Moon, T. J., Cunningham, C. C., and Kas, J. (2001) The optical stretcher: a novel laser tool to micromanipulate cells, *Biophys. J.*, **81**, 767–784.

50. Guck, J., Schinkinger, S., Lincoln, B., Wottawah, F., Ebert, S., Romeyke, M., Lenz, D., Erickson, H. M., Ananthakrishnan, R., Mitchell, D., Kas, J., Ulvick, S., and Bilby, C. (2005) Optical deformability as an inherent cell marker for testing malignant transformation and metastatic competence, *Biophys. J.*, **88**, 3689–3698.

51. Lautenschlager, F., Paschke, S., Schinkinger, S., Bruel, A., Beil, M., and Guck, J. (2009) The regulatory role of cell mechanics for migration of differentiating myeloid cells, *Proc. Natl. Acad. Sci. USA*, **106**, 15696–15701.

52. Lincoln, B., Wottawah, F., Schinkinger, S., Ebert, S., and Guck, J. (2007) High-throughput rheological measurements with an optical stretcher, *Methods Cell Biol.*, **83**, 397–423.

53. Remmerbach, T. W., Wottawah, F., Dietrich, J., Lincoln, B., Wittekind, C., and Guck, J. (2009) Oral cancer diagnosis by mechanical phenotyping, *Cancer Res.*, **69**, 1728–1732.

54. Charras, G. T., and Horton, M. A. (2002) Single cell mechanotransduction and its modulation analyzed by atomic force microscope indentation, *Biophys. J.*, **82**, 2970–2981.

55. Radmacher, M., Fritz, M., Kacher, C. M., Cleveland, J. P., and Hansma, P. K. (1996) Measuring the viscoelastic properties of human platelets with the atomic force microscope, *Biophys. J.*, **70**, 556–567.

56. Rotsch, C., and Radmacher, M. (2000) Drug-induced changes of cytoskeletal structure and mechanics in fibroblasts: an atomic force microscopy study, *Biophys. J.*, **78**, 520–535.

57. Vinckier, A., and Semenza, G. (1998) Measuring elasticity of biological materials by atomic force microscopy, *FEBS Lett.*, **430**, 12–16.

58. Wu, H. W., Kuhn, T., and Moy, V. T. (1998) Mechanical properties of l929 cells measured by atomic force microscopy: effects of anticytoskeletal drugs and membrane crosslinking, *Scanning*, **20**, 389–397.

59. Lehenkari, P. P., Charras, G. T., Nesbitt, S. A., and Horton, M. A. (2000) New technologies in scanning probe microscopy for studying molecular interactions in cells, *Expert Rev. Mol. Med.*, **2**, 1–19.

60. Lehenkari, P. P., Charras, G. T., Nykanen, A., and Horton, M. A. (2000) Adapting atomic force microscopy for cell biology, *Ultramicroscopy*, **82**, 289–295.

61. Haupt, B. J., Pelling, A. E., and Horton, M. A. (2006) Integrated confocal and scanning probe microscopy for biomedical research, *ScientificWorldJournal.*, **6**, 1609–1618.

62. Stolz, M., Gottardi, R., Raiteri, R., Miot, S., Martin, I., Imer, R., Staufer, U., Raducanu, A., Duggelin, M., Baschong, W., Daniels, A. U., Friederich, N. F., Aszodi, A., and Aebi, U. (2009) Early detection of aging cartilage and osteoarthritis in mice and patient samples using atomic force microscopy, *Nat. Nanotechnol.*, **4**, 186–192.

63. Cross, S. E., Jin, Y. S., Rao, J., and Gimzewski, J. K. (2007) Nanomechanical analysis of cells from cancer patients, *Nat. Nanotechnol.*, **2**, 780–783.

64. Muller, D. J. (2008) Afm: a nanotool in membrane biology, *Biochemistry*, **47**, 7986–7998.

65. Ludwig, T., Kirmse, R., Poole, K., and Schwarz, U. S. (2008) Probing cellular microenvironments and tissue remodeling by atomic force microscopy, *Pflugers Arch.*, **456**, 29–49.

66. Heidemann, S. R., Kaech, S., Buxbaum, R. E., and Matus, A. (1999) Direct observations of the mechanical behaviors of the cytoskeleton in living fibroblasts, *J. Cell Biol.*, **145**, 109–122.

67. Helmke, B. P., Rosen, A. B., and Davies, P. F. (2003) Mapping mechanical strain of an endogenous cytoskeletal network in living endothelial cells, *Biophys. J.*, **84**, 2691–2699.

68. Wang, N., and Suo, Z. G. (2005) Long-distance propagation of forces in a cell, *Biochem. Biophys. Res. Commun.*, **328**, 1133–1138.

69. Hu, S., Chen, J., Fabry, B., Numaguchi, Y., Gouldstone, A., Ingber, D. E., Fredberg, J. J., Butler, J. P., and Wang, N. (2003) Intracellular stress tomography reveals stress focusing and structural anisotropy in cytoskeleton of living cells, *Am. J. Physiol. Cell Physiol.*, **285**, C1082-C1090.

70. Hu, S. H., Chen, J. X., Butler, J. P., and Wang, N. (2005) Prestress mediates force propagation into the nucleus, *Biochem. Biophys. Res. Comm.*, **329**, 423–428.

71. Rosenbluth, M. J., Crow, A., Shaevitz, J. W., and Fletcher, D. A. (2008) Slow stress propagation in adherent cells, *Biophys. J.*, **95**, 6052–6059.

72. Pelling, A. E., Nicholls, B. M., Silberberg, Y. S., and Horton, M. A. (2007) *Modern Research and Educational Topics on Microscopy*, 1st edn, Formatex, Badajoz.

73. Silberberg, Y. R., Felling, A. E., Yakubov, G. E., Crum, W. R., Hawkes, D. J., and Horton, M. A. (2008) Tracking displacements of intracellular organelles in response to nanomechanical forces, in *5th IEEE International Symposium on Biomedical Imaging: From Nano to Macro, 2008. ISBI 2008*, 1335–1338
74. Silberberg, Y. R., Yakubov, G. E., Horton, M. A., and Pelling, A. E. (2009) Cell nanomechanics and focal adhesions are regulated by retinol and conjugated linoleic acid in a dose-dependent manner, *Nanotechnology*, **20**, 285103.
75. Bereiter-Hahn, J., and Voth, M. (1994) Dynamics of mitochondria in living cells: shape changes, dislocations, fusion, and fission of mitochondria, *Microsc. Res. Tech.*, **27**, 198–219.
76. Drubin, D. G., Jones, H. D., and Wertman, K. F. (1993) Actin structure and function: roles in mitochondrial organization and morphogenesis in budding yeast and identification of the phalloidin-binding site, *Mol. Biol. Cell*, **4**, 1277–1294.
77. Morris, R. L., and Hollenbeck, P. J. (1995) Axonal transport of mitochondria along microtubules and f-actin in living vertebrate neurons, *J. Cell Biol.*, **131**, 1315–1326.
78. Suelmann, R., and Fischer, R. (2000) Mitochondrial movement and morphology depend on an intact actin cytoskeleton in aspergillus nidulans, *Cell Motil. Cytoskeleton*, **45**, 42–50.
79. Alenghat, F. J., and Ingber, D. E. (2002) Mechanotransduction: all signals point to cytoskeleton, matrix, and integrins, *Sci STKE*, 2002, PE6.
80. Blumenfeld, R. (2006) Isostaticity and controlled force transmission in the cytoskeleton: a model awaiting experimental evidence, *Biophys. J.*, **91**, 1970–1983.
81. Wang, N., Butler, J. P., and Ingber, D. E. (1993) Mechanotransduction across the cell surface and through the cytoskeleton, *Science*, **260**, 1124–1127.
82. Heggeness, M. H., Simon, M., and Singer, S. J. (1978) Association of mitochondria with microtubules in cultured cells, *Proc. Natl. Acad. Sci. USA*, **75**, 3863–3866.
83. Brady, S. T., Lasek, R. J., and Allen, R. D. (1982) Fast axonal transport in extruded axoplasm from squid giant axon, *Science*, **218**, 1129–1131.
84. Napoli, J. L. (1996) Retinoic acid biosynthesis and metabolism, *Faseb J.*, **10**, 993–1001.
85. Chambon, P. (1996) A decade of molecular biology of retinoic acid receptors, *Faseb J.*, **10**, 940–954.
86. Balmer, J. E., and Blomhoff, R. (2002) Gene expression regulation by retinoic acid, *J. Lipid Res.*, **43**, 1773–1808.
87. Rozzo, C., Chiesa, V., Caridi, G., Pagnan, G., and Ponzoni, M. (1997) Induction of apoptosis in human neuroblastoma cells by abrogation of integrin-mediated cell adhesion, *Int. J. Cancer*, **70**, 688–698.
88. Varani, J., Nickoloff, B. J., Dixit, V. M., Mitra, R. S., and Voorhees, J. J. (1989) All-trans retinoic acid stimulates growth of adult human keratinocytes cultured in growth factor-deficient medium, inhibits production of thrombospondin and fibronectin, and reduces adhesion, *J. Invest. Dermatol.*, **93**, 449–454.

89. Darling, E. M., Zauscher, S., and Guilak, F. (2006) Viscoelastic properties of zonal articular chondrocytes measured by atomic force microscopy, *Osteoarthr. Cartil.*, **14**, 571–579.

90. Harris, A. K., Wild, P., and Stopak, D. (1980) Silicone-rubber substrata—new wrinkle in the study of cell locomotion, *Science*, **208**, 177–179.

91. Lee, J., Leonard, M., Oliver, T., Ishihara, A., and Jacobson, K. (1994) Traction forces generated by locomoting keratocytes, *J. Cell Biol.*, **127**, 1957–1964.

92. Oliver, T., Dembo, M., and Jacobson, K. (1995) Traction forces in locomoting cells, *Cell Motil. Cytoskeleton*, **31**, 225–240.

93. Dembo, M., Oliver, T., Ishihara, A., and Jacobson, K. (1996) Imaging the traction stresses exerted by locomoting cells with the elastic substratum method, *Biophys. J.*, **70**, 2008–2022.

94. Dembo, M., and Wang, Y. L. (1999) Stresses at the cell-to-substrate interface during locomotion of fibroblasts, *Biophys. J.*, **76**, 2307–2316.

95. Doyle, A., Marganski, W., and Lee, J. (2004) Calcium transients induce spatially coordinated increases in traction force during the movement of fish keratocytes, *J. Cell Sci.*, **117**, 2203–2214.

96. Doyle, A. D., and Lee, J. (2002) Simultaneous, real-time imaging of intracellular calcium and cellular traction force production, *Biotechniques*, **33**, 358–364.

97. du Roure, O., Saez, A., Buguin, A., Austin, R. H., Chavrier, P., Siberzan, P., and Ladoux, B. (2005) Force mapping in epithelial cell migration, *Proc. Natl. Acad. Sci. USA*, **102**, 2390–2395.

98. Ganz, A., Lambert, M., Saez, A., Silberzan, P., Buguin, A., Mege, R. M., and Ladoux, B. (2006) Traction forces exerted through n-cadherin contacts, *Biol. Cell.*, **98**, 721–730.

99. Munevar, S., Wang, Y. L., and Dembo, M. (2001) Traction force microscopy of migrating normal and h-ras transformed 3t3 fibroblasts, *Biophys. J.*, **80**, 1744–1757.

100. Pelham, R. J., and Wang, Y. L. (1997) Cell locomotion and focal adhesions are regulated by substrate flexibility, *Proc. Natl. Acad. Sci. USA*, **94**, 13661–13665.

101. Wang, N., Ostuni, E., Whitesides, G. M., and Ingber, D. E. (2002) Micropatterning tractional forces in living cells, *Cell Motil. Cytoskeleton*, **52**, 97–106.

102. Tan, J. L., Tien, J., Pirone, D. M., Gray, D. S., Bhadriraju, K., and Chen, C. S. (2003) Cells lying on a bed of microneedles: an approach to isolate mechanical force, *Proc. Natl. Acad. Sci. USA*, **100**, 1484–1489.

103. Curtze, S., Dembo, M., Miron, M., and Jones, D. B. (2004) Dynamic changes in traction forces with dc electric field in osteoblast-like cells, *J. Cell Sci.*, **117**, 2721–2729.

Chapter 19

THE ROLE OF ATOMIC FORCE MICROSCOPY IN ADVANCING DIATOM RESEARCH INTO THE NANOTECHNOLOGY ERA

Michael J. Higgins[a] **and Richard Wetherbee**[b]

[a] *ARC Centre of Excellence for Electromaterials Science, Intelligent Polymer Research Institute, AIIM Facility, Innovation Campus, University of Wollongong, Wollongong NSW 2522, Australia*

[b] *Botany Department, University of Melbourne, Victoria, 3000, Australia*

mhiggins@uow.edu.au

19.1 INTRODUCTION TO GENERAL DIATOM BIOLOGY

Diatoms are unicellular, micro-sized algae abundant in most of the world's marine and freshwater habitats. When observed under a light microscope, diatoms are strikingly beautiful organisms because of the transmission of brilliant yellow-green to golden-brown colours from their intracellular photosynthetic pigments. They come in diverse shapes and sizes ranging from five to hundreds of microns and are easily distinguished by their highly elaborate, mineralized cell walls composed of micro- and nanostructured segments and appendages (Fig. 19.1a). Planktonic diatoms live free-floating in open water, while benthic diatoms reside at the water–sediment interface or adhere to any submerged substrate, including sand and rocks, the surface of larger organisms and man-made structures.[1]

The cell wall of diatoms, termed the frustule, is composed of silica and consists of two overlapping halves or thecae that fasten together like a Petri dish.[2] Each theca is composed of a valve and one or more rings of silica called girdle bands that run around the circumference of the frustule and permit cell growth following division (Fig. 19.1b). A major valve feature, called

Life at the Nanoscale: Atomic Force Microscopy of Live Cells
Edited by Yves Dufrêne

www.panstanford.com

the raphe, provides an opening for adhesives involved in the motility of benthic diatoms.[3] Development of the frustule involves several processes[4] thought to be genetically encoded, including silicon uptake and metabolism, biomineralization and morphogenesis, which together lead to species-specific morphologies upon which diatom taxonomy is based. The whole frustule, typically consisting of uniformly patterned pores, spine-like processes, organic material and other nanostructured components, provides an avenue for nutrient and gas transport and secretion of adhesives.[1,2] Although the frustule structure conveys a profound level of intricacy, it has remarkable material strength to withstand external environmental forces.[5]

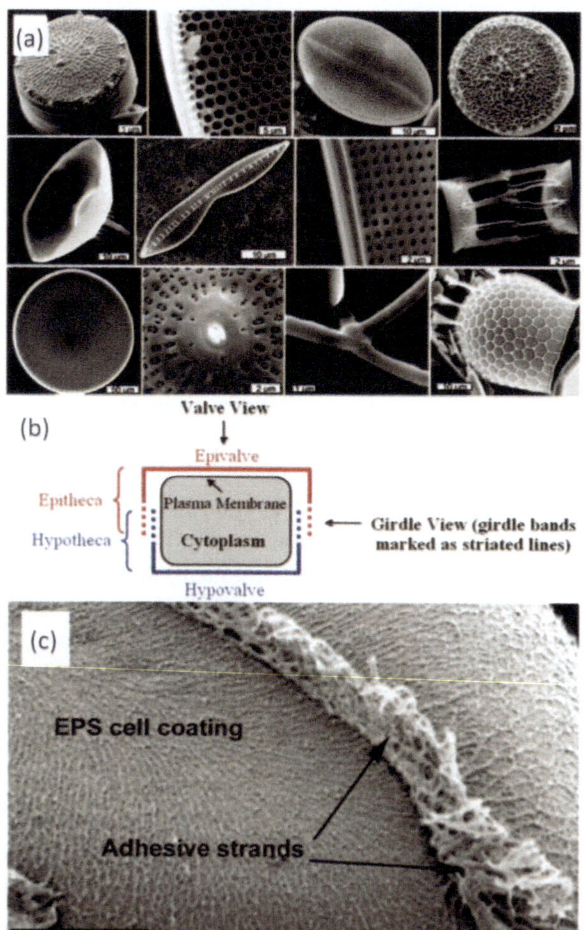

Figure 19.1. (a) SEM images of different diatoms highlighting the structures of the cell wall.[31] (b) Schematic of frustule comprising valve and girdle components.[32] (c) SEM image of outer frustule with EPS coating and adhesive strands.[24] Scale bar, 3 μm.

Another conspicuous feature of diatoms is their production of extracellular polymeric substances (EPS), a key survival strategy that provides energy production, habitat stabilization, colony formation, mechanical protection, adhesion and motility.[1] EPS mainly consists of complex carbohydrates and glycoproteins and can be secreted externally to form various structures just as elaborate as the silica cell wall. Some EPS forms are intimately associated with the frustule as coatings, whereas others such as strands, tethers, pads and stalks serve primarily as adhesive structures (Fig. 19.1c).[1] The diversity of EPS structure and function underpins their ability in seeking out nutrient-rich and suitable photosynthetic conditions and subsequent colonization of most of the world's aquatic habitats. Their ecological success is epitomized by diatoms accounting for an estimated 40% of marine primary productivity, 20% of the total photosynthetic CO_2 fixation as well as being predominant contributors to silicon cycling in oceans.[6] As a major group of organisms controlling the world's CO_2 levels, the importance of diatoms on future trends of climate change is well stated.[7]

19.2 CURRENT TRENDS IN DIATOM RESEARCH: INFLUENCES FROM NANOTECHNOLOGY

Diatom research in recent years has seen a significant shift in the motivation behind fundamental aspects of their biology. An emphasis on nanotechnology research and related applications has certainly been a major factor in shaping the context of the research. Perhaps the biggest revolution in recent times has undoubtedly been in research on understanding the formation of the silica frustule. During this process, the cells convert the soluble form of silicic acid in the aqueous environment into solid silica. The phenomenon that follows involves the nanostructuring and moulding of the silica in synchrony with self-assembly processes to form a new valve for each daughter cell during replication. Several proposed models provide an overview of the process,[4] though critical aspects still remain a mystery. Diatoms undergo rapid logarithmic growth rates (>10^6 cells in 3–5 days), thus formation of the valves occurs at unprecedented speeds, densities and under ambient conditions. It is no wonder that this process, usually referred to as "diatom biomineralization and morphogenesis", has been gripped by the current nanotechnology wave and grabbed the attention of nanotechnologists and multidisciplinary researchers alike. It is also the case that numerous recent reviews have used diatom cell wall formation as a case study for the three-dimensional (3-D) self-assembly of nanostructures,[8–10] making them synonymous with nanotechnology practices.

Current research into diatom EPS on the other hand is looking towards nanotechnology to advance its field. This is related to the tenacity of diatoms to adhere to artificial marine surfaces (ships, pipes, and filters), producing slime layers and instigating biofilm formation that is problematic and costly for the marine industry. Studies on the mechanisms of diatom adhesion and chemical composition of their adhesives have sought to provide clues for possible genetic and molecular targets for prevention of their detrimental attachment to surfaces.[3] An applied approach to the problem has been to perform cell adhesion assays to assess the potential of different materials to act as "non-stick" surfaces or coatings. There has been a recent emergence in designing dynamic, multifaceted surfaces by way of nanostructuring with nanomaterials and tailored chemistries to gain finer control over the cell-surface interactions.[11] It is hoped that through nanotechnology approaches, the design of these "smart" surfaces will address the complexity and diversity of diatom adhesion and adhesives and enhance antifouling surface properties. With new worldwide environmental legislation prohibiting the use of toxic antifouling coatings and tightening restrictions on biocides, nanotechnology will be one of the sciences relied upon to come up with environmentally friendly solutions.

The idea of learning from, or mimicking, diatoms to assemble and synthesize new materials, structures or adhesives on the same scale has been around since the early electron microscopy structural studies observing cell wall formation and EPS production.[1,2] The recent excitement surrounding "diatom inspired nanotechnology" can be attributed to current research trends, greater awareness by researchers outside the field and emergence of tangible diatom-based nanotechnology applications,[10] including gas sensors, photonic crystals and solar cells. The exhaustive work in elucidating the mechanistic origins and genetic and molecular processes[3,4,12] has also brought nanotechnology researchers closer to an understanding of the cell wall and EPS biology and their potential applications. Much of this work has required the novel application and development of new techniques, capable of probing diatoms at sub-micron length scales. Genetic and molecular tools have been important, as well as microscopy techniques for morphological characterization. In terms of the latter, atomic force microscopy (AFM) and its application to study the diatom cell wall has played a significant role in advancing our understanding of biomineralization and morphogenesis at the nanometre scale. Its unique ability to measure nanoscale forces has also provided discoveries on the design, mechanical properties and function of diatom adhesives. The purpose of this chapter is to emphasize the impetus AFM has provided in placing diatoms under the nanotechnology spotlight by highlighting some of the research in this field.

19.3 THE DIATOM CELL WALL

19.3.1 The Living Outer Frustule

The first AFM images of the frustule were taken on the surface of living diatoms.[13,14] To enable imaging of the motile, pennate diatoms, *Pinnularia viridis*, *Craspedostauros australis* and *Nitzchia navis-varingica*, cells in artificial media were settled onto an adhesive polymer surface (poly-L-lysine or polyethylenimine) for immobilization and the AFM cantilever tip brought directly into contact with living cells positioned either on their girdle or valve face. The original intention of this approach was to probe the outer EPS layers; however it was established that contact mode imaging at higher forces easily removed the EPS coating to reveal the underlying frustule structures.[13,14] After "sweeping" away the EPS, the large, flat valve face of *P. viridis* was amenable for observing common microstructures such as the raphe opening and endings (Fig. 19.2a), while the nanostructure of other valve components exhibiting small changes in their surface height, including foramen chambers, raised circular nodules and the surrounding silica wall, were more clearly resolved in AFM images than in scanning electron microscopy (SEM) images of chemically cleaned frustules.[13] Live *C. australis* cells positioned on their valve could not be imaged because of their instability, though imaging of the flatter girdle region to observe their silica bands and 30–50 nm pores was possible (Fig. 19.2b).[14] Exposing the girdle region subsequently allowed the direct visualization of EPS secretion emanating from the pores. The girdle regions of live *N. navis-varingica* cells in logarithmic growth phase were void of an EPS coating and were shown to consist of numerous 50–100 nm spherical particles (Fig. 19.2c),[15] confirming previous SEM reports of "silica warts" for this species. The silica particles were only weakly connected to the frustule, as they could be removed by nanonewton lateral forces imposed by the cantilever tip, suggesting that particle formation occurred through the fine, nanoscale deposition, or bottom-up assembly, of silica at the distal girdle surface. When in their stationary growth phase, *N. navis-varingic* produced an EPS coating on the girdle region and silica particles, but not the valve mantle openings, which instead had branching polymer strands adhering to the surface. Studies on live *Phaeodactylum tricornutum* revealed that the triradiate form had a clean, smooth surface morphology, in contrast to the rougher, streaky appearance of the ovoid form indicating the presence of EPS. Further studies on *P. viridis* and *C. australis* aimed at preserving the EPS coatings using low amplitudes to reduce the tapping force on the cells revealed the EPS coatings had distinct nanostructure specific to each species.[15] The EPS coating for *C.*

australis had a grooved surface topography, while *P. viridis* had a spherical particulate structure (Fig. 19.2d). Until this study, the EPS coating had only been observed as dried strand-like material under SEM,[1,16] or interpreted as an amorphous mucilage when hydrated that is generally sloughed off the cell surface. AFM showed that the EPS coating in reality is a discrete, structured polymer layer that maintains its integrity and association with silica frustule.

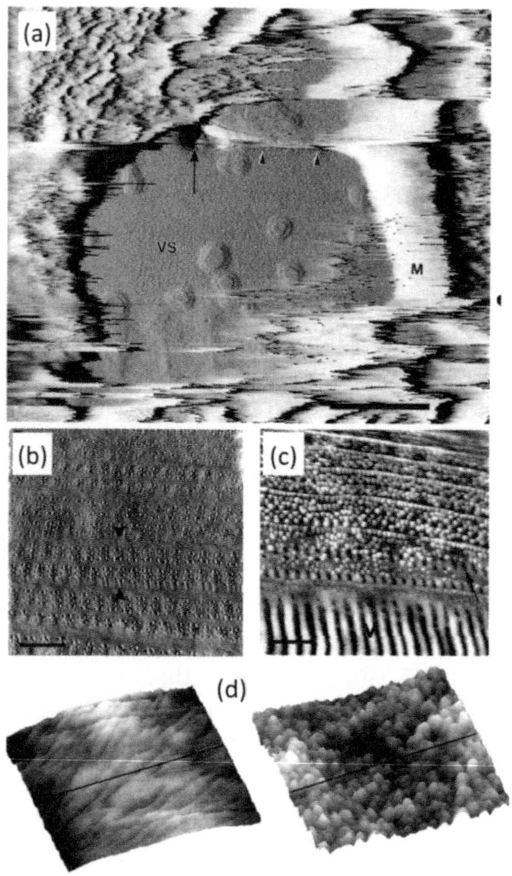

Figure 19.2. (a) Outer frustule surface of living *Pinnularia viridis*. EPS coating (M) has been removed after scanning to reveal the valve surface (vs), raphe (arrowheads), raphe ending (large arrow) and other frustule structures.[13] Scale bar, 3 µm. (b) Outer girdle region of living *Craspedostauros australis* showing rows of pores.[14] Scale bar, 1 µm. (c) Outer surface of living *Nitzchia navis-varingica* showing the silica particles (bright spots).[15] Scale bar, 1 µm. (d) 3-D height images show nanostructure of hydrated EPS coating of living *Craspedostauros australis* (left) and *Pinnularia viridis* (right).[15] Scan areas, 5 µm.

With respect to the outer frustule surface, a common finding to all these studies on living diatoms was that a purported tightly bound organic sheath covering the silica wall and situated beneath the EPS coating was not evident, suggesting a lack of an additional protective layer, or residual organic component involved in valve formation. If such a layer were to exist, it would have to be of a molecular layer thickness covering the silica topography for it to go undetected by AFM, which has the capability of resolving sub-nanometre changes in height. It is more likely that the organic sheath visualized in previous SEM studies[16] results from residual EPS coating after cell preparation (e.g. chemical fixation and drying). Thus, a clear advantage of AFM studies is that observations on the structure and properties of the outer frustule surface can be made under natural physiological conditions. The integrity of the whole frustule structure is retained, rather than its disassembly into separate components as is sometimes the case when frustules undergo chemical treatment and drying. This allows nanoscale silica structures and frustule components to be observed in relation to one another and without potential modification from any prior harsh chemical treatments, mechanical perturbations or disassembly. The approach will be of particular use for species such as *C. australis* whose delicate frustules collapse and deform under hydrostatic pressure in ambient air conditions.

A clearer representation of the outer living frustule emerging from AFM imaging of live diatoms is one of a smooth or particulate silica wall, comprising various nano- and micromorphologies, generally encased within a structured, visoelastic polymer layer, expect at major openings in the cell wall. Although parts of this description have always been the "status quo", this area of diatom research has possibly redefined our thinking of the frustule as not just silica with traces of organics but a complete silica-polymer composite layered structure.

19.3.2 Nanoscale Silica Structures

Preparing acid cleaned frustules provides another method for imaging of diatom silica structures with AFM. Although the cells are not alive, the EPS is removed to better expose the silica and provide greater access to different areas of the frustule, including both distal and proximal surfaces. Early studies on air-dried *Navicula pelliculosa* showed the capability of imaging the whole ellipsoidal frustule structure, including the distal raphe and pores.[17] For chemically cleaned and dried *P. viridis* frustules, it was found that the outer surface of the siliceous valve when imaged by SEM or AFM in contact mode was identical to that of the living cells, whose EPS coating had been removed by scanning,[13,14] as described earlier. This provided a good indication that in this case the acid treatment did not modify the true, smooth silica outer surface. Since these studies, relatively few diatoms had similarly been characterized until a recent survey of 16 diatom species was undertaken by imaging all the major cell wall components (valves, girdle bands and setae) using AFM.[18] The main conclusion from this study was that diatom nanoscale silica structure is highly diverse between species, within a single species, and even within a single frustule component (Fig. 19.3a–f). This provided an indication that no direct correlation necessarily exists between the nanoscale silica morphology and the frustule component that contains it. The authors also summarized that at the mesoscale level (defined as intermediate structures between the nanoscale and microscale), the prevalence of linear structures, even within different frustule components (e.g. girdle bands), suggested that an organization of linear organic molecules or subcellular features play a conserved role in templating structure formation on that scale.

In addition to nanoscale silica structures on the proximal and distal surfaces, there is great interest in understanding the structural details and composition embedded within the silicified structures, as this may shed more light on the principal nanostructures, or proposed organic template, involved in biomineralization and silica deposition. A very innovative sample design was specifically developed for this purpose so that the cross-sectional nanostructure of the frustule could be observed.[19] The method involved attaching a single chemically cleaned diatom to an optical fibre by embedding the cell in a bead of epoxy resin. The fibre was then cleaved at the mid-region of the frustule, and then threaded vertically into an aperture holder with the cleaved face of the frustule positioned upwards for imaging. High-resolution images of *P. viridis* and *Hantzschia amphioxys* frustules cleaved in cross-section revealed the presence of individual silica particles in the valves and girdle bands ranging from 30 to 50 nm in diameter (Fig. 19.3g,h).[13] Statistical analysis revealed no significant difference in particle size from major structures (i.e. girdle bands and valves) within a frustule, indicating for the first time that these nanoparticles represented the primary silica building blocks of a fully constructed cell wall.[13] In particular, this highlighted that a formless silica structure of the frustule, typically perceived from the smooth proximal and distal valve and girdle surfaces, was in fact composed of individual particles. A significant difference observed between species indicated a species-specific dependence and was mentioned to reflect differences in organic molecules embedded within the silica, such as long chain polyamines and silaffins, proposed to play a regulatory role in silica polymerization.[12] The study also reported nanoscale silica particles in the frustules of other species, including *Sureilla, Neidium* and *Pleurosigma*.

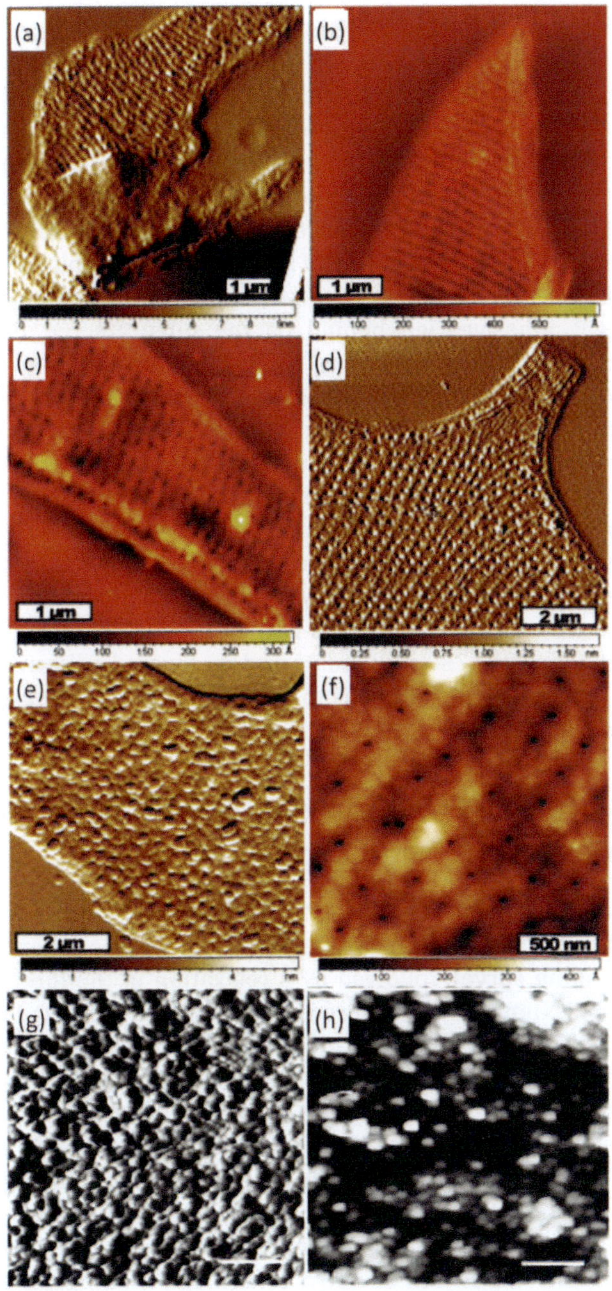

Figure 19.3. (a–f) Distinct silica morphologies on different side of girdle bands from the same species.[18] (a) *Chaetoceros laciniosis*. (b–c) *Chaetoceros decipiens*. (d–f) *Ditylum brightwelli*. (g–h) Deflection and height images of *Pinnularia viridis* valve in cross-section showing silica nanoparticles.[13] Scale bars, 250 nm.

19.4 DIATOM EPS AND ADHESIVES

SEM approaches have dominated studies on diatom EPS over the last four decades, and while it has been possible to observe large, conspicuous EPS macrostructures (e.g. EPS stalks and pads), the fixation, drying and vacuum operating environment has severely limited the ability to characterize the vast majority of EPS with confidence that dehydration effects have not altered the material. With the ability of AFM to probe soft biological systems in fluids, it has been possible to observe the true hydrated morphology of EPS coatings,[15] adhesives left behind on the substrate (i.e. diatom trails and biofilms) by motile diatoms[20] as well as adhesives from other algae species.[21] EPS adhesives that are too sticky, or in the form of an unsupported 3-D structure (e.g. adhesives strand protruding vertically from the cell) are difficult to image. Applying AFM force measurements in these cases has been especially important for elucidating the elastic and adhesive properties of the EPS, in particular at the single cell and molecular level. Localized regions on the cell surface can be targeted because of the nanometre size (10–20 nm) and lateral positioning of the tip over the desired EPS region. Most of the work done so far on measuring forces with the AFM has distinguished non-adhesive and adhesive EPS components and discovered adhesive properties and designs that give explanation as to why diatoms have the great tenacity to attach to surfaces.

19.4.1 Non-Adhesive Components: Cell Coatings and Outer Frustule Surface

AFM force measurements on the EPS coating have shown a non-linear increase in the force acting on the cantilever tip as it is indented into the surface (extending curve), followed by a relaxation in the force (retracting curve), which is not fully recovered, as the tip is retracted away (Fig. 19.4a). This force profile indicates the properties of a viscoelastic polymer, which is compressible but does not fully recover its form on the timescale of the measurement.[13,14] By analysing the approaching part of the force measurement (i.e. as the tip is pushed into the surface) with mechanical models such as Hertz theory, Young's modulus values ranging from 250 to 750 KPa for the EPS coating and outer living cell surface have been obtained.[15] Such measurements have highlighted the diverse polysaccharide and glycoprotein composition and structure of EPS coatings, as inferred by significant variations in the Young's modulus between species.[15] Similar measurements have also been used to distinguish the extent of silica composition in ovoid and triradiate forms of *P. tricornutum*.[22] Stiffer ovoid forms (500 KPa) confirmed a higher silica content compared with fusiform and triradiate forms (100 KPa). The girdle region of both fusiform and ovoid forms was five times softer than the valve, suggesting that this region is poor in silica and enriched in organic material (Fig. 19.4b,c). In addition to being a low modulus, viscoelastic polymer, the EPS coating has been shown to be non-adhesive. Force measurements on *P. viridis* showed no adhesion.[13] For *C. australis*, an adhesion force of 13 nN recorded on the EPS valve coating was five times less than that over the position of the raphe.[14] Even though the

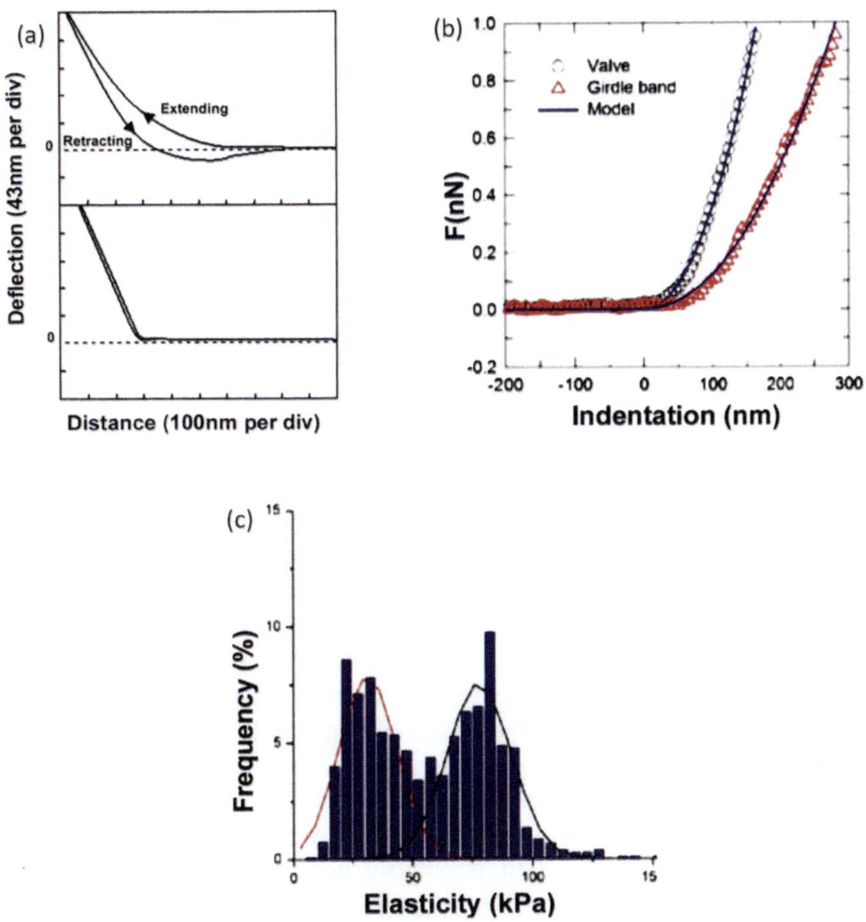

Figure 19.4. (a) Top graph shows an AFM force measurement on the EPS coating of living *Pinnulari viridis* cells. Bottom graph shows a force measurement on the silica cell wall after removal of the EPS. The slope of the cantilever deflection signal (force) is steeper, indicating a stiffer material.[13] (b–c) Mechanical properties of the fusiform girdle and valve interface.[22] (b) Force measurements indicate a stiffer material for the valve compared with the girdle. (c) Histogram of the Young's modulus values, valve (black gaussian fit), girdle (red gaussian fit).

EPS adhesion values were significantly less, they still may have been affected by residual adhesives from the nearby raphe and/or high loading forces (25 nN) applied to polymer, which has the effect of increasing the contact area of tip with the polymer. Measurements on the EPS girdle coating using loading forces < 1 nN showed adhesion forces of only ≈200 pN corresponding to the picking up and subsequent detachment of 50–200 nm individual polymer chains with similar elastic properties.[15]

19.4.2 From Microscale to Single-Molecule Adhesives: Pads, Tethers, Strands and Nanofibres

To study diatom adhesive interactions with surfaces, researchers have prepared live "bioprobes" by attaching an individual living cell to a tipless cantilever.[23] Using these probes, forces measured against a mica substrate and antifouling coating, Intersleek™, showed comparable cell adhesion strength for *Navicula sp.* on the two surfaces, indicating cells secreted an adhesive consisting of both hydrophilic and hydrophobic motifs. To more directly probe diatom adhesives involved in such interactions, fly fishing measurements on *C. australis* and *P. viridis*, whereby the tip was hovered above the surface, enabled single adhesive strands protruding from the non-driving raphe of living cells to be "caught" by the tip.[24] Their subsequent detachment from the tip recorded forces of ≈150 pN. To enable strong adhesion to the surface, the cells secreted a conglomerate of these strands in the form of a single micro-sized tether that extended for ≈40 μm and terminated in a holdfast-like attachment to the surface (Fig. 19.5a).[23] When the AFM tip was brought into direct contact with the raphe, these tethers recorded an adhesion force > 20 nN and, because of their high extensibility, could remain attached to the tip even when the z-height limit of the piezo had been reached (Fig. 19.5b, ii). Force profiles for these measurements revealed an irregular sawtooth pattern (Fig. 19.5b, i), indicating the successive unbinding of domains (i.e. inter- and intra-bonds within strands and tethers) when the raphe tether was placed under stress. These unbinding domains had previously been explained as "sacrificial bonds" which give way under force before the backbone of the adhesive breaks, effectively increasing its lifetime.[25] Rises and falls in the force (i.e. sawtooths) over long extension distances also greatly increased the area under the curve, or energy required to break the adhesives. This imparted extra fracture toughness into the adhesive material.[25] Similar sawtooth patterns were observed on regions of a glass slide, presumably the location of residual adhesive, where a chain-forming species, *Eunotia sudetica*, had been mechanically removed.[26] The cell samples were cultured with another diatom species (*Sellphora seminulum*) and the force measurements were not pinpointed to an observable adhesive structure, so it is difficult to rule out interactions from the conditioning film, adhesive material from the other species, or general biofilm formed during culture. Seminal AFM studies on the adhesive pads of living *Toxarium undulatum* cells (Fig. 19.5c) discovered an amazing new adhesive structure which the authors termed adhesive nanofibres (ANFs).[27,28] Unlike previous studies where the "sawtooth" profile was irregular because of the random breaking of inter- and intrachain bonds, Dugdale and co-workers showed for the first time a natural adhesive or composite material specifically engineered with modular domains whose purpose was to successively unbind under stress, giving rise to a *regular* sawtooth profile and enhanced mechanical toughness (Fig. 19.5d).

Several remarkable attributes of the ANFs were shown: (1) their modular domains reversibly unbind and refold upon hundreds of stretch-relax cycles, indicating self-healing properties, (2) they are composed of

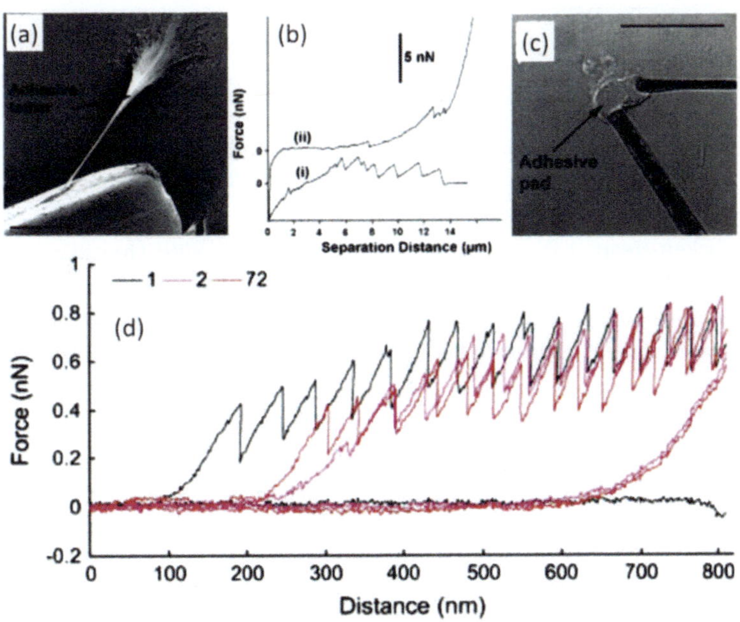

Figure 19.5. (a) Adhesive tether of *Pinnulari viridis.*[24] Scale bar, 15 µm. (b) Force measurements on the adhesive tether, where the tether does not detach from the tip (ii) and a sawtooth profile is observed (i).[24] (c) Optical microscope image of the adhesive pad of *Toxarium undulatum.*[27] Scale bar, 50 µm. (d) Force measurements on the adhesive pad showing the reversible unbinding and refolding of domains of the same adhesive nanofibres (ANFs) attached to the tip after 72 cycles, 1st cycle (black), 2nd cycle (magenta), 72nd cycle (red).[27]

supramolecular assemblies of different nanofibres (ANFs I, II, II), each with their own modular properties, and aligned domains which can all unfold–refold in registry and (3) they have an additional flexible polymer region and adhesive motif to enable the ANF to extend beyond the cell surface and adhere to surfaces. Energy dispersive X-ray analysis and Fourier transform infrared spectroscopy showed that the adhesive contained mainly protein, carbohydrate, sulphate, calcium and magnesium.[29] Further analysis of soluble EDTA extracts suggested that the ANFs composed sulphated high-molecular-mass glycoproteins cross-linked by calcium and magnesium ions. The cross-linking was proposed to enable domains of the adjacent protein backbones to unbind and refold in register. Although the exact modular structure of the ANFs was unknown, the force profile had the characteristic fingerprint of a true modular protein such as the muscle protein titin. Synthesis of a titin assembly consisting of several modular constructs in parallel was proven to unfold in registry, further supporting the proposed ANF supramolecular modular mechanism.[30]

References

1. Hoagland, K. D., Rosowski, J. R., Gretz, M. R., and Roemer, S. C. (1993) Diatom extracellular polymeric substances: function, fine structure, chemistry and physiology, *J. Phycol.*, **29**, 537–566.
2. Pickett-Heaps, J. D., Schmid, A. M., and Edgar, L. A. (1990) The cell biology of the diatom valve formation, *Prog. Phycol. Res.*, **7**, 1–168.
3. Wetherbee, R., Lind, J. L., Burke, J., and Quatrano, R. S. (1998) The first kiss: establishment and control of initial adhesion by raphid diatoms, *J. Phycol.*, **34**, 9–15.
4. Hildebrand, M. (2008) Diatoms, biomineralization processes, and genomics, *Chem. Rev.*, **108**, 4855–4874.
5. Hamm, C. E., Merkel, R., Springer, O., Jurkojc, P., Maier, C., Prechtel, K., and Smetacek, V. (2003) Architecture and material properties of diatom shells provide effective mechanical protection, *Nature*, **421**, 841–843.
6. Nelson, D. M., Tréguer, P., Brzezinski, M. A., and Leynaert, A. (1995) Production and dissolution of biogenic silica in the ocean: revised global estimates, comparison with regional data and relationship to biogenic sedimentation, *Global Biogeochem. Cycles*, **9**, 359–372.
7. Falkowski, P. G., and Oliver, M. J. (2007) Mix and match: how climate selects phytoplankton, *Nat. Rev. Microbiol.*, **5**, 813–819.
8. Sterrenburg, F. A. S. (2005) Crystal palaces—diatoms for engineers, *J. Nanosci. Nanotechnol.*, **5**, 100–107.

9. Drum, R. W., and Gordon, R. (2003) Star trek replicators and diatom nanotechnology, *Trends Biotechnol.*, **21**, 325–328.

10. Gordon, R., Losic, D., Tiffany, M. A., Nagy, S. S., and Sterrenburg, F. A. S. (2008) The glass menagerie: diatoms for novel applications in nanotechnology, *Trends Biotechnol.*, **27**, 116–127.

11. Akesso, L., Pettitt, M. E., Callow, J. A., Callow, M. E., Stallard, J., Teer, D., Liu, C., Wang, S., Zhao, Q., D'Souza, F., Willemsen, P. R., Donnelly, G. T., Donik, C., Kocijan, A., Jenko, M., Jone, L. A., and Guinaldo, P. C. (2009) The potential of nano-structured silicon oxide type coatings deposited by PACVD for control of aquatic biofouling, *Biofouling*, **25**, 55–67.

12. Kröger, N., and Poulsen, N. (2008) Diatoms—from cell wall biogensis to nanotechnology, *Annu. Rev. Genet.*, **42**, 83–107.

13. Crawford, S. A., Higgins, M. J., Mulvaney, P., and Wetherbee, R. (2001) Nanostructure of the diatom frustule as revealed by atomic force and scanning electron microscopy, *J. Phycol.*, **37**, 543–554.

14. Higgins, M. J., Crawford, S. A., Mulvaney, P., and Wetherbee, R. (2002) Characterization of the adhesive mucilages secreted by live diatom cells using atomic force microscopy, *Protist*, **153**, 25–38.

15. Higgins, M. J., Sader, J. E., Mulvaney, P., and Wetherbee, R. (2003). Probing the surface of living diatoms with atomic force microscopy: the nanostructure and nanomechanical properties of the mucilage layer, *J. Phycol.*, **39**, 722–734.

16. Edgar, L. A., and Pickett-Heaps, J. D. (1984) Ultrastructural localization of polysaccharides in the motile diatom *Navicula pelliculosa*, *Protoplasma*, **113**, 10–22.

17. Almqvist, N., Delamo, Y., Smith, B. L., Thomson, N. H., Bartholdson, A., Lal, R., Brzezinski, M. A., and Hansma, P. K. (2001) Micromechanical and structural properties of a pennate diatom investigated by atomic force microscopy, *J. Microsc.*, **202**, 518–532.

18. Hildebrand, M., Holton, G., Joy, D. C., Doktycz, M. J., and Allison, D. P. (2009) Diverse and conserved nano- and mesoscale structures of diatom silica revealed by atomic force microscopy, *J. Microsc.*, **235**, 172–187.

19. Egerton-Warbuton, L. M., Huntington, S. T., Mulvaney, P., Griffin, B. J., and Wertherbee, R. (1998) A new technique for preparing biominerals for atomic force microscopy, *Protoplasma*, **204**, 34–37.

20. Higgins, M. J., Crawford, S. A., Mulvaney, P., and Wetherbee, R. (2000) The topography of soft, adhesive diatom "trails" as observed by atomic force microscopy, *Biofouling*, **16**, 133–139.

21. Callow, J. A., Crawford, S. A., Higgins, M. J., Mulvaney, P., and Wetherbee, R. (2000) The application of atomic force microscopy to topographical studies and force measurements on the secreted adhesive of the green alga *enteromorpha*, *Planta*, **211**, 641–647.

22. Francius, G., Tesson, B., Dague, E., Martin-Jézéquel, V., and Dufrêne, Y. F. (2008) Nanostructure and nanomechanics of live *Phaeodactylum tricornutum* morphotypes, *Environ. Microbiol.*, **10**, 1344–1356.

23. Terán Arce, F., Avci, R., Beech, I. W., Cooksey, K. E., and Wigglesworth-Cooksey, B. (2004) A live bioprobe for studying diatom-surface interactions, *Biophys. J.*, **87**, 4284–4297.

24. Higgins, M. J., Molino, P., Mulvaney, P., and Wetherbee, R. (2003). The structure and properties of the adhesive mucilage that mediates diatom-substratum adhesion and motility, *J. Phycol.*, **39**, 1181–1193.

25. Smith, B. L., Schaffer, T. E., Viani, M., Thompson, J. B., Fredrick, N. A., Kindt, J., Belcher, A., Stucky, G. D., Morse, D. E., and Hansma, P. K. (1999) Molecular mechanistic origin of the toughness of natural adhesives, fibres and composites, *Nature*, **399**, 761–763.

26. Gebeshuber, I. C., Thompson, J. B., Del Amo, Y., Stachelberger, H., and Kindt, J. H. (2002) In vivo nanoscale atomic force microscopy investigation of diatom adhesive properties, *Mater. Sci. Tech.*, **18**, 763–766.

27. Dugdale, T. B., Dagastine, R., Chiovitti, A., Mulvaney, P., and Wetherbee, R. (2005) Single adhesive nanofibres from alive diatoms have the fingerprint of modular proteins, *Biophys. J.*, **89**, 4252–4260.

28. Dugdale, T. B., Dagastine, R., Chiovitti, A., Mulvaney, P., and Wetherbee, R. (2006) Diatom adhesive mucilage contains distinct supramolecular assemblies of a single modular protein, *Biophys. J.*, **90**, 2987–2993.

29. Chiovitti, A., Heraud, P., Dugdale, T. M., Hodson, O. M., Curtain, R. C. A., Dagastine, R., Wood, B. R., and Wetherbee, R. (2008) Divalent cations stabilize the aggregation of sulphated glycoproteins in the adhesive nanofibres of the biofouling diatom *Toxarium undulatum*, *Soft Matter*, **4**, 811–820.

30. Sarkar, A., Caamano, S., and Fernandez, J. (2007) The mechanical fingerprint of a parallel polyprotein dimer, *Biophys. J.*, **92**, L36–L38.

31. Hildebrand, M., Doktycz, M., and Allison, D (2008) Application of AFM in understanding biomineral formation in diatoms, *Eur. J. Physiol.*, **456**, 127–137.

32. Molino, P. J., and Wetherbee, R. (2009) The biology of biofouling diatoms and their role in the development of microbial slimes, *Biofouling*, **24**, 365–379.

Chapter 20

ATOMIC FORCE MICROSCOPY FOR MEDICINE

Shivani Sharma[a,b] **and James K. Gimzewski**[a,b,c]

[a] *Department of Chemistry and Biochemistry, University of California, Los Angeles, CA, USA*

[b] *California NanoSystems Institute, University of California, Los Angeles, CA, USA*

[c] *International Center for Materials Nanoarchitectonics Satellite (MANA), National Institute for Materials Science (NIMS), Tsukuba, Japan*

gim@chem.ucla.edu

20.1 INTRODUCTION

Because of increasing healthcare costs, changing demographics and rapid growth in chronic illnesses, it is very likely that many healthcare systems around the world will become unsustainable by 2015. Worldwide healthcare spending is expected to grow from 9% of worldwide Gross Domestic Product to 15% by 2015, and by 2050 the world's population older than 60 years will triple from 600 million to over 2 billion. Moreover, the number of people in US only with a chronic illness will grow from 118 million in 1995 to 157 million in 2020 (World Health Organization). Therefore, new technologies will be needed to overcome these challenges such as implementation of nanotechnology applications for healthcare (www.OECD.org). In particular, the development of a wide spectrum of emerging nano-enabled technologies may hold great promise for medicine and healthcare benefits by complementing and enhancing the current diagnostic and therapeutic capabilities of existing healthcare systems. Indeed, nanotechnology could be the crucial enabling technology that will turn the promise of theranostics[1] into reality, i.e., personalized therapy customized to serve patient needs based on their exact genetic and molecular diagnostics.

Life at the Nanoscale: Atomic Force Microscopy of Live Cells
Edited by Yves Dufrêne

www.panstanford.com

It is advantageous to use nanotechnology for medical applications since most biological processes, including those processes leading to cancer and other diseases, occur at the nanoscale (1–100 nm). Nanotechnology allows the understanding and manipulation of these biological processes at the cellular, sub-cellular and single-molecule level. Rapid interest in the medical applications has led to the emergence of a new field called nanomedicine.[2] Nanomedicine refers to the specialized application of nanotechnology for diagnosing, treating and preventing disease and improving human health. A bibliographic analysis of research articles in the Pubmed Citation Index shows that nanomedicine has seen a surge in research activity over the past decade, with publication numbers rising from 25 in year 2000 by a factor of 10 up to 2009 (Fig. 20.1).

The overall goal of nanomedicine is to achieve accurate and early diagnosis, effective treatment with minimal or no side effects and rapid and non-invasive monitoring of treatment efficacy. Traditionally, medicine takes a generalized approach to treat diseases, though the response may vary dramatically among individuals. The development of nanotechnology-based theranostic tests involving cellular, proteomic and genomic level testing platforms such as microchips represents a paradigm shift in patient care. It provides unique, individualized medications for each patient, being more targeted and cost-effective. Based on unique capabilities, nanoscale science probes cells and biomolecules in their physiological states at forces, displacement resolutions and concentrations at the piconewton, nanometre and picomolar scales, respectively. Studying human diseases from a nanoscale perspective may lead to better understanding of the

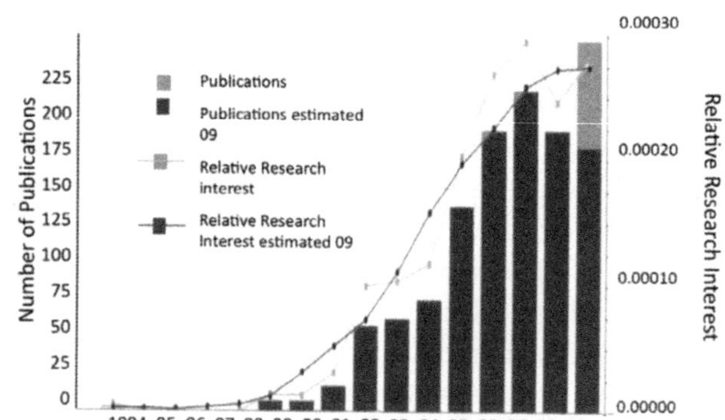

Figure 20.1. Trends in the number of research articles on nanotechnology and medicine published during the last 15 years (Pubmed Citation Index) and relative research interest in the field.

pathophysiology and pathogenesis of a variety of human diseases by correlating changes occurring at the molecular and cellular levels to changes in patient physiology. This will provide an alternative and better approach to assess the onset or progression of diseases as well as to identify targets for therapeutic interventions. Such measurements, previously not technically achievable, facilitate quantitative studies on the morphological, biophysical and biochemical nano and microscale properties of biological cells and their organization. Several recently developed nanotechnological tools and probe techniques that have influenced healthcare research and development include the following: nanoparticles for imaging and drug delivery, atomic force microscopy (AFM), molecular force spectroscopy, nanomaterials and microfluidics in management of major diseases such as cardiovascular diseases, cancer, diabetes and other diseases. Table 20.1 outlines some uses of molecules, molecular assemblies, materials and devices in the range of 1–100 nm, and the exploitation of the unique properties and processes at this dimensional scale.

Table 20.1. Nanotechnology-based medical tools for diagnostics and therapeutics

	Benefits	Examples
Drug delivery		
Nanoparticles liposomes, virosomes, polymerosomes, nanosuspensions	Greater affectivity, biocompatibility, low toxicity	*Abraxane™* against advanced breast cancer; 130 nm albumin-bound paclitaxel particles[+] *Doxil®* for ovarian cancer and Kaposi's sarcoma; polyethylene glycol (PEG)-coated lipid nanoparticles evade the potential impact of the immune system[+]. *Emend®* Anti-nausea drug for chemotherapy patients containing aprepitant; colloidal suspension of surface stabilized NanoCrystal particles (<1000 nm)
Therapeutics		
Fullerenes, dendrimers, nanoshells	Low side effects, multiple drug therapy, targeted drug release to tumour cells reducing toxic side effects; thermal drug release	*BrachySil™* 30 mm BioSilicon particles encapsulating radioactive 32P bonded within silicon microcrystalline shell, remain localized and deliver targeted dose of beta radiation *Aurimune* Recombinant human tumour necrosis factor alpha-coated pegylated colloidal Au nanoparticles *Gold Nanorods* *Nanobomb* Laser heating of hydrated carbon nanotubes kills tumour cells

Medical imaging		
Nanoparticles for MRI, ultrasound contrast Supermagnetic iron oxide nanoparticles	Detection of small tumours	*Magnetic Nanoclinic* is a thin silica bubble, the surface of which can be customized using a peptide carrier group to selectively target cancer cells. Inside the bubble are ferromagnetic nanoparticles that exhibit a strong inclination to align in the direction of a magnetic field
***In vitro* diagnostic devices/sensors**		
Nanotubes, nanowires, nanocantilevers, AFM	High sensitivity detection of analytes, infectious agents, pathophysiology of single cells, biomolecules	*Nanomix* Carbon nanotube-based sensors for monitoring respiratory functions *Bioforce* Uses AFM for detecting whole viruses *Cell Tracks* Ferrofluids for cancer cell detection *Nanochips* DNA/RNA microarray technology; Implantable Personal ID device*
Biomaterials		
Dental fillers, nano-hydroxyapatite implant coating	Self-assembling particles or nanomaterials with improved biocompatibility and mechanical properties	Nanoscale CAP dual acid etched surfaces improves bone healing Retina Implantat AG Bone replacement materials: Hydroxyapatite (HA) tricalcium phosphate-Ostim®, VITOSS® Nanostructured HA for hip, knee implant coating and dental prostheses
Bioimplants		
	Long-term detection and assessment	EKG monitor glucose sensors, implantable cardioverter defibrillator*

*FDA approved

20.2 AFM FOR CANCER DIAGNOSIS

Cancer is a complex disease involving multiple molecular and cellular processes arising from a gradual accumulation of genetic changes in individual cells (Fig. 20.2). It continues to be the leading cause of death worldwide and is presently responsible for about 25% of all deaths.[3] It is estimated that there will be about 15 million new cases of cancer annually by 2010.[4] National Cancer Institute committed $114 billion for nanotechnology research on cancer detection. Earlier diagnosis has been shown to be the most important factor in prognostic outcome[5] highlighting the critical need

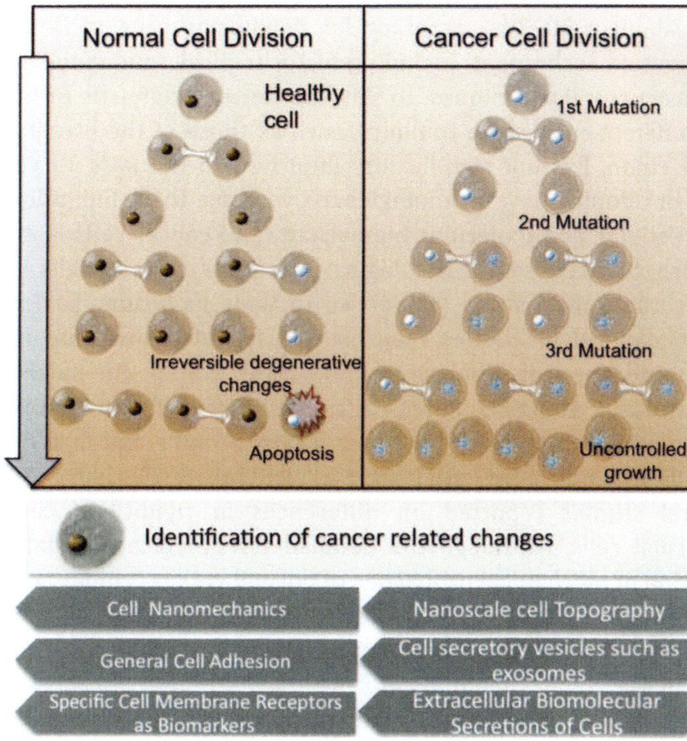

Figure 20.2. Schematics of normal (healthy) and abnormal cancer cell division showing genetic aberrations and novel techniques to identify cancer-related changes in cells.

for developing novel approaches for early cancer detection. Additionally, quantification of cancer cell physiology is required for customization of drug therapies based on specific characteristics and extent of abnormality of cancer cell populations as well as the response of cancer cell populations to therapies. Cancer diagnosis has been widely and aggressively pursued by various nanotechnology research initiatives. In particular, this chapter will highlight some recent AFM techniques developed for cancer diagnostics based on unique nanoscale properties and/or structure of cancer cells or tumour-associated biomolecules.

20.2.1 Cellular Nanomechanics: Using AFM for Cancer Detection

Conventionally, detection of cancerous cells is based on morphological analysis though it is realized that diagnosis based on morphological examination can be difficult,[6] with cyto-morphological analysis alone showing about a 50–70% accuracy for diagnosing cancer.[7] Despite using complementary techniques, including histochemical, immunohistochemical and ultrastructural techniques, to develop better diagnostic protocols, the ability to detect early-stage tumours, such as those of the breast, prostate, cervix or colon, has not significantly improved in the past 30 years.[7] The need for developing new technologies to overcome these limitations is thus evident. Cellular and molecular biomechanics of cancer cells is an exciting area, which characterizes the rheological properties of cancer cells and relates the measurable mechanical properties to their molecular basis. Changes in the rheological properties may provide useful information for cancer diagnosis and physical evidence to understand therapeutic mechanisms of various anti-cancer agents. Recent advances in experimental biomechanics have enabled direct and real-time mechanical probing and manipulation of single cells and molecules with nano and picoscale resolutions.

Several studies reported on differences in rigidity of cancer cells from normal cells.[8] Although the detailed physiological mechanisms and propagation of mechanical properties of normal versus tumour cells are still being investigated, AFM-based cytological analysis provides an entirely new technological platform for cancer diagnosis and evaluation by quantitatively measuring the Young's modulus of cells.[9] Low stiffness of cancer cells may be caused by a partial loss of actin filaments and/or microtubules, and therefore lowers the density of the cellular scaffold.[10] In general, malignant cells respond either more elastically (softer) or less viscously to the applied stress since metastatic cells must squeeze to go through the surrounding tissue matrix when they make their way into the circulatory systems where they are directed to establish distant settlements.[11]

AFM has emerged as an important instrument for the investigation of mechanical properties associated with live cells.[12] It is considered as a powerful tool for probing biological samples with sub-nanometre resolution thus providing tremendous insight regarding the surface features and cellular nanomechanics,[13] or cellular processes based on the mechanical properties of living cells.[14] An AFM consists of a cantilever (with tip mounted to the soft cantilever spring), a sample stage and an optical beam deflection system which consists of a laser diode and a position-sensitive photodiode. A schematic diagram of an AFM tip interacting with an individual cell is shown in Fig. 20.3.

Mechanical measurements acquired using AFM rely on measuring the force as the tip is pushed towards (Fig. 20.3a), indented into (Fig. 20.3b) and retracted from the sample or cell surface in this case (Fig. 20.3c). The cantilever is mounted on the end of a piezoelectric tube scanner which is used to bring the tip into contact with the surface. The force is measured

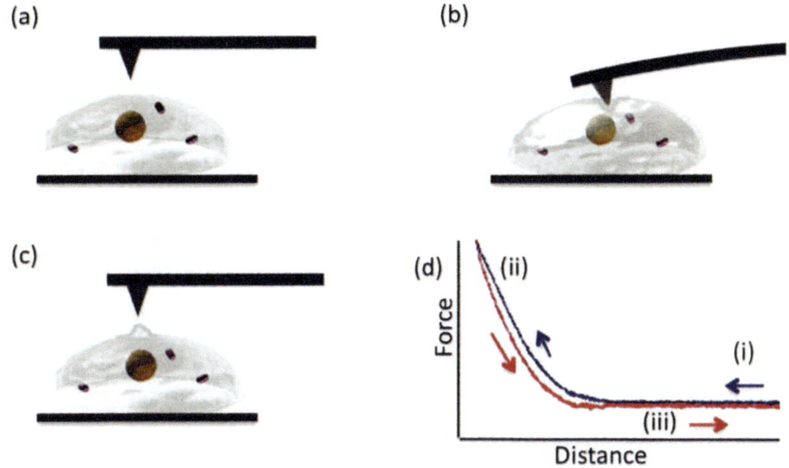

Figure 20.3. Schematic of an AFM probing a cell surface (a) AFM tip approaching cell surface, (b) indented into cell surface and (c) retracted from cell surface. (d) Typical force–displacement curve ((i–iii) correspond to the positions described earlier), recorded as the "approach" and "retract" curves of the cantilever as it moves towards and away from the surface. The force acting on the cantilever is recorded as a function of the piezoelectric crystal displacement. Mechanical properties, such as the Young's modulus (E) or cell stiffness, can be calculated from force curves using a Hertz model.

by recording the deflection (vertical bending) of the cantilever. As described earlier, the cantilever deflection is usually detected by a laser beam focused on the free end of the cantilever and reflected into a photodiode; this deflection is directly proportional to the force. Force–displacement curves are obtained by monitoring the deflection of the cantilever (Fig. 20.3d). The microcantilever-based system allows us to probe the local Young's modulus (E) or "stiffness" of living cells, performs force spectroscopy measurements with piconewton resolution and provides a sensor to record *in vivo* measurements of the cell wall at sub-nanometre resolution. In particular, AFM is a key tool in acquiring kinetic information, and real-time signals of living cells, and is capable of offering *in vivo* single-cell diagnostics. AFM measurements provide a greater understanding of structure, function and relationships of biological macromolecules, thus generating characteristics inherent to specific biological cells.[15] These emerging concepts aid in the development of new types of nanomechanical sensors, which may contribute significantly to the understanding of changes in cytoarchitecture, which are characteristic of cellular de-differentiation, malignant transformation, growth activation, cell motility and disease states.

Cross *et al.*[9] reported a novel approach to identify and diagnose cancerous cells based on cellular nanomechanical behaviour through the implementation of AFM. Quantitative measurements showed that the metastatic tumour cells obtained from human patients were about five times softer than normal mesothelial cells despite showing similar morphology. The use of AFM to probe and study single biological systems on the nanoscale can yield information about the integrity and local nanomechanical properties of these cells.[16] The AFM approach may help to understand the mechanics inherent to changes in cytoarchitecture and dynamics under *in vitro* conditions and elucidate the mechanisms and related biological alterations associated with tumour phenotype. AFM-based cytological analysis can quantify the pathophysiology and potential aggressiveness of individual tumours. It could also be used for customization/monitoring of drug therapies based on specific cancer cell characteristics in near future.

20.2.2 Sub-Cellular Vesicles for the Detection of Novel Cancer Markers from Biological Fluids

The development of nanotechnology-based methods for early cancer detection from easily assessable body fluids such as blood,[17] urine[18] or saliva[19] can be highly beneficial for diagnostics and monitoring treatment response and remain of paramount importance. One such class of biomarkers that has gained renewed interest is a unique type of sub-100 nm membrane-bound secretory vesicles called "exosomes". Exosomes are secreted by a wide range of normal mammalian cell types[20] and released into body fluids such as epididymal fluid, seminal plasma, broncoalveolar fluid, pleural effusions, ascites, amniotic fluid, blood and urine via exocytosis.[21] Malignancy and other diseases cause elevated exosome secretion and tumour-antigen enrichment of exosomes associated with cancer cells.[22,23] Their physiological functions are unclear; however, exosomes possess cell type-specific membrane and proteins enclosed in a lipid bilayer, and serve to signal the physiological state of various distant cells without direct access to the originating tissue or cells themselves. Previous studies have identified populations of various types of normal and tumour-derived exosomes. These vesicles hold tremendous promise as biomarkers for several types of cancers. Yet, because of their small size, sensitive and quantitative detection tools are needed for their individual characterization. Currently, exosomes characterization includes electron microscopy-based morphological analysis and semi-quantitative proteomic and transcriptional analysis of exosome populations.[24] Single vesicle structural and surface molecular details on human saliva exosomes considered as potential non-invasive biomarker resource for oral cancer[19] have been studied recently using AFM.[25] Single exosomes vesicle ultrastructure, quantitative surface molecular constitution and nanomechanical characteristics of exosomes may be helpful for understanding the role of exosomes in intercellular communication and delivery of genetic components through the extracellular domain (Fig. 20.4a).

AFM has developed as a useful single-molecule tool for sensing and mapping molecular recognition interactions on biological cell interfaces.[26,27] Cell type-specific markers such as CD63 receptors on individual exosomes can be analysed using force spectroscopy. Force spectroscopy relies on measuring the interaction force with piconewton sensitivity as the tip is pushed towards the sample and retracts from it in the *z* direction. The force is monitored by measuring the deflection (vertical bending) of the cantilever. Measuring molecular receptors on the exosome surface requires recording force curves between the modified tips (antiCD63 antibody) and the exosomes surface. At large tip–sample separation distances, the force experienced by the tip is zero. As the tip approaches the surface, the cantilever may bend upwards owing to repulsive forces until the tip jumps into contact with the exosome surface (Fig. 20.4b). Upon retracting the tip from the surface, in the event of successful binding of the antiCD63 antibody to the complementary receptors on the vesicle surface, the curve shows an unbinding event calculated as the adhesion "pull-off" force. The rupture force represents the unbinding force between complementary antiCD63 IgG receptors and ligand molecules borne on the vesicle outer membrane. The recognition of single receptor molecules on biological fluid-derived exosomes, such as saliva, can potentially detect surface tumour-antigen-enriched cancer exosomes, and thereby enable early cancer diagnosis where conventional methods may prove ineffective because of sensitivity limitations.

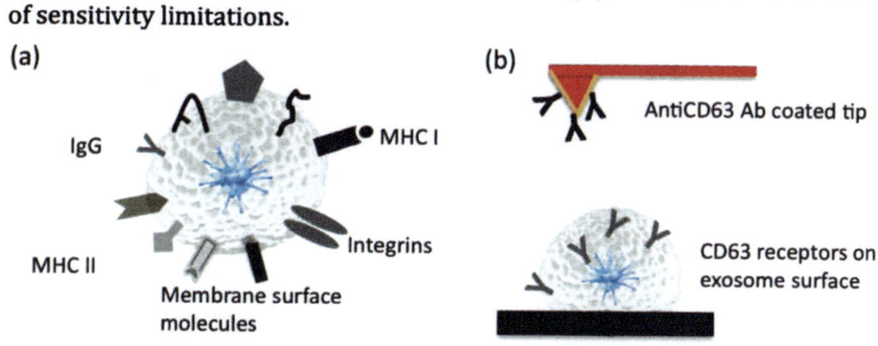

Figure 20.4. (a) Schematic showing exosome vesicle and surface receptors. (b) Schematic of receptor recognition spectroscopy via adhesion force measurements between AntiCD63 IgG-functionalized AFM tips and exosome surface.

20.2.3 *Ex vivo* Molecular Recognition for Early Cancer Detection

Several emerging nanodevices can provide rapid and sensitive detection of cancer-related molecules by enabling detection of biomolecular changes in diseased states. It is increasingly evident that the single-molecule detection sensitivities of nanodevices hold tremendous advantages for early detection of cancer—a critical step in improving cancer treatment. Miniaturization of such diagnostic tools, as in the case of nanocantilevers and nanowires, also enables possible screening for multiple cancer markers on a single device, thereby allowing cancer screening being faster and more cost-efficient.

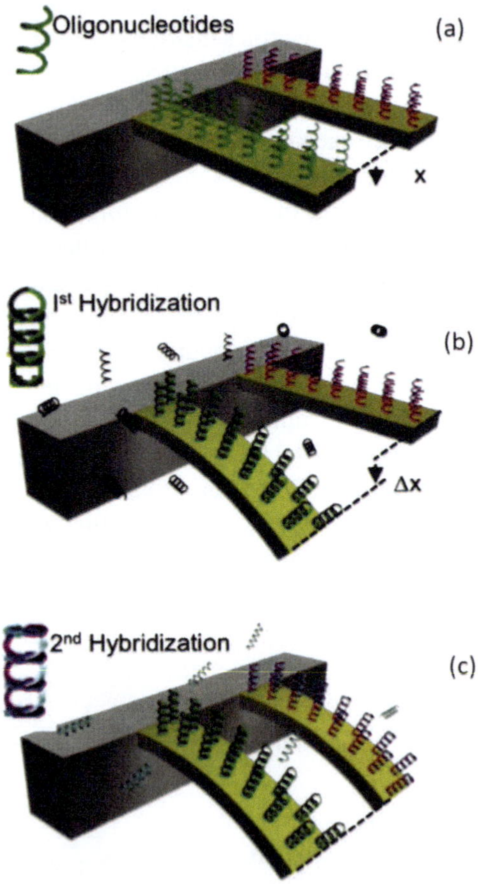

Figure 20.5. Scheme illustrating hybridization of different oligonucleotides functionalized over each cantilever. (a) Differential signal is set to zero. (b) Hybridization of first (black) complementary sequence resulting in increased differential signal, □x. (c) Second oligonucleotide (cyan)-functionalized cantilever bent because of binding of second matching oligonucleotide (magenta).

Gimzewski *et al.*[28] pioneered the concept that biomolecular binding events yield forces and deformations that might be detected and recognized by appropriately selective sensing nanostructures, leading to new approaches to multiplexed molecular recognition (Fig. 20.5).

Primary examples of such devices are micro- or nanocantilevers, which deflect and change resonant frequencies as a result of affinity binding e.g., nucleic acid hybridization or proteomic binding events occurring on their free surfaces. A nanocantilever is a thin silicon nitride (typically 1 μm thick, 500 μm long, 100 μm wide) projection attached to a microchip. The cantilevers' surfaces are covered with a layer of receptor probes with a specific binding affinity to target sequence. Because of the extreme thinness of the probe, any adjacent bindings of probes to target molecules would cause the cantilever to locally bend at those binding sites through steric and charge interactions. The resulting bending could then be measured dynamically through the change in resonance frequency in response to the added mass, or statically through the deflection of a laser beam in response to bending. The biochemically induced surface stress was shown to directly and specifically transduce molecular recognition into nanomechanical responses in a cantilever array. Cantilevers in an array were functionalized with a selection of oligonucleotides of variable lengths. The differential deflection of the cantilevers was found to provide a true molecular recognition signal, despite large nonspecific responses of individual cantilevers. Hybridization of complementary oligonucleotides shows detection of a single-base mismatch between two 12-mer oligonucleotides (Fig. 20.5). The nanometre-sized cantilevers, being extremely sensitive and able to detect single molecules of DNA or proteins, also provide fast and sensitive detection for cancer-related molecules. Other applications include microcantilevers to detect single nucleopeptides in a 10-mer DNA target oligonucleotide without the use of extrinsic fluorescent or radioactive labeling.[29,30] Quantitation of prostate serum albumin at clinically significant concentrations has also been demonstrated.[29] Nanocantilevers possess extraordinary multiplexing capability.[31] In future, fabrication of arrays of cantilevers may allow the simultaneous reading of proteomic profiles or the entire proteome.

Based on similar concept, silicon nanowires can be engineered to detect molecular markers of cancer cells in microfluidic channel devices. Because of their tiny size (20–100 nanometres wide), they exhibit special properties such as superconductivity, and extremely high sensitivity to outside electric fields. Nanowires can be coated with a probe such as an antibody that binds to a target protein. Proteins that bind to the antibody can change the nanowire's electrical conductance, and this can be measured by a detector.[32] Each nanowire may bear a different antibody or oligonucleotide, a short stretch of DNA that can be used to recognize specific RNA sequences or proteins secreted by cancer cells.[33] Self-assembled carbon nanotubes and probe DNA oligonucleotides are immobilized by covalent binding to the nanotubes.[34] When hybridization between the probe and the target DNA sequence occurs, the change is noted as a voltage peak.[35] The nano-based biosensors being developed are more efficient and more selective than current detectors and may be utilized as alternative and complementary cancer detection probes.

20.2.4 Cell Nanomechanical Motion

Cells are dynamic structures that display nanometre to micrometre scale motions at their cell membranes. The AFM can investigate the nanomechanical motion of the cell surface ranging from yeast[16] to cardiomyocytes.[36] If the AFM tip is held stationary over a cell surface that is vibrating or moving, the tip will bend and follow these motions. The AFM can thus be used as an ultra-sensitive, high-resolution motion detector. AFM may be used to probe the surface of cells under a variety of external and internal environmental conditions to obtain an oscillatory signal of nanomechanical origin. This oscillatory, periodic signal can be converted into sound and used as an indicator of cell health. The process termed as "sonocytology" enables cell damage detection; for example, when microtubule and actin dissociating agents used in chemotherapy are added, a change in cell elasticity is discernable much earlier than biochemical measurements of cell death.[16] Thus, by observing their motion, the healthy and cancerous cells can be distinguished. Sonocytology may thus be used as a diagnostic tool by analyzing variations in cell nanomechanical motions. In future, sonocytology may be incorporated as a complementary tool into medical disciplines such as cancer research and make cancer detection possible before a tumour forms and significant cellular biophysical and biochemical changes manifest.

20.3 SUMMARY

Nanomedicine is an emerging field with significant potential to yield new generation of scientific and technological approaches and advance clinical tools and devices. Diagnostics and biosensors are among the earliest applications of nanotechnology rapidly translating from research to clinical environments. They provide alternative and better approaches to assess the onset or progression of diseases. Diseases such as cancer can be quantified based on morphological, biophysical and biochemical nanoscale properties of cells and subcellular structures. In particular, AFM techniques developed for cancer diagnostics enable detection of nanoscale properties and/or structure of cancer cells or tumour-associated biomolecules. The next generation of AFMs can be integrated with complementary methodologies, including ionic conductance, total internal reflection fluorescence, fluorescence resonance energy transfer and fluorescence imaging, microfluidics, and physico-chemical measurements, thereby enabling a detailed structure–function studies of biological tissues.[37] Cancer-associated rapid quantitative changes in whole-cell morphology, motion and mechanical rigidity via live cell interferometry[38] can also be combined with the dynamic capability of AFM. The flexibility and high-resolution capability of these integrated tools will invariably provide new and exciting information from multiscale biological systems. An approximate 20-year generic gestation period exists between any science discovery and its implementation in the market.[39] However, rapid growth of nano-enabled products and devices in the healthcare market during the last five years suggests imminent emergence of nano-enabled technologies.[2] As the capability to detect and measure nano-dimensional changes in cells and their environment and using novel platforms to derive physiological information progresses, medical applications of nano-enabled and nano-enhanced products and technologies including AFM are bound to rise in the coming years.

References

1. Warner, S. (2004) Diagnostics + therapy = theranostics, *The Scientist*, **18**, 38–39.
2. Wagner, V., Dullaart, A., Bock, A. K., and Zweck, A. (2006) The emerging nanomedicine landscape, *Nat. Biotechnol.*, **24**, 1211–1217.
3. Jemal, A., Murray, T., Ward, E., Samuels, A., Tiwari, R. C., Ghafoor, A., Feuer, E. J., and Thun, M. J. (2005) Cancer statistics, 2005, *CA Cancer J. Clin.*, **55**, 10–30.
4. Lee, G. Y. H., and Lim, C. T. (2007) Biomechanics approaches to studying human diseases, *Trends Biotechnol.*, **25**, 111–118.
5. Christofori, G. (2006) New signals from the invasive front, *Nature*, **441**, 444–450.
6. Osterheld, M. C., Liette, C., and Anca, M. (2005) Image cytometry: an aid for cytological diagnosis of pleural effusions, *Diagn. Cytopathol.*, **32**, 173–176.
7. Bedrossian, C. W. (1994) Cytopathology: in search of a new identity, *Diagn. Cytopathol.*, **10**, 1–2.
8. (a) Goldmann, W. H., and Ezzell, R. M. (1996) Viscoelasticity in wild-type and vinculin-deficient (5.51) mouse F9 embryonic carcinoma cells examined by atomic force microscopy and rheology, *Exp. Cell Res.*, **226**, 234–237; (b)

Lekka, M., Laidler, P., Gil, D., Lekki, J., Stachura, Z., and Hrynkiewicz, A. Z. (1999) Elasticity of normal and cancerous human bladder cells studied by scanning force microscopy, *Eur. Biophys. J.*, **28**, 312–316; (c) Weisenhorn, A. L., Khorsandi, M., Kasas, S., Gotzos, V., and Butt, H. J. (1993) Deformation and height anomaly of soft surfaces studied with an AFM, *Nanotechnology*, **4**, 106–113.

9. Cross, S. E., Jin, Y.-S., Rao, J., and Gimzewski, J. K. (2007) Nanomechanical analysis of cells from cancer patients, *Nat. Nanotechnol.*, **2**, 780–783.
10. Daisuke, Y., Shusaku, K., and Tadaomi, T. (2005) Regulation of cancer cell motility through actin reorganization, *Cancer Sci.*, **96**, 379–386.
11. Farina, K. L., Wyckoff, J. B., Rivera, J., Lee, H., Segall, J. E., Condeelis, J. S., and Jones, J. G. (1998) Cell motility of tumor cells visualized in living intact primary tumors using green fluorescent protein, *Cancer Res.*, **58**, 2528–2532.
12. (a) Rotsch, C., Jacobson, K., and Radmacher, M. (1999) Dimensional and mechanical dynamics of active and stable edges in motile fibroblasts investigated by using atomic force microscopy, *Proc. Natl. Acad. Sci. USA*, **96**, 921–926; (b) Pelling, A. E., Li, Y., Shi, W., and Gimzewski, J. K. (2005) Nanoscale visualization and characterization of *Myxococcus xanthus* cells with atomic force microscopy, *Proc. Natl. Acad. Sci. USA*, **102**, 6484–6489.
13. Sharma, S., Cross, S. E., French, S., Gonzalez, O., Petzold, O., Baker, W., Wanda, W., Yougsunthon, R., Baker, D., and Gimzewski, J (2009) Influence of substrates on hapatocyes: a nanomechanical study, *J. Scanning Probe Microsc.*, **4**, 7–16.
14. Suresh, S. (2007) Biomechanics and biophysics of cancer cells, *Acta Biomater.*, **3**, 413–438.
15. Zullo, S. J., Srivastava, S., Looney, J. P., and Barker, P. E. (2002) Nanotechnology: emerging developments and early detection of cancer. A two-day workshop sponsored by the National Cancer Institute and the National Institute of Standards and Technology, August 30–31, 2001, on the National Institute of Standards and Technology Campus, Gaithersburg, MD, USA, *Dis. Markers*, **18**, 153–158.
16. Pelling, A. E., Sehati, S., Gralla, E. B., Valentine, J. S., and Gimzewski, J. K. (2004) Local nanomechanical motion of the cell wall of *Saccharomyces cerevisiae*, *Science*, **305**, 1147–1150.
17. Skog, J., Wurdinger, T., van Rijn, S., Meijer, D. H., Gainche, L., Sena-Esteves, M., Curry, W. T., Jr., Carter, B. S., Krichevsky, A. M., and Breakefield, X. O. (2008) Glioblastoma microvesicles transport RNA and proteins that promote tumour growth and provide diagnostic biomarkers, *Nat. Cell Biol.*, **10**, 1470–1476.
18. Pisitkun, T., Shen, R., and Knepper, M. A. (2004) Identification and proteomic profiling of exosomes in human urine, *Proc. Natl. Acad. Sci. USA*, **101**, 13368–13373. Epub 23 August 2004.
19. Ogawa, Y., Kanai-Azuma, M., Akimoto, Y., Kawakami, H., and Yanoshita, R. (2008) Exosome-like vesicles with dipeptidyl peptidase IV in human saliva, *Biol. Pharm. Bull.*, **31**, 1059–1062.
20. Keller, S., Sanderson, M., Stoeck, A., and Altevogt, P. (2006) Exosomes: from biogenesis and secretion to biological function, *Immunol. Lett.*, **107**, 102–108.

21. Simpson, R. J., Jensen, S. S., and Lim, J. W. (2008) Proteomic profiling of exosomes: current perspectives, *Proteomics*, **8**, 4083–4099.

22. Andre, F., Schartz, N. E. C., Movassagh, M., Flament, C., Pautier, P., Morice, P., Pomel, C., Lhomme, C., Escudier, B., Le Chevalier, T., Tursz, T., Amigorena, S., Raposo, G., Angevin, E., and Zitvogel, L. (2002) Malignant effusions and immunogenic tumour-derived exosomes, *Lancet*, **360**, 295–305.

23. Wolfers, J., Lozier, A., Raposo, G., Regnault, A., Thery, C., Masurier, C., Flament, C., Pouzieux, S., Faure, F., Tursz, T., Angevin, E., Amigorena, S., and Zitvogel, L. (2001) Tumor-derived exosomes are a source of shared tumor rejection antigens for CTL cross-priming, *Nat. Med.*, **7**, 297–303.

24. Thery, C., Zitvogel, L., and Amigorena, S. (2002) Exosomes: composition, biogenesis and function, *Nat. Rev. Immunol.*, **2**, 569–579.

25. (a) Palanisamy, V., Sharma S., Deshpande, A., Zhou, H., Gimzewski, J., and Wong, D. (2010) Nanostructural and transcriptomic analyses of human saliva derived exosomes, *PLoS One*, **5**, e8577. (b) Sharma, S., Rasool, H., Palanisamy, V., Wong, D. T. and Gimzewski, J. K. (2010) Structural-mechanical characterization of nanoparticle exosomes in human saliva, using correlative AFM, FESEM, and force spectroscopy, *ACS Nano*, **4**, 1921–1926.

26. Moy, V. T., Florin, E. L., and Gaub, H. E. (1994) Intermolecular forces and energies between ligands and receptors, *Science*, **266**, 257–259.

27. Hinterdorfer, P., Baumgartner, W., Gruber, H. J., Schilcher, K., and Schindler, H. (1996) Detection and localization of individual antibody-antigen recognition events by atomic force microscopy, *Proc. Natl. Acad. Sci. USA*, **93**, 3477–3481.

28. Fritz, J., Baller, M. K., Lang, H. P., Rothuizen, H., Vettiger, P., Meyer, E., Guntherodt, H., Gerber, C., and Gimzewski, J. K. (2000) Translating biomolecular recognition into nanomechanics, *Science*, **288**(5464), 316–318.

29. Wu, G., Datar, R. H., Hansen, K. M., Thundat, T., Cote, R. J., and Majumdar, A. (2001) Bioassay of prostate-specific antigen (PSA) using microcantilevers, *Nat. Biotechnol.*, **19**, 856–860.

30. Majumdar, A. (2002) Bioassays based on molecular nanomechanics, *Dis. Markers*, **18**, 167–174.

31. Yue, M., Stachowiak, J. C., and Majumdar, A. (2004) Cantilever arrays for multiplexed mechanical analysis of biomolecular reactions, *Mech. Chem. Biosyst.*, **1**, 211–220.

32. McAlpine, M. C., Agnew, H. D., Rohde, R. D., Blanco, M., Ahmad, H., Stuparu, A. D., Goddard, W. A., III, and Heath, J. R. (2008) Peptide-nanowire hybrid materials for selective sensing of small molecules, *J. Am. Chem. Soc.*, **130**, 9583–9589.

33. Bunimovich, Y. L., Shin, Y. S., Yeo, W. S., Amori, M., Kwong, G., and Heath, J. R. (2006) Quantitative real-time measurements of DNA hybridization with alkylated nonoxidized silicon nanowires in electrolyte solution, *J. Am. Chem. Soc.*, **128**, 16323–16331.

34. Cai, H., Cao, X., Jiang, Y., He, P., and Fang, Y. (2003) Carbon nanotube-enhanced electrochemical DNA biosensor for DNA hybridization detection, *Anal. Bioanal. Chem.*, **375**, 287–293.

35. Williams, K. A., Veenhuizen, P. T., de la Torre, B. G., Eritja, R., and Dekker, C. (2002) Nanotechnology: carbon nanotubes with DNA recognition, *Nature*, **420**, 761.

36. Domke, J., Parak, W. J., George, M., Gaub, H. E., and Radmacher, M. (1999) Mapping the mechanical pulse of single cardiomyocytes with the atomic force microscope, *Eur. Biophys. J.*, **28**, 179–186.

37. Lal, R., and Arnsdorf, M. F. (2010) Multidimensional atomic force microscopy for drug discovery: a versatile tool for defining targets, designing therapeutics and monitoring their efficacy, *Life Sci.*, **86**, 545–562.

38. Reed, J., Troke, J. J., Schmit, J., Han, S., Teitell, M. A., and Gimzewski, J. K. (2008) Live cell interferometry reveals cellular dynamism during force propagation, *ACS Nano*, **2**, 841–846.

39. The 2006 Nanomedicine, Device & Diagnostics Report (2006) *Nanobiotech News*, National Health Information, Atlanta, USA.

Index

For Product Safety Concerns and Information please contact our EU representative GPSR@taylorandfrancis.com Taylor & Francis Verlag GmbH, Kaufingerstraße 24, 80331 München, Germany